COMBINATORIAL REASONING

COMBINATORIAL REASONING

An Introduction to the Art of Counting

DUANE DeTEMPLE
WILLIAM WEBB
Department of Mathematics
Washington State University
Pullman, WA

WILEY

Published by John Wiley & Sons, Inc., Hoboken, New Jersey.
Published simultaneously in Canada.

Library of Congress Cataloging-in-Publication Data:

DeTemple, Duane W.
Combinatorial reasoning : an introduction to the art of counting / Duane DeTemple, William Webb.
pages cm
Includes bibliographical references and index.
ISBN 978-1-118-65218-3 (cloth)
1. Combinatorial analysis–Textbooks. 2. Mathematical analysis–Textbooks. I. Webb, William, 1944– II. Title.
QA164.D48 2013
511′.6–dc23

2013035522

Printed in the United States of America

10 9 8 7 6 5 4 3 2 1

CONTENTS

PREFACE

Counting problems, or more precisely enumerative combinatorics, are a source of some of the most intriguing problems in mathematics. Often these problems can be solved using ingenious and creative observations, what we call *combinatorial reasoning*. It is this kind of thinking that we stress throughout the descriptions, examples, and problems in this text.

Combinatorics has many important applications to areas as diverse as computer science, probability and statistics, and discrete optimization. But equally important, the subject offers many results of beautiful mathematics that are enjoyable to discover and problems that are simply fun to think about and solve in innovative ways.

Each of us has over 40 years of experience teaching combinatorics, as well as other mathematics courses, at both the undergraduate and graduate levels. We think that we have learned some effective ways to present this subject. Early versions of the notes for the book were used in both undergraduate and graduate courses, and our students found the approach both easy to understand and quite thorough.

FEATURES OF THIS TEXT

Chapter 1 introduces the reader to combinatorial thinking by considering topics of existence, construction, and enumeration that lead, by the end of the chapter, to general principles of combinatorics that are employed throughout the remainder of the text. The problems solved in this chapter, often involving dot patterns and tilings of rectangular boards, are easily described and visualized, and foreshadow much of what comes later.

More formal considerations of combinatorics are taken up in Chapter 2, where selections, arrangements, and distribution are treated in detail. Special attention is given to combinatorial models—block walking, tiling of rectangular boards, committee selection, and others. It is shown how general results can be derived by combinatorial reasoning based on an appropriate model. Most often, only the simplest of algebraic calculations are necessary.

An unusually complete discussion of generating functions—both ordinary and exponential—is given in Chapter 3, where the binomial series is shown to be a prototype of a much larger collection of generating functions. It is also shown how enumeration problems can often be solved with generating functions.

Chapter 4 begins with the DIE method (Description—Involution—Exceptions), which is shown to be a powerful combinatorial approach to the evaluation of alternating series. This leads naturally to the Principle of Inclusion/Exclusion. The chapter then turns to a section on rook polynomials that combines generating functions and inclusion–exclusion to solve an interesting class of restricted arrangement problems. The chapter concludes with an optional section on the Zeckendorf representation of integers and its application to creating a winning strategy for the game Fibonacci Nim.

Recurrence sequences are treated in detail in Chapters 5, using what we call the operator method. By employing the readily understood successor operator E, which simply replaces n by $n + 1$, the properties of recurrence sequences seem natural and easy to understand. This approach not only deepens comprehension but also simplifies many calculations.

Chapter 6 enlarges the library of special numbers that often answer combinatorial questions. Since the sections within this chapter are largely independent, there is freedom to pursue whatever topics seem of most interest—Stirling numbers, harmonic numbers, Bernoulli numbers, Eulerian numbers, partition numbers, or Catalan numbers.

Chapter 7 returns to the operator approach for solving linear recurrence relations that was introduced earlier in Chapter 5. Here, by viewing recurrence sequences as vector spaces, additional methods to solve recurrence relations become available. Moreover, we discover a powerful new approach to both discover and verify combinatorial identities.

Pólya-Redfield counting—the enumeration of arrangements that take symmetries into account—is the subject of Chapter 8, the final chapter. Here, abstraction is minimized by showing how general formulas can be derived from the consideration of carefully chosen simple figures and arrangements.

FLEXIBILITY FOR COURSES

A beginning course for undergraduates can be easily constructed using selected sections from Chapters 1–5. A course for advanced undergraduates and beginning graduate students might give quicker coverage to early chapters and include material chosen from Chapters 6, 7, and 8. For example, Sections 6.1–6.6, on special numbers, are largely independent, and any of these sections can be covered in any order. There are no special prerequisites for this material beyond a little exposure to power series. An elementary introduction to linear algebra is needed for Chapter 7. A little background in group theory is helpful for Chapter 8, but in our experience this chapter can provide a good introduction to this algebraic topic.

We would like to thank Ken Davis (Hardin-Simmons University) for his very helpful suggestions and comments.

Duane DeTemple
William Webb

Washington State University
Pullman

PART I

THE BASICS OF ENUMERATIVE COMBINATORICS

1

INITIAL EnCOUNTers WITH COMBINATORIAL REASONING

1.1 INTRODUCTION

Although this text is devoted largely to enumerative combinatorics, Section 1.2 presents a brief encounter with a simple yet surprisingly versatile method to prove existence, the pigeonhole principle. Section 1.3 discusses some combinatorial construction problems associated with covering a chessboard with dominoes. In Section 1.4, we consider some number sequences that often arise in combinatorial problems such as *triangular* numbers 1,3,6,10,...; *square* numbers 1,4,9,16,...; and other *figurate numbers*, where the terminology alludes to the representation of these numbers by geometric patterns of dots. In Section 1.5, we count the number of ways a $1 \times n$ rectangle can be tiled with either unit squares of two contrasting colors or with a mixture of 1×1 squares and 1×2 dominoes. By counting the number of dots in a pattern or the number of tilings of a chessboard, we will discover several general principles of counting that are fundamental to enumerative combinatorics. In particular, we will encounter the addition and multiplication principles, which are explored in detail in Section 1.6, which concludes the chapter.

1.2 THE PIGEONHOLE PRINCIPLE

The pigeonhole principle was first applied in 1834 by Peter Dirichlet (1805–1859) to solve a problem in number theory. Soon, other mathematicians found his idea equally

Combinatorial Reasoning: An Introduction to the Art of Counting, First Edition. Duane DeTemple and William Webb.

useful and referred to it as *Dirichlet's box principle* (*Schubfachprinzip* in German). Later, in the nineteenth century, the term *pigeonhole* was used in reference to the small boxes or drawers common in desks of that century. (It may be comforting to know that envelopes, and not pigeons, are placed in the pigeonholes.)

Dirichlet's idea is simply stated as follows.

Theorem 1.1 (Pigeonhole Principle) If $n + 1$ or more objects are placed into n boxes, then at least one of the boxes contains two or more of the objects.

Proof (by contradiction). Suppose, to the contrary, that each of the n boxes contains no more than one object. Then the n boxes together contain no more than n objects, a contradiction. ■

The examples that follow show how the pigeonhole principle provides the basis for an existence proof. It is helpful to think of pigeons and pigeonholes as metaphorical terms for the objects and boxes of the theorem.

Example 1.2 In a family of seven, there must be two family members for which either the sum or difference in their ages can be given in decades, that is, as a multiple of 10.

Solution. A multiple of 10 is easy to identify by the digit 0 in the units position. If two family members have ages ending with the same digit, the difference in their ages is a multiple of 10. Also, if someone in the family has an age ending in the digit 1 and another family member's age ends with the digit 9, their sum of ages ends with the digit 0. Continuing with this type of reasoning suggests that we define the following pigeonholes:

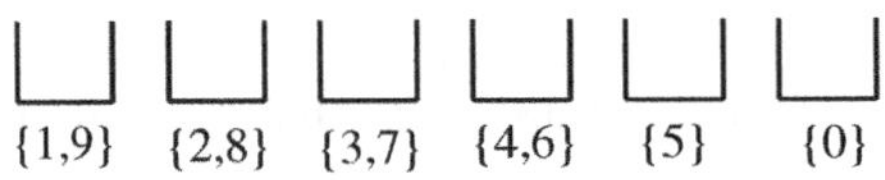

The *pigeons* are the seven ages of the family member. When these are placed in the box labeled with the set containing the last digit of the age, the pigeonhole principle guarantees that at least one of the six boxes contains at least two people with one of the labeled ages. If these two ages happen to have the same unit's digit, then their difference is a multiple of 10. If the two ages have different last digits, these must be 1 and 9, or 2 and 8, or 3 and 7, or 4 and 6. In each case, the sum of the two ages is a multiple of 10. ■

The pigeonholes set up for Example 1.2 show why seven numbers were needed. There is no pair of numbers from the six members of the set $\{1, 2, 3, 4, 5, 10\}$ whose sum or difference is divisible by 10.

Example 1.3 There are five people in a 6 mile by 8 mile rectangular forest, each carrying a walkie-talkie with a range of 5 miles. Show that at least two of the five people can talk with one another on their walkie-talkies.

Solution. Divide the forest into four rectangular 3×4-mi plots; these are the *pigeonholes*. There are five people (the *pigeons*), so by the pigeonhole principle, at least two people are in one of the four plots. Since the maximum distance in a 3×4 rectangular plot is the 5-mi-long diagonal, these two people are within talking range. ∎

The solution of a problem by means of the pigeonhole principle requires us to carry out these steps:

1. Recognize that the pigeonhole principle can be helpful.
2. Identify the pigeons and the pigeonholes.
3. Show that there are more pigeons than pigeonholes.
4. Show why the existence of two pigeons in the same pigeonhole solves the given problem.

These steps are carried out to solve the following problem.

Example 1.4 The *lattice plane* is the set of points in the Cartesian plane with integer coordinates. Given any five points of the lattice plane, show that the midpoint of some pair of points is a point in the lattice plane.

Solution. If (a, b) and (c, d) are two lattice points, their midpoint is $((a + c)/2, (b + d)/2)$, the average of the x and y coordinates. This will be a point in the lattice plane if, and only if, both $a + c$ and $b + d$ are even; that is, a and c must have the same parity, and b and d must have the same parity. This observation suggests that we use parity to define these four pigeonholes:

Box 1: x even, y even

Box 2: x even, y odd

Box 3: x odd, y even

Box 4: x odd, y odd

Since five points are placed in the four boxes, there is some box with at least two members. Both of these points have x and y coordinates with the same parity, so the midpoint of these two points has integer coordinates; that is, their midpoint is a point of the lattice plane. ∎

1.2.1 Applications to Ranges and Domains of Functions

There are a number of useful variations and interpretations of the pigeonhole principle. For example, suppose that A is a set of objects and B is a set of distinct boxes. Then any placement of the objects into the boxes describes a function $f : A \to B$. If no two objects are assigned to the same box, then the function is *one-to-one*, or *injective*. The pigeonhole principle requires there to be at least as many pigeonholes as pigeons, so if $|A|$ and $|B|$ denote the respective number of elements in set A and B, we have the following theorem.

Theorem 1.5 If the function $f : A \to B$ is one-to-one, then $|A| \leq |B|$.

In the opposite direction, suppose that we have a placement of objects from set A into the boxes of set B that leaves no box empty. In other words, we have a function $f : A \to B$ that is *onto* or *surjective*, meaning that its range is all of set B. Since there must be at least as many objects as boxes, we have this theorem.

Theorem 1.6 If $f : A \to B$ is a surjective function from A onto B, then $|A| \geq |B|$.

A function that is both one-to-one and onto (i.e., is both injective and surjective), is said to be *bijective*. When the two theorems above are combined, we get the following result.

Theorem 1.7 If $f : A \to B$ is a bijective function of A onto B, then $|A| = |B|$.

Finally, we have the following result.

Theorem 1.8 If $f : A \to B$ and $|A| = |B|$, then f is one-to-one if and only if f is onto.

Proof. Let $|A| = |B|$ and first suppose that f is one-to-one. Then $|A| = |\text{range}(f)|$ and therefore $|\text{range}(f)| = |B|$. This shows us that $\text{range}(f) = B$, and we see that f is onto. Similarly, if f is not one-to-one, then $|A| > |\text{range}(f)|$, so $\text{range}(f)$ is a proper subset of B and f is not onto. ■

Bijections are immensely useful in combinatorial reasoning. Suppose that we have a difficult problem counting the number of elements in set A, but we can find a bijection of set A onto a set B, and B is more easily counted. Since $|A| = |B|$, our difficulties are over! This strategy is known as a *bijective proof*, and we'll see several examples of this type of combinatorial reasoning later.

1.2.2 An Application to the Chinese Remainder Theorem

Theorem 1.8 will be useful for our next application. Readers with a background or interest in number theory may find the following approach to this topic interesting. However, it is optional and can be skipped since it will not be used later.

Example 1.9 Suppose that a *torus* (i.e., the surface of a doughnut) is divided into quadrangular regions by 11 circles in one direction that are crossed by 16 circles in the orthogonal direction. If the surface is cut along one circle of each type, the surface can be unrolled and stretched to form an 11×16 rectangle partitioned into $11 \cdot 16 = 176$ squares. As shown in the diagram below, a path of squares numbered 0,1,2, . . . has been initiated that spirals around the torus, starting with square 0 in row 0 and column 0. Prove that a continuation of the spiral path covers the entire torus, so that each square is assigned a unique number 0,1,2, . . .,175.

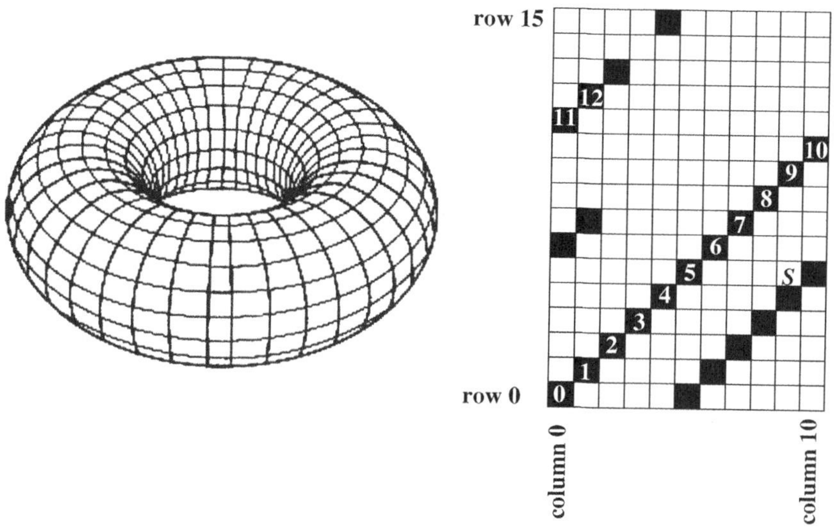

Solution. It suffices to show that the spiral path includes all 16 of the squares located in column 0, the far left column. These are the squares numbered by the entries in the set

$$A = \{0 \cdot 11, 1 \cdot 11, 2 \cdot 11, 3 \cdot 11, \ldots, 15 \cdot 11\} \tag{1.1}$$

We still need to know that no number in the left column is repeated, where the rows are numbered by the set

$$B = \{0,1,2, \ldots ,15\} \tag{1.2}$$

The row of square $k \cdot 11 \in A$ is given by its remainder when divided by 16. For example, square $2 \cdot 11 = 22$ has the remainder of 6 when divided by 16, and we see from the figure that square 22 is in row 6. This suggests we consider the function $f : A \rightarrow B$ defined by mapping each element of set A to the corresponding remainder r in set B. To see why f is one-to-one, suppose that two values, say, $j \cdot 11$ and $k \cdot 11$, $j, k \in \{0,1,2, \ldots .15\}$, have the same remainder r; that is, suppose that

$$j \cdot 11 = p \cdot 16 + r \text{ and } k \cdot 11 = q \cdot 16 + r \tag{1.3}$$

for the quotients p and q. When these equations are subtracted from one another, we see that

$$(j - k) \cdot 11 = (p - q) \cdot 16 \tag{1.4}$$

Since 16 divides the right side of (1.4) it must also divide the left side $(j-k)\cdot 11$. Since 16 is relatively prime to 11 (i.e., 11 and 16 have no common prime divisors), we see that 16 divides $j-k$. However, $0 \leq |j-k| < 15$, and only $j-k=0$ is divisible by 16. Therefore, f is a one-to-one function that maps set A to the set B, and $|A| = |B| = 16$. We now see that each of the 16 squares in column 0 is along the spiral path, and so every square of the entire rectangle is assigned a unique number along the spiral path of squares. ■

Except for its geometric interpretation as a spiral path on a torus, Example 1.8 gives a proof of a special case of the *Chinese remainder theorem*, which first appeared in a third-century (A.D.) book *Sun Zi Suanjing* written by the Chinese mathematician Sun Tzu. It was important that $m = 11$ and $n = 16$ have no positive common divisor other than 1. More generally, we say that two integers m and n are *relatively prime* when their largest common integer divisor is 1. The following theorem can be proved similarly to the approach followed in Example 1.9.

Theorem 1.10 (Chinese Remainder Theorem) Let m and n be relatively prime positive integers, and let a and b be integers with $0 \leq a < m$ and $0 \leq b < n$. Then there is a unique k, $0 \leq k < mn$, for which $k = mj + a$ and $k = nj' + b$ for some j and j'.

1.2.3 Generalizations of the Pigeon Principle

For some problems, we want to know that there are not just two but some larger number of pigeons in some box. In these cases, we can turn to a generalized version of the pigeonhole principle.

Theorem 1.11 (Generalized Pigeonhole Principles)

(a) If $nk + 1$ or more pigeons are put into n pigeonholes, then at least one pigeonhole has $k + 1$ or more pigeons.

(b) If $p_1 + p_2 + \cdots + p_n + 1$ or more pigeons are put into n pigeonholes numbered 1 through n, then, for some r, pigeonhole r has $p_r + 1$ or more pigeons.

Proof. Since part (a) follows from (b) by setting $p_r = k$, $1 \leq r \leq n$, it suffices to prove part (b). As before, it is easy to give a proof by contradiction. If each hole r has at most p_r pigeons, then all n holes together contain at most $p_1 + p_2 + \cdots + p_n$ pigeons. This contradicts the hypothesis that more than $p_1 + p_2 + \cdots + p_n$ pigeons were placed in the holes. ■

Example 1.12 An equilateral triangle has sides of length 6. Show that (a) if there are 10 points inside the triangle, then at least 2 of them are within 2 units of each other; and (b) if there are 9 points inside the triangle, there are at least 3 of them for which each is at most 3 units from each of the others.

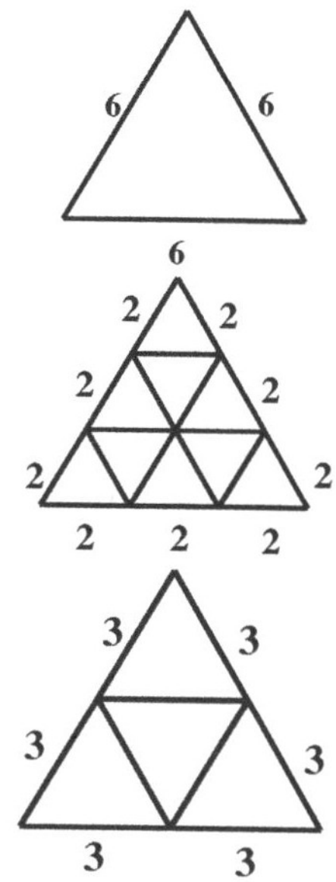

Solution

(a) The triangle can be partitioned into nine equilateral triangles with sides of length 2. Given any 10 points within or on the large triangle, there must be 2 in the same small triangle, and these are at a distance of at most 2 units.

(b) Again using the partitioning of the large triangle into four congruent equilateral triangles of side length 3, not each of the 4 small triangles can contain just two of the nine points. Thus, some small triangle has three or more of the nine points, and each of these is at a distance of no more than 3 units from the other two points. ■

For the next example, it will be helpful to introduce some notations and terminologies that are often used in combinatorics and elsewhere in mathematics:

- The set of the first n natural numbers will be denoted by $[n] = \{1,2,\dots,n\}$.
- A *sequence* of length n is an ordered list $(a_1, a_2, \dots, a_n)$. Equivalently, a sequence of length n is any function f: $[n] \to A$, where $f(j) = a_j$ is the jth *term* of the sequence.
- A *permutation* of $[n]$ is an ordered arrangement of the n elements of set $[n]$. Equivalently, a permutation is a bijection $\pi : [n] \to [n]$.
- A sequence $(a_1, a_2, \dots, a_n)$ of real numbers in *monotone increasing* if $a_1 < a_2 < \cdots < a_n$ and *monotone decreasing* if $a_1 > a_2 > \cdots > a_n$.

- A *subsequence* of $(a_1, a_2, \ldots, a_n)$ is a sequence formed by deleting some of the terms of the given sequence but preserving the order in which the remaining terms are listed.

For example, (8,3,5,2,6,1,4,10,9,7) is a permutation of [10]. The subsequence (3,5,6,10) is monotone increasing. According to the following result, we can always find a monotone sequence of length 4 for any permutation of [10].

Example 1.13 Ten students, all of different heights, are standing shoulder to shoulder in a line. Prove that four of the students can take a step forward so that they form a line of students whose heights either decrease or increase from left to right.

Solution. Suppose that the students' heights are $h_1, h_2, \ldots, h_{10}$ from left to right. Assume that there is no subsequence of four of the students (in the same left-to-right order, of course) with increasing heights. We will then show that a subsequence of four students with decreasing heights can be found. Starting with student 1 at the left, let s_1 be the largest number of students that can step forward to form a row with increasing height with student 1 at the left of the row. Similarly, let s_2 be the largest number of students that can take a step forward to form a row with increasing height with student 2 at the far left. More generally, let s_k be the largest number of students that can step forward to form a row of increasing height and with student k at the left. Since we cannot find a subsequence of four students with increasing height, we know that $1 \le s_k \le 3$, $k = 1, 2, \ldots, 10$; that is, we have 10 numbers that have one of the values 1, 2, or 3. By Theorem 1.11 part (a), at least 4 of the 10 numbers are the same, say, $s_a = s_b = s_c = s_d$, $a < b < c < d$. We see that student a must be taller than student b, since otherwise student a could be added to the left of the longest increasing subsequence starting with student b, and then s_a would be larger than s_b. Thus, we have $h_a > h_b$. By the same reasoning, $h_b > h_c$ and $h_c > h_d$, so that altogether $h_a > h_b > h_c > h_d$. We see that if students a, b, c, and d take a step forward, they have decreasing heights from left to right. ■

The reasoning used in Example 1.13 can be extended to prove this theorem of Erdös and Szekeres [1].

Theorem 1.14 Every sequence of $n^2 + 1$ distinct real numbers contains a subsequence of $n + 1$ numbers that is either increasing or decreasing.

PROBLEMS

1.2.1. A bag contains seven blue, four red, and nine green marbles. How many marbles must be drawn from the bag without looking to be sure that we have drawn

(a) a pair of red marbles?

(b) a pair of marbles of the same color?

(c) a pair of marbles with different colors?

(d) three marbles of the same color?

(e) a red marble, a blue marble, and a green marble?

1.2.2. Given 10 French books, 20 Spanish books, 8 German books, 15 Russian books, and 25 Italian books, how many books must be chosen to guarantee there are

(a) twelve books of the same language?

(b) a book of each language?

1.2.3. There are 10 people at a dinner party. Show that at least two people have the same number of acquaintances at the party.

1.2.4. Given 10 distinct numbers chosen from the arithmetic sequence $1, 4, 7, \ldots, 1+3k, \ldots, 40, 43, 46$, prove there is at least one pair of the 10 chosen integers that has the sum 50.

1.2.5. Given any five points in the plane, with no three on the same line, show that there exists a subset of four of the points that form a convex quadrilateral.[1] [*Hint:* Consider the *convex hull* of the points; that is, consider the convex polygon with vertices at some or all of the given points that encloses all five points. This scenario can be imagined as the figure obtained by bundling the points within a taut rubber band that has been snapped around all five points. There are then three cases to consider, depending on whether the convex hull is a pentagon, a quadrilateral containing the fifth point, or a triangle containing the other two given points.]

1.2.6. Given four points on a circle, show that some three of the points lie in some closed semicircle (a closed semicircle includes its two endpoints).

1.2.7. Given five points on a sphere, show that some four of the points lie in a closed hemisphere [2].
(*Note:* A closed hemisphere includes the points on the bounding great circle.)

1.2.8. A *graph* is a set of points known as *vertices* together with a set of line segments called *edges* that connect some of the pairs of vertices. If every pair of vertices are joined by an edge, the graph is said to be *complete.* The complete graphs on five and six vertices, K_5 and K_6, are shown below:

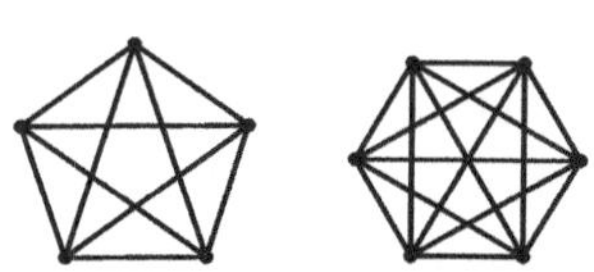

K_5 K_6

[1]This well-known problem is called the "happy ending" problem, since two of its first investigators, Esther Klein and George Szekeres, would later be married to one another.

(a) Show that all of the edges of K_5 can be colored blue or red so that no triangle exists in K_5 with its three edges having the same color.

(b) Show that every red and blue coloring of the edges of K_6 contains a triangle with all of its edges of the same color.

1.2.9. Suppose that 51 numbers are chosen randomly from $[100] = \{1,2,\ldots,100\}$. Show that two of the numbers have the sum 101.

1.2.10. Assuming that there are 48 different pairs of people who know each other at a party of 20 people, show that some person has four or fewer acquaintances.

1.2.11. Choose any 51 numbers from $[100] = \{1,2,\ldots,100\}$. Show that two of the chosen numbers are relatively prime (i.e., have no common divisor other than 1).

1.2.12. Show that any subset of eight distinct integers between 1 and 14 contains a pair of integers m and n such that m divides n.

1.2.13. Choose any 51 numbers from $[100] = \{1,2,\ldots,100\}$. Show that there are two of the chosen numbers for which one divides the other.

1.2.14. State and prove a theorem that generalizes the results of Problems 1.2.12 and 1.2.13.

1.2.15. Consider a string of $3n$ consecutive natural numbers. Show that any subset of $n+1$ of the numbers has two members that differ by at most 2.

1.2.16. Let $(a_1, a_2, \ldots, a_n)$ be any sequence of n natural numbers. Show that there is a subsequence of consecutive members of the sequence that is divisible by n. [*Hint:* Consider the sums $s_k = a_1 + a_2 + \cdots + a_k$.]

1.2.17. Suppose that the numbering of the squares along the spiral path shown in Example 1.9 is continued. What number k is assigned to the square S whose lower left corner is at the point (9, 5)?

1.2.18. Suppose that a torus is divided into mn quadrangular regions by m circles crossed orthogonally by n circles, as in Example 1.9. By the Chinese remainder theorem 1.10, if m and n are relatively prime, then each of the regions is reached by the unique spiral path on the torus.

(a) Using Example 1.9 as a model, draw and number the squares along the spiral path in the case that $m = 4$ and $n = 5$.

(b) How many distinct spiral paths can be found when $m = 4$ and $n = 6$?

(c) Repeat part (b) for $m = 3$ and $n = 6$.

1.2.19. Generalize the results of Problem 1.2.18.

(a) How many spiral paths exist on the torus if $m = n$?

(b) Suppose that $d \geq 2$ is the largest common divisor of m and n. How many distinct spiral paths exist on the torus?

1.2.20. **(a)** Find a permutation of $[9] = \{1, 2, \ldots, 9\}$ for which no subsequence of length 4 is either monotone increasing or monotone decreasing (see Example 1.13).

(b) Place 10 at the right end of your sequence from part (a) and underline the four terms of an increasing subsequence.

(c) Place a 10 at the left end of your sequence from part (a) and underline the four terms of a decreasing subsequence.

1.3 TILING CHESSBOARDS WITH DOMINOES

In this section, our attention turns to combinatorial *construction*. A construction settles the question of existence in a very satisfying way, since the constructed object provides an explicit example with the required properties. Sometimes it can be shown that an object cannot possibly be constructed, so that the existence question is answered in the negative.

In each of the examples that follow, we consider how a shape formed with unit squares can be completely covered with nonoverlapping dominoes. A *domino* is a 1×2 rectangle that is viewed simply as a tile with no attention given to the dots, known as *pips*, that are imprinted on actual dominoes. For example, the 3×6 rectangle in Figure 1.1(a) can be tiled with nine dominoes in many ways, such as the tiling constructed in Figure 1.1(b). Note that a tiling must cover the entire figure and the dominoes can touch along their edges but can never overlap one another.

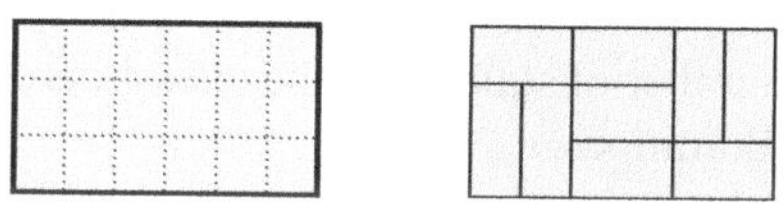

(a) (b)

FIGURE 1.1 A 3×6 rectangle (a) and one way to tile it with dominoes (b).

In enumerative combinatorics, we would ask "in how many ways" can a given shape be tiled with dominoes, but here we will be content to consider only the existence question:

> Given a chessboard, possibly with some of its squares deleted, can we construct a tiling of the board with dominoes? If no construction is found, can it be explained why no tiling exists?

It will soon become apparent why we consider chessboards, since we will be able to take advantage of the alternating pattern of the colors of the unit squares.

Insights into the general case are often given by the examination of special cases. In the examples that follow we will consider the five chessboards shown in Figure 1.2.

(a) (b) (c) (d) (e)

FIGURE 1.2 Five examples of chessboards.

Example 1.15 Chessboard (a) is standard chessboard, with 8 unit squares in each row that are colored with an alternating pattern of black and gray colors. The bottom row can be tiled easily with 4 horizontal dominoes laid end to end. Indeed, each row can be tiled in this way, so the entire 8×8 chessboard can be tiled with horizontally aligned dominoes. Of course, there are many other ways to construct a tiling as well.

Example 1.16 If the last column of an 8×8 standard chessboard is removed, this leaves board (b) with 7 unit squares in each row. Horizontally aligned dominoes no longer can be used to tile the rows as for board (a). However, each column is 8 units high, and therefore the 8×7 chessboard can be tiled with vertically oriented dominoes.

Example 1.17 Board (c) is a 7×7 board, so it has 49 unit squares, an odd number. But each domino in a tiling covers 2 unit squares. This means that any tiling by dominoes covers an even number of squares of the chessboard, so there is no possible way to tile board (c) with dominoes.

Our analysis of the first three chessboards can be generalized to rectangular chessboards of any size.

Theorem 1.18 A rectangular $m \times n$ chessboard can be tiled with dominoes if and only if at least one of its dimensions m or n is an even number.

Example 1.19 Chessboard (d) is obtained by removing two opposite corner squares from the 8×8 standard chessboard, leaving a trimmed board with 62 unit squares. It might seem, since 62 is even, that a tiling with dominoes exists. However, a closer look reveals that the 2 unit squares that were removed were both black, leaving a board with 32 gray and 30 black unit squares. But a domino, whether vertical or horizontal, simultaneously covers both a gray unit square and an adjacent black one. Thus, any trimmed chessboard cannot be tiled if it has an unequal number of gray and black unit squares. In particular, chessboard (d) cannot be tiled with dominoes.

Example 1.20 Chessboard (e) is a 7×7 board has one of its black squares removed, leaving a board with 48 unit squares, 24 gray and 24 black. The reasoning that we have used to examine boards (c) and (d) is not applicable, so it may yet be possible to tile the board with dominoes. As with boards (a) and (b), we can see if a particular tiling can be constructed. Rather than try brute force on the entire board, it may be best to consider first a simpler related problem. For example, what if the center black unit square is removed from a 3×3 chessboard? The following diagram shows that the trimmed 3×3 board can tiled with four dominoes:

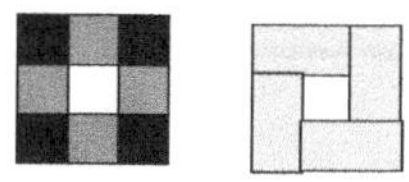

This example suggests that we look for a tiling that combines both horizontally and vertically aligned dominoes. Nicely enough, we quickly find a tiling of board (e):

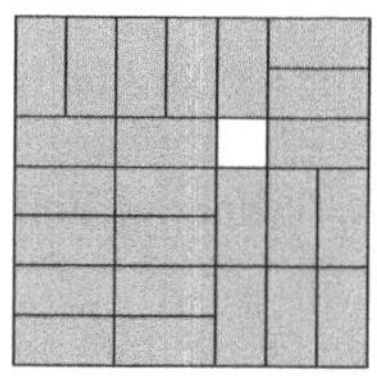

More generally, we can prove the next theorem.

Theorem 1.21 Suppose that an $m \times n$ chessboard, m and n odd, has black corners. If any black square is removed from the board, the trimmed board can be tiled with dominoes.

Proof. Suppose that the black square that has been removed was in column i and row j. Then i and j are either both odd or both even. We consider the two cases separately.

Case 1: i and j are odd. This means the remainder of the board is a collection of rectangular boards each with at least one even dimension, so it can be tiled with dominoes by Theorem 1.18. Here is a typical example

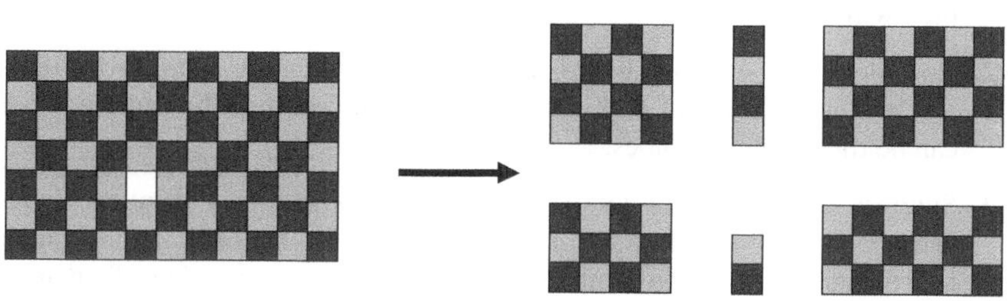

Case 2: i and j even. This means that the 3×3 square with a deleted black center can be tiled, and that the remainder of the board is a collection of rectangular boards each with at least one even dimension, so it can be tiled with dominoes by Theorem 1.18. Here is a diagram illustrating this case:

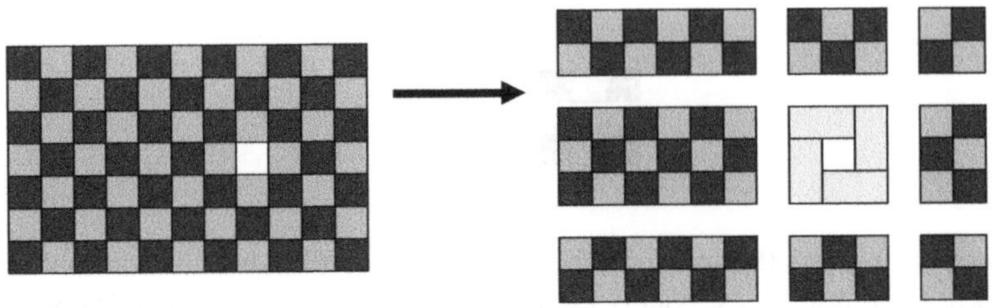

■

It may seem curious that we have avoided a consideration of the enumerative question: How many ways can an $m \times n$ board be tiled with dominoes?

This question, although easy to ask, is not easy to answer! In 1961, the following result was derived independently by Temperley and Fisher [3] and Kasteleyn [4].

Theorem 1.22 An $m \times n$ board can be tiled by $mn/2$ dominoes in

$$\prod_{j=1}^{m}\prod_{k=1}^{n}\left(4\cos^2\frac{j\pi}{m+1}+4\cos^2\frac{k\pi}{n+1}\right)^{1/4} \tag{1.5}$$

ways.

Fortunately, the problem is much easier when m is small. Later we will introduce enumerative methods to determine the number of ways to tile the $2 \times n$ and $3 \times n$ boards with dominoes, and in Section 1.5 we will discuss the number of ways to tile $1 \times n$ boards with either colored squares or a mixture of squares and dominoes.

PROBLEMS

1.3.1. Consider an $m \times n$ chessboard, where m is even and n is odd. Prove that if two opposite corners of the board are removed, the trimmed board can be tiled with dominoes.

1.3.2. Consider an $m \times n$ chessboard, where both m and n are even. Prove that if any two unit squares of opposite color are removed, then the trimmed board can be tiled with dominoes.

1.3.3. Suppose that the lower left $j \times k$ rectangle is removed from an $m \times n$ chessboard, leaving an angle-shaped chessboard. Prove that that angular board can be tiled with dominoes if it contains an even number of squares.

1.3.4. In three dimensions, suppose that a domino consists of two unit cubes joined at their faces. What conditions on l, m, and n are necessary and sufficient for a rectangular solid of size $l \times m \times n$ to be filled with solid dominoes?

1.3.5. Consider a rectangular solid of size $l \times m \times n$, where l, m, and n are all odd positive integers. Imagine that the unit cubes forming the solid are alternately colored gray and black, with a black cube at the corner in the first column, first row, and first layer.

(a) What is the color of each of the remaining corner cubes of the solid?

(b) How can the color of the cube in column j, row k, and layer h of the solid be determined?

(c) Prove that removing any black cube leaves a trimmed solid that can be filled with solid dominoes.

1.3.6. A *tromino* is formed from three unit squares joined along common edges. There are two different trominoes, the 1×3 rectangular I tromino and the angular L tromino.

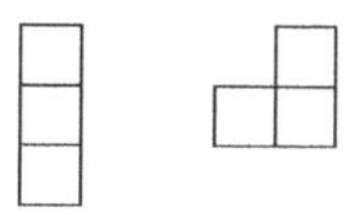

(a) Give a necessary and sufficient condition for when an $m \times n$ rectangular board can be tiled with the I tromino.

(b) Suppose that three corners are removed from a 9×9 board. Can the trimmed board be tiled with trominoes?

[*Hint:* Color the board with three colors in a pattern of alternating colors.]

1.3.7. A *tetromino* is formed with four squares joined along common edges. For example, the O and the Z tetrominoes are shown here:

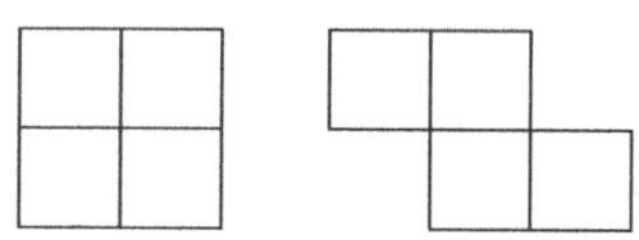

(a) find the three other tetrominoes, called the I, J, and T tetrominoes.

(b) the set of five tetrominoes has a total area of 20 square units. Explain why it is not possible to tile a 4×5 square with a set of tetrominoes.

(c) show that a 4×10 rectangle can be tiled with two sets of tetrominoes.

(d) show that a 5×8 rectangle can be tiled with two sets of tetrominoes.

1.3.8. **(a)** Construct all of the tilings of a 2×4 chessboard with dominoes. [If you have skill with a computer algebra system, you may be interested in verifying that the number of tilings that you obtain is given by formula (1.5).]

(b) Repeat part (a), but for a 3×4 chessboard.

1.4 FIGURATE NUMBERS

In the fifth and sixth centuries BCE, numbers were the essence of the Pythagorean universe. Indeed, for them all objects were composed of whole numbers. A unit was not an abstraction, but was viewed as a very tiny geometric sphere that could be represented by a dot. Numbers were classified by the shape of the patterns that can be formed by the corresponding number of dots. For example, the numbers 1,3,6,10,15,... were called *triangular numbers*, because each of these numbers of dots could be arranged into a triangular pattern as shown in Figure 1.3.

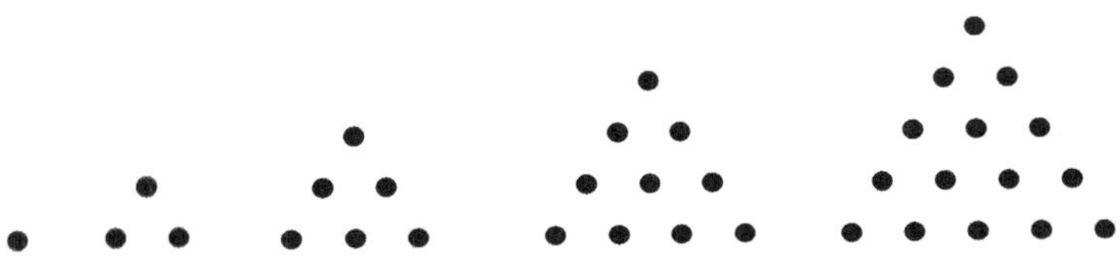

FIGURE 1.3 The first five triangular numbers 1, 3, 6, 10, and 15.

To investigate the triangular numbers in more detail, it is helpful to let t_n denote the nth triangular number. For example, $t_1 = 1$, $t_2 = 3$, $t_3 = 6$, $t_4 = 10$, and $t_5 = 15$. The triangular patterns in Figure 1.3 then make it clear, by counting the dots by rows from top to bottom, that

$$t_n = 1 + 2 + 3 + \cdots + n \tag{1.6}$$

The pattern that represents t_n consists of a triangle of t_{n-1} dots above the bottom row of n dots, so it follows that

$$t_n = t_{n-1} + n \tag{1.7}$$

This equation is a *recurrence relation* for the sequence of triangular numbers, since it is a formula for the nth triangular number that depends on the value of an earlier term in the sequence. Recurrence relations often arise in combinatorial analysis, and they will be considered in detail in Chapter 5.

Equations (1.6) and (1.7) are of interest, but neither one provides a *closed-form* expression for t_n. To obtain a simple algebraic formula for the triangular numbers, place two dot patterns side by side, with one turned upside down. This results in the pattern shown in Figure 1.4.

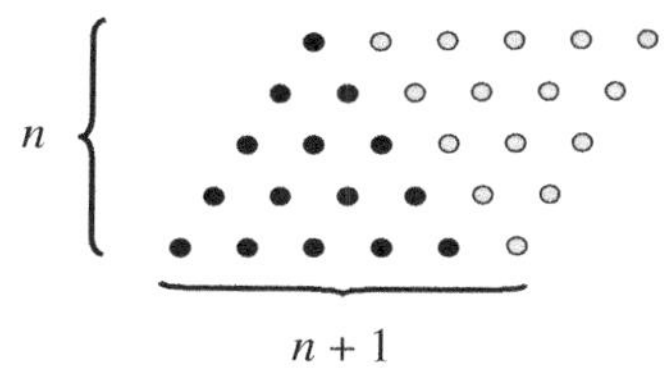

FIGURE 1.4 Dot pattern showing $t_n + t_n = n(n+1)$.

Since each of the n rows in the combined pattern contains $n + 1$ dots, there are $n(n+1)$ dots in all. Thus, we obtain the following theorem that provides a closed form expression for the nth triangular number.

Theorem 1.23 The nth triangular number $t_n = 1 + 2 + \cdots + n$ is given by

$$t_n = \frac{1}{2}n(n+1) \tag{1.8}$$

For example, we see from equation (1.8) that

$$t_{100} = 1 + 2 + \cdots + 100 = \frac{1}{2} \cdot 100 \cdot 101 = 50 \cdot 101 = 5050$$

This sum has become well known because of its connection with a story concerning the mathematical abilities shown by Carl Friedrich Gauss (1777–1855) as a young boy. When about 10 years of age, his teacher Master Büttner asked the class to sum the numbers 1 through 100. Gauss very quickly wrote his answer on his slate, whereas his classmates continued to calculate for another hour. Master Büttner was surprised to discover that only Gauss had given the correct answer, and asked him how he obtained his result. Gauss explained that $1 + 100 = 101, 2 + 99 = 101, \ldots, 50 + 51 = 101$, so there are 50 such sums that are each 101. Therefore, the answer is $50 \cdot 101 = 5050$.

Let's now turn to the *square numbers* $s_1 = 1, s_2 = 4, s_3 = 9, s_4 = 16, s_5 = 25, \ldots, s_n = n^2$, which are represented by the dot patterns shown Figure 1.5.

FIGURE 1.5 The first five square numbers 1, 4, 9, 16, and 25.

Once again, the dot patterns quickly reveal several properties of the square numbers. For example, the dots can be counted by diagonal rows or with angular shapes known to the Pythagoreans as *gnomens*. We also see that the square pattern that represents s_n is formed with two successive triangular patterns representing t_n and t_{n-1}.

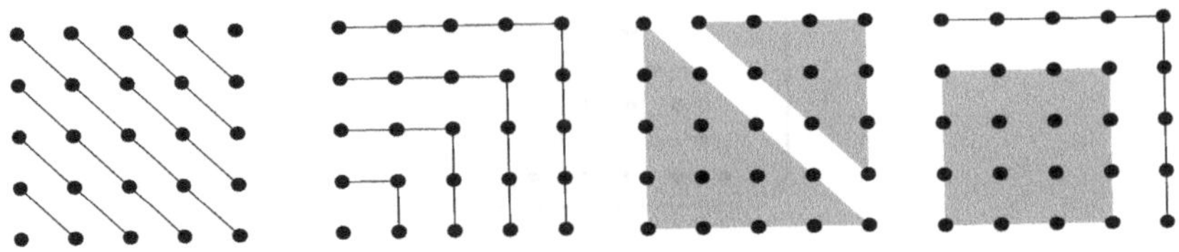

FIGURE 1.6 Discovering properties of the square numbers.

The following formulas are suggested by Figure 1.6:

$$s_n = 1 + 2 + 3 + \cdots + n + \cdots + 3 + 2 + 1 \tag{1.9}$$

$$s_n = 1 + 3 + 5 + \cdots + (2n - 1) \tag{1.10}$$

$$s_n = t_n + t_{n-1} \tag{1.11}$$

$$s_n = s_{n-1} + (2n - 1) \tag{1.12}$$

1.4.1 More General Polygonal Numbers

The Pythagoreans saw no reason to stop with the triangular and square numbers, since it was nearly as easy to arrange dots into pentagonal, hexagonal, and other polygonal patterns. For example, the pentagonal and hexagonal number patterns are shown in Figure 1.7.

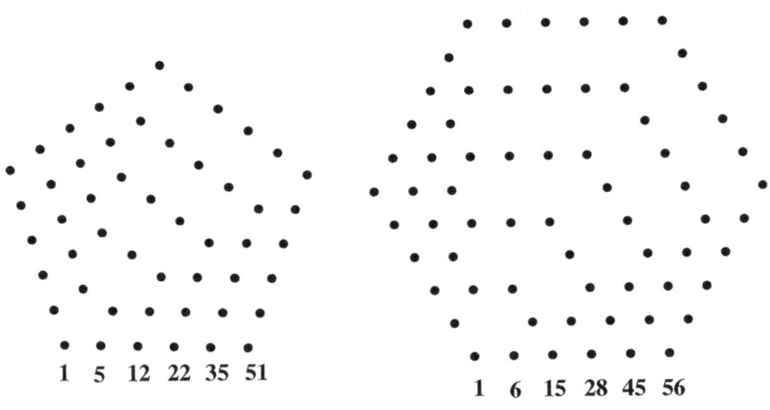

FIGURE 1.7 The pentagonal numbers $1, 5, 12, 22, 35, 51, \ldots$ and the hexagonal numbers $1, 6, 15, 28, 45, 56, \ldots$.

Let's denote the *n*th *r-gonal number* by $p_n^{(r)}$. For example, Figure 1.7 shows us that $p_3^{(5)} = 12$ is the third pentagonal number, and $p_5^{(6)} = 45$ is the fifth hexagonal number.

To obtain a closed-form expression for $p_n^{(r)}$, we can subdivide the dot pattern into a row of n dots together with $r - 2$ triangular arrays. There are $r - 2$ triangular patterns of dots that each contain t_{n-1} dots. The hexagonal case of this subdivision is shown in Figure 1.8.

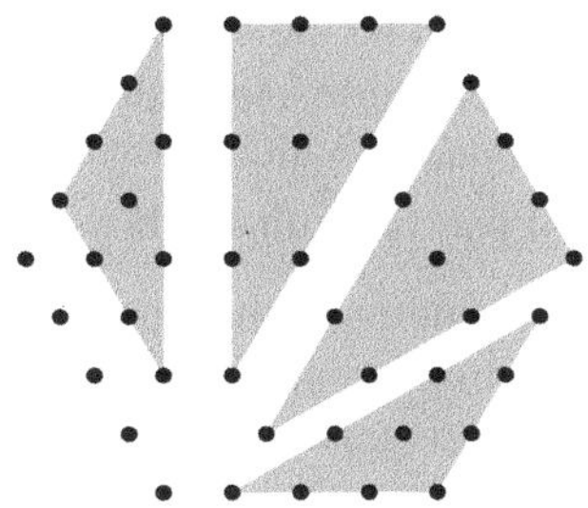

FIGURE 1.8 A hexagonal dot pattern shows that $p_5^{(6)} = 5 + (6-2)t_{5-1}$.

In general, the following theorem expresses the nth r-gonal number in terms of triangular numbers.

Theorem 1.24 The nth r-gonal number is given by

$$p_n^{(r)} = n + (r-2)t_{n-1}, n \geq 1 \tag{1.13}$$

where t_{n-1} is the $(n-1)$st triangular number.

In view of (1.8), some algebraic rearrangement gives us the equivalent closed-form expressions in the next theorem.

Theorem 1.25 The nth r-gonal number is given by the closed-form expressions

$$\begin{aligned} p_n^{(r)} &= \frac{n}{2}[n(r-2) - (r-4)] \\ &= \frac{n}{2}[(n-1)r - 2(n-2)] \end{aligned} \tag{1.14}$$

These two formulas suggest that we define $p_0^{(r)} = 0$.

There are several useful variations of dot patterns beyond the polygonal numbers. In particular, the *centered polygonal numbers* are introduced in the following problem set. Dot patterns will also be used extensively in Chapter 6 when we investigate partition numbers.

PROBLEMS

1.4.1. The following diagram illustrates that $t_{2m} = m(2m+1)$:

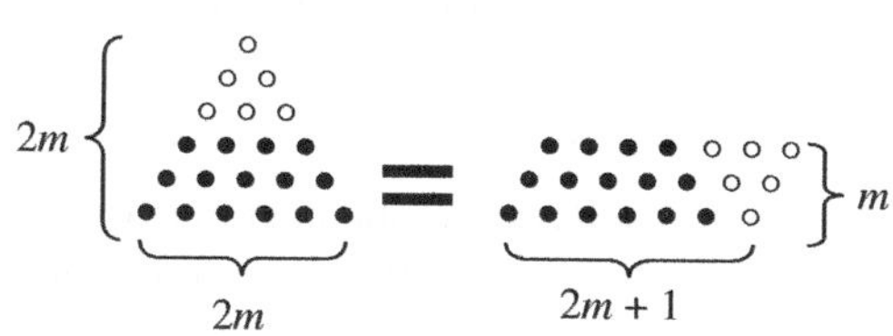

Create a similar diagram that illustrates the formula $t_{2m+1} = (2m+1)(m+1)$.

1.4.2. **(a)** Use dot patterns to show that the square of an even integer is a multiple of 4. [In number-theoretic terms, the square of any even integer is said to be *congruent to* 0 *modulo* 4, which is written as $(2n)^2 \equiv 0 \pmod 4$.]

(b) Verify your result of part (a) with algebra.

1.4.3. Use both algebra and dot patterns to show that the square of an odd integer is congruent to 1 modulo 8; that is, show that $s_{2n+1} = 8u_n + 1$ for some integer u_n. Be sure to identify the integer u_n by its well-known name.

1.4.4. The *centered triangular numbers* are obtained by starting with a single dot and then surrounding it by triangles with 2, 3, 4, 5, ... dots per side. The following diagram shows that the first five centered triangular numbers are 1, 4, 10, 19, 31, and 46:

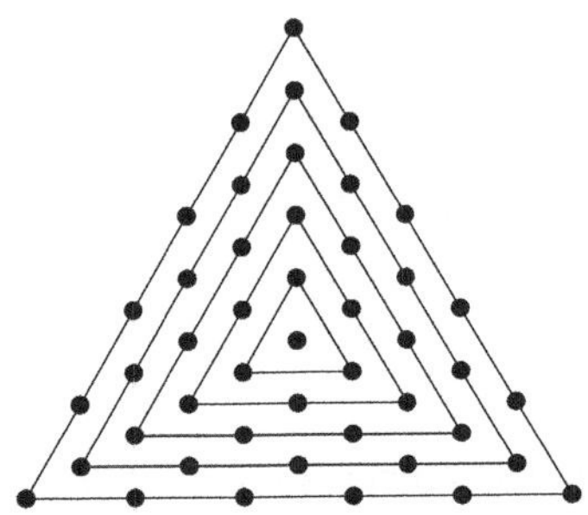

(a) use the diagram to show that the nth centered triangular number is given by $1 + 3t_n$.

(b) derive a closed form expression for the nth centered triangular number.

1.4.5. The *centered square numbers* are obtained much like the centered triangle numbers of Problem 1.4.4, except that squares with an increasing number of dots per side surround a center dot.

(a) Create a diagram that shows the sequence of centered square numbers begining with 1, 5, 13, 25, and 41.

(b) Color the dots in the diagram from part (a) to show that the nth centered square number is given by $(n+1)^2 + n^2$.

(c) Shade your diagram from part (a) to shows that every centered square number is congruent to 1 modulo 4.

(d) Verify part (c) with algebra.

1.4.6. The *centered polygonal numbers* are obtained much like the centered triangle and centered square numbers of Problems 1.4.4 and 1.4.5, except that polygons with an increasing number of dots per side surround a center dot:

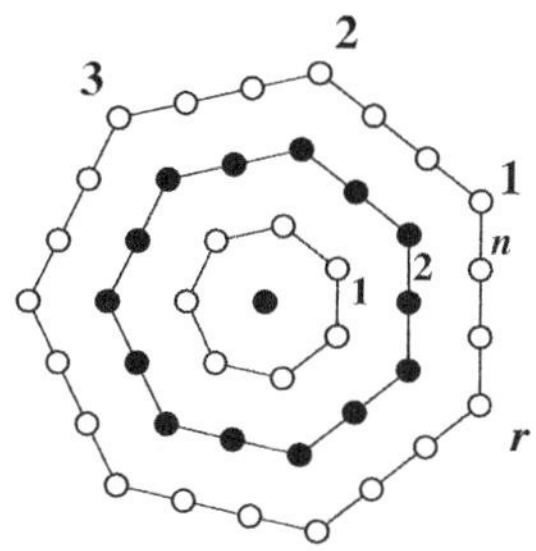

Explain why the nth centered r-gonal number can be mathematically represented by $c_n^{(r)} = rt_n + 1$.

1.4.7. The first three *trapezoidal numbers* are 1, 5, and 12, as shown by the dot pattern here.

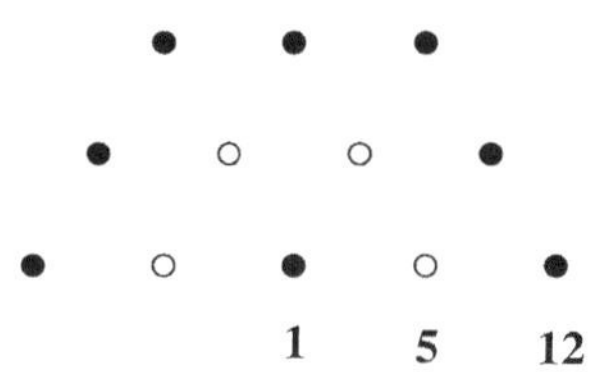

(a) Continue the trapezoidal pattern to find the next three trapezoidal numbers.

(b) Draw some lines on your diagram from part (a) to explain why the trapezoidal numbers are simply an alternative pattern for the pentagonal numbers $p_1 = 1, p_2 = 5, p_3 = 12, \ldots$.

(c) Use the trapezoidal diagram to show why each pentagonal number is the sum of a triangular number and a square number. Give an explicit formula for p_n in terms of the triangular and square numbers.

(d) The trapezoidal diagram shows that each pentagonal number is the difference of two triangular numbers. Determine the two triangular numbers corresponding to p_n and express this result in a formula.

(e) Construct a diagram showing that each pentagonal number is one-third of a triangular number. Give an explicit formula of this property.

1.4.8. A domino is a 1×2 rectangular tile, divided by a centerline into two unit squares. In a double-6 set, each square is imprinted with a pattern ranging from 0 to 6 dots that determines the *suit* of the domino. For example, the 2–5 domino has 2 pips on one square and 5 pips on the other, so it belongs to both the two suit and the five suit. A *double* domino, such as the 6–6 domino, belongs to just one suit.

(a) How many dominoes are in a double-6 set? Note that there is just one domino for each pair of suits, since a $p-q$ domino is the same as the $q-p$ domino.

(b) How many total pips are on the complete double-6 set?

1.4.9. Dominoes, as described in Problem 1.4.8, also come in double-9, double-12, double-15, and even double-18 sets. Consider, more generally, a double-n set, so each half-domino is imprinted with 0 to n pips.

(a) Derive a formula for the number of dominoes in a double-n set. Use the formula to determine the number of dominoes in a double-n set for $n = 6, 9, 12, 15$, and 18.

(b) Derive a formula for the total number of pips in a double-n set. Use the formula to determine the total number of pips in a double-n set for $n = 6, 9, 12, 15$, and 18.

1.4.10. Trace the octagonal dot pattern and provide shading (see Figure 1.8) to show that $p_4^{(8)} = 4 + (8-2)t_{4-1} = 4 + 6 \cdot 6 = 40$:

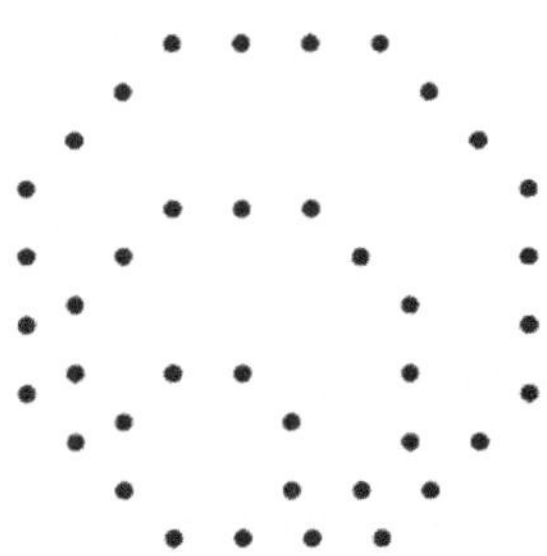

1.5 COUNTING TILINGS OF RECTANGLES

In this section, we will investigate the number of ways that a $1 \times n$ rectangular board can be tiled with either squares of two colors or a mixture of 1×1 squares and 1×2 dominoes. We will discover several useful principles of combinatorial reasoning.

1.5.1 Tiling a Rectangle with Squares of Two Colors

For small values of n it is not unreasonable to draw all of the tilings with squares of two colors and then directly count the number of ways this can be done. Our hope is to discover and utilize patterns to obtain general answers. Figure 1.9 shows all of the tilings of a $1 \times n$ rectangular board with gray and white square tiles in the cases $n = 1, 2$, and 3.

If we let $T^{(n)}$ denote the number of ways to tile a board of length n with tiles of two colors, we see that $T^{(1)} = 2, T^{(2)} = 4, T^{(3)} = 8$. This suggests that the number of

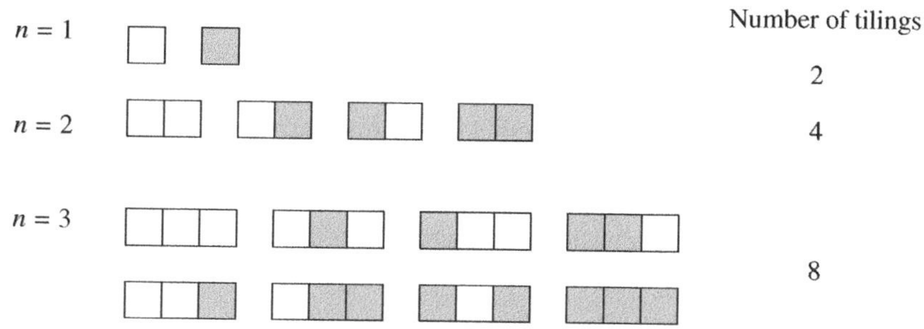

FIGURE 1.9 Tilings by squares of two colors.

tilings doubles each time the length of the board increases by one; thus it appears that we have the recurrence relation

$$T^{(n)} = 2T^{(n-1)} \tag{1.15}$$

Note that this equation will hold for all integers $n \geq 1$ if we define $T^{(0)} = 1$.

To understand why equation (1.15) holds, we see that there are two ways to extend a tiling of a board of length $n-1$ to become a tiling of a board of length n, namely, these, where we add either a white or a gray square to the right end:

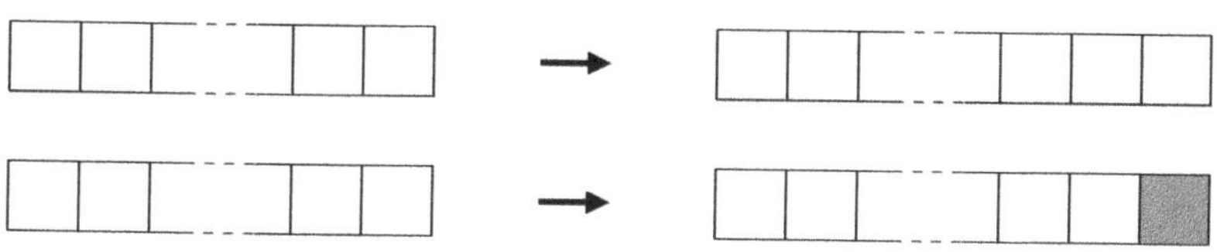

The two methods never form the same tiling, since one extension gives a tiling that ends with a white square and the other extension ends with a gray square. Moreover, *all* tilings of a board of length n are obtained in this way, since given any tiling of length n, we can delete its rightmost square to create a tiling of a board of length $n-1$. In summary, *all* of the tilings of length n can be separated into two disjoint classes depending on the color of the rightmost square—either white or gray—and the two classes together contain all of the $T^{(n)}$ tilings of a board of length n. Thus, we have derived the recursion relation $T^{(n)} = T^{(n-1)} + T^{(n-1)} = 2T^{(n-1)}$ that we had guessed earlier.

In later chapters, we will consider a variety of methods to solve recurrence relations, but the doubling relation can be solved easily by iterating the recurrence until we arrive at an expression containing the known value $T^{(0)} = 1$:

$$T^{(n)} = 2T^{(n-1)} = 2\left(2T^{(n-2)}\right) = 2\left(2\left(2T^{(n-3)}\right)\right) = \cdots = 2^n T^{(0)} = 2^n. \tag{1.16}$$

We have proved the following theorem.

Theorem 1.26 There are

$$T^{(n)} = 2^n \tag{1.17}$$

ways to tile a rectangular board of length n with white and gray squares.

This answer is unexpectedly simple, which suggests there may be a more direct derivation of the result. Let's consider the board of length 4, as shown in Figure 1.10.

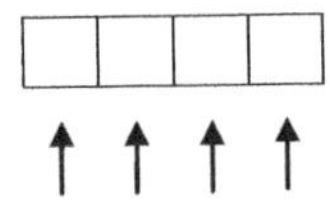

FIGURE 1.10 There are 4 two-way choices to make: cover each of the four cells of the board with a white or gray tile.

The arrows in the figure highlight where we can decide what color square to use, white or gray. Since each of the four choices can go either of two ways, there are $2 \times 2 \times 2 \times 2 = 2^4 = 16$ ways to tile a board of length 4. The same reasoning can be applied to count the number of tilings of a board of any length n; we have n choices that can each be made in two ways, so the number of tilings of a $1 \times n$ board with squares of two colors is

$$2 \times 2 \times 2 \times \cdots \times 2 = 2^n$$

The reasoning we have used is called the *multiplication principle* of combinatorics, and will be explored in detail in the next section of this chapter. This principle can be used to solve some related tiling problems, as demonstrated with these two questions:

Question 1. In how many ways can a $1 \times n$ rectangular board be tiled so that each cell of the board is covered with a red, green, or blue 1×1 square tile?

Answer 1. There is a three-way choice to be made n times, so the board can be tiled in $3 \times 3 \times \cdots \times 3 = 3^n$ ways.

Question 2. In how many ways can a $1 \times n$ rectangular board be tiled so that the first cell of the board is covered with a red square tile, the second cell is covered with either a green or blue tile, and more generally the jth cell is covered by a square tile with j choices of its color?

Answer 2. There are j choices for the color of the tile that covers the jth cell, $1 \leq j \leq n$, so the board can be tiled in $1 \times 2 \times 3 \times \cdots \times n$ ways.

The product of the first n positive integers occurs frequently in combinatorics, so it is given a special name and notation:

$$n! = n \times (n-1) \times \cdots \times 3 \times 2 \times 1 \text{ is } n \textit{ factorial}$$

1.5.2 Tiling a $1 \times n$ Rectangle with Exactly r Gray Squares

As before, we will count the number of ways to tile a board with white and gray squares, but now we will restrict the number of squares of each color. Let $C(n, r)$ denote the number of ways to tile a $1 \times n$ board using r gray squares and $n - r$ white squares. Some simple drawings should provide the entries in the following table. We have set $C(0,0) = 1$ to provide a more complete table:

$C(n, r)$	Number of Gray Squares in Tiling r				
Length of board n	$r = 0$	$r = 1$	$r = 2$	$r = 3$	$r = 4$
$n = 0$	1				
$n = 1$	1	1			
$n = 2$	1	2	1		
$n = 3$	1	3	3	1	
$n = 4$	1	4	6	4	1

It follows from Theorem 1.26 that the sum of the entries in row n is 2^n, and so we have the identity

$$\sum_{r=0}^{n} C(n, r) = 2^n \tag{1.18}$$

We might also note from the table that the sum of two adjacent values in a row is the entry in the next row just below the right-hand summand. This could have been anticipated, since there are two distinct ways to create a tiling of length n that has r gray squares:

1. Given any of the $C(n - 1, r - 1)$ tilings of length $n - 1$ with $r - 1$ gray tiles, add a gray tile at the right to create a tiling of length n with r gray squares:

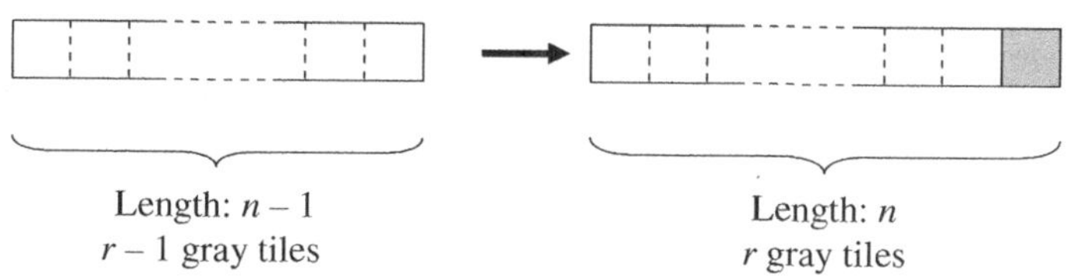

2. Given any of the $C(n - 1, r)$ tilings a tiling of length $n - 1$ and r gray tiles, add a white tile at the right to create a tiling of length n still with r gray tiles:

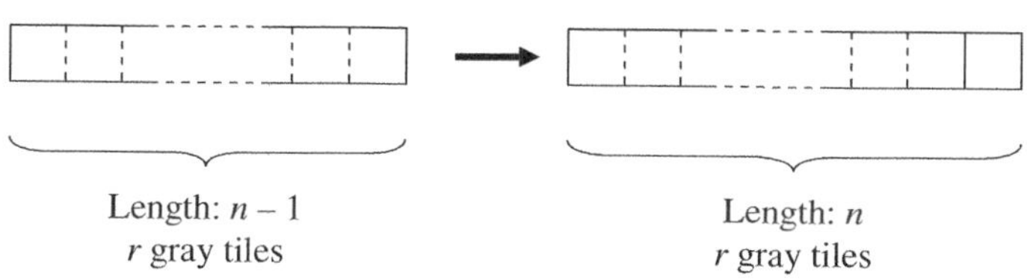

The two methods never form the same tiling, since we create tilings that end in squares of different colors. Moreover, all of the tilings of length n with r gray squares are formed in this way, since any tiling necessarily ends with either a gray or a white tile. Therefore, the set of all tilings of length n with r gray squares splits into two disjoint subsets, one with $C(n-1, r-1)$ elements and the other with $C(n-1, r)$ elements. We have proved the following recursive identity:

$$C(n, r) = C(n-1, r-1) + C(n-1, r) \tag{1.19}$$

This identity makes it easy to extend the table of values for as many rows as we wish, as shown here:

				r					
	1								
	1	1							
	1	2	1						
	1	3	3	1					
n	1	4	6	4	1				
	1	5	10	10	5	1			
	1	6	15	20	15	6	1		
	1	7	21	35	35	21	7	1	
	1	8	28	56	70	56	28	8	1

This tabulation is the famous *Pascal triangle*,[2] for which the entry in row n and column r is the *binomial coefficient* denoted by $C(n,r)$ or by $\binom{n}{r}$. For example, $\binom{8}{3} = 56$ is the entry in row 8 and column 3 of the tabulation above. Both of the indices n and r begin with 0.

The C symbol refers to its meaning as a *combination*; that is, $C(n,r)$ is the number of ways in which a subset of r objects can be chosen from a given set of n distinct objects, where the order in which objects are selected is unimportant. It is helpful to use the letter C as a reminder that we are counting the ways to *choose* a subset of a certain size, and the symbols $C(n,r)$ and $\binom{n}{r}$ are usefully read as "n choose r." Combinations and the binomial coefficients play a major role in combinatorial analysis, and their study will be taken up in detail in later chapters. In particular, later we will obtain a formula for the binomial coefficients, namely

$$C(n, r) = \binom{n}{r} = \frac{n!}{r!(n-r)!} \tag{1.20}$$

[2]Blaise Pascal (1623–1662) was a French mathematician, physicist, inventor, essayist, philosopher, and theologian. As a teenager, he invented a calculating machine called the *Pascaline.* He wrote fundamental works on projective geometry and probability, and later in his a life a treatise on the triangular array that we now call *Pascal's triangle.*

where

$$0! = 1, \quad k! = k(k-1)(k-2)\cdots 3 \cdot 2 \cdot 1 \tag{1.21}$$

are the *factorials* introduced earlier.

For our tiling problem, $C(n,r)$ is the number of ways to choose the r cells to be covered with gray tiles on a $1 \times n$ board, leaving the remaining $n - r$ squares of the board to be covered with white tiles. For example, to tile a 1×7 board, there are $\binom{7}{1} = 7$ (i.e., 7 choose 1) ways to choose one cell covered by a gray tile. To form a tiling with 2 gray tiles, there are $\binom{7}{2}$ (i.e., 7 choose 2) ways to choose the 2 cells covered by gray tiles. Since $\binom{7}{2} = 21$, as seen from Pascal's triangle or from the factorial expression

$$\binom{7}{2} = \frac{7!}{2!5!} = \frac{7 \cdot 6}{2 \cdot 1} = 21$$

we see that a 1×7 board can be tiled in 21 ways using 2 gray and 5 white tiles. Similarly, there are 35 ways to tile a board of length 7 with 3 gray tiles and 4 white tiles.

The reasoning above can be applied generally to boards of any length, giving us the following theorem.

Theorem 1.27 The number of tilings of a $1 \times n$ board with r gray tiles and $n - r$ white tiles is given by the binomial coefficient (n choose r):

$$C(n,r) = \binom{n}{r} = \frac{n!}{r!(n-r)!} \tag{1.22}$$

Counting tilings has provided us with two identities, (1.18) and (1.19) for the binomial coefficients that will be encountered again later. Combinatorialists generally favor writing "n choose r" with the $\binom{n}{r}$ style of notation, so the identities take the form

$$\sum_{r=0}^{n} \binom{n}{r} = 2^n \tag{1.23}$$

$$\binom{n}{r} = \binom{n-1}{r-1} + \binom{n-1}{r} \qquad \text{(Pascal's identity)} \tag{1.24}$$

Both of these identities were derived with combinatorial reasoning; that is, they were obtained using the principles of counting rather than, say, algebraic calculation. It would be more difficult to prove identity (1.23) algebraically using formulas (1.20) and (1.21), although Pascal's identity can be verified in a few lines of calculation. To the combinatorialist, a combinatorial proof is much the preferred approach, with algebra used sparingly. Combinatorial proofs will be used frequently throughout this text.

1.5.3 Tiling a $1 \times n$ Rectangle with Squares and Dominoes

Figure 1.11 shows the ways that a 1×1, 1×2, 1×3, and 1×4 board can be tiled with 1×1 squares and 1×2 dominoes.

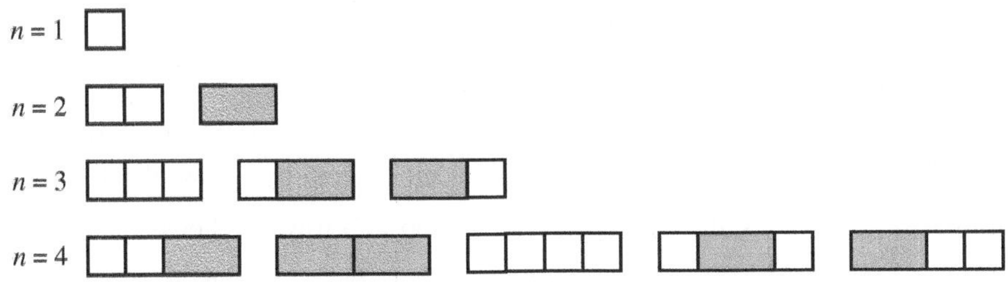

FIGURE 1.11 Tilings by squares and dominoes.

To see a pattern, it helps to check that there are 8 ways to tile a board of length 5. If we let f_n denote the number of tilings of a $1 \times n$ board, we see that $f_1 = 1, f_2 = 2$, $f_3 = 3, f_4 = 5$, and $f_5 = 8$.

Although the evidence is minimal, so far it appears that the sum of two successive terms of the sequence gives the next term in the sequence. To verify that this property holds in general, we can proceed by analogy with the approach that was successful for the tiling of a board with colored squares. That is, we want to see how the tilings of boards of length $n - 2$ and $n - 1$ can be modified to form all of the tilings of a board of length n. As before, it seems helpful to add an additional tile at the right of the board in one of these two distinct ways:

1. Given a tiling of length $n - 2$, add a domino at the right to create a tiling of length n:

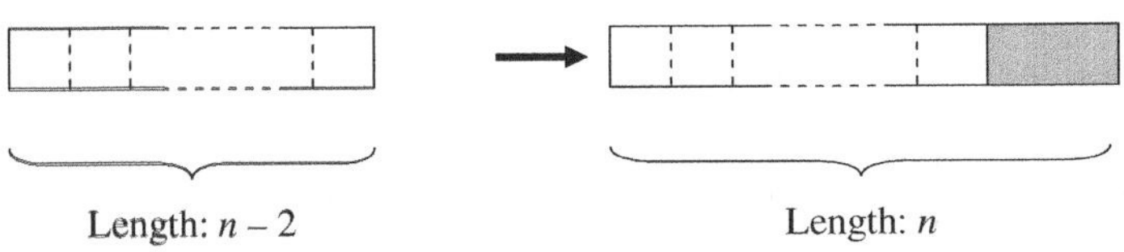

2. Given a tiling of length $n - 1$, add a square at the right to create a tiling of length n:

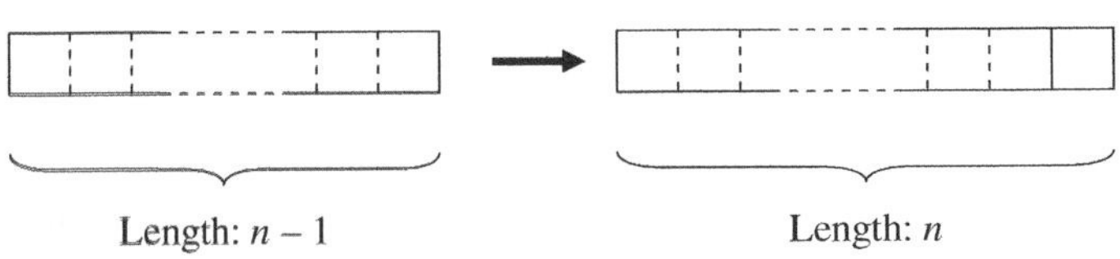

The two methods never form the same tiling, since we create a tiling ending in either a domino or a square. Moreover, any tiling must end with either a square or domino, so *all* of the tilings of length n are created in *exactly one* of these two ways. Thus, we have the following recurrence relation:

$$f_n = f_{n-2} + f_{n-1} \tag{1.25}$$

Using the beginning values, $f_1 = 1, f_2 = 2$, we can use the recurrence formula (1.25) to extend the sequence as far as we wish. For example, the first 10 terms of the sequence are given in the following table:

n	1	2	3	4	5	6	7	8	9	10
f_n	1	2	3	5	8	13	21	34	55	89

This is the famous sequence of *Fibonacci numbers*,[3] which are most often defined by the following recurrence relation and initial conditions:

$$F_0 = 0, F_1 = 1, F_n = F_{n-1} + F_{n-2}, n \geq 2 \tag{1.26}$$

We see that $f_1 = F_2 = 1, f_2 = F_3 = 2$, and more generally that $f_n = F_{n+1}$. The numbers in the sequence f_n are often called the *combinatorial Fibonacci numbers,* since they frequently appear in counting situations.

The following theorem relates the number of tilings by squares and dominoes to the Fibonacci numbers.

Theorem 1.28 The number of tilings f_n of a $1 \times n$ board with squares and dominoes is given by the Fibonacci number F_{n+1}.

Often combinatorial problems come in different disguises, but they are the same problems beneath the disguises. For example, suppose that we wish to know the number of sequences of length n where each term is a 0 or a 1. Such a sequence—known as a *binary sequence*—is just a way to describe how a $1 \times n$ board can be tiled with gray and white tiles, where each 0 corresponds to a white tile and each 1 corresponds to a gray one. Thus, by Theorem 1.26, there are 2^n binary sequences of length n. Similarly, to determine how many binary sequences of length n have r ones and $n - r$ zeros, we would use the same method as used to find the number of tilings of a $1 \times n$ board with r gray tiles and $n - r$ white tiles. In Problem 1.5.11, you will discover how to answer this question: *How many binary sequences of length n have no two consecutive ones?*

[3]Leonardo Fibonacci (ca. 1170–1250), also known as Leonardo of Pisa, is most famous for his book *Liber Abaci*, which introduced Indo-Arabic mathematics and notations to European scholars. The number sequence that bears his name was first introduced in a combinatorial problem related to breeding rabbits under some idealized conditions.

PROBLEMS

1.5.1. **(a)** Extend Figure 1.9 to depict the set of 16 tilings of a board of length 4, where each tile is either gray or white.

(b) Explain why it is easy to use the 8 tilings of boards of length 3 to draw all of the tiled boards of length 4.

1.5.2. Explain how to obtain the tilings of a 1×4 board with 2 gray squares by modifying the tilings of the 1×3 boards that have either 1 or 2 gray squares.

1.5.3. Use formulas (1.20) and (1.21) to prove Pascal's identity (1.24).

1.5.4. Show that there are 8 ways to tile a 1×5 rectangular board with squares and dominoes.

1.5.5. **(a)** Find all of the ways that a 2×4 rectangular board can be tiled with 1×2 dominoes. Here is one way to tile the board:

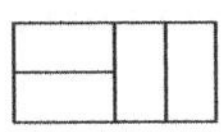

Draw all of the ways to tile a 2×4 board with dominoes.

(b) How many ways can a $2 \times n$ board be tiled with dominoes?

1.5.6. Suppose that a *train* of length n is a tiling of a $1 \times n$ rectangle by $1 \times r$ rectangles called *cars*, where the length of a car is any positive integer r. For example, here is a train of length 13 formed with 6 cars of lengths 1, 3, 1, 1, 4, and 3, with total length $1 + 3 + 1 + 1 + 4 + 3 = 13$:

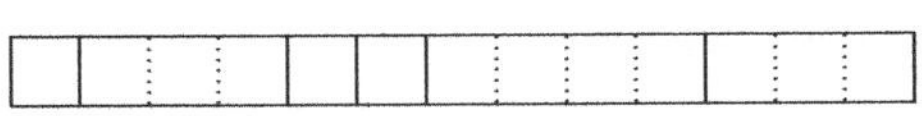

(a) Let $t^{(n)}$ denote the number of trains of length n. Determine $t^{(1)}, t^{(2)}, t^{(3)}, t^{(4)}$ by constructing a figure showing all of the trains.

(b) Guess and then prove a formula for $t^{(n)}$.

1.5.7. The following train (see Problem 1.5.6 for the definition of a train) has just one car of length 13:

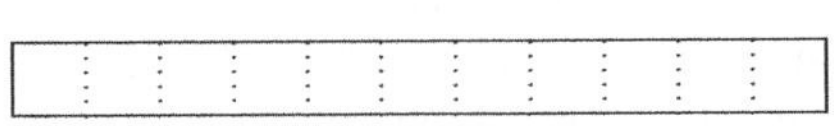

(a) in how many ways can a train of length 13 be formed with 2 cars?

(b) why are there $\binom{12}{4}$ trains of length 13 that can be formed with 5 cars?

(c) generalize your answer to part (b) to give a binomial coefficient that expresses the number of trains of length n with r cars.

1.5.8. There are 4 ways to express 3 as an ordered sum of positive integers, namely, $3, 1 + 2, 2 + 1$, and $1 + 1 + 1$.

(a) In how many ways can 4 be expressed as an ordered sum?

(b) In how many ways can a positive integer n be expressed as an ordered sum of positive integers? (*Suggestion:* See Problem 1.5.6 and Problem 1.5.7.)

1.5.9. There are 4 ways to express 5 as a sum of 2 ordered summands, namely $4 + 1, 3 + 2, 2 + 3$, and $1 + 4$.

(a) In how many ways can 5 be expressed as a sum of 3 ordered summands? (See Problem 1.5.8.)

(b) In how many ways can a positive integer n be expressed as a sum of k summands?

1.5.10. Suppose that a *g-tiling* is a tiling of a rectangular board by squares, dominoes, or trominoes, where a tromino is a 1×3 tile. Let $G^{(n)}$ denote the number of g-tilings of a board of size $1 \times n$.

(a) Determine $G^{(1)}, G^{(2)}, G^{(3)}, G^{(4)}$.

(b) Explain how to associate the g-tilings of length 4 with the tilings of length 1, 2, and 3, and obtain an expression for $G^{(4)}$ in terms of $G^{(1)}, G^{(2)}, G^{(3)}$.

(c) Derive a recurrence relation for $G^{(n)}$, where $n \geq 4$.

1.5.11. How many binary sequences of length n have no two consecutive ones? (A binary sequence in an ordered list of ones and zeros, such as 100101001.) For example, there are five binary sequences of length 3 with no two consecutive ones, namely, 000, 100, 010, 001, and 101.

1.6 ADDITION AND MULTIPLICATION PRINCIPLES

In enumerative combinatorics, the goal is to determine the number of members of a set S, as determined by the properties that determine membership in the set. For example, S may be the set of (1) dots that can be arranged into a geometric figure of a certain shape, (2) tilings of a $1 \times n$ rectangle by squares of two colors, or (3) tilings of a chessboard with dominoes. The number of elements of set S is denoted by $|S|$.

1.6.1 Addition Principles

In several of the problems addressed in the last section, it was helpful to separate the members of a set S into two subsets, say, sets A and B. For example, we separated the tilings of a rectangle into two subsets; one subset contained those tilings ending with a white tile and the second subset contained the tilings ending with a gray tile. To account for every member of S, it is important that *every* member of S either belong to either A or B. In other words, we required $S = A \cup B$. But we also want to count every element just once, so no element of S can belong to both A and B; that is, we require A and B to be disjoint sets, so that $A \cap B = \emptyset$. When these two conditions are met, we say that the subsets A and B are a *partition* of set S. The number of elements in S is simply the sum of the number of elements in the subsets A and B. This important idea is known as the *addition principle* of combinatorics.

Theorem 1.29 (Addition Principle) Let $S = A \cup B$, where $A \cap B = \emptyset$. Then

$$|S| = |A \cup B| = |A| + |B|. \tag{1.27}$$

Example 1.30 The traditional soccer ball, first made in Denmark around 1950, is a spherical polyhedron known as a *truncated icosahedron*. The surface of the ball has 20 white hexagonal panels and 12 black pentagonal panels (see diagram). What is the total number of panels of the ball?

Solution. Each panel of the ball is either hexagonal or pentagonal, so the set of hexagonal panels and the set of pentagonal panels partition the set of all of the panels of the ball. By the addition principle, there are $20 + 12 = 32$ panels in all. ■

For some problems, the addition principle is inverted. As a quick example, if there are 15 members of the math club and 9 members are women, then there are $15 - 9 = 6$ male members. More generally, we have this theorem, where $S - A$ denotes the difference of sets defined by $S - A = \{x \in S \mid x \notin A\}$.

Theorem 1.31 (Subtraction Principle) Let $A \subseteq S$ and $B = S - A$. Then

$$|B| = |S - A| = |S| - |A| \tag{1.28}$$

Proof. Since A and $B = S - A$ are a partition of the set S, we have $|S| = |A| + |B| = |A| + |S - A|$. Solving this equation for $|B|$ gives equation (1.28). ■

Example 1.32 The figure shown is called the *gyroelongated square bicupola*, a polyhedron with 34 faces that are either squares or equilateral triangles. If there are 10 square faces, how many faces are triangles?

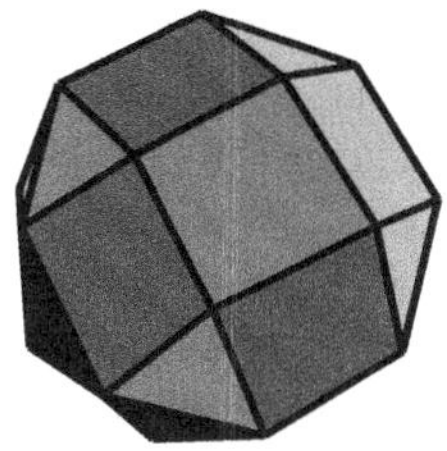

Solution. Let F, S, and T denote the sets of all faces, square faces, and triangular faces, respectively. Thus $T \subset F$ and $T = F - S$. According to the subtraction principle, there are $|T| = |F - S| = |F| - |S| = 34 - 10 = 24$ triangular faces. ■

In the next example, the addition principle is extended to a union of two sets that are not necessarily disjoint. If $S = A \cup B$, where $A \cap B \neq \varnothing$, then $|S|$ is no longer given by the sum $|A| + |B|$ since each element $x \in A \cap B$ is counted once as a member of A and a second time as a member of B. To compensate, we can subtract $|A \cap B|$ from the sum $|A| + |B|$ so that each member of S is counted exactly once. As a result, we have this theorem, which we also prove more formally by applying the subtraction principle.

Theorem 1.33 (Addition Principle for Unions of Two Sets) Let $S = A \cup B$. Then

$$|S| = |A \cup B| = |A| + |B| - |A \cap B| \tag{1.29}$$

Proof (via the Subtraction Principle). The Venn diagram below makes it clear that $S = A \cup B = A \cup (B - A)$, so S is partitioned by the disjoint sets A and $B - A$:

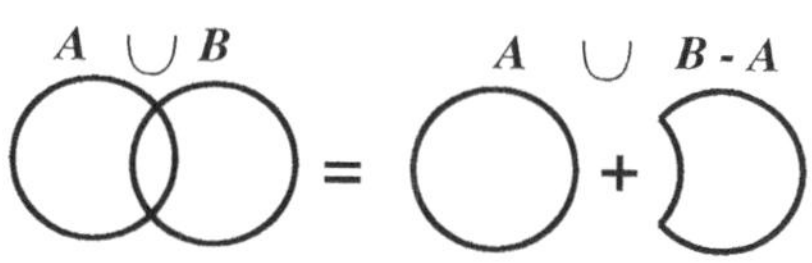

Therefore $|S| = |A| + |B - A|$ by the addition principle (1.27). But $A \cap B \subseteq B$, so by the subtraction principle $|B - A| = |B| - |A \cap B|$. Combining equations, we get $|S| = |A \cup B| = |A| + |B - A| = |A| + |B| - |A \cap B|$. ■

Example 1.34 Let H be the set of hearts, and F be the set of face cards from an ordinary 52-card deck. How many cards in the deck are either a heart or a face card?

Solution. We need to determine $|S|$, where $S = H \cup F$. There are $|H| = 13$ hearts, and $|F| = 12$ face cards with 3 face cards in each of the 4 suits. In particular, there are 3 hearts that are also face cards, so $|H \cap F| = 3$. We can now apply formula (1.29) to get $|H \cup F| = |H| + |F| - |H \cap F| = 13 + 12 - 3 = 22$. ■

In Chapter 4, we will consider the principle of inclusion–exclusion (PIE), which extends Theorem 1.33 to a formula that counts the number of elements in a set given as a union $S = A_1 \cup A_2 \cup \cdots \cup A_n$ of arbitrarily many sets. For now, however, suppose that each element of S belongs to exactly one of the subsets. In this case we say that $A_1, A_2, \ldots, A_n$ are *pairwise disjoint*, so that $A_j \cap A_k = \varnothing, j \neq k$. When $S = A_1 \cup A_2 \cup \cdots \cup A_n$ is a union of pairwise disjoint sets, we say that the sets $A_1, A_2, \ldots, A_n$ are a *partition* of set S. We can again find the number of members of S by summing the number of members of each set of the partition, giving us this theorem.

Theorem 1.35 (Extended Addition Principle) Let $S = A_1 \cup A_2 \cup \cdots \cup A_n$, where $A_j \cap A_k = \varnothing, j \neq k$. Then

$$|S| = |A_1| + |A_2| + \cdots + |A_n| \tag{1.30}$$

Example 1.36 The *rhombicosidodecahedron* is a polyhedron with 20 triangular faces, 30 square faces, and 12 pentagonal faces. Therefore the total number of faces is 20 + 40 + 12 = 62 faces:

Example 1.37 In Section 1.5, we saw that $C(n,r)$ is the number of the ways to tile a rectangular $1 \times n$ board with n tiles, r of which are gray and the rest white. Use the tiling model to show that

$$C(n + r + 1, r) = C(n, 0) + C(n + 1, 1) + \cdots + C(n + r, r) \tag{1.31}$$

Solution. The left side of (1.31) counts the number of ways to tile a $1 \times (n + r + 1)$ board with r gray tiles, and $n + 1$ white tiles. We must show the sum on the right side of (1.31) gives the same count. Suppose that the rightmost white square, call it S, is in position $n + k + 1$, where $k = 0, 1, \ldots, r$. To the right of S is a string of $r - k$ consecutive gray tiles, and the k other gray tiles must be located among the $n + r$ tiles to the left of S:

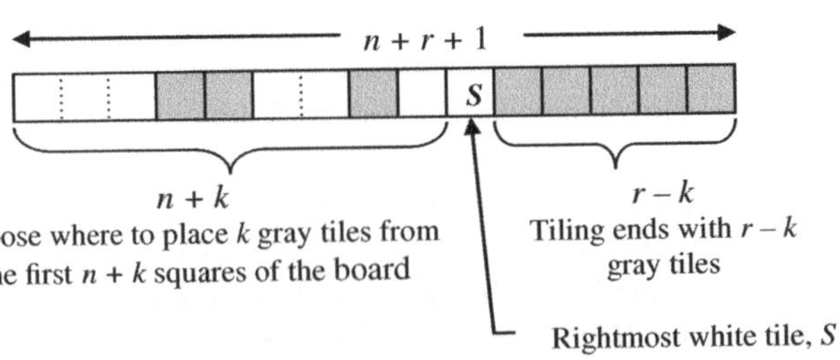

There are $C(n + k, k)$ ways to tile the first $n + k$ squares of the board using k gray tiles, so this is the number of tilings for which the S has position $n + k + 1$. By noting the position of the rightmost white square, we have partitioned the tiling into disjoint sets. By the extended addition principle, we see that the number of tilings is given by $\sum_{k=0}^{r} C(n + k, k)$, which is just the sum on the right of (1.31). ■

Identity (1.31) is most often written using the notation $C(n, r) = \binom{n}{r}$, acquiring the form shown in the following theorem.

Theorem 1.38 (Hockey Stick Identity 1)

$$\binom{n+r+1}{r} = \binom{n}{0} + \binom{n+1}{1} + \binom{n+2}{2} + \cdots + \binom{n+r}{r} \tag{1.32}$$

The name *hockey stick* becomes evident if a loop is drawn in Pascal's triangle that tightly encloses the binomial coefficients that appear in identity (1.32). Here is the hockey stick identity that corresponds to the case $n = 2$, $r = 5$:

$$56 = \binom{8}{5} = \binom{2}{0} + \binom{3}{1} + \binom{4}{2} + \binom{5}{3} + \binom{6}{4} + \binom{7}{5} = 1 + 3 + 6 + 10 + 15 + 21$$

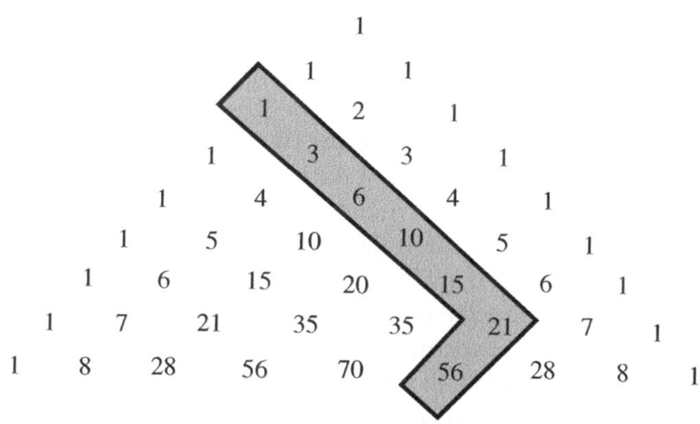

A second hockey stick identity can be proved by partitioning the set of tilings according to the position of the rightmost gray tile of any tiling of a board of length $n + 1$ with $r + 1$ gray squares (see Problem 1.6.8).

1.6.2 Multiplication Principle

In how many ways can you flip a coin and roll a six-sided die? The answer is certainly not 8, since there are six outcomes for the die when the coin lands heads and 6 additional outcomes for the die when the coin lands tails. Altogether there are

$12 = 2 \cdot 6$ ways to flip the coin and roll the die. We see that the number of outcomes is given by the *product* of choices, not the sum.

If $C = \{\text{H,T}\}$ is the set of outcomes of a coin flip, and $D = \{1,2,3,4,5,6\}$ the set of outcomes of a roll of the die, then the set of outcomes of the coin-and-die experiment is described by the set of ordered pairs in the set

$$S = C \times D = \{(\text{H},1),(\text{H},2),(\text{H},3),(\text{H},4),(\text{H},5),(\text{H},6)\\ (\text{T},1),(\text{T},2),(\text{T},3),(\text{T},4),(\text{T},5),(\text{T},6)\} \quad (1.33)$$

The set shown in (1.33) is the Cartesian product of the sets C and D. More generally, we have the following definition.

Definition 1.39 The *Cartesian product* of sets A and B, denoted by $S = A \times B$, is the set of all ordered pairs whose first coordinate is a member of set A and whose second coordinate is from set B:

$$A \times B = \{(a,b) \mid a \in A \text{ and } b \in B\}$$

The number of ordered pairs is the product of the number of choices $|A|$ for the first entry and the number of choices $|B|$ for the second entry of the ordered pair. Therefore, we have the following theorem.

Theorem 1.40 (Multiplication Principle for a Cartesian Product) For any two sets A and B, the number of ordered pairs in the Cartesian product $S = A \times B$ is given by

$$|S| = |A \times B| = |A||B| \quad (1.34)$$

It is very important to note that the number of choices available for the second entry in each ordered pair is independent of the choice made for the first entry of the ordered pair. For example, we were *not* asked to roll a 12-sided die if the coin landed tails and otherwise roll a six-sided die if the coin landed heads.

To form the Cartesian product, it was helpful to create a sequence of two choices: first, choose the first coordinate and then choose the second coordinate. In other words, we may formulate a two-stage experiment that results in all of the possibilities. The total number of possibilities is then the product of the number of choices available at each stage, *where the number of choices available for the second stage is independent of which choice was made in the first stage.*

The following example illustrates this type of reasoning, where a diagram known as a *possibility tree* is a useful way to visualize the succession of stages.

Example 1.41 A bag contains 4 balls, labeled A, B, C, and D. In how many ways can two balls be drawn from the bag, where the first drawn ball is not replaced into the bag?

Solution. Four possible balls can be drawn first: ball A, B, C, or ball D. If ball A is drawn first, there are 3 ways—B or C or D—to draw the next ball. Indeed, there are always 3 ways to draw the second ball independently of which ball was drawn first.

The $12 = 4 \cdot 3$ ways to draw the two balls without replacement is illustrated with this possibility tree:

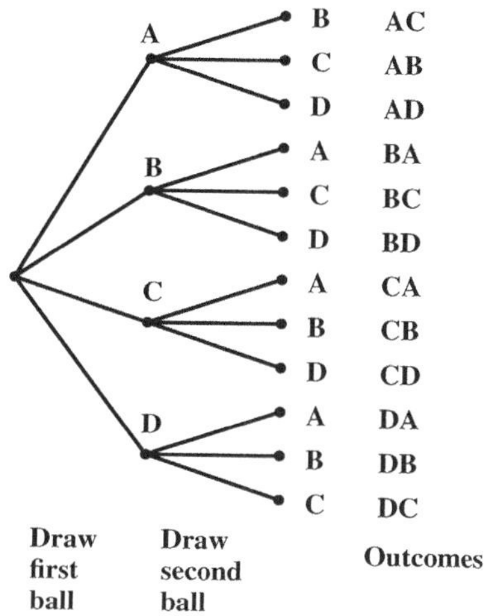

If the first ball were replaced before the second ball is drawn, there would always be four choices for the second ball and $16 = 4 \cdot 4$ ways to draw the two balls with replacement. Therefore, $S = \{A, B, C, D\}$ is the set of choices for both the first and second ball drawn, and the number of ways to draw the balls with replacement is $|S \times S| = |S||S| = 4^2 = 16$. ■

The following example is solved with a three-stage process.

Example 1.42 At Henry's Ice Cream Shop, ice cream can be ordered in two sizes—one or two scoops, in three flavors—vanilla, peach, or mint—and in two containers—a dish or a cone. How many ways can ice cream be purchased at Henry's? See the following diagram:

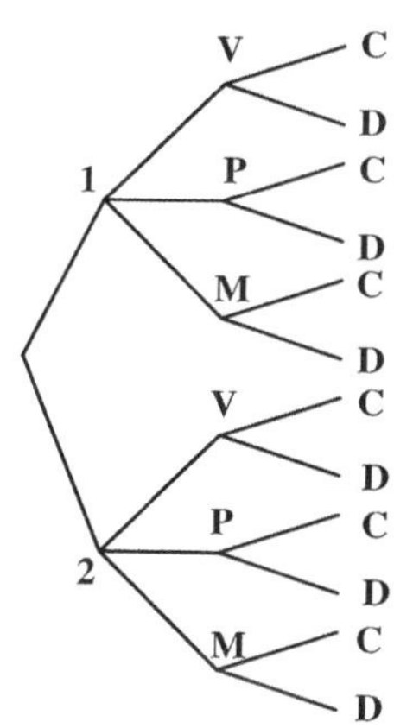

Solution. Let $R = \{1, 2\}$ be the set of available sizes; $S = \{V, P, M\}$, the set of flavor choices; and $T = \{D, C\}$, the set of containers. An order is placed by choosing the size, flavor, and container, and can be visualized by this possibility tree that shows there are 12 ways to place an order.

Alternatively, each order is described by a ordered triple in the Cartesian product $R \times S \times T$, and $|R \times S \times T| = |R||S||T| = 2 \cdot 3 \cdot 2 = 12$. For example, (2,P,D) corresponds to two scoops of peach ice cream in a dish. ■

The reasoning we have used for two- and three-stage processes extends to multistage processes. However, it is important to observe that the number of choices at each successive stage cannot depend on which choices have been made in any of the earlier stages. The particular set of choices at any stage is allowed to vary according to previous choices, but the number of choices at each stage must be independent of the choices made earlier.

Theorem 1.43 (Multiplication Principle for a Multistage Process) Suppose that a multistage experiment consists of a sequence of n steps, for which there are

1. p_1 ways to perform step 1
2. p_2 ways to perform step 2 (independently of how step 1 was performed)
3. p_3 ways to perform step 3 (independently of how steps 1 and 2 were performed)

   ...

n. p_n ways to perform step n (independently of how the preceding steps were performed)

Then there are $p_1 p_2 \cdots p_n$ ways to carry out all n steps of the multistage experiment.

The following examples illustrate how to apply the multiplication principle.

Example 1.44 In Washington State, a car license is an arrangement of three letters followed by four digits.

(a) What is the number of possible licenses?

(b) What is the number of possible licenses if no digit or letter can be repeated?

Solution. Any license plate can be viewed as a seven-step process in which the leftmost symbol is chosen, then the second symbol, and so on, finishing with the choice made for the seventh symbol at the far right.

(a) There are 26 letters and 10 digits, so the number of possible license plates is

$$26 \cdot 26 \cdot 26 \cdot 10 \cdot 10 \cdot 10 \cdot 10 = 175{,}760{,}000$$

by the multiplication principle.

(b) The second letter cannot repeat the first, so there are 25 choices available. Likewise, the third letter cannot repeat either of the two letters that have been selected already, so there are 24 remaining choices. Similar considerations apply to the selection of the digits.

By the multiplication principle, there are $26 \cdot 25 \cdot 24 \cdot 10 \cdot 9 \cdot 8 \cdot 7 = 78{,}624{,}000$ license plates with no repeated symbol. ■

Example 1.45 Eight people—Alice, Bob, Carly, Dave, . . . , Hal—arrive at the tennis courts. In how many ways can they

(a) form pairs to play four singles matches?

(b) divide up to play two games of doubles?

Solution. Imagine that you are the match organizer. Formulate a sequence of steps that creates all of the possible matches for either singles or doubles.

(a) Let's choose the pairs of opponents in succession, beginning with the choice of an opponent for Alice. There are seven possible opponents for Alice, any one of Bob, Carly, . . . , or Hal. After Alice and her opponent leave to start their match, six people remain. Choose any one of them to pick his or her opponent in 5 ways. This leaves a group of four, and you can ask one of them to choose an opponent in 3 ways. This leaves just two people, with one way for them to become opponents. Altogether, the four-stage process shows that there are $7 \cdot 5 \cdot 3 \cdot 1 = 105$ ways to divide eight people into four pairs.

(b) The four steps used in part (a) to form the singles matches can just as well be used to form four pairs of doubles partners. However, as a fifth step, we can ask Alice and her partner to choose one of the three other teams as their opposing team. The two remaining teams will then form the other doubles match. We now see that there are $7 \cdot 5 \cdot 3 \cdot 3 \cdot 1 = 315$ ways to form the two doubles matches. ■

The reasoning in Example 1.45(a) applies to any even number of players, say, $2n$, to show that the number of ways to arrange singles matches among $2n$ players is

$$(2n-1)!! = (2n-1)(2n-3)\cdots(3)(1) \tag{1.35}$$

where the decreasing product of positive odd positive integers is sometimes called a *double factorial.* The double factorial of an even integer is given by

$$(2n)!! = (2n)(2n-2)\cdots(4)(2) \tag{1.36}$$

so that the double factorial $m!!$ is defined for all positive integers. Note that $(n!)! \neq n!!$.

In the next example, we consider a sequence with terms that are each either a 0 or a 1. Since only two symbols appear, we recall that this is called a *binary sequence*. For example, 101011 is a binary sequence of length 6.

Example 1.46 What is the number of binary sequences

(a) of length 6?
(b) of length n?

Solution. There is a two-way choice—either 0 or 1—for each term of the sequence, and each choice is independent of the choices made for the previous terms. This gives us the answers

(a) $2 \cdot 2 \cdot 2 \cdot 2 \cdot 2 \cdot 2 = 2^6$
(b) $2 \cdot 2 \cdot \cdots \cdot 2 = 2^n$ ■

Binary sequences arise frequently as a way to convey or code information. For example, they may code a binary (base two) numeral such as

$$101011_2 = 1 \cdot 2^5 + 0 \cdot 2^4 + 1 \cdot 2^3 + 0 \cdot 2^2 + 1 \cdot 2^1 + 1 = 32 + 0 + 8 + 0 + 2 + 1 = 43_{10}$$

The binary sequences of length 6 correspond to the $2^6 = 64$ integers $0, 1, \ldots, 63$.

In the next example, binary sequences provide a way to count the number of subsets of a finite set.

Example 1.47 What is the number of subsets of the six-element set $S = \{a, b, c, d, e, f\}$, including the empty subset $\varnothing$ and S itself?

Solution. Since $|S| = 6$, any subset of S can be associated with a binary sequence of length 6 by assigning a 1 if the element is included in the subset and a 0 if not. For example, the binary sequence 101011 corresponds to the subset $\{a, c, e, f\}$. Similarly, 000000 corresponds to the empty subset and 111111 to S itself. This association describes a bijection (i.e., a one-to-one matching) of binary sequences of length 6 and the subsets of a six-element set, so, from Example 1.46, we see there are 2^6 subsets of the six element set S. ■

The reasoning in Example 1.47 applies equally well to any finite set with n members, giving us the following theorem.

Theorem 1.48 A set S with $|S| = n$ members has 2^n subsets.

Just as the addition principle can be inverted to become the subtraction principle, the multiplication principle can be inverted to become the *division principle*.

Theorem 1.49 (Division Principle) Let set S have the partition $S = A_1 \cup A_2 \cup \cdots \cup A_k$, where the pairwise disjoint sets $A_1, A_2, \ldots, A_k$ each have the same number of members, say, $r = |A_1| = |A_2| = \cdots = |A_k|$. Then there are $k = |S|/r$ subsets in the partition, or equivalently, there are $r = |S|/k$ members of each set of the partitioning family.

Proof. Applying the extended addition principle in Theorem 1.35, we obtain

$$|S| = |A_1| + |A_1| + \cdots + |A_k| = rk$$ ■

Example 1.50 A child has five plastic letters and has written PANDA. How many words[4] can be written with different arrangements of the same set of plastic letters? For example, AANPD and DNAPA are two more words.

Solution. Suppose, temporarily, that all of the letters have different colors. For example, one A is red and the other A is blue. To form a word, we have a five-step process as we choose letters from left to right; that is, we can form a set S of words with $5 \cdot 4 \cdot 3 \cdot 2 \cdot 1 = 5!$ members. Suppose that A_1 is the subset of words with the red A to the left of the blue A, and A_2 is the subset with the reversed order. By switching the two As, we see that $|A_1| = |A_2|$. Thus, by the division principle, there are $5!/2 = 120/2 = 60$ different words that can be formed if the two As have the same color. ■

1.6.3 Combining the Addition and Multiplication Principles

Many combinatorial problems are solved using both the addition and multiplication principles.

Example 1.51 Mischa is spending a long weekend at a beach house, and wants to watch a video each evening on Friday and Saturday. The beach house has a video library containing three Westerns, five dramas, and six comedies from which to choose. How many ways can Mischa choose to watch two movies of different genres?

Solution. There are three cases, depending on which genres are chosen. Once the two genres are chosen, she can perform a three-stage process in choosing a video of each genre and choosing the order in which they will be watched. Here is a useful

[4] In the present context, we use the term *word* either to refer to a specific English word (e.g., panda) or to a permutation of letters forming a nonword (e.g., dnapa).

way to show both the three cases and the choices made in each case:

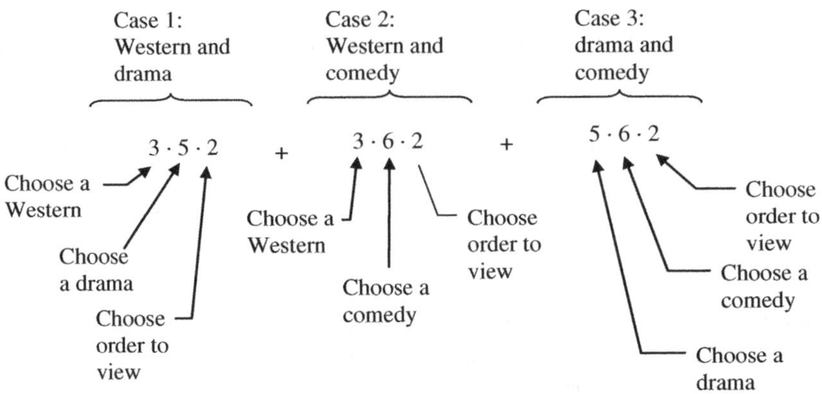

It is now clear that Mischa has $3 \cdot 5 \cdot 2 + 3 \cdot 6 \cdot 2 + 5 \cdot 6 \cdot 2 = 30 + 36 + 60 = 126$ ways to watch a video each night. ■

Example 1.52 The diagram below shows a system of roads from Abbottsville (A) to Dusty (D), some of which pass through Baker (B) or Cavendish (C). How many routes, always following the arrows, extend from Abbotsville to Dusty?

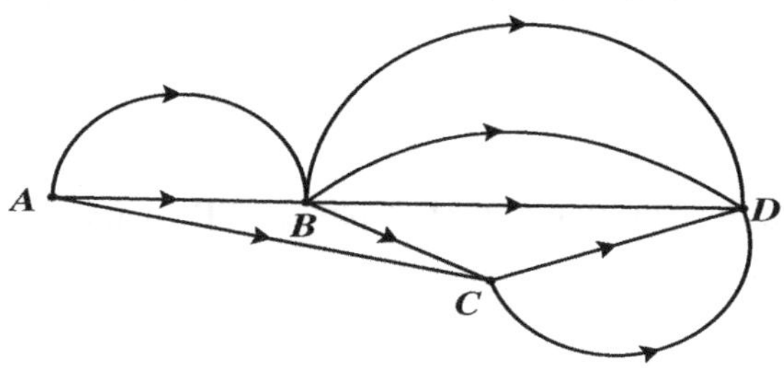

Solution. There are three cases: ABD, ACD, and $ABCD$, with two routes from A to B, one direct route from A to C, one direct route from B to C, three direct routes from B to D, and two direct routes from C to D. Therefore, there are $2 \cdot 3 + 1 \cdot 2 + 2 \cdot 1 \cdot 2 = 12$ ways to go from Abbottsville to Dusty. ■

Example 1.52 solves a problem in *path counting*, a topic of considerable interest and usefulness in combinatorial reasoning. Path counting will be revisited later, when we choose a variety of different patterns for the edges that a path must follow.

PROBLEMS

1.6.1. The 2006 World Cup used a soccer ball called the *+teamgeist* (*team spirit*, with the + sign to allow the German word to be copyrighted). Mathematically, the ball is a spherical analog of the truncated octahedron, obtained by starting with the octahedron of eight triangles as shown, then replacing (truncating) each corner with a square, and finally rounding the faces to become spherical. What counting principle can be used to determine the number of panels on the +teamgeist soccer ball?

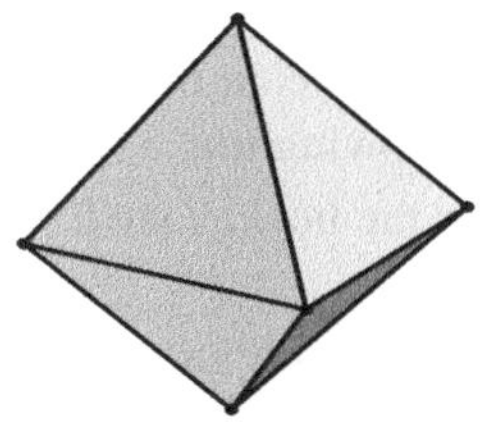

1.6.2. The 2010 World Cup held in South Africa used the *jabulani* soccer ball, where the name means *celebrate* in the Zulu (Bantu) language. There are exactly eight panels on the ball, including four rounded triangular panels. What counting principle can be used to determine the number of nontriangular panels of a jabulani ball?

1.6.3. There are two types of seams of the traditional soccer ball shown in Example 1.30: those that separate two hexagonal panels and those that separate a pentagonal panel from a hexagonal panel. Find the total number of seams on the ball, and the number that separate one hexagonal panel from another.

1.6.4. How many ways can you put on a pair of socks and a pair of shoes?

1.6.5. Maria likes to order double-scoop ice cream cones, with chocolate or strawberry on the bottom, and chocolate, vanilla, or mint on the top. Describe the ways in which Maria can order her ice cream cones with a Cartesian product, and count the number of types of cones that she likes.

1.6.6. Suppose that 13 players arrive at the tennis courts. In how many ways can they

(a) split up to form 6 games of singles, with one player sitting out?

(b) split up to for 3 games of doubles, with one player sitting out?

1.6.7. Generalize the results of Example 1.45 and Problem 1.6.6; that is, provide formulas for the number of ways to split up m people into singles and doubles matches. For doubles, create as many matches as possible and then set up a singles match if enough people remain not already playing doubles.

1.6.8. Consider the tilings of a $1 \times (n + 1)$ board with $r + 1$ gray tiles.

(a) Explain why $\binom{r+k}{r}$ tilings have the rightmost gray tile in cell $r + k + 1$, where $k = 0, 1, 2, \ldots, n - r$.

(b) Use part (a) to prove the following (hockey stick identity 2):

$$\binom{n+1}{r+1} = \binom{r}{r} + \binom{r+1}{r} + \binom{r+2}{r} + \cdots + \binom{n}{r}$$

1.6.9. Construct Pascal's triangle showing rows 0 through 6.

(a) Draw a loop around the terms in the hockey stick identity (1.32) for the case $n = 3$ and $r = 2$, showing that a hockey stick shape in formed.

(b) Repeat part (a) but for the hockey stick identity of Problem 1.6.8 in the case that $n = 5$ and $r = 3$.

(c) How do the terms in the handle of the hockey stick relate to the term in the blade of the hockey stick?

1.6.10. The Math Club needs to choose a president and treasurer, where the two officers cannot be from the same year in school. The club has four seniors, seven juniors, and six sophomores. In how many ways can the two officers be chosen?

1.6.11. There are three roads from Sylvan to Tacoma, four roads from Tacoma to Umpqua, and two roads from Sylvan directly to Umpqua. How many routes, with no backtracking, can be taken from Sylvan to Umpqua?

1.7 SUMMARY AND ADDITIONAL PROBLEMS

- *The Pigeonhole Principle.* The simplest statement of the principle is as follows: If more than n pigeons are placed into n pigeonholes, at least one hole is occupied by two or more pigeons. More generally, if more than $p_1 + p_2 + \cdots + p_n$ pigeons are placed into holes $1, 2, \ldots, n$, then some hole k contains more than p_k pigeons. The pigeonhole principle is an important strategy to prove the existence of some mathematical object, but little if any information about how to construct the object is given.
- *Chessboard Tilings with Dominoes.* The approach taken was constructive. For example, since each domino covers two unit squares, an $m \times n$ board must have an even number of unit squares, thus requiring that one or both of m and n be even. Indeed, this condition is also sufficient since it is easy to construct the tiling in which every domino is aligned with the side of even dimension. For a board with odd dimensions, with alternate squares colored black and gray, trimming any square with the color of the corner squares leaves a board that can be tiled with dominoes. The proof was again constructive; the trimmed board can always be partitioned into rectangles each with an even dimension.
- *Figurate Numbers.* This section investigated enumerative problems revolving around the question "How many dots are in this sequence of patterns?" Several important principles of counting were found to be helpful:

The nth term of the sequence may be related in a simple way to earlier terms in the sequence; that is, the sequence was described by a recurrence relation.

The complex dot pattern could be partitioned into simpler dot patterns with known formulas; therefore, a formula could be obtained for the nth term of the complex pattern.

- *Counting Tilings of Rectangles.* The central problem was to count the number of ways that a $1 \times n$ board can be tiled with tiles of a specified type. In particular, if r of the unit squares were to be covered with a gray tile, and the remaining $n - r$ unit square covered with a white tile, it was critical to know how many ways r units squares can be chosen from the set of n unit squares. This number was called the *combination* $C(n,r)$ that is read as "n choose r" and often written in the alternative notation $\binom{n}{r}$. Using the tiling model, it was shown that there are 2^n ways to tile a $1 \times n$ board with any number $r = 0, 1, 2, \ldots, n$ gray tiles; that is, $\sum_{r=0}^{n} \binom{n}{r} = 2^n$. Also, any tiling of a $1 \times n$ board with r gray tiles end with either a white or a gray tile, so the $1 \times (n - 1)$ boards to the left of the rightmost have either r or $r - 1$ gray tiles. Indeed, all such $1 \times (n - 1)$ boards are accounted for, and it follows that

$$\binom{n}{r} = \binom{n-1}{r} + \binom{n-1}{r-1}$$

a relation known as *Pascal's identity.* Finally, it was shown that the number ways to tile an $1 \times n$ board with a sequence of squares and dominoes is given by the *combinatorial Fibonacci number* f_n, where $f_n = F_{n+1}$ and $F_0 = 0, F_1 = 1, F_2 = 1, F_3 = 2, F_4 = 3, F_5 = 5, \ldots$ are the standard Fibonacci numbers.

- *Addition and Multiplication Principles.* To count the number of elements in a set S, it is helpful to partition S into pairwise disjoint subsets $A_1, A_2, \ldots, A_k$ for which $S = A_1 \cup A_2 \cup \cdots \cup A_k$, $A_i \cap A_j = \emptyset$, $i \neq j$. Then $|S| = |A_1| + |A_2| + \cdots + |A_k|$, a result known as the *addition principle.* If each element of S is obtained as the result of a n-stage process in which there are p_j ways to carry out stage j independently of the way any previous stage was carried out, then the number of outcomes in S is given by $p_1 p_2 \cdots p_n$, a result known as the *multiplication principle.* Often, both the addition and the multiplication principle are used together to solve a combinatorial problem.

PROBLEMS

1.7.1. **(a)** Let A be any set of 51 numbers chosen from [100]. Show that two members of A differ by 50.

(b) State and prove a generalization of the result of part (a).

1.7.2. Let m be an odd number. Show that m is a divisor of at least one of the terms in the sequence $(1,3,7,15,\ldots, 2^m - 1)$ whose jth term is $a_j = 2^j - 1$.

1.7.3. Show that any set of 10 natural numbers, each between 1 and 100, contains two disjoint subsets with the same sum of its members.

1.7.4. Given $nk + 1$ or more pigeons placed into n pigeonholes, show that the average number of pigeons per hole is larger than k.

1.7.5. A traditional dartboard divides the circular board into 20 sectors that are numbered clockwise from the top with the sequence

20 - 1 - 18 - 4 - 13 - 6 - 10 - 15 - 2 - 17 - 3 - 19 - 7 - 16 - 8 - 11 - 14 - 9 - 12 – 5

There is considerable variation in the sum of three successive numbers, from $23 = 1 + 18 + 4$ to $42 = 19 + 7 + 16$. Can the numbers 1 through 20 be rearranged so that the sum of each group of three successive numbers is smaller than 32?

1.7.6. Imagine large dots represent stacked logs as seen from their ends. It is required that the stack be either one or two rows high, that the logs in the lower row have no gap between adjacent logs, and that each of the logs in the upper row must touch two logs in the lower row. The diagram that follows shows all five ways to stack five logs:

Investigate the sequence of numbers that gives the number of ways to stack n logs into piles that meet the criteria described above. Justify your answer.

1.7.7. The centered hexagon numbers (or hex numbers) H_n are obtained by starting with a single dot and then surrounding it by hexagons with 6,12,18,... dots on its sides. The diagram shows that $H_0 = 1$, $H_1 = 7$, and $H_2 = 19$:

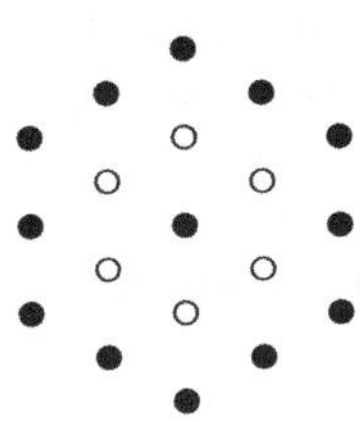

(a) extend the diagram with two more surrounding hexagons to determine H_3 and H_4.

(b) derive a formula that gives H_n in terms of the triangular numbers.

(c) obtain an expression for H_n as a function of n.

(d) suppose that *tridominoes* are formed with a pair of equilateral triangles joined along a common edge, and are colored gray, white, or black according to their orientation. The figure here shows the three types of tridominoes and one way they can be used to tile the hex pattern for the H_2 array of dots.

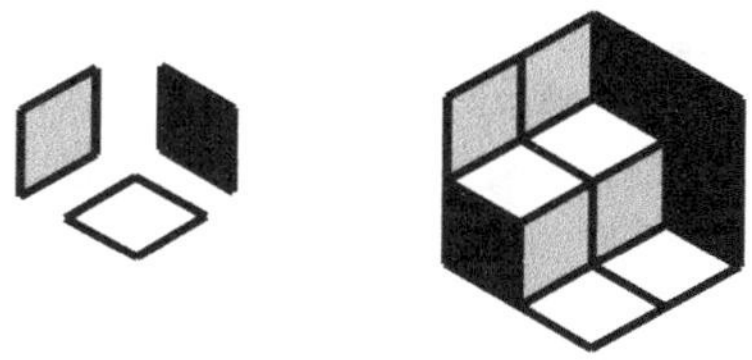

Show that every hex pattern of H_n dots can be tiled with tridominoes, and give the number of tridominoes of each color that are used in the tiling. (*Suggestion:* The hex numbers might also be called the *corner* numbers!)

1.7.8. Recall that $\binom{8}{3}$ is the number of ways to tile a 1×8 board with 3 gray and 5 white squares.

(a) Using the tiling model, explain why $\binom{8}{3} = \binom{8}{5}$.

(b) Generalize your answer to part (a) to explain why

$$\binom{n}{r} = \binom{n}{n-r}$$

1.7.9. Use the result of Problem 1.7.8 to show that hockey stick identity 1 (1.32) can be rewritten to become

$$\binom{n+1}{r+1} = \binom{r}{r} + \binom{r+1}{r} + \binom{r+2}{r} + \cdots + \binom{n}{r}$$

(hockey stick identity 2).

1.7.10. A $2 \times n$ rectangular board is to be tiled with three gray squares and $2n - 3$ white squares.

(a) How many tilings have all three gray squares in the same row?

(b) How many tilings have no row that is all white?

(c) Use your answers to the parts above to give an identity involving the binomial coefficient $\binom{2n}{3}$.

1.7.11. **(a)** How many paths extend from point P_1 to each of the points $P_2, P_3, \ldots, P_{12}$ in the following directed graph? Each step along any path must be in the direction indicated by the arrow. For example, there are two paths from

P_1 to P_3:

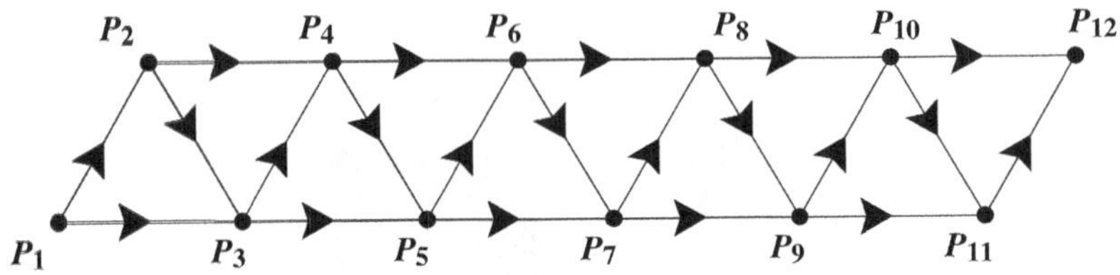

(b) Imagine extending this graph to an arbitrary number of points $P_1, P_2, \ldots, P_n$. What famous number sequence gives the number of paths to the points $P_1, P_2, \ldots, P_n$? Provide a justification for your answer.

1.7.12. How many ways can some coins be removed from a coin purse containing 3 pennies, 7 nickels, 4 dimes, and 2 quarters? The coins of the same denomination are considered to be identical, and the order of removal is not important.

1.7.13. There are six positive divisors of $12 = 2^2 \times 3^1$, namely, $1 = 2^0 \times 3^0$, $2 = 2^1 \times 3^0$, $4 = 2^2 \times 3^0$, $3 = 2^0 \times 3^1$, $6 = 2^1 \times 3^1$, and $12 = 2^2 \times 3^1$. What is the number of positive divisors of these integers?

(a) 660 **(b)** $2^5 \times 3^7 \times 11^2 \times 23^4$ **(c)** 10^{100}

REFERENCES

1. *The Man Who Loved Only Numbers: The Story of Paul Erdös and the Search for Mathematical Truth*, Paul Hoffman, Hyperion, 1998.
2. Sixty-third Annual William Lowell Putnam Mathematical Competition, 2002, problem A2.
3. H. N. V. Temperley and M. E. Fisher, *Philosophical Magazine* (1961) 6, 1061–1063.
4. P. W. Kasteleyn, The statistics of dimers on a lattice. I. The number of dimer arrangements on a quadratic lattice, *Physica* 27 (1961), 1209–1225.

2

SELECTIONS, ARRANGEMENTS, AND DISTRIBUTIONS

2.1 INTRODUCTION

In this chapter we will investigate three types of problems that occur frequently in combinatorial analyses:

Selections A *selection* is the creation of a subset of a given set. The subset formed is called a *combination*. No consideration is given to the order in which elements are chosen to belong to the subset; it is only the unordered listing of the elements in the subset that is of importance. *Example:* How many ways can we select three toppings for a pizza at a shop that offers a dozen toppings?

Arrangements An *arrangement* is a selection *and an ordering* of some or all of the objects from a set of objects, so that each selected object has a specified position in an ordered list. Each resulting ordered arrangement is called a *permutation*. *Example:* How many ways can the seven finalists in a piano competition be scheduled to perform in the final round?

Distributions A *distribution* is an assignment of objects given to recipients or, equivalently, the placement of objects into containers. *Example:* How many ways can we contribute $100 to five charities?

Combinatorial Reasoning: An Introduction to the Art of Counting, First Edition. Duane DeTemple and William Webb.

Most often, a problem is accompanied by side conditions that must be taken into account. For example, at the pizza shop, we may be able to order a "double cheese" and black olives, in which the cheese option can be repeated. In the piano competition, what if the three men finalists are to be interspersed with the four women finalists? In the contributions to charities, what if each charity is to be given an amount between \$10 and \$50?

The most common side condition is whether repetition is allowed.

Repetition allowed—multiple choices of identical objects are available, so objects of the same type can be chosen repeatedly.

Repetition not allowed—all of the objects are distinct, so that any particular object is either selected or omitted.

We will see the significance of whether repetition is allowed.

Combinatorial results are often best understood by interpreting their meaning with a *combinatorial model.* For example, in Chapter 1, the binomial coefficients $C(n, r)$ were viewed as the number of ways to choose r cells from a $1 \times n$ board that are to be covered with a gray square tiles. Similarly, the number of ways to tile a $1 \times n$ board with squares and dominoes gave us the combinatorial Fibonacci numbers $f_1 = 1, f_2 = 2, f_3 = 3, f_4 = 5, \ldots$. The tiling model is just one of many combinatorial models that will be explored in more depth throughout the chapter. Section 2.6 discusses two special types of permutations: *circular permutations* and *derangements*.

2.2 PERMUTATIONS AND COMBINATIONS

2.2.1 Permutations

The following example illustrates how permutations arise and how the number of permutations can be counted.

Example 2.1 There are 11 members of the Math Club. In how many ways can the slate of four officers—president, vice president, secretary, and treasurer—be filled?

Solution. Any slate of officers is a permutation, since it is a list of names written in the order president first, vice president second, and so on. We assume that no person can hold more than one office, so only permutations without repetition are allowed. Any permutation is the result of a four-stage process: namely, pick the president, then pick the vice president, then the secretary, and finally pick the treasurer. Therefore, the multiplication principle described in Chapter 1 allows us to count the number of permutations—there are 11 choices for president, 10 choices remaining for vice president, 9 choices left for secretary, and finally 8 choices for treasurer. Altogether, there are

$$11 \times 10 \times 9 \times 8 \tag{2.1}$$

slates of 4 officers that can be chosen from the 11 members of the club. ■

More generally, we introduce the following terminology and notation, where an *n-set* is a set with n distinct (i.e., all different) elements.

Definition 2.2 A *permutation* is an ordered arrangement of objects chosen from a set of distinct elements. An *r-permutation* is a permutation of r distinct objects taken from an n-set. The number of r-permutations from an n-set is denoted by $P(n, r)$.

Example 2.1 shows that any slate of officers is a 4-permutation of the 11-set of club members, and expression (2.1) shows there are $P(11, 4) = 11 \cdot 10 \cdot 9 \cdot 8$ permutations of 4 members chosen from the 11-member club.

The same reasoning can be generalized to count the number of r-permutations from an n-set.

Theorem 2.3 The number of r-permutations from a set of n distinct elements is given by

$$P(n, r) = n(n-1) \cdots (n-r+1) \tag{2.2}$$

Proof. We must create all possible ordered lists of r distinct objects chosen from a set of n elements. The first element in the list can be selected in n ways. This leaves $n - 1$ ways to select the second element of the list different from the element that was selected first. The number of ways to select the remaining entries in the list is determined in the same way, ending with $n - (r - 1)$ ways to select the rth element of the list. It is helpful to note that there are r factors in the product (2.2). ■

The notation $P(n, r)$ is used widely, but sometimes alternate notations are encountered, including these:

$$P(n, r) = {}_{\mathrm{n}}\mathrm{P}_{\mathrm{r}} = (n)_r$$

The notation nPr is especially common on calculators. For example, the entry string `11 nPr 4` ENTER on a graphing calculator computes the value $P(11, 4) = 7920$. The notation $(n)_r$ is known as the *Pochhammer symbol,* named for the Prussian mathematician Leo August Pochhammer, 1841–1920. For example, $(11)_4 = 11 \times 10 \times 9 \times 8$. Later we will have occasion to use the polynomials given by $(x)_r = x(x-1) \cdots (x-r+1)$ known as the *falling factorial*, where the variable x is no longer confined to an integer value.

If all 11 members of the Math Club are asked to line up for a photo, this is an 11-permutation of all 11 club members. Thus, there are

$$P(11, 11) = 11 \times 10 \times 9 \times 8 \times 7 \times 6 \times 5 \times 4 \times 3 \times 2 \times 1 = 11!$$

ways to line up for the photo, where we recall (from Chapter 1) that 11! is *eleven factorial*. This special case is noted in the following corollary to Theorem 2.3.

Corollary 2.4 The number of n-permutations of all of the elements of an n-set is

$$P(n,n) = n \times (n-1) \times (n-2) \times \cdots \times 3 \times 2 \times 1 = n!$$

Example 2.5 Jenny has 10 different mathematics books and 8 different chemistry books. In how many ways can she

(a) put all 18 books on a shelf?
(b) put 6 books on a shelf?
(c) put half of the math and half of the chemistry books on a shelf, with the math books all to the left of the chemistry books?

Solution

(a) $P(18, 18) = 18!$
(b) $P(18, 6) = 18{\cdot}17{\cdot}16{\cdot}15{\cdot}14{\cdot}13{\cdot}$ (note that there are six factors in the 6-permutation)
(c) $P(10, 5){\cdot}\ P(8, 4) = (10{\cdot}9{\cdot}8{\cdot}7{\cdot}6)(8{\cdot}7{\cdot}6{\cdot}5)$ ■

2.2.2 Combinations

We now turn our attention to combinations, where the following notation and terminology will be helpful.

Definition 2.6 A *combination* is a subset whose elements have been chosen from a given set of distinct elements. An *r-combination* is an r-subset of an n-set. The number of r-combinations in an n-set is denoted either by $C(n,r)$ or by $\binom{n}{r}$, and is read as "n choose r."

This definition gives $C(n,r)$ the same meaning in which it was used in Chapter 1, where $C(n,r)$ was the number of ways to choose r of the cells in a $1 \times n$ board that were to be covered with gray square tiles. In Chapter 3, we will see why $C(n,r)$ is also called a *binomial coefficient.*

The following example illustrates how to count the number of combinations. Since no element is chosen more than once, we are counting *combinations without repetitions.*

Example 2.7 The Math Club needs to choose a four-person team to staff its display booth for freshman orientation. How many ways can this team be chosen from the club's 11 members?

Solution 1 (via the Division Principle). If the team members are chosen in a specific order, we have already seen in Example 2.1 that there are $P(11,4) = 11 \times 10 \times 9 \times 8$ permutations. But there are 4! ways to permute these same four persons, and each permutation includes the same subset of four team members. According to the division principle, the number of combinations is

$$C(11,4) = \frac{P(11,4)}{4!} \tag{2.3}$$

Solution 2 (via the Multiplication Principle). A permutation is the result of a two-stage process—first select a subset of elements, and then arrange these elements into an order. Thus, we can count the number of ways to choose a slate of four officers from the 11 club members as follows: (1) choose the four-element subset of officers in $C(11,4)$ ways and then (2) choose which person occupies which office in 4! ways. Thus, $P(11,4) = C(11,4) \times 4!$, which is equivalent to formula (2.3). ■

The reasoning just given can be generalized to derive a formula for the number of r-combinations from a set of n distinct members.

Theorem 2.8 The number of r-combinations selected from a set of n distinct elements is given by

$$C(n,r) = \binom{n}{r} = \frac{P(n,r)}{r!} = \frac{n(n-1)\cdots(n-r+1)}{r!} = \frac{n!}{r!\,(n-r)!} \tag{2.4}$$

Proof. All of the $P(n,r)$ permutations result by first choosing an r-subset in $C(n,r)$ ways and then permuting the r elements in $r!$ ways. Thus $P(n,r) = C(n,r)\,r!$, which is equivalent to (2.4). ■

It follows from (2.4) that $\binom{n}{0} = 1$, all $n \geq 0$ and $\binom{n}{r} = 0$, all $r > n$. It is also convenient to define

$$\binom{n}{r} = 0, \text{ all } r < 0 \leq n \tag{2.5}$$

We will see that this notational convention is helpful when manipulating sums since the limits of a summation variable need not be exact provided that all of the nonzero terms are included in the domain of the summation variable. For example, all of these summations are the same:

$$\sum_{k=0}^{\min(m,n)} \binom{m}{k}\binom{n-k}{r-k} = \sum_{k\geq 0} \binom{m}{k}\binom{n-k}{r-k} = \sum_{k} \binom{m}{k}\binom{n-k}{r-k}$$

It is usually acceptable to express answers to combinatorial questions with simple expressions involving the binomial coefficients, factorials, permutations, and so on. It is seldom necessary to evaluate the coefficients numerically using formula (2.4). If a numerical value is of interest, it may be easiest to use a calculator or computer. For example, on a graphing calculator the entry string `11 nCr 4` [ENTER] shows that $\binom{11}{4} = 330$. Similarly, in *Maple* both commands `binomial(11,12)` and `binomial(11,-3)` return a 0, with corresponding commands in *Mathematica, MatLab,* and other computer algebra systems (CASs). A convenient way to evaluate combinatorial expressions is available at the Internet site *WolframAlpha.*

We now consider some examples of combinatorial situations in which combinations must be counted.

Example 2.9 A standard deck of cards contains 52 cards divided into four *suits:* the red suits, hearts and diamonds, and the black suits, clubs and spades. Each suit, in turn, is divided in 13 ordered *ranks:* ace through 10 followed by the face cards jack, queen, and king. The ace can act as either the lowest or highest card in the ranking. What is the number of 5-card hands in a 52-card deck that contain

(a) two pairs (i.e., two pairs from different ranks, and a fifth card of a third rank)?

(b) three of a kind (i.e., exactly three cards of the same rank and two other cards that are not a pair)?

Solution

(a) The two ranks of the pairs are a 2-combination of the 13 ranks, so there are $\binom{13}{2}$ ways to choose the two ranks of the pairs. There are 4 cards in a rank, so there are $\binom{4}{2}$ ways to choose a pair from the lower rank and another $\binom{4}{2}$ ways to choose a pair from the higher rank The fifth card in the hand must be one of the 44 cards from the remaining 11 ranks. Altogether, we see that there are

$$\binom{13}{2}\binom{4}{2}^2\binom{44}{1} = 123{,}552$$

hands containing exactly two pairs.

(b) There are four cards in each of the 13 ranks, so there are $\binom{13}{1}\binom{4}{3}$ ways to choose a rank and then three cards with that rank. The remaining two cards must not be a pair or include a card of the rank of the three of a kind, since otherwise we would have a full house or four of a kind. Thus, the ranks of the other two cards can be chosen in $\binom{12}{2}$ ways, and there are $\binom{4}{1}$ ways to select a card of a given rank; that is, the remaining two cards can be chosen in $\binom{12}{2}\binom{4}{1}^2$ ways. The number of three-of-a-kind hands is therefore

$$\binom{13}{1}\binom{4}{3}\binom{12}{2}\binom{4}{1}^2 = 54{,}912$$

This shows that three of a kind beats two pair since three of a kind occurs less frequently than two pair. ■

The following example provides a connection of the binomial coefficients with the number of paths in a square grid.

Example 2.10 How many paths, always moving north (N) or east (E), connect point $A(0,0)$ with point $B(7,4)$ in the square grid shown below? One path is shown by the heavy lines in the diagram:

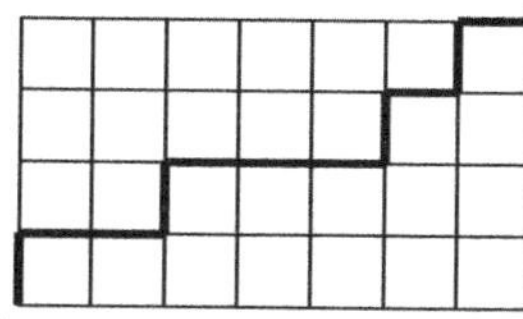

Solution. Each path from A to B must cover four blocks to the north and seven blocks to the east. For example, the path shown in the diagram can be described by NEENEEENENE. More generally, in the sequence of 11 blocks that must be covered, we can choose four positions for the N, leaving the remaining seven positions for a move E to the east. We see that there are $\binom{11}{4}$ paths covering 11 blocks, with four to the north and the remaining seven to the east. ■

2.2.3 Applications to Probability

Combinatorics is of particular importance in the calculation of probabilities. For example, there are $6 \times 6 = 36$ ways to roll of a pair of dice, and two ways to roll an 11 (i.e., a 5 and a 6 or a 6 and a 5), so the probability of rolling an 11 is 2 in 36. More generally, a probability is determined by counting the number of outcomes considered favorable and counting the total number of outcomes, both favorable and unfavorable. If every outcome is equally likely to occur, the ratio between the two counts is the probability of obtaining a favorable outcome.

Often, the number of outcomes is expressed as combinations or permutations. For instance, in Example 2.9 combinations were evaluated to show that there are 123,552 five-card hands with two-pairs, and 54,912 hands with three-of-a-kind arrangements. The total number of hands is the number of 5-combinations in a 52-card deck, so there are a total of $\binom{52}{5} = 2{,}598{,}960$ possible hands. We therefore have the following probabilities P, where $\doteq$ indicates a decimal approximation:

$$P\ (\text{two pair}) = \frac{123{,}552}{2{,}598{,}960} \doteq 0.048$$

$$P\ (\text{three of a kind}) = \frac{54{,}912}{2{,}598{,}960} \doteq 0.021$$

We can see why the less likely hand three of a kind beats a hand with two pair.

Example 2.11 In the game of bridge, all of the 52 cards are dealt out to the four players, so each player has a hand of 13 cards. What is the probability that a bridge hand has

(a) all 13 cards of the same suit?
(b) no face cards?
(c) four aces, four kings, four queens, and a jack?

Solution

(a) There are $\binom{52}{13} \doteq 6.35 \times 10^{11}$ possible bridge hands, and just 4 of these are of a single suit. The probability of a hand of a single suit is therefore $\frac{4}{\binom{52}{13}} \doteq 6.3 \times 10^{-12}$

(b) There are 40 non-face cards, so the probability is $\frac{\binom{40}{13}}{\binom{52}{13}} \doteq 0.019$.

(c) There is just one way to get all of the aces, kings, and queens, and four ways to get a jack, so the probability is the same as for part (a). ■

Example 2.12 Randy has 13 loose keys to his house in his right pocket. One dark and rainy night, he randomly draws a key from his pocket, tries it, and if it doesn't work puts it in his left pocket and draws another key from his right pocket. What is the probability that Randy takes i tries before getting the correct key?

Solution. There is just one correct key among the 13, so the probability of drawing the correct key on the first attempt is 1 in 13. Now assume that the correct key is drawn on the ith attempt where $i \geq 2$. The sample space S (i.e., the set of *all* ways to try i keys without replacement) is the set of i-permutations of the 13 keys, so the number of ways to try i keys is

$$|S| = P(13, i) = 13 \times 12 \times \cdots \times (13 - i + 1) = 13 \times 12 \times \cdots \times (14 - i)$$

If the correct key is taken on the ith attempt, the first $i - 1$ keys are an $(i - 1)$-permutation of the 12 incorrect keys. Thus, event E, in which the ith key is correct, has the count

$$|E| = P(12, i - 1) = 12 \times 11 \times \cdots \times (12 - (i - 1) + 1) = 12 \times 11 \times \cdots \times (14 - i)$$

The probability that the ith key is correct is therefore

$$P(E) = \frac{|E|}{|S|} = \frac{12 \times 11 \times \cdots \times (14 - i)}{13 \times 12 \times 11 \times \cdots \times (14 - i)} = \frac{1}{13}$$

Interestingly, the probability is the same that the correct key will be drawn on each of the possible trials.

More generally, if Randy had any number of keys in his pocket, say, n, the probability that the correct key is found at trial i is $1/n$ for any i, $1 \leq i \leq n$. ■

Example 2.13 In organized sports, the champion is often determined by a "final" or "world" series of $2n + 1$ games, with the team to first win $n + 1$ games declared champion. Consider a playoff series between two evenly matched teams.

(a) Determine the probability p_{n+1+k} that the series ends after $n + 1 + k$ games, where $k = 0, 1, \ldots, n$.

(b) Use part (a) to prove the identity

$$\sum_{k=0}^{n} \frac{1}{2^k} \binom{n+k}{n} = 2^n$$

Solution

(a) There are 2^{n+k} outcomes of the first $n + k$ games, and the series ends with game $n + k + 1$ if the team winning the last game has already won exactly n games. Since there are $\binom{n+k}{n}$ ways to choose which of the first $n + k$ games were won by the champion team, we see that the probability the series ends with game $n + 1 + k$ is

$$p_{n+1+k} = \frac{\binom{n+k}{n}}{2^{n+k}}, k = 0, 1, \ldots, n$$

(b) It is certain with probability 1 that the series ends by game $2n + 1$, so

$$1 = \sum_{k=0}^{n} p_{n+1+k} = \sum_{k=0}^{n} \frac{\binom{n+k}{n}}{2^{n+k}}$$

When multiplied by 2^n, the desired identity is obtained. ■

PROBLEMS

Unless the problem specifically asks for a numerical evaluation, it suffices to answer with an expression involving factorials, permutations, and/or combinations. However, your solution must clearly explain the reasoning that you have used to obtain the expression. When asked to use *combinatorial reasoning*, base your reasoning on the combinatorial meaning of combinations, permutations, and the like. In particular, avoid algebraic calculation and numerical calculation. To give a combinatorial proof of an identity, propose a counting question and then answer it in two ways, so that equating the expressions derived in your two answers yields the identity.

2.2.1. What is the number of permutations of the 26 letters of the alphabet

(a) with no restrictions?

(b) with all five vowels in the consecutive group *aeiou*?

(c) with all five vowels in a group of 5, although in any order within the group?

(d) with the five vowels appearing in the natural order *a,e,i,o,u,* but not necessarily in a group of consecutive letters?

2.2.2. How many permutations of the 26 letters of the alphabet

(a) contain either the consecutive string *abc* or *xyz,* but not both strings?

(b) contain either the consecutive string *abc* or *cde,* but not both strings?

2.2.3. What is the number of permutations of the digits 0, 1, 2, ... , 9

(a) with no restrictions?

(b) if the odd digits are adjacent to one another?

(c) if no two odd digits are adjacent?

(d) if every prime is to the left of every nonprime?

2.2.4. A van has 12 seats, arranged in three rows. In each row, there are two pairs separated by the aisle. In how many ways can nine people be seated, where three insist on sitting left of the aisle, four insist on sitting right of the aisle, and two people don't care where they are seated?

2.2.5. There are three flagpoles and 11 different flags, where each pole can fly up to 4 flags. In how many ways can all 11 flags be flown?

2.2.6. In how many ways can 10 freshmen each be assigned to work with a mentor chosen from a group of 20 upper-class volunteers, if each freshman is assigned to

(a) one mentor and each mentor works with one freshman?

(b) a pair of mentors?

2.2.7. In how many ways can a two-scoop ice cream cone be ordered from 10 flavors if

(a) the two flavors are different and it matters which flavor is on top?

(b) the two flavors are different and it doesn't matter which flavor is on top (e.g., chocolate over strawberry is the same as strawberry over chocolate)?

(c) both scoops can be the same flavor, but it still matters which flavor is on top?

(d) both scoops can be the same flavor, but if two different flavors are ordered it makes no difference which flavor is on top?

2.2.8. In how many ways can a five person committee be chosen from an 11-member group, where two members of the committee are designated as the chair and secretary? Give three answers, depending on whether

(a) the officers are chosen after the committee selection

(b) the committee members other than the two officers are selected first

(c) the two officers are selected first and then the rest of the committee

Numerically check that the three answers are the same.

2.2.9. Every week, Jane rearranges the six pictures in her den, using the same hooks on the walls. What is the maximum number of weeks that can pass until some arrangement is repeated?

2.2.10. In how may ways can a seven member committee be chosen from a group of five men and eight women

(a) with no restrictions?

(b) if the committee must have exactly three men?

(c) if the committee must have at least three men?

(d) if the committee cannot have both Mr. and Mrs. Smith?

2.2.11. Give a combinatorial proof of the identity $n^2 = 2\binom{n}{2} + n$ by asking a question that can be answered in two ways.

2.2.12. How many rectangles can be found in an $m \times n$ grid of unit squares? For example, three rectangles are shown in this grid. Note that the rectangles can overlap one another:

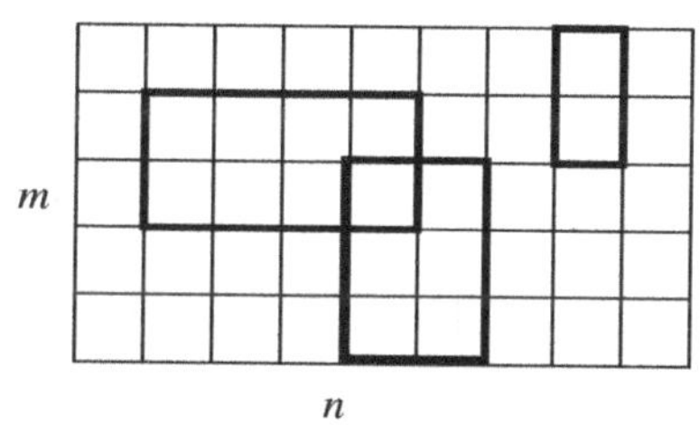

2.2.13. The nth Catalan number (named after Belgian mathematician Eugène Charles Catalan, 1814–1894; also see the next section for its meaning in the block walking model) is given by $C_n = [1/(n+1)]\binom{2n}{n}$. Prove that another formula is $C_n = (1/n)\binom{2n}{n-1}$, using both

(a) combinatorial reasoning

(b) algebraic calculation.

2.2.14. In Example 2.7, a four person committee was to be chosen from the 11 members of the Math Club. Suppose that exactly three seniors belong to the club, and at least two of them are to be chosen for the committee. Decide which of the following approaches to counting the number of possible committees is correct, and, if incorrect, explain what error of reasoning has been made:

(a) first choose two seniors to be on the committee, and then choose the other two committee members from the nine club members not yet chosen, giving $\binom{3}{2}\binom{9}{2}$ possible committees.

(b) take cases according to whether two or three seniors are on the committee, giving $\binom{3}{2}\binom{8}{2}+\binom{3}{3}\binom{8}{1}$ committees.

(c) remove committees with just one or no seniors, giving $\binom{11}{4}-\binom{3}{1}\binom{8}{3}-\binom{8}{4}$.

2.2.15. The United States Post Office automates mail handling with the PostNET (postal numeric encoding technique) barcode. There are 62 vertical bars of two sizes, long and short. Between the two long guard bars at the beginning and end, there are 12 five-symbol blocks that each contain two long and three shorts bars:

7 4 8 0 5 3 7 5 8 1 3 9

For example, the first block of five symbols that follows the leftmost guard bar is long–short–short–short–long, which encodes a 7.

(a) Explain why blocks of two long and three short symbols work well.

(b) What blocks do you expect code the digits 2 and 6?

2.2.16. In how many ways can 3 numbers from $[n] = \{1,2, \ldots, n\}$ be selected so that

(a) at least two are adjacent? [*Hint.* First choose the middle number.]

(b) no two are adjacent?

2.2.17. Place n points on a circle and then construct all of the chords that join pairs of points. Assume that your n points are in "general position," so that no more than two chords intersect at a single point within the circle. The following diagram depicts the case $n = 7$, and shows that there are 21 chords and 35 intersections of chords within the circle:

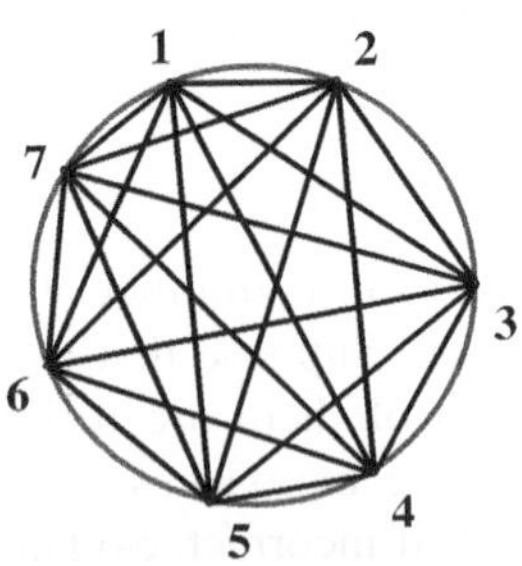

For n points, what is

(a) the number of chords?

(b) the number of points of intersection of chords within the circle?

2.2.18. How many paths, always moving to the right or upward, join A to B in the following street pattern (see diagram) when

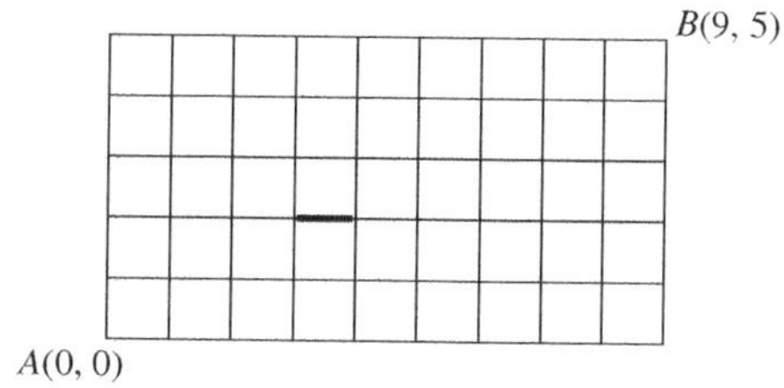

(a) there are no restrictions?

(b) the path cannot cross the street in bold, which is under construction?

2.2.19. How many paths connect A to B in the system of blocks shown (see diagram) that surround the lake in the center?

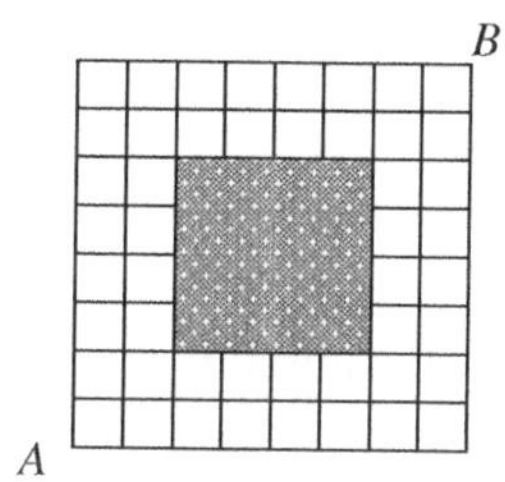

2.2.20. **(a)** Give a combinatorial proof of the identity

$$P(m+n,r) = \sum_{k\geq 0} P(m,k)P(n,r-k)\binom{r}{k}$$

by answering this question in two ways: in how many ways a line of r people can be formed, chosen from a group of m men and n women?

(b) Show algebraically that this identity is equivalent to the identity

$$\binom{m+n}{r} = \sum_{k\geq 0}\binom{m}{k}\binom{n}{r-k}$$

2.2.21. A bridge hand consists of 13 cards selected from a 52-card deck. How many bridge hands

(a) are there?

(b) do not contain a pair (i.e., with no two cards having the same rank)?

(c) contain *exactly* one pair?

2.2.22. Determine the number of five-card poker hands of the indicated type:

(a) one pair

(b) four of a kind

(c) full house (a pair and three of a kind)

(d) royal flush (10-J-Q-K-A, all of same suit)

(e) straight but not a flush (sequence of cards not all of the same suit with no gap in rank, such as 3-4-5-6-7; an ace-high straight 10-J-Q-K-A is a *broadway*, and a five-high straight A-2-3-4-5 is a *wheel*).

(f) straight flush other than a royal flush (straight of the same suit, but not ace high)

(g) flush, but not a royal flush or a straight flush (five cards of the same suit, not a straight)

(h) high card (no pair, no straight, and no flush)

2.2.23. Consider the identity

$$(2n-1)!! = \frac{n!}{2^n}\binom{2n}{n} \quad \text{where} \quad (2n-1)!! = (2n-1)\cdot(2n-3)\cdot\cdots\cdot 3\cdot 1$$

is a double factorial.

(a) Prove the identity combinatorially, by counting the number of ways to set up singles tennis matches for $2n$ players.

(b) Prove the identity algebraically.

2.2.24. A rook moves any distance along a row or column of a chessboard. How many ways can two white and three black nonattacking rooks be placed on an 8×8 chessboard?

2.2.25. Would you want to bet for or against getting 9, 10, or 11 heads in 20 flips of a fair coin?

2.2.26. If the five letters a,b,c,d,e are randomly arranged, what is the probability that

(a) the two-letter string ab occurs somewhere in the arrangement?

(b) the two vowels are side by side in either order?

2.3 COMBINATORIAL MODELS

It is often useful to interpret a combinatorial quantity in a context that is easily understood. We have already encountered some of the most fruitful combinatorial models such as committee selections, block walking and path counting, tiling patterns, and dot patterns. These models provided us a way to obtain several important identities. For example, since the binomial coefficient $C(n, r)$ counts the number of ways to tile a $1 \times n$ board with r gray square tiles, we then derived Pascal's identity $C(n, r) = C(n-1, r) + C(n-1, r-1)$ by partitioning the tilings into those ending with a white tile and those ending with a gray tile. Similarly, since the combinatorial Fibonacci number f_n counts the number of ways to tile a $1 \times n$ board with squares and dominoes, partitioning the tilings into those ending with a square and those ending with a domino gave us the Fibonacci recursion $f_n = f_{n-1} + f_{n-2}$.

Our goal in this section is to explore how these and other models can be utilized to prove other identities for permutations, combinations, Fibonacci numbers, and so on. Typically, it is best to start with the simpler side of the identity, and choose or create a model that counts some set S of objects (tilings, committees, patterns, etc.) that is evident in the model. Next, an argument it given that explains why the other side of the identity counts the same number $|S|$. Often, one side of the identity is a sum, and it can be shown that each term in the sum counts the number of objects in some partitioning of the set into disjoint cases.

The examples that follow illustrate how this strategy is carried out using a variety of combinatorial models.

2.3.1 Tiling Models

Example 2.14 Recall that the combinatorial Fibonacci numbers f_n are given by the recursively defined sequence $f_0 = 1, f_1 = 1, f_2 = 2, f_3 = 3, f_4 = 5, \ldots,$ $f_n = f_{n-1} + f_{n-2}$. Prove these identities:

(a)
$$f_{m+n} = f_m f_n + f_{m-1} f_{n-1} \tag{2.6}$$

(b)
$$f_0 + f_1 + f_2 + \cdots + f_n = f_{n+2} - 1 \tag{2.7}$$

(c)
$$\sum_{k \geq 0} \binom{n-k}{k} = f_n \tag{2.8}$$

Solution

(a) The left side of (2.6) counts the number of ways to tile a $1 \times (m+n)$ board with squares and dominoes. The right side suggests that we consider two cases. If there is a break between cells n and $n+1$ of the board, then there are f_n ways to tile the $1 \times n$ board to the left of the break and f_m ways to tile the $1 \times m$ board to the right of the break. Therefore, a *breakable board* can be tiled in $f_n f_m$ ways. The board is not breakable when a domino spans cells n and $n+1$, and then there are $f_{n-1} f_{m-1}$ ways to tile the boards to the left and right of the domino.

(b) The easier side of (2.7) is on the right, which counts all except one of the tilings of a board of length in $n+2$. What tiling has been omitted? The strongest possibility is the one tiling by all squares. Thus, the right side can be viewed as the ways to tile a board of length $n + 2$ with at least one domino. We can assume that the left side of (2.7) is a sum taken over cases. Since we know the tilings of interest have at least one domino, let's assume that the rightmost domino is followed by k squares, where $k = 0, 1, \ldots, n$. The board of length $n - k$ to the left of the rightmost domino can be tiled in f_{n-k} ways. Summing over all of the cases $k = 0, 1, \ldots, n$ shows that there are $f_n + f_{n-1} + \cdots + f_0$ tilings with at least one domino. Equating the two counts proves identity (2.7).

(c) The right side of (2.8) counts the number of tilings of an n-board with squares and dominoes. If the tiling uses k dominoes, there are then $n - 2k$ squares and $k + (n - 2k) = n - k$ tiles altogether. There are $\binom{n-k}{k}$ permutations of these tiles. ■

The interpretation of the Fibonacci numbers as tilings with squares and dominoes can be applied to prove a large number of identities. The interested reader is invited to consult Benjamin and Quinn's book [1] for a thorough look at this approach.

The next example illustrates how nearly any combinatorial question about tiling leads to an identity that typically involves the binomial coefficients.

Example 2.15 Prove the identity

$$\binom{n}{r} 2^{n-r} = \sum_{k=r}^{n} \binom{n}{k} \binom{k}{r}, \qquad 0 \le r \le n \tag{2.9}$$

by giving two expressions that each provide an answer to the following question: "In how many ways can a $1 \times n$ board be tiled with r gray tiles and an assortment of black and white tiles?"

Solution

Answer 1. There are $\binom{n}{r}$ ways to position the r gray tiles and 2^{n-r} ways to tile the remaining $n - r$ squares with either a white or a black tile.

Answer 2. Take cases according to the number k of cells that will *not* be covered with a black tile. There are $\binom{n}{k}$ ways to choose the "not black" cells, and then $\binom{k}{r}$ ways to choose r of these k cells that will be covered with a gray tile. This means there are $\binom{n}{k}\binom{k}{r}$ tilings with $k - r$ white tiles. Summing over k gives the total number of tilings, which is the right side of (2.9). ■

2.3.2 Block Walking Models

Another productive combinatorial model is provided by *block walking*. Here we start with a system of edges on a square grid and then count the number of paths in the system that join a starting point A to a second point B. Example 2.10 provided an example of block walking, connecting the binomial coefficients to the number of paths in a square grid. The connection to Pascal's triangle is made more evident by placing point A at the uppermost point of the rotated pattern shown in Figure 2.1. A path then is required to always move downward, or south, and there is a choice at each intersection to take the downward path to the east or to the west.

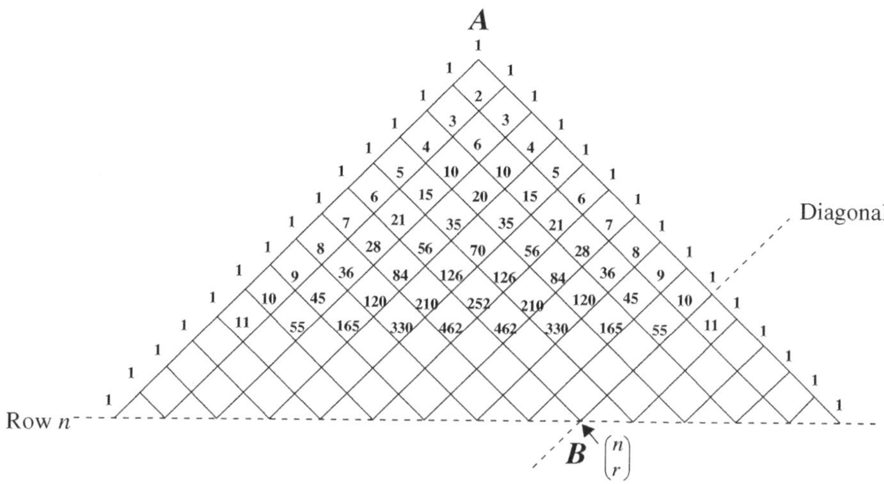

FIGURE 2.1 Block walking model for Pascal's triangle.

It is important to observe that the rows are numbered $n = 0,1,2,\ldots$ and the diagonals are numbered $r = 0,1,2,\ldots$. There are then $\binom{n}{r}$ paths from $A = A(0, 0)$ to $B = B(n,r)$, since any path is a list of n steps with r steps to the east. Note also that points such as $B(n,r)$ are described by their row index n and their diagonal index r.

Identities and properties of the binomial coefficients are often evident from the block walking diagram shown in Figure 2.1. For example, the vertical symmetry of the pattern makes it evident that $\binom{n}{r} = \binom{n}{n-r}$. Block walking is a useful model both suggesting and proving identities involving the binomial coefficients. Here are two easy examples:

- $\sum_{r=0}^{n} \binom{n}{r} = 2^n$, since any path from A to a point of row n is a sequence of n two-way decisions—move downward to either east or west.
- $\binom{n}{r} = \binom{n-1}{r-1} + \binom{n-1}{r}$, since any path reaching point $B(n, r)$ had to pass through exactly one of the points $B(n-1, r-1)$ or $B(n-1, r)$.

Example 2.16 Use block walking to prove the following identities:

(a) $$\binom{n+1}{r+1} = \binom{r}{r} + \binom{r+1}{r} + \binom{r+2}{r} + \cdots + \binom{n}{r} \tag{2.10}$$

(b) $$\binom{2n}{n} = \sum_{k=0}^{n} \binom{n}{r}^2 \tag{2.11}$$

Solution

(a) The left side of (2.10) gives the number of paths that reach row $n + 1$ at diagonal $r + 1$. But any path must have passed through diagonal r at exactly one of its intersections with a row above $n + 1$. Thus the number of paths is given by the sum $\sum_{k=r}^{n} \binom{k}{r}$, which is on the right side of (2.10).

(b) Any path ending at point B on diagonal n of row $2n$ must pass through row n at some unique point P_r of diagonal r. There are $\binom{n}{r}$ paths from A to P_r and then $\binom{n}{n-r} = \binom{n}{r}$ paths from P_r to B. Therefore, there are $\binom{n}{r}^2$ paths from A to B that pass through row n at diagonal r. Summing over r gives the right side of identity (2.11). ■

Example 2.17 Use a block walking argument to show that the number of paths from $A(0,0)$ to $B(2n, n)$ in Figure 2.1 that stay on or to the west of the vertical line l through A and B is given by

$$C_n = \frac{1}{n+1}\binom{2n}{n} \tag{2.12}$$

Such numbers are known as the *Catalan numbers* (named after Belgian mathematician Eugène Charles Catalan, 1814–1894), and will be examined further in Chapter 6.

Solution. There are $\binom{2n}{n}$ paths from A to B, each with n steps to the east and n steps to the west. Our answer can then obtained by subtracting the number of these paths that cross to the east of line l. Given a path that crosses l, suppose that it first crosses l along the edge from point $F(2r, r)$ to point $G(2r + 1, r + 1)$. Since the path from A to G used one more edge to the east than to the west, the remaining tail of the path from G to $B(2n,n)$ covers one more westward edge than eastward edge before ending at B. Now suppose that the edges in the tail are swapped, with each east edge in the path from G to B replaced by a west edge and each west edge replaced by an east edge. The new tail-swapped path starting at A still ends in row $2n$ but now at point $H(2n, n + 1)$. Conversely, any path from A to H can be tail swapped to provide a path from A to B that necessarily crosses l. In the figure that follows, the dotted part of path from A to H is the swapped tail of the crossing path:

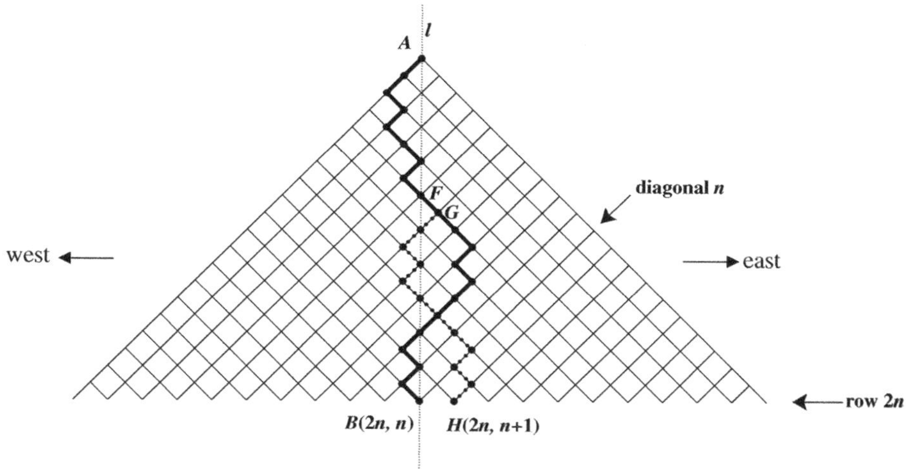

Since there are $\binom{2n}{n+1}$ paths from A to B that cross line l, this leaves $\binom{2n}{n} - \binom{2n}{n+1}$ paths that never pass through a point to the east of line l. To obtain the form shown in (2.12), just note that

$$n\binom{2n}{n} = \binom{2n}{n+1}(n+1) \qquad (2.13)$$

This can be shown algebraically, but (2.13) follows even more easily with path counting; any path from A to B can be mapped to a path from A to H by replacing one of its n westward blocks with an eastward block. Conversely, any path from A to H can be mapped to a path from A to B by replacing one of its $n + 1$ eastward blocks with a westward block. Using (2.13), we see that the number of paths from A to B that are never to the east of line l is given by

$$\binom{2n}{n} - \binom{2n}{n+1} = \binom{2n}{n} - \frac{n}{n+1}\binom{2n}{n} = \frac{1}{n+1}\binom{2n}{n}$$

thus completing the proof. ■

There are endless variations on the block walking model. For example, the square grid can be replaced with any system of directed edges. A proof based on this more general setting is known as *path counting*.

2.3.3 The Committee Selection Model

The key idea of the *committee selection model* is to count the number of committees that can be created under various constraints. If the same count can be made in two (or even more) ways, then equating the resulting expressions derives an identity (or identities). Most often this argument is far more direct than an algebraic calculation. For example, it is easy to prove identity (2.13) with a committee selection model. If asked how many ways a committee of n can be created from a club of $2n$ members, as well as a chair of the committee, we can answer in (either of) two ways.

Answer 1. Pick the committee in $\binom{2n}{n}$ ways, and then choose one of the committee members as chair in n ways. Therefore, the number of ways to form a committee with chair is $\binom{2n}{n}n$.

Answer 2. Choose $n + 1$ people who will *not* be ordinary committee members but might be chair in $\binom{2n}{n+1}$ ways. From the $n + 1$ people, choose the chair in $n + 1$ ways. Therefore the committee with chair can be formed in $\binom{2n}{n+1}$ $(n + 1)$ ways.

Since both answers count the same number of possibilities, they can be equated to derive identity (2.13).

Identity (2.13) is one case of this general result.

Theorem 2.18 (Committee with Chair Identities)

$$r\binom{n}{r} = (n-r+1)\binom{n}{r-1} = n\binom{n-1}{r-1} \tag{2.14}$$

Proof. First, ask this question: "In how many ways can a committee of r members be chosen from a group of n people, together with a choice of one of the r committee members as its chair?" Next, give three different answers to the question:

Answer 1. There are $\binom{n}{r}$ ways to select the committee, and then r ways to select the chair from among the r people on the committee; that is, there are $\binom{n}{r}r$ ways to select the committee and its chair.

Answer 2. First select the members of the committee other than the chair in $\binom{n}{r-1}$ ways. This leaves $n-(r-1) = n-r+1$ people from which to choose the chair of the committee and bring the number of committee members up to r. In other words there are $\binom{n}{r-1}$ $(n - r + 1)$ ways to choose a committee of r and the committee chair.

Answer 3. First select the chair from the entire membership in n ways, and then choose the remaining $r - 1$ members of the committee in $\binom{n-1}{r-1}$ ways, which shows there are $n\binom{n-1}{r-1}$ committee with chair selections.

Equating the expressions in the three answers to the same question gives the identities shown in (2.14). ∎

Example 2.19 Give a committee selection argument to verify the identity

$$\binom{m+n}{3} = \binom{m}{3} + \binom{n}{3} + \binom{m}{2} n + \binom{n}{2} m$$

Solution. In how many ways can a committee of three be chosen from a club with m men and n women?

Answer 1. The three members of the committee can be selected with no regard to gender in $\binom{m+n}{3}$ ways.

Answer 2. There are four disjoint cases to consider. All members of the committee are men, all are women, two are men, and one is a woman, or one is a man and two are women. The respective number of ways to choose the four types of committees are

$$\binom{m}{3}, \binom{n}{3}, \binom{m}{2} n, m \binom{n}{2}$$

By the addition principle, there are

$$\binom{m}{3} + \binom{n}{3} + \binom{m}{2} n + m \binom{n}{2}$$

ways to select the committee.

The desired identity results by equating the expressions derived in the two answers. ■

2.3.4 The Flagpole Model

Tiling, block walking, path counting, and committee selections are perhaps the most commonly used combinatorial models. But other models can be devised as well. One such possibility is the *flagpole model.* The idea is simple to visualize; there is a row, or rows, of blocks that can be used to anchor a flagpole and that can also be used to anchor guywires that support the pole. Conditions can be imposed about the location, number, and color of the guywires, and whether one or more wires can be anchored to the same block. Counting the number of permissible arrangements—in two ways, of course—then derives combinatorial identities.

Here are two examples of the flagpole model.

Example 2.20 Prove the identity

$$1 \cdot n + 2 \cdot (n-1) + 3 \cdot (n-2) + \cdots + n \cdot 1 = \binom{n+2}{3} \tag{2.15}$$

by answering this question in two ways: "In how many ways can a flagpole be erected on a row of $n + 2$ blocks, supported by two guywires, with one wire anchored to a block to the left of the pole and another wire anchored to a block to the right of it?" See diagram:

Solution. It's best to start with the simpler side of the identity, and determine rules for the arrangements showing that the number of allowable arrangements is given by that side of the identity. For the opposite side of the identity, look for a new way to count the same number of arrangements. Often this is a sum, as in the present example, and we must determine how the terms in the sum count those arrangements that have some distinguishing characteristic.

Answer 1. Any choice of 3 of the $n + 2$ blocks corresponds to the arrangement with the pole on the middle block and the guywires on the other two blocks. Thus, there are $\binom{n+2}{3}$ arrangements.

Answer 2. We can take cases according to which block supports the flagpole. Suppose that there are k blocks to the left of the block supporting the pole, and therefore $n + 1 - k$ blocks to the right of the pole. The guywire to the left of the pole can be anchored in k ways, and the guywire to the right of the pole can be anchored in $n + 1 - k$ ways. We see that if the pole is erected on block $k + 1$, then the two guywires can be anchored in $k(n + 1 - k)$ ways. Summing over k gives us $\sum_{k=1}^{n} k(n + 1 - k)$, which is the left side of (2.15).

Equating the expressions obtained in the two answers gives the desired identity. ∎

The next example illustrates a useful combinatorial principle. If a 1-to-n mapping from a set A to a set B can be found, then $|A| = (1/n)|B|$.

Example 2.21 Prove the identity

$$\sum_{k=1}^{n} k^2 = \frac{1}{4}\binom{2n+2}{3} \tag{2.16}$$

by answering this question in two ways: "In how many ways can a flagpole be erected on a row of $n + 1$ blocks, with a red and blue guywire attached to either the same or two different blocks to the left of the pole?" See diagram:

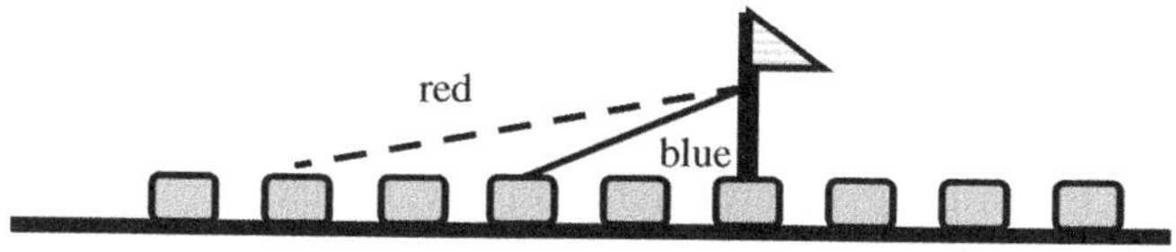

Solution. The division by 4 makes the right side of (2.16) the more difficult side, so this time we start with the sum on the left side of the identity:

Answer 1. Let the flagpole be erected on block $k + 1$, where $1 \le k \le n$. There are then k choices of the block or blocks to which the red and blue guywires are anchored. In other words, there are k^2 ways to anchor the two guywires when the pole is on block $k + 1$. Summing over k shows that there are $\sum_{k=1}^{n} k^2$ flagpole arrangements.

Answer 2. The right side of the identity suggests that we need to consider the $\binom{2n+2}{3}$ ways that a three element subset can be chosen from a set of $2n + 2$ elements. Since there are $n + 1$ blocks, let's try the set $S = \{1, 1', 2, 2', \ldots, n+1, n'+1\}$, ordered by $1 < 1' < 2 < 2' < \cdots < n+1 < n'+1$. The division by 4 tells as that a flagpole arrangement must correspond to exactly 4 of the three-element subsets. For example, consider the arrangement in the diagram above for which the red wire is on block 2, the blue wire is on block 4, and the pole is on block 6. We can associate this arrangement with the four subsets $\{2, 4, 6\}$, $\{2', 4, 6\}$, $\{2, 4', 6\}$, and $\{2', 4', 6\}$. Similarly, if the blue wire is to the left of the red wire, we can let each of the subsets $\{2, 4, 6'\}$, $\{2', 4, 6'\}$, $\{2, 4', 6'\}$, and $\{2', 4', 6'\}$ correspond to the arrangement with the blue wire on block 2, the red wire on block 4, and the pole on block 6. Finally, the arrangement with both the red and blue wires on block 2 and the pole on block 6 can be associated with the four subsets $\{2, 2', 6\}$, $\{2, 2', 6'\}$, $\{2, 6, 6'\}$, and $\{2', 6, 6'\}$. The figures below give the one-to four mapping in which each flagpole arrangement corresponds to exactly 4 of the three element subsets of S. Moreover, each three element subset of S corresponds to a unique flagpole arrangement. Since there is a 1-to-4 mapping between flagpole arrangements and the 3-subsets of S, we see there are $\frac{1}{4}\binom{2n+2}{3}$ flagpole arrangements.

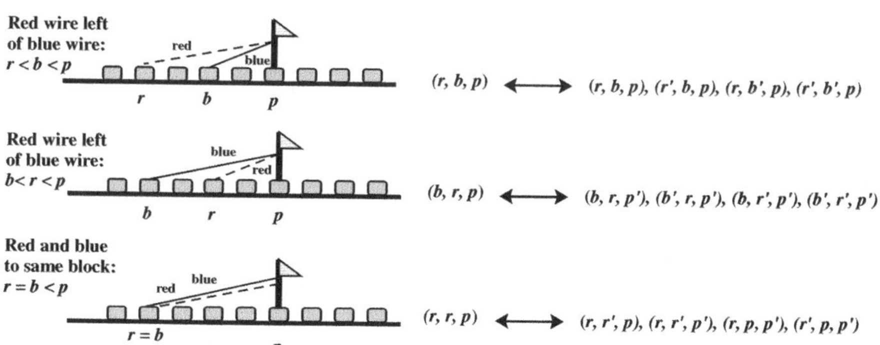

Equating the expression derived in the two answers gives us identity (2.16). ■

PROBLEMS

2.3.1. Recall that the combinatorial Fibonacci number f_n is the number of ways to tile a $1 \times n$ board with squares and dominoes. Define $f_0 = 1$ and use the tiling model to prove the identity $f_{2n+1} = f_0 + f_2 + f_4 + \cdots + f_{2n}$, $n \geq 0$.

2.3.2. Use the tiling model to prove the identity $f_{2n} - 1 = f_1 + f_3 + f_5 + \cdots + f_{2n-1}$, $n \geq 1$.

2.3.3. **(a)** Prove that

$$f_{2n+1} = \sum_{k=1}^{n+1} \binom{n+1}{k} f_{k-1}, n \geq 0$$

using the tiling model. [*Hint:* Any tiling of a $1 \times (2n + 1)$ board must include at least one square and at least n other tiles.]

(b) Restate the identity of part (a) in terms of the standard Fibonacci numbers $F_m = f_{m-1}$.

2.3.4. Use the tiling model to prove that $f_n^2 + f_{n+1}^2 = f_{2n+2}$.

2.3.5. Use the tiling model to prove that $F_{n+1}^2 - F_{n-1}^2 = F_{2n}$. [*Hint:* Consider tilings of two boards each of length n, not both ending with a domino.]

2.3.6. Prove the identity below by providing two ways to answer this question: "In how many ways can a $2 \times n$ board be tiled with n red and n blue square tiles?"

$$\binom{2n}{n} = \sum_{k \geq 0} \binom{n}{k} \binom{n}{n-k}$$

2.3.7. Prove the identity below by providing two ways to answer this question: "In how many ways can a $2 \times n$ board be tiled with n red and n blue square tiles?"

$$\binom{2n}{n} = \sum_{k \geq 0} \binom{n}{k} \binom{n-k}{k} 2^{n-2k}$$

2.3.8. Prove the identity

$$\binom{2n}{n} = \sum_{k \geq 0} \binom{n}{2k} \binom{2k}{k} 2^{n-2k}$$

by providing two ways to answer this question: "In how many ways can a $2 \times n$ board be tiled with n red and n blue square tiles?"

2.3.9. Use the block walking model to prove that

$$\binom{2n}{n} = \sum_{k=0}^{n} \binom{n}{k}^2$$

2.3.10. Use block walking to prove the hockey stick identity

$$\binom{n+1}{r+1} = \binom{r}{r} + \binom{r+1}{r} + \binom{r+2}{r} + \cdots + \binom{n}{r}$$

2.3.11. There have been 100 East-versus-West bowl games, with 50 wins for each. What is the probability that West never had fewer wins than East throughout the series?

2.3.12. The *Delannoy number* $D(a, b)$ is the number of paths from point A to point $B(a, b)$ in the system of edges shown here, where each edge is southwest, southeast, or directly south:

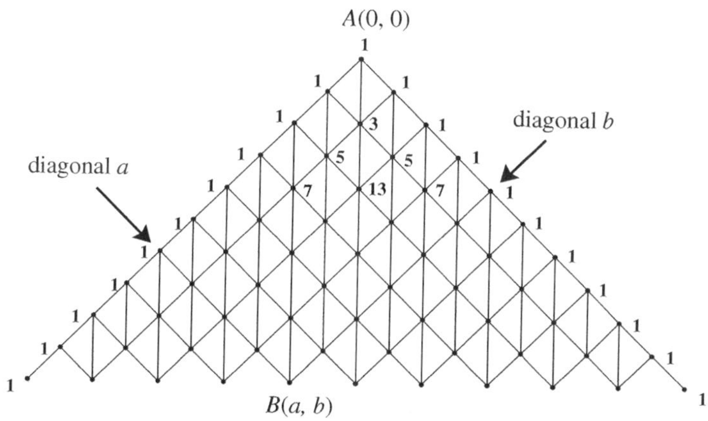

For example, $D(1, 1) = 3$, $D(1, 2) = 5$, and $D(2, 2) = 13$.

(a) Draw and label a system of edges through $a + b = 10$ and check that $D(6, 4) = 1289$ and $D(5, 5) = 1683$.

(b) Use path counting to derive a Pascal-like identity for the Delannoy numbers.

(c) Use path counting to prove that

$$D(a, b) = \sum_{s=0}^{a} \binom{b}{s} \binom{a+b-s}{b}$$

[*Hint*: Let s represent the number of edges due south along a path.]

2.3.13. Give a block walking proof that the Catalan numbers described in Example 2.17 satisfy the recursion relation $C_{n+1} = C_0C_n + C_1C_{n-1} + \cdots + C_{n-1}C_1 + C_nC_0$, where $C_0 = 1$.

2.3.14. Give a combinatorial proof that the number of compositions of a positive integer n into parts each larger than one is the combinatorial Fibonacci number f_{n-2}.

2.3.15. Use the committee selection model to prove the identity

$$\binom{2n}{n} = \sum_{k=0}^{n} \binom{n}{k}^2$$

2.3.16. Use the committee selection model to derive this identity:

$$\binom{m+n}{r} = \sum_{k\geq 0} \binom{m}{k}\binom{n}{r-k}$$

2.3.17. Prove that the binomial coefficients in row n of Pascal's triangle are *unimodal,* meaning that they strictly increase to a maximum $\binom{2m}{m}$ when $n = 2m$ is even and then strictly decrease, or, strictly increase to the maximum $\binom{2m-1}{m} = \binom{2m-1}{m+1}$ when $n = 2m - 1$ is odd, and then strictly decrease.

2.3.18. Use the flagpole model to prove the identity

$$\sum_{k=q}^{n} \binom{k}{q}\binom{n-k}{r} = \binom{n+1}{q+r+1}$$

by giving two answers to this question: "How many arrangements of a flagpole on a row of $n + 1$ blocks can be made if there are q supporting guywires anchored to distinct blocks to the left of the pole and r guywires anchored to distinct blocks to the right of the pole?"

2.3.19. Use the flagpole model to prove the identity $\sum_{k=1}^{n} k\binom{n}{k} = n\,2^{n-1}$, where $n \geq 1$, by giving two answers to this question: "In how many ways can a flagpole be placed on one of n blocks, with any subset (possibly empty) of the remaining blocks used to anchor one guywire each?"

2.3.20. Give a combinatorial tiling proof the identity $\sum_{j=2}^{n-1} (n-j)f_{j-2} + n + 1 = f_{n+1}$ for the combinatorial Fibonacci numbers f_n.

2.4 PERMUTATIONS AND COMBINATIONS WITH REPETITIONS

2.4.1 Multisets

Since a set either does or does not contains an element, it was natural to view both permutations and combinations without repetition as arrangements and selections of elements taken from a set. However, if repetition is allowed, we need to extend the concept of a set to allow for multiple copies of elements; that is, we introduce the notion of a *multiset.* For example, if a bakery has 30 glazed, 45 sugar, and 28 jelly doughnuts in their display case, then its stock of doughnuts is neatly described by the multiset $\{30 \cdot \mathrm{G}, 45 \cdot \mathrm{S}, 28 \cdot \mathrm{J}\}$.

Definition 2.22 A *multiset* is a collection $S = \{n_1 \cdot a_1, n_2 \cdot a_2, \ldots, n_k \cdot a_k\}$ containing n_1 copies of the element a_1, n_2 copies of the element a_2, and so on. If an unlimited number of copies of element a is available in S, this is indicated by the notation $\infty \cdot a$.

We see that there are k types of elements in the multiset $S = \{n_1 \cdot a_1, n_2 \cdot a_2, \ldots, n_k \cdot a_k\}$, and the number of elements in S is $|S| = n_1 + n_2 + \cdots + n_k$ when counted by multiplicity.

2.4.2 Permutations with Repetition

In Example 2.10, each path in the 7×4 system of square blocks was uniquely described by a permutation of 7 Es and 4 Ns, where an E denoted a move to the east and an N a move to the north; thus, the paths through the grid are just the permutations of the multiset $S = \{7 \cdot \mathrm{E}, 4 \cdot \mathrm{N}\}$. The number of permutations was counted by choosing the seven positions of the Es from the 11 total moves, so we see that there are $\binom{11}{7}$ permutations of the multiset S.

We can extend this reasoning to count the paths through a three-dimensional grid of cubes.

Example 2.23 How many paths extend from point $A(0,0,0)$ to the point $B(7,4,3)$, where each segment of any path is a unit-long step in the x (east), y (north), or z (upward) direction?

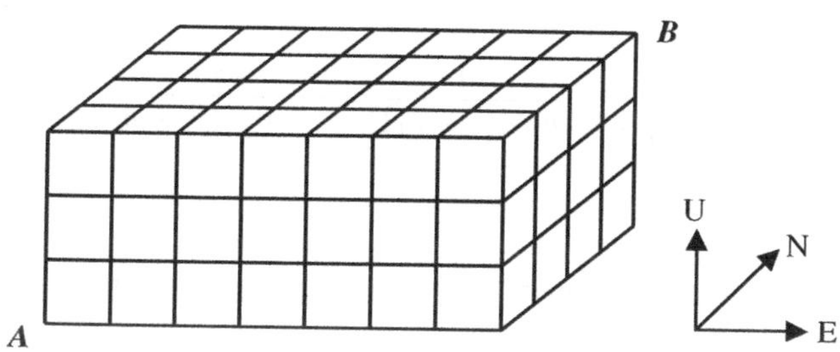

Solution. Each path is a permutation of all of the elements of the multiset $\{7 \cdot \text{E}, 4 \cdot \text{N}, 3 \cdot \text{U}\}$, where E, N, and U represent, respectively, a unit step in the east, north, or upward direction. For example, EEUNNNEEUUENEE describes one such path. Each path corresponds to a unique list of 14 moves in which there are $\binom{14}{7}$ ways to choose the positions of the eastward moves, then $\binom{7}{4}$ ways to choose where to take northward moves, and finally $\binom{3}{3}$ ways to choose where to move upward. Thus, there are $\binom{14}{7}\binom{7}{4}\binom{3}{3}$ permutations of the multiset $\{7 \cdot \text{E}, 4 \cdot \text{N}, 3 \cdot \text{U}\}$, and this is the number of paths from A to B. ■

In the previous example, it shouldn't make a difference if we had first selected the three upward edges, then the seven edges to the east, and finally the four edges to the north. To see that the order makes no difference, we note that

$$\binom{14}{7}\binom{7}{4}\binom{3}{3} = \frac{14!}{7!7!} \cdot \frac{7!}{4!3!} \cdot \frac{3!}{3!0!} = \frac{14!}{7!4!3!} \tag{2.17}$$

The same reasoning used in Example 2.23, together with the same calculation that gives formula (2.17), can be generalized to give us the following result.

Theorem 2.24 Let

$$\binom{n}{n_1, n_2, \dots, n_k}$$

denote the number of permutations of all $n = n_1 + n_2 + \cdots + n_k$ elements of the multiset $\{n_1 \cdot a_1, n_2 \cdot a_2, \dots, n_k \cdot a_k\}$. Then

$$\binom{n}{n_1, n_2, \dots, n_k} = \frac{n!}{n_1!n_2! \cdots n_k!} = \frac{(n_1 + n_2 + \cdots + n_k)!}{n_1!n_2! \cdots n_k!} \tag{2.18}$$

Proof. Each permutation is a sequence of all n elements of the multiset. There are then $\binom{n}{n_1}$ to choose where the n_1 copies of element a_1 are placed. This leaves $n - n_1$

positions for the n_2 copies of element a_2, so these can be placed in $\binom{n-n_1}{n_2}$ ways. Continuing until all of the elements are positioned, we see that there are

$$\begin{aligned}
&\binom{n}{n_1}\binom{n-n_1}{n_2}\binom{n-n_1-n_2}{n_3}\cdots\binom{n-n_1-n_2-\cdots-n_{k-1}}{n_k}\\
&\quad=\frac{n!}{(n_1)!\,(n-n_1)!}\frac{(n-n_1)!}{(n_2)!\,(n-n_1-n_2)!}\\
&\qquad\times\frac{(n-n_1-n_2)!}{(n_3)!\,(n-n_1-n_2-n_3)!}\cdots\frac{n!}{(n_k)!\,(n-n_1-n_2-\cdots-n_{k-1})!}\\
&\quad=\frac{n!}{(n_1)!\,(n_2)!\,(n_3)!\cdots(n_k)!}
\end{aligned}$$

permutations of the multiset. ■

The term

$$\binom{n}{n_1, n_2, \ldots, n_k}$$

is called a *multinomial coefficient* or *k-nomial coefficient* for reasons that will become clear in Chapter 3. It is seen from (2.18) that if $k = 2$, then

$$\binom{n}{r, n-r} = \frac{n!}{r!\,(n-r)!} = \binom{n}{r} \tag{2.19}$$

so that a 2-nomial coefficient is the same as a binomial coefficient.

Example 2.25 How many words (i.e., actual words or other letter combinations) can be formed with permutations of all of the letters in PANAMABANANAMAN?

Solution. Each word is a permutation of the 15 letters in the multiset $\{1 \cdot \mathsf{P}, 7 \cdot \mathsf{A}, 4 \cdot \mathsf{N}, 2 \cdot \mathsf{M}, 1 \cdot \mathsf{B}\}$. Applying Theorem 2.24, we see that

$$\binom{15}{1, 7, 4, 2, 1} = \frac{15!}{1!7!4!2!1!} = 5{,}405{,}500$$

words that can be formed. ■

It is important to understand that a multinomial coefficient counts the number of permutations that use *all* of the elements of a multiset. Later, we will develop methods to determine the number of permutations when only some of the elements of the multiset are selected and arranged into a specific order.

The division principle offers an alternative approach to counting permutations from a multiset. Suppose, for example, that we temporarily distinguish the letters in the multiset $\{1 \cdot \mathsf{P}, 7 \cdot \mathsf{A}, 4 \cdot \mathsf{N}, 2 \cdot \mathsf{M}, 1 \cdot \mathsf{B}\}$ by using subscripts. This creates the set $\{\mathsf{P}, \mathsf{A}_1, \mathsf{A}_2, \ldots, \mathsf{A}_7, \mathsf{N}_1, \mathsf{N}_2, \mathsf{N}_3, \mathsf{N}_4, \mathsf{M}_1, \mathsf{M}_2, \mathsf{B}\}$ which has 15! permutations of its 15 distinct letters. But the 7 As can be permuted in 7! ways, the 4 Ns can be permuted in 4! ways, and the 2 Ms can be permuted in 2! ways, all with no change made to the word if the subscripts are overlooked. We conclude there are 15!/(1!7!4!2!1!) permutations of the multiset.

2.4.3 Combinations with Repetition

Our interest now is counting the number of unordered selections taken from a multiset $\{\infty \cdot a_1, \infty \cdot a_2, \ldots, \infty \cdot a_k\}$. If the number of objects that we intend to select is less than the number of elements of each type in the multiset, we may reasonably assume that there are unlimited copies of each type of element.

Example 2.26 In how many ways can we purchase a dozen donuts at a bakery that sells glazed, sugar, and jelly doughnuts?

Solution. Our doughnut order is a 12-combination from the multiset $\{\infty \cdot \mathrm{G}, \infty \cdot \mathrm{S}, \infty \cdot \mathrm{J}\}$. Any order of a dozen donuts chosen from three kinds can be neatly represented by a permutation of 12 *stars* and $3 - 1$ *bars*, where the number of stars to the left of the bars is the number of glazed doughnut, the number of stars between the bars is the number of sugar doughnuts, and the number of stars to the right of the bars is the number of jelly doughnuts. Here are two examples of how an order is coded by a permutation of 12 stars and 2 bars:

$$\star\star\star \mid \star\star\star\star\star\star\star \mid \star\star \leftrightarrow \text{order of 3 glazed, 7 sugar, and 2 jelly}$$
$$\star\star\star\star\star\star\star\star \mid \star\star\star\star \mid \leftrightarrow \text{order of 8 glazed, 4 sugar, and 0 jelly}$$

In general, each permutation of 12 stars and 2 bars corresponds to a unique order, and conversely, any order of a dozen doughnuts corresponds to a unique permutation of 12 stars and 2 bars. That is, each order of a dozen doughnuts uniquely corresponds to a permutation of the multiset $\{12 \cdot \star, 2 \cdot \mid\}$. By Theorem 2.24, there are $\binom{12+3-1}{12} = \binom{14}{12} = \binom{14}{2} = 91$ permutations, so this is the number of ways to order a dozen doughnuts of the three types. ■

The following theorem is proved by generalizing the reasoning in Example 2.26.

Theorem 2.27 Let $\left(\!\binom{k}{n}\!\right)$ denote the number of combinations of n elements chosen from a multiset $\{\infty \cdot a_1, \infty \cdot a_2, \ldots, \infty \cdot a_k\}$ with k types of elements each of unlimited number. Then

$$\left(\!\!\binom{k}{n}\!\!\right) = \binom{n+k-1}{n} \tag{2.20}$$

Proof. Each combination corresponds to a permutation of n stars and $k-1$ bars, where the bars serve as dividers separating the k types of elements and the stars designate the numbers of elements of each type to include in the combination; that is, the number of n-combinations of a multiset of k types is the number of permutations of the multiset $\{n\cdot\bigstar, (k-1)\cdot|\}$. By Theorem 2.24, there are

$$\binom{n+k-1}{n, k-1} = \binom{n+k-1}{n}$$

permutations of the multiset. ■

The term $\left(\!\binom{k}{n}\!\right)$ is known as the *multichoose coefficient*. It expresses the number of ways to choose a multiset with n elements chosen from a multiset with k types of elements. The term $\left(\!\binom{k}{n}\!\right)$ is read "k multichoose n." It will be seen in Chapter 3 why the symbol can correctly be called a *coefficient*.

Example 2.28 The coach has decided to buy ice cream cones at "Jerry's 35 Flavors" for the nine members of the softball team. She knows that 3 players always want chocolate and 2 others always want cookie dough, but the other 4 players like to be surprised. In how many ways can the coach place the order for the 9 ice cream cones?

Solution. The coach knows she can start by buying 3 chocolate and 2 cookie dough cones, so there are $9-3-2=4$ cones left to buy. She still has 35 flavors from which to choose, since additional chocolate and/or cookie dough cones can be ordered. From the result of Theorem 2.27, the number of ways to order the cones is

$$\left(\!\!\binom{35}{4}\!\!\right) = \binom{4+35-1}{4} = \binom{38}{4}$$

Alternatively this can be expressed as

$$\binom{9-3-2+35-1}{9-3-2}$$

■

The multichoose coefficients satisfy a number of identities analogous to identities for the binomial coefficients. Indeed, any multichoose coefficient identity can be rewritten in terms of the binomial coefficients by means of equation (2.20), so identities with binomial coefficients can be recast as identities with the multichoose coefficients. However, it is usually best to reason directly from the combinatorial meaning of the multichoose coefficients. This is illustrated in the following theorem.

Theorem 2.29 The multichoose coefficients satisfy the identity

$$\left(\!\!\binom{k}{n}\!\!\right) = \left(\!\!\binom{k-1}{n}\!\!\right) + \left(\!\!\binom{k}{n-1}\!\!\right) \tag{2.21}$$

Proof. There are $\left(\!\binom{k}{n}\!\right)$ ways to choose an n-multiset from a multiset with k types $a_1, a_2, \dots, a_k$, giving us the left side of (2.21). But these multisets are of two distinct types: those with no element of type a_k and those with at least one element of type a_k. There are $\left(\!\binom{k-1}{n}\!\right)$ ways to choose the n-multiset with no element of type a_k, and $\left(\!\binom{k}{n-1}\!\right)$ ways to choose one element of type a_k and $n-1$ additional elements of any of the k types. Summing the two quantities gives right side of (2.21). ■

There is a useful algebraic interpretation of the meaning of the multichoose coefficients. Let x_i be the number of elements selected of type a_i, $1 \le i \le k$. Then any n-combination with repetition is a solution in nonnegative integers of the equation

$$x_1 + x_2 + \cdots + x_k = n, \quad x_i \ge 0. \tag{2.22}$$

In the opposite direction, any solution of (2.22) corresponds to an n-combination with repetition from a multiset with k types, where x_i is the number of objects of type a_i.

Example 2.30 Find the number of solutions in integers of these equations.

(a) $x_1 + x_2 + x_3 + x_4 = 11,\ x_1 \ge 0, x_2 \ge 0, x_3 \ge 0, x_4 \ge 0$
(b) $t_1 + t_2 + t_3 + t_4 = 13,\ t_1 \ge 3, t_2 \ge -2, t_3 \ge 0, t_4 \ge 1$

Solution.

(a) This equation corresponds to buying 11 doughnuts, available in four types. Thus, the equation has $\left(\!\binom{4}{11}\!\right) = \binom{11+4-1}{11} = \binom{14}{11} = \binom{14}{3} = 364$ solutions in nonnegative integers.
(b) Make the change of variables $x_1 = t_1 - 3, x_2 = t_2 + 2, x_3 = t_3, x_4 = t_4 - 1$, so that

$$\begin{aligned} x_1 + x_2 + x_3 + x_4 &= (t_1 - 3) + (t_2 + 2) + t_3 + (t_4 - 1) \\ &= t_1 + t_2 + t_3 + t_4 - 2 = 13 - 2 = 11, \quad x_1 \ge 0, \\ &\quad x_2 \ge 0, x_3 \ge 0, x_4 \ge 0. \end{aligned}$$

We see that the same equation as in part (a) has emerged with the change of variable, so equation (b) also has 364 solutions. ■

Many combinatorial problems can be rephrased as equations of the type found in Example 2.30, and can therefore be solved in terms of the multichoose coefficients. Additional interpretations and applications of the multichoose coefficients will be examined in Section 2.5 covering distributions.

PROBLEMS

2.4.1. How many words (i.e., words or nonword letter combinations) can be formed using all of the letters of the word combinatorics if

(a) there are no restrictions?

(b) the three-letter sequence bat must appear in the word?

(c) the two-letter sequence cc does not appear?

2.4.2. The state fish of Hawaii is the humuhumunukunukuapuaa (reef trigger fish).

(a) What multiset is formed by the letters of the Hawaiian state fish?

(b) How many words can be formed with the letters in the Hawaiian name for the state fish?

2.4.3. **(a)** How many paths join (0,0,0) to (3,3,3) where each step along the path has the form (1,0,0), (0,1,0) or (0,0,1)?

(b) How many paths that join (0,0,0) to (3,3,3) include the opposite corners (1,1,1) and (2,2,2) of the inner unit cube?

2.4.4. How many paths, making unit steps in the positive x, y, or z direction, connect the origin to the point (3,5,4)?

2.4.5. **(a)** Show, by numerical evaluations, that

$$\binom{12}{3,5,4} = \binom{11}{2,5,4} + \binom{11}{3,4,4} + \binom{11}{3,5,3}$$

(b) Explain why the result in part (a) holds by combinatorial reasoning.

(c) Prove the identity

$$\binom{n}{a,b,c,\ldots,z} = \binom{n-1}{a-1,b,c,\ldots,z} + \binom{n-1}{a,b-1,c,\ldots,z} + \cdots + \binom{n-1}{a,b,\ldots,z-1}$$

2.4.6. An ice cream shop has 15 flavors. In how many ways can you order

(a) a triple scoop cone, where it matters which order the scoops are taken?

(b) seven single-scoop cones, where there is only enough vanilla left in the shop for five scoops?

2.4.7. A well-stocked bakery sells four kinds of doughnuts—maple, glazed, jelly, and chocolate. In how many ways can you buy a dozen of the items

(a) with no restrictions?

(b) provided there are at least two glazed and three jelly donuts?

(c) provided there are no more than two maple bars?

2.4.8. **(a)** Find the number of integer solutions of the equation $x_1 + x_2 + x_3 = 10, x_1 \geq 0, x_2 \geq 0, x_3 \geq 0$

(b) Create a selection problem corresponding to the equation given in part (a).

2.4.9. Find the number of integer solutions of the equation $x_1 + x_2 + x_3 + x_4 = 15, x_1 \geq 2, x_2 \geq 0, x_3 \geq 0, x_4 \geq 0$

2.4.10. **(a)** Find the number of integer solutions of the equation $x_1 + x_2 + x_3 + x_4 + x_5 = 17, x_1 \geq 3, x_2 \geq 1, x_3 \geq 0, x_4 \geq 0, x_5 \geq 0$

(b) Create a selection problem corresponding to the equation given in part (a).

2.4.11. Find the number of integer solutions of the equation $e_1 + e_2 + e_3 + e_4 = n, 2 \leq e_1 \leq 5, 0 \leq e_2, 0 \leq e_3, 0 \leq e_4$.

2.4.12. You have 20 books and a bookcase with four shelves. In how many ways can the books be placed on shelves if the books are

(a) identical?

(b) distinct and the order of the books on a shelf is of no importance?

(c) distinct and the order of the books on a shelf is of importance?

2.4.13. Apple and berry pies each call for a cup of sugar, but a rhubarb pie requires 3 cups of sugar. In how many ways can pies be prepared if all 10 cups of sugar that are available must be used?

2.4.14. **(a)** How many words (or other letter combinations) can be formed using all of the letters of MISSISSIPPI?

(b) How many of the words in part (a) do not contain the pair PP?

(c) How many of the words in part (a) do not contain the pair SS?

2.4.15. How many permutations of the multiset $S = \{3 \cdot \mathsf{A},\ 5 \cdot \mathsf{B},\ 2 \cdot \mathsf{C},\ 2 \cdot \mathsf{D}\}$ have no adjacent As?

2.4.16. Recall that the combinatorial Fibonacci number f_n counts the number of ways to tile a board of length n with squares and dominoes.

(a) How many tilings of a board of length n use exactly d dominoes?

(b) Prove that $f_n = \sum_{d \geq 0} \binom{n-d}{d}$.

(c) Draw three parallel diagonal lines through the binomial coefficients in Pascal's triangle that are summed by the identity of part (b) in the cases $n = 4$, 5, and 6. Why is it now clear that $f_4 + f_5 = f_6$?

2.4.17. Let u_n denote the number of binary sequences of As and Bs of length n with no adjacent As. For example, there are $u_2 = 3$ sequences of length 3: AB, BA, and BB.

(a) How many of these sequences have precisely k As? Use the observation that any A must be followed with a B unless it is at the end of the sequence.

(b) Solve part (a) by counting the number of integer solutions of an equation.

(c) Derive a recursion formula for the u_n sequence.

(d) Identify the well-known sequence given by the recursion.

(e) What identity follows from parts (a) and (d)?

2.4.18. The *tribonacci numbers* are defined by the recurrence relation $T_{n+3} = T_{n+2} + T_{n+1} + T_n$ with the initial values $T_0 = 1,\ T_1 = 1, T_2 = 2$. It can be verified that T_n gives the numbers of ways to tile a $1 \times n$ board with unit squares, dominoes, and 1×3 trominoes. For example, simple drawings show that there are $T_3 = 4$ tilings of a 1×3 board. Prove that

$$T_n = \sum_{d,t \geq 0} \binom{n - d - 2t}{n - 2d - 3t, d, t}$$

2.4.19. The Delannoy numbers (see Problem 2.3.12) of the form $D(n, n)$ are also known as the "king's walk" numbers, since they give the number of paths that the king in chess can move from one corner to the opposite corner of a square chess board. Prove that

$$D(n,n) = \sum_{k \geq 0} \binom{n+k}{k} \binom{n}{k}$$

2.4.20. Show that the number of permutations of m As and at most n Bs is $\binom{m+n+1}{m+1}$, where $m \geq 0$ and $n \geq 0$. [*Hint:* How can permutations of $m + 1$ As and n Bs be "pruned"?]

2.4.21. Show that the number of permutations of at most m As and at most n Bs is $\binom{m+n+2}{m+1} - 2$, where $m \geq 0$, $n \geq 0$ and $m + n > 0$.

2.4.22. Prove that $\left(\!\binom{k}{n}\!\right) = \left(\!\binom{n+1}{k-1}\!\right)$. [*Hint*: Switch bars and stars.]

2.4.23. Prove that $n\left(\binom{k}{n}\right) = k\left(\binom{k+1}{n-1}\right)$ [*Hint*: Replace the Δ in any permutation of the multiset $S = \{(n-1)\cdot *, (k-1)\cdot |, 1\cdot \Delta\}$) with either a star or a bar.]

2.4.24. **(a)** Use combinatorial reasoning to prove that

$$\left(\binom{k+1}{n}\right) = \sum_{r=0}^{n} \left(\binom{k}{r}\right)$$

(b) Identify the identity proved in part (a) when it is written in terms of binomial coefficients.

2.4.25. **(a)** Use combinatorial reasoning to prove that

$$\left(\binom{k+1}{n+1}\right) = \sum_{r=0}^{k} \left(\binom{r+1}{n}\right)$$

[*Hint*: The left side of the identity counts the number of ways to order $n + 1$ ice cream cones chosen from $k + 1$ flavors.]

(b) Identify the identity proved in part (a) when it is written in terms of binomial coefficients.

2.4.26. Use combinatorial reasoning to prove the identity

$$\left(\binom{k}{n}\right) = \sum_{r=0}^{k} \binom{k}{r} \left(\binom{r}{n-r}\right)$$

2.5 DISTRIBUTIONS TO DISTINCT RECIPIENTS

A *distribution* is an assignment of objects to recipients. It continues to be important to know whether the objects are distinct or identical. Likewise, it is important to know whether the recipients are distinct or indistinguishable. For example, if m people are formed into n discussion groups, this is a distribution of m distinct objects—people—into n indistinguishable *containers*—the discussion groups. Or, as a second example, suppose that m dollars are donated to n charities; this is a distribution of m identical objects—dollars—to n distinct recipients—charities.

A particularly nice way to model a distribution is to imagine a sack of balls that are to be placed into boxes. The balls are distinct if they are of different colors or have different labels. For example, m different balls can be numbered with labels 1, 2, ..., m from the set $[m]$. Similarly, if the boxes are distinct, they can be labeled or numbered, say, with the integers 1,2, ..., n.

In this section, we will consider only distributions made to distinct recipients. In Chapter 6 we will consider distributions to indistinguishable recipients, with combinatorial situations that introduce the partition numbers, the Stirling (after James Stirling) numbers, and the Bell (after Eric T. Bell) numbers.

2.5.1 Distributions of Distinct Objects to Distinct Recipients

We will first discuss the distribution of distinct objects to distinct recipients, starting with the unrestricted distributions, then distributions in which no recipient receives more than one object, and finally distributions that assign at least one object to each recipient.

Example 2.31 In how many ways can 10 different toys—a doll, a baseball, a puzzle, and so on—be given to four children?

Solution. The doll can be given to one of four children, the baseball to one of four children, and so on. Since no restrictions have been imposed on the distribution, each toy has four possible assignments. The number of unrestricted distributions of the 10 distinct toys to the four distinct children is therefore

$$4 \cdot 4 \cdot \cdots \cdot 4 = 4^{10}$$ ■

It is easy to extend the reasoning in Example 2.31 to the general case.

Theorem 2.32 There are n^m unrestricted distributions of a set of m distinct objects to a set of n distinct recipients.

Proof. There are n choices for the recipient of each of the m objects, giving $n \cdot n \cdot \cdots \cdot n = n^m$ distributions. ■

Many of the distributions counted in Example 2.31 seem a little heartless, certainly including the four distributions in which all 10 toys are given to just one of the four children. Surely it would be nicer distribute the toys a little more evenly. Indeed, if there are just m toys for n children, where $m \leq n$, then each child should receive at most one toy. A distribution is called *injective,* or *one-to-one* if no two objects are assigned to the same recipient. The following theorem counts the number of injective distributions.

Theorem 2.33 There are $P(n, m) = n(n-1) \cdots (n-m+1)$ injective distributions of m distinct objects to n distinct recipients. In other words, $P(n, m)$ is the number of ways to distribute m distinct objects onto n distinct recipients so that no recipient is assigned more than one object.

Proof. We can assume that we are distributing m distinct toys numbered 1 through m to n children, and no child is given more than one toy. First, there are n ways to choose which child gets toy 1. This leaves $n-1$ ways to choose the remaining child that is given toy 2, and so on. Effectively, we are forming an m-permutation of the

n children, and then giving toy 1 to the first child, toy 2 to the second child, and so on. ■

A happier situation occurs when there are more toys than children. If m toys are distributed to n children and $m \geq n$, we can seek a distribution that leaves no child without a toy. This is a *surjective,* or *onto,* distribution. The number of surjective distributions is given by the distribution numbers that are defined this way.

Definition 2.34[1] The *distribution number* $T(m,n)$ is the number of surjective distributions of m distinct objects onto n distinct recipients; that is, $T(m,n)$ is the number of ways to distribute m distinct objects to n distinct recipients so that each recipient is assigned at least one object.

Some values of $T(m,n)$ are easy to determine. For example, $T(m,1) = 1$, since there is just one way to give all m objects to a single recipient. Also, $T(n,n) = n!$, and $T(m,n) = 0$ for $m < n$. If $m = n + 1$, we can first form a pair of objects in $\binom{n+1}{2}$ ways, and then distribute the one pair and the $n - 1$ single objects in $n!$ ways. Therefore

$$T(n+1,n) = \binom{n+1}{2} n!$$

The identity derived in the next theorem gives a formula to compute additional distribution numbers.

Theorem 2.35 For $m \geq n > 1$, the distribution numbers $T(m,n)$ satisfy

$$T(m,n) = n\,[T(m-1,n-1) + T(m-1,n)] \tag{2.23}$$

Proof. Assume we are placing m balls all of different colors into n distinct boxes. If no box is left empty, this can be done in $T(m,n)$ ways, which is the left side of (2.23). Now suppose that one ball is pink, so it can be placed in a box in n ways. We can then consider two cases; namely, the pink ball is either alone in a box or accompanied by other balls in its box. If no other balls are placed in the box with the pink ball, the remaining $m - 1$ balls are distributed onto the remaining $n - 1$ boxes in $T(m-1,n-1)$ ways. In the other case, the nonpink balls are distributed by one of the $T(m-1,n)$ distributions onto all n boxes since we must assign at least another ball to the box with the pink ball. Together, the two cases show us there are $n\,[T(m-1,n-1) + T(m-1,n)]$ onto distributions of the m balls to the n distinct boxes. ■

[1] This terminology is not standardized. In Chapter 6 we will see that $T(m,n) = n!S(m,n)$, where $S(m,n)$ is the Stirling number of the second kind that gives the number of ways to distribute m distinct objects to n *identical* recipients.

Equation (2.23) continues to hold for all nonnegative integers if we define $T(0,0) = 1, T(m,0) = 0$ all $m > 0$, and $T(0,n) = 0$ all $n > 0$. We then obtain the values shown in Table 2.1.

TABLE 2.1 Distribution Numbers $T(m,n)$ for $0 \le m, n \le 6$

	n						
m	0	1	2	3	4	5	6
0	1	0	0	0	0	0	0
1	0	1	0	0	0	0	0
2	0	1	2	0	0	0	0
3	0	1	6	6	0	0	0
4	0	1	14	36	24	0	0
5	0	1	30	150	240	120	0
6	0	1	62	540	1560	1800	720

For example, the entry $T(6,3) = 3(30 + 150) = 540$ in the bottom row of Table 2.1 follows from (2.23) when $m = 6$ and $n = 3$. Since two adjacent entries in a row determine an entry in the next row, the formula given in (2.23) is sometimes called a *triangle identity*. We have seen that the binomial and multichoose coefficients each satisfy a triangle identity, and in later chapters we will encounter even more arrays that satisfy a triangle identity.

Example 2.36 Mike has an inkpad and 6 rubber stamps that print the letters A, B, C, D, E, F. How many five-letter words or nonwords can Mike create where each of them uses exactly three of his stamps? For example, he could write DBEBD.

Solution. First choose the three stamps that are each to be used at least once in $\binom{6}{3} = 20$ ways. Use these letters to label three boxes. Next, distribute the five objects 1, 2, 3, 4, and 5 onto the three boxes in $T(5,3) = 150$ ways, so that each box is assigned at least one of the numbers. The numbers give the order in which to use the stamp to form the word or nonword. For example, the surjective distribution $\mathsf{B} = \{2,4\}, \mathsf{D} = \{1,5\}, \mathsf{E} = \{3\}$ corresponds to the word DBEBD since the D stamp is used first, the B stamp is used second, and so on, with the D as fifth stamp used to finish the word or nonword. Altogether we see that there are $\binom{6}{3}T(5,3) = 20 \cdot 150 = 3000$ words/nonwords of length 5 that each use three of the six stamps. ■

A surjective distribution of toys to children guarantees no child is without a toy, but it may be more equitable to decide in advance the number of toys, say m_k, that will be given to child c_k. All of the toys will be distributed if $m_1 + m_2 + \cdots + m_n = m$, and each child gets at least one toy if $m_k \ge 1$ for $k = 1, 2, \ldots, n$. To count the number of distributions of the n toys, note that any permutation of the multiset $\{m_1 \cdot c_1, m_2 \cdot c_2, \ldots, m_n \cdot c_n\}$ corresponds to a desired distribution if the toy t_k is given to the kth child listed in the permutation. For example, the list c_3, c_1, c_3, c_2, c_2 describes the distribution of five toys to three children for which toy t_1 is given to child c_3, toy t_2 is given child c_1, toy t_3 is given child c_3, toy t_4 is given child c_2, and toy t_5 is given child c_2. By Theorem 2.24, the number of permutations of a multiset is given by a multinomial coefficient, giving us the following theorem.

Theorem 2.37 The number of distributions $f: X \to Y$ for which m_k objects of X are distributed to the kth recipient in set Y is given by the multinomial coefficient

$$\binom{m}{m_1, m_2, \ldots, m_n} = \frac{m!}{m_1!m_2! \cdots m_n!}, \quad m_1 + m_2 + \cdots + m_n = m$$

Example 2.38 In how many ways can five different prizes be awarded to three contest winners, where no winner goes without a prize? Give two answers, one using a distribution number and the other a sum of multinomial coefficients.

Solution. The quick answer is $T(5,3) = 150$. A second approach is to consider cases. If five prizes are given to three people, either one winner gets three prizes and the others one apiece, or else one person gets one prize and the other two winners each get two prizes. According to Theorem 2.37, the prizes can be awarded in

$$3\binom{5}{3,1,1} + 3\binom{5}{2,2,1} = \frac{3 \cdot 5!}{3!} + \frac{3 \cdot 5!}{2!2!} = 60 + 90 = 150$$

ways. ■

Example 2.38 is easily generalized, and we see that

$$T(m,n) = \sum_{\substack{m_1+m_2+\cdots+m_n=m \\ m_1\geq 1, m_2\geq 1, \cdots, m_n\geq 1}} \binom{m}{m_1, m_2, \ldots, m_n} \tag{2.24}$$

2.5.2 Distributions of Identical Objects to Distinct Recipients

Often the objects to be distributed are identical. For example, if we are donating dollar bills to charities, or giving cherry gumdrops to children, it is irrelevant which particular dollar bill or which cherry gumdrop is given to a recipient. However, it continues to be important to distinguish recipients. For example, it makes a difference whether charity A receives \$100 in one distribution and \$5 in another distribution.

As before, imagine that the recipients are n distinct boxes, which are distinguishable by their label. However, since m identical objects are being placed into the boxes, it only matters how many objects are placed in a particular box. A distribution for which no box is given more that one object is injective, and a distribution for which no box is left empty is surjective.

Example 2.39 In how many ways can 6 identical cherry gumdrops be given to 10 children so no child receives more than one candy?

Solution. There are $\binom{10}{6}$ injective distributions, since this is the number of ways to select the subset of children who get one cherry gumdrop apiece. ■

This example has an obvious generalization.

Theorem 2.40 There are $\binom{n}{m}$ injective distributions of m identical objects to n distinguishable recipients.

Suppose next that we have complete freedom to place any number of identical balls into whichever boxes we like; that is, we consider unrestricted distributions. For example, suppose we have ten \$100 bills that we will donate to six charities. This is equivalent to distributing 10 identical balls into 6 distinguishable boxes. One possible distribution is shown in Figure 2.2(a).

FIGURE 2.2 An unrestricted distribution of 10 identical objects distributed to six distinct recipients corresponds to a permutation of 10 stars and 6 – 1 = 5 bars.

Figure 2.2(b) shows how the distribution can be uniquely represented with a "star and bar" permutation of 10 stars and $6 - 1 = 5$ bars that separate the six charities. We see that the two balls in box 1 corresponds to the selection of two objects of type 1, the single ball in box 2 corresponds to the selection of one object of type 2, and so on. By Theorem 2.27, the number of unrestricted distributions of the 10 balls to six distinct boxes is given by the multichoose coefficient

$$\left(\!\!\binom{6}{10}\!\!\right) = \binom{10+6-1}{10} = \binom{15}{10} = 3003$$

The reasoning above easily generalizes to show that the number of unrestricted distributions is equivalent to an m-combination of a multiset with n types: the distribution is represented by a permutation of m stars and $n - 1$ bars, and from

Theorem 2.27 there are $\binom{m+n-1}{m} = \left(\!\binom{n}{m}\!\right)$ such permutations. This proves the following theorem.

Theorem 2.41 There are $\binom{m+n-1}{m} = \left(\!\binom{n}{m}\!\right)$ unrestricted distributions of m identical objects to n distinct recipients.

We now have two interpretations of the multichoose coefficient: $\left(\!\binom{n}{m}\!\right)$ is the number of unrestricted distributions of m identical objects to n distinct recipients, and it is the number of m-combinations from a multiset with n types.

So far, we have counted the number of injective unrestricted distributions. Next we will count the number of surjective distributions, those for which each of the n distinct boxes is assigned at least one of the m identical balls. This turns out to be easy; first place one ball into each of the n boxes, and then make an unrestricted distribution of the remaining $m - n$ balls into the n boxes. Using Theorem 2.41, we can do this in $\left(\!\binom{n}{m-n}\!\right)$ ways. Since

$$\left(\!\binom{n}{m-n}\!\right) = \binom{m-n+n-1}{m-n} = \binom{m-1}{m-n} = \binom{m-1}{n-1}$$

we have the following theorem.

Theorem 2.42 There are $\binom{m-1}{n-1}$ surjective distributions of m identical objects to n distinct recipients.

Example 2.43 Suppose that a *train* of length m is the $1 \times m$ rectangle formed by an abutting sequence of n *cars*, where a car is any rectangle of the form $1 \times k$, where $k \geq 1$ For example, the diagram below shows a train of length 11 composed of 4 cars of length 2, 4, 2, and 3 from left to right:

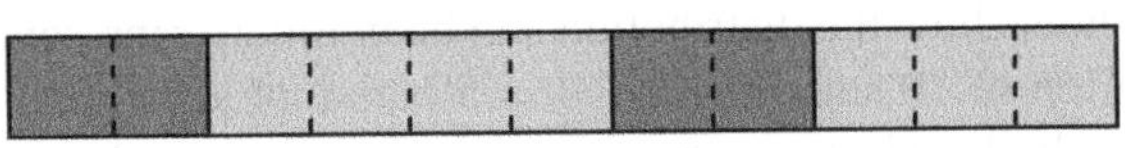

(a) determine the number of trains of length m with n cars.
(b) what is the total number of trains of length m with any number of cars?
(c) what identity is proved by combining the results of parts (a) and (b)?

Solution

(a) The train shown above can be considered to be a distribution of $m = 11$ identical unit squares into a sequence of $n = 4$ boxes reading from left to right, where at least one square is placed in each box to create a car of positive length. By Theorem 2.42, there are $\binom{11-1}{4-1}$ trains of length 11 with 4 cars. In the general case, there are $\binom{m-1}{n-1}$ trains of length m composed of n cars.

(b) To form a train of length n, there are $n - 1$ divisions between successive cells of a $1 \times n$ board at which we can either continue the car to the left or else end the car at the left and begin a new car to the right. That is, there are $m - 1$ two-way choices to make to form any train, showing there are 2^{m-1} trains of length m.

(c) There are $\binom{m-1}{n-1}$ trains of length m with n cars, where $n = 1,2, \ldots, m$. Since there are 2^{m-1} trains altogether, we see that $2^{m-1} = \sum_{n=1}^{m} \binom{m-1}{n-1}$. This is the familiar result that the coefficients in row $m - 1$ of Pascal's triangle sum to 2^{m-1}.

■

The train problem considered in Example 2.43 shows that there are 2^{m-1} ways to write a positive integer m as an ordered sum of positive integer summands. These sums are known as the *compositions* of m. For example, there are $2^{4-1} = 8$ compositions of $m = 4$, namely 4, $1 + 3$, $3 + 1$, $2 + 2$, $1 + 1 + 2$, $1 + 2 + 1$, $2 + 1 + 1$, and $1 + 1 + 1 + 1$.

2.5.3 Mixed Distribution Problems

Until now, we have considered distributions of objects that are either all different or all alike. Sometimes, however, a mixed collection of objects is to be distributed, with some objects distinct and others repeated.

Example 2.44 In how many ways can four distinct toys, six identical cherry gumdrops, and 14 identical oranges be distributed to three children so that each child receives eight items, including at least one toy and one gumdrop per child?

Solution. First distribute the four toys in $T(4,3) = 36$ ways that leave no child without a toy, and note that no child receives more than two toys. Next, distribute the six identical gumdrops in $\binom{6-1}{3-1} = \binom{5}{2} = 10$ ways so that each child receives at least one gumdrop, and note that no child gets more than four gumdrops. Since no child has more that six items at this point, simply distribute the 14 oranges so that each child has the required eight items. Altogether, there are $36 \cdot 10 = 360$ distributions.

■

Example 2.45 In how many ways can three blue flags, four green flags, a yellow flag, and a red flag be flown on three distinct flagpoles? The order of the flags from top to bottom is important, so a yellow flag over a red flag is a different distribution from one that places a red flag over a yellow flag on the same pole. Flags of the same color are identical. One possible distribution is shown here:

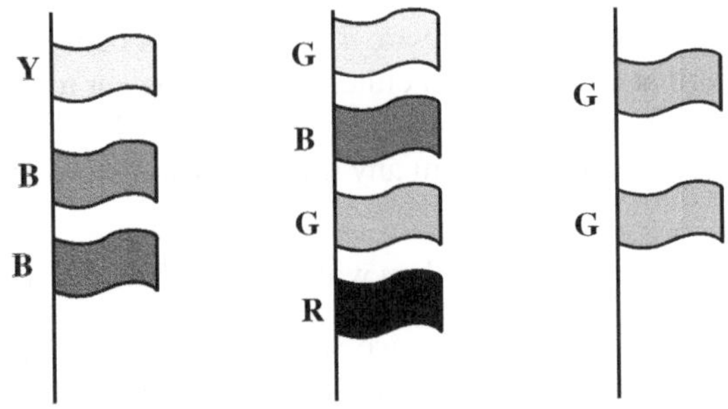

Solution. By Theorem 2.41 there are

$$\binom{9+3-1}{9} = \frac{11!}{9!2!}$$

ways to fly nine identical flags, say, all white, on the three poles. Next, arrange the colored flags in any of their $9!/(3!4!1!1!)$ permutations. Now replace the white flags in the order of the permutation, starting with the flags from top to bottom on the first pole and moving on to the subsequent poles in order. This shows there are

$$\frac{11!}{9!2!}\frac{9!}{3!4!1!1!} = \binom{11}{2,3,4,1,1}$$

ways to fly the flags. This answer reveals another approach to solve the distribution problem: any distribution of the flags is equivalent to one of the $\binom{11}{2,3,4,1,1}$ permutations of the multiset $\{2 \cdot |,\ 3 \cdot \mathrm{B},\ 4 \cdot \mathrm{G},\ 1 \cdot \mathrm{Y},\ 1 \cdot \mathrm{R}\}$, where the two bars are separators to indicate when to move to the next pole. For example, the permutation Y B B | G B G R | G G gives the distribution of the nine flags on three distinct poles shown in the preceding diagram above. ■

2.5.4 Equations for Distributions

The following example illustrates how counting distributions can be viewed as counting the number of solutions in integers of a related equation. This type of question arises frequently in many mathematical problems, so it is of great interest to have a way to count the number of solutions that can exist.

Example 2.46 Apples, bananas, and pears each weigh 4 oz (ounces) but melons weigh 12 oz each. A bag of mixed fruit weighs 5 lb, and contains at least two melons, at least three apples, and at least one pear. How many different contents of the bag can exist, where each piece of fruit is identical to any other of the same type?

Solution. We can suppose that the unit of weight is 4 oz, so that a melon weighs 3 units, apples, bananas, and pears weigh 1 unit each, and the entire bag of fruit weighs 20 units. If we let y_1, y_2, y_3, and y_4 denote the respective number of apples, bananas, pears, and melons in the bag, we have the equation

$$y_1 + y_2 + y_3 + 3y_4 = 20,\ y_1 \geq 3,\ y_2 \geq 0,\ y_3 \geq 1,\ y_4 \geq 2$$

It is useful to make the change of variables $x_1 = y_1 - 3,\ x_2 = y_2,\ x_3 = y_3 - 1,\ x_4 = y_4 - 2$ so that, in the new variables, we have the equation

$$\begin{aligned} x_1 + x_2 + x_3 + 3x_4 &= (y_1 - 3) + y_2 + (y_3 - 1) + 3(y_4 - 2) = 20 - 3 - 1 - 6 \\ &= 10,\ x_1 \geq 0,\ x_2 \geq 0,\ x_3 \geq 0,\ x_4 \geq 0 \end{aligned}$$

This gives us four equations, $x_1 + x_2 + x_3 = 10 - 3x_4,\ x_1 \geq 0,\ x_2 \geq 0,\ x_3 \geq 0$, since it is evident that $x_4 \in \{0,1,2,3\}$. If $x_4 = 0$, we get $x_1 + x_2 + x_3 = 10,\ x_1 \geq 0,\ x_2 \geq 0,\ x_3 \geq 0$, whose solutions correspond to the ways that 10 identical objects can be distributed into three distinct boxes. Therefore, we see that there are

$$\binom{10+3-1}{10} = \binom{12}{10} = \binom{12}{2}$$

ways to fill the bag with exactly two melons. If $x_4 = 1$, we get $x_1 + x_2 + x_3 = 7,\ x_1 \geq 0,\ x_2 \geq 0,\ x_3 \geq 0$, which is the equation for a distribution of seven identical objects into three distinct boxes, so there are $\binom{9}{2}$ ways to fill the bag with exactly three melons. Similarly, there are, respectively, $\binom{6}{2}$ and $\binom{3}{2}$ ways to fill the bag with exactly four or five melons. As a check of our reasoning $\{4 \cdot \text{A}, 1 \cdot \text{P}, 5 \cdot \text{M}\}$, $\{3 \cdot \text{A}, 2 \cdot \text{P}, 5 \cdot \text{M}\}$, and $\{3 \cdot \text{A}, 1 \cdot \text{P}, 1 \cdot \text{B}, 5 \cdot \text{M}\}$ are the 3 ways in which the 5-lb bag contains five melons. Altogether, the bag can be filled in $\binom{12}{2} + \binom{9}{2} + \binom{6}{2} + \binom{3}{2}$ ways. ■

2.5.5 Counting Functions

A distribution of m distinct objects to n distinct recipients is a function $f : X \to Y$, where the domain X is a set of m elements and the codomain Y is a set with n elements. The function is either injective or surjective precisely when the same is true of the distribution. The theorems obtained earlier that counted the number of unrestricted, injective, and surjective distributions of distinct objects can now be restated to count functions.

Theorem 2.47 Consider the set of all functions $f : X \to Y$ when $m = |X|$ and $n = |Y|$.

(a) There are n^m functions in all.
(b) There are $P(n, m)$ injective (one-to-one) functions.

(c) There are $T(m,n)$ surjective (onto) functions.

(d) There are $n!$ bijective (one-to-one and onto) functions when $m = n$, and none otherwise.

Next suppose that X and Y are ordered sets, so that we may as well consider the functions $f : [m] \to [n]$. It then becomes of interest to consider those functions that are strictly increasing, so that $f(j) < f(k)$ whenever $j < k$, and those functions that are nondecreasing, so that $f(j) \leq f(k)$ whenever $j < k$. The next two theorems count the number of strictly increasing functions and the number of nondecreasing functions.

Theorem 2.48 There are $\binom{n}{m}$ strictly increasing functions $f : [m] \to [n]$.

Proof. According to Theorem 2.40, there are $\binom{n}{m}$ injective distributions of m identical balls to n distinct boxes. Therefore, it seems that there must be a correspondence between injective distributions and strictly increasing functions. Indeed, this is the case, as can be seen from the example in this diagram.

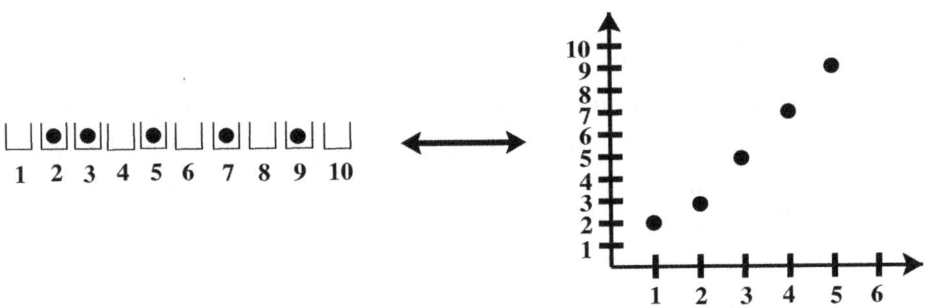

Injective distribution

Strictly increasing function

To describe the correspondence precisely, let's start with an injective distribution in which m identical balls have be placed in boxes numbered 1 through n. This distribution corresponds to the function $f : [m] \to [n]$ for which $f(x) = y$ if x balls have been distributed to boxes 1 through y. The function is strictly increasing since any distribution of more than x balls will require more boxes because any box can contain at most one ball. In the opposite direction, let's start with a strictly increasing function $f : [m] \to [n]$. Now place one ball in the boxes labeled by the range of f. This is an injective distribution since f is a strictly increasing function and therefore injective. ■

The next theorem gives us yet another meaning of the multichoose coefficients.

Theorem 2.49 There are

$$\binom{m+n-1}{m} = \left(\!\binom{n}{m}\!\right)$$

nondecreasing functions $f : [m] \to [n]$.

Proof. Comparing this result with Theorem 2.41 suggests that there is a correspondence between nondecreasing functions and unrestricted distributions. The diagram below illustrates a distribution of $m = 10$ balls to $n = 6$ boxes and its correspondence to a nondecreasing function $f : [10] \to [6]$:

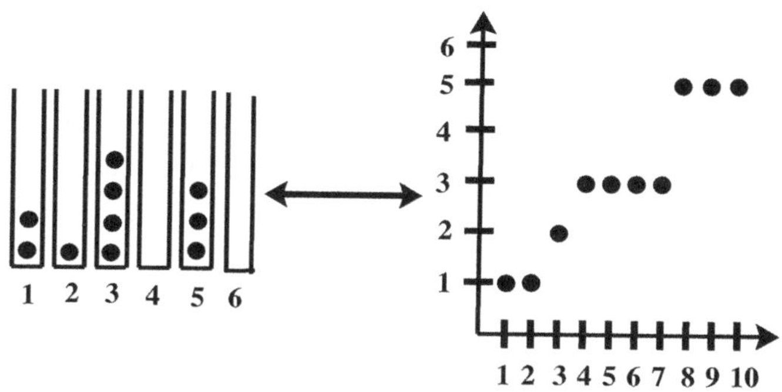

Distribution Nondecreasing function

To describe the correspondence, start with any nondecreasing function $f : [m] \to [n]$. If $f^{-1}(y)$ denotes the set of preimage values of $y \in [m]$, place $|f^{-1}(y)|$ balls in box y. For example, in the figure above, $|f^{-1}(3)| = |\{4, 5, 6, 7\}| = 4$, so four balls are placed in box 3. Similarly, $f^{-1}(4) = \emptyset$, so no balls are placed in box 4. This rule describes a distribution of m identical balls into n distinct boxes. In the opposite direction, start with any distribution. Imagine the balls are distributed into the boxes from left to right. If ball x is placed into bin y, set $f(x) = y$. For example, balls 1 and 2 are placed into bin 1, so $f(1) = f(2) = 1$, ball 3 is placed into bin 2 so $f(3) = 2, \ldots,$ and finally ball 10 is placed into bin 5 so $f(10) = 5$. ■

PROBLEMS

2.5.1. How many ways can 15 identical candies be given to three children so that each receives an odd number of candies?

2.5.2. Give the number of solutions of the equation $x_1 + x_2 + \cdots + x_n = m$ when the integers $x_1, x_2, \ldots, x_n$ satisfy

(a) $x_j \in \{0, 1\}$ **(b)** $x_j \geq 1$ **(c)** $x_j \geq 0$

2.5.3. **(a)** In how many ways can six identical balls be distributed to 10 people?

(b) Propose a selection problem at the ice cream store that is equivalent to the problem of part (a).

(c) What equation (in integers) is equivalent to the problems in parts (a) and (b)?

2.5.4. **(a)** What is the number of solutions (in integers) of the following equation under these conditions?

$$x_1 + x_2 + \cdots + x_{10} = 18$$
$$x_1 \geq 3,\ x_2 \geq 5,\ x_3 \geq 4$$
$$x_i \geq 0, i = 4, 5, \ldots, 10$$

(b) Create a distribution problem that is answered by the equation of part (a).

(c) Create a selection problem that is answered by the equation of part (a).

2.5.5. Use Table 2.1 and the triangle identity (2.23) to calculate the distribution number $T(8, 4)$.

2.5.6. In how many ways can the 13 hearts and the 13 spades in a deck of cards be distributed to four players if each player receives at least

(a) one card?

(b) one heart and one spade?

(c) one heart?

2.5.7. For each expression listed below, discuss whether it answers the question "In how many ways can four different books be placed in a bookcase with three shelves, where no shelf remains empty?" Give the reasoning that led to the answer, and explain why it is correct or incorrect.

(a) $T(4, 3) \cdot 4!$ **(b)** $T(4, 3) \cdot 2!$ **(c)** $\left(\!\binom{3}{1}\!\right) \cdot 4!$

2.5.8. In how many ways can five runners finish a race? Count all possible outcomes, including all of the possible ways in which some or even all of the runners can be tied.

2.5.9. Use combinatorial reasoning to prove that $T(m, 2) = 2^m - 2$.

2.5.10. Use combinatorial reasoning to prove that

$$T(n+2, n) = n! \left[\binom{n+2}{3} + 3\binom{n+2}{4} \right]$$

2.5.11. Use combinatorial reasoning to prove that

$$T(n+3,n) = n!\left[\binom{n+3}{4} + 10\binom{n+3}{5} + 5\cdot 3\binom{n+3}{6}\right]$$

2.5.12. A gourmet club with eight members meets for dinner each month. In how many ways can a new host be chosen for each dinner so that every club member hosts at least one dinner in one year's time?

2.5.13. How many permutations of the letters I, N, F, I, N, I, T, I, E, S do not contain any adjacent pair of Is?

2.5.14. Use combinatorial reasoning to prove that

$$T(m, a+b) = \sum_{k=0}^{m} \binom{m}{k} T(k,a)\, T(m-k,b)$$

2.5.15. **(a)** Given any polynomial $p(x) = a_0 + a_1x + a_2x^2 + \cdots + a_nx^n$, prove that there are constants $b_0, b_1, \ldots, b_n$ so that $p(k) = b_0\binom{k}{0} + b_1\binom{k}{1} + b_2\binom{k}{2} + \cdots + b_n\binom{k}{n}$ for all positive integers k.

(b) Find the constants b_0, b_1, b_2, b_3 for which

$$k^3 = b_0\binom{k}{0} + b_1\binom{k}{1} + b_2\binom{k}{2} + b_3\binom{k}{3}$$

[*Note*: It can be shown (see Problem 2.5.18) that $b_r = \sum_{k=r}^{n} T(k,r)\, a_k$.]

2.5.16. Use combinatorial reasoning to prove that $k^m = \sum_{n=0}^{m} T(m,n)\binom{k}{n}$, where k and m are nonnegative integers, not both 0.

2.5.17. Use the result of Problem 2.5.16 to prove the polynomial identity

$$x^m = \sum_{n=0}^{m} T(m,n)\, \frac{(x)_n}{n!}$$

where m is a nonnegative integer, x is the variable of the polynomials, and $(x)_r = x(x-1)\cdots(x-r+1)$.

2.5.18. **(a)** Show that any polynomial $p(x) = a_0 + a_1x + a_2x^2 + \cdots + a_nx^n$ can also be written in the form $p(x) = b_0 + b_1(x)_1 + b_2(x)_2 + \cdots + b_n(x)_n$, where

$$b_r = \frac{1}{r!}\sum_{k=r}^{n} T(k,r)\, a_k$$

(b) Show that $p(x) = 2 + 4x - x^2 + 2x^3 + 5x^4 = 2 + 10(x)_1 + 40(x)_2 + 32(x)_3 + 5(x)_4$.

2.6 CIRCULAR PERMUTATIONS AND DERANGEMENTS

For many applications, the permutations of interest must satisfy additional conditions or restrictions. For example, we know that there are $n!$ ways to seat n people in a row, but what if the people are to be seated in a circle of n identical chairs? Such a permutation is known as a *circular permutation* to distinguish it from a *linear permutation* already considered. As a second example, in how many ways can n people currently seated in a row of chairs be reseated if each person must move to a different chair? This permutation, leaving no element fixed in position, is known as a *derangement.* In this section, we will count the number of circular permutations and the number of derangements.

2.6.1 Circular Permutations

Definition 2.50 A *circular n-permutation* is an arrangement of n distinct objects around a fixed circle, where the term *fixed* indicates the circle can be freely rotated but not flipped upside down.

For example, Figure 2.3 shows the six circular permutations of the four-element set $[4] = \{1,2,3,4\}$.

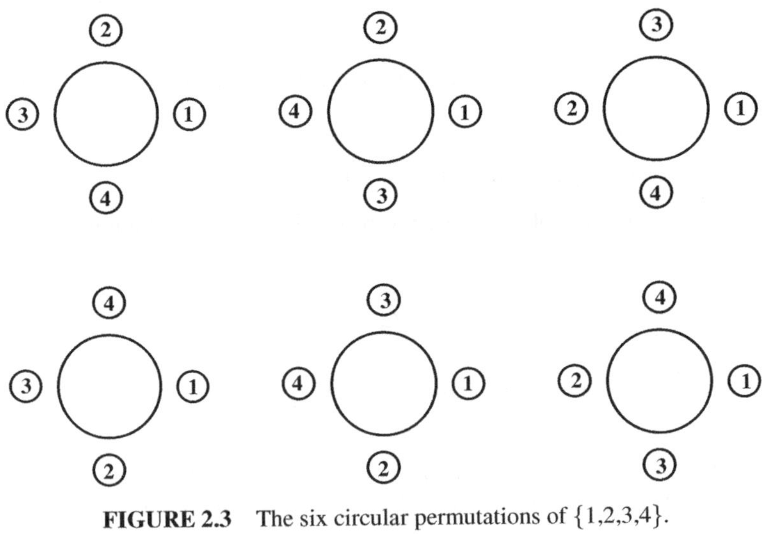

FIGURE 2.3 The six circular permutations of $\{1,2,3,4\}$.

It is important to understand that the permutation is unchanged by any rotation of the circle, which allows us to place element 1 always at the right in Figure 2.3. We also note that any circular permutation corresponds to a linear permutation and its *cyclic permutations*. For example, the circular permutation (1,4,2,3) can be obtained by the four cyclic permutations of the linear permutations (1,4,2,3), (4,2,3,1), (2,3,1,4), and (3,1,4,2). Although cyclic permutations do not change a circular permutation, a flip of the circle does give a new circular permutation if there are at least three elements in the permutation. For example, the permutations in the second row of Figure 2.3 are the flips of the circular permutations in the top row. If flips and rotations are both allowed, we obtain the *free circular permutations*. If $n \geq 3$, there are half as many free circular permutations as circular permutations, since each permutation is now considered to be the same as its flipped permutation.

To see why there are $6 = 3!$ circular permutations of $[4] = \{1,2,3,4\}$, just observe that there are 3! ways to form the linear permutation of the elements $\{2,3,4\}$ that extend counterclockwise around the circle starting with element 1. Extending this reasoning, there are $(n-1)!$ circular permutations of any n element set, since this is the number of linear permutations of the objects $\{2,3,\ldots,n\}$ extending counterclockwise from element 1. This proves part (a) of the following theorem.

Theorem 2.51

(a) The number of circular n-permutations of all of the elements of an n-set is

$$(n-1)! \tag{2.25}$$

(b) The number of circular r-permutations taken from an n-set is

$$\frac{P(n,r)}{r} = \frac{n(n-1)\cdots(n-r+1)}{r} \tag{2.26}$$

Proof. Formula (2.25) of part (a) has already been proved. To prove part (b), we can first choose an r-element subset in

$$\binom{n}{r} = \frac{n(n-1)\cdots(n-r+1)}{r!}$$

ways and next form a circular r-permutation of these r elements in $(r-1)!$ ways by part (a). Altogether, there are

$$\frac{n(n-1)\cdots(n-r+1)}{r!}(r-1)! = \frac{n(n-1)\cdots(n-r+1)}{r}$$

circular r-permutations. ■

Example 2.52 There are three circular tables in a room, one seating 10 people, another seating 8 people, and the third seating 7 people. In how many ways can

(a) 25 people be seated, so that all of the chairs at the tables are occupied?
(b) 25 out of a group of 30 people be seated at the three tables?
(c) 20 people be seated, leaving five of the chairs unoccupied?

Solution

(a) By Theorem 2.51, there are

$$\left(\frac{25 \cdot 24 \cdot \cdots \cdot 16}{10}\right)\left(\frac{15 \cdot 14 \cdot \cdots \cdot 8}{8}\right)\left(\frac{7 \cdot 6 \cdot \cdots \cdot 1}{7}\right) = \frac{25!}{10 \cdot 8 \cdot 7}$$

ways to make a seating, which is seen by first seating people at the 10-person table, then at the 8-person table, and finally at the 7-person table. Another way to see this is to form any of the 25! linear permutations and seat the people in that order counterclockwise around the tables. However, the 10 people at the largest table can all simultaneously move any of 1 through 10 seats to their right, and there will be no change to the circular permutation at that table. Similar considerations for the other two tables means 25! overstates the number of seating arrangements by a factor of $10 \cdot 8 \cdot 7$. Using the division principle, we conclude that there are $25!/(10 \cdot 8 \cdot 7)$ seatings at the circular tables, just as before.

(b) There are $\binom{30}{25}$ ways to choose the 25 people that can be seated, and $25!/(10 \cdot 8 \cdot 7)$ ways to seat the selected people. This gives

$$\binom{30}{25}\frac{25!}{10 \cdot 8 \cdot 7} = \frac{30 \cdot 29 \cdot \cdots \cdot 26}{5!} \cdot \frac{25!}{10 \cdot 8 \cdot 7} = \frac{30!}{5! \cdot 10 \cdot 8 \cdot 7}$$

seatings of a subset of 25 of the 30 people.

(c) Suppose that five extra people, $U_1, U_2, \ldots, U_5$ are temporarily appended to the set of 20, so by part (a) there are now 25 people who can be seated in $25!/(10 \cdot 8 \cdot 7)$ ways. If $U_1, U_2, \ldots, U_5$ people are asked to leave their chairs unoccupied, we see there are $25!/(5! \cdot 10 \cdot 8 \cdot 7)$ seatings that leave five chairs empty. Another way to see this is to first permute, in $25!/5!$ ways, the multiset of the 20 people and five Us. In the order of the permutation, seat the elements of the multiset around the table of 10, then around the table of 8, and then around the table of 7. As before, this overstates the count by the factor $10 \cdot 8 \cdot 7$, so once again we see there are $25!/5! \cdot 10 \cdot 8 \cdot 7$ seatings. ■

Example 2.53 How many necklaces can be made by stringing 18 spherical beads, all of different colors into a loop, if

(a) there are no restrictions?

(b) the white and black beads cannot be side by side?

(c) the white and black beads are on opposite sides of the necklace, with 8 beads between them on each side of the necklace?

Solution. Since the necklace can be worn flipped over, we are counting the free circular permutations in which the circle is allowed to both rotate and flip over.

(a) There are $\frac{17!}{2}$ different ways to string the 18 beads, since the 17! circular rotations occur in pairs that are the flip of each other.

(b) There are $\frac{16!}{2}$ ways to make a string of 17 beads that exclude the white bead. Then the white bead can be inserted in 15 ways that avoid the two positions that are adjacent to the black bead. Altogether, there are $15 \cdot \frac{16!}{2}$ necklaces with the black and white beads separated by other beads.

(c) There are $\frac{16!}{2}$ necklaces that exclude the white bead, and just one way to insert the white bead on the opposite side from the black bead of the necklace. ■

2.6.2 Derrangements

A linear permutation of $[n]$ is a bijection (i.e., a one-to-one–onto mapping) $\pi : [n] \to [n]$. There are several ways to describe a permutation π. For example

$$\pi = \begin{pmatrix} 1 & 2 & 3 & 4 & 5 \\ 3 & 5 & 4 & 1 & 2 \end{pmatrix} = 3\ 5\ 4\ 1\ 2 \tag{2.27}$$

gives two ways to write the permutation $\pi(1) = 3, \pi(2) = 5, \pi(3) = 4, \pi(4) = 1, \pi(5) = 2$.

Permutations of more general sets can be described similarly. For example,

$$\begin{pmatrix} a_1 & a_2 & a_3 & a_4 & a_5 \\ a_3 & a_5 & a_4 & a_1 & a_2 \end{pmatrix}$$

is the permutation that maps a_1 to a_3, a_2 to a_5, a_3 to a_4, a_4 to a_1, and a_5 to a_2.

The permutation π given in (2.27) never maps an element to itself, so it is an example of a *derangement.*

Definition 2.54 A *derangement* is a permutation that maps every element of a set (or sequence) to a different element of the set (or sequence). The *number of derangements* of an n element set or sequence is denoted by $D_{n.}$.

A permutation $\pi : [n] \to [n]$ is a derangement if and only if $\pi(k) \neq k$ for all k, $1 \leq k \leq n$. A permutation is not a derangement if there is some element j for which $\pi(j) = j$, in which case j in known as a *fixed point* of the permutation.

It is of interest to obtain formulas and relationships for the derangement numbers, and see how these numbers arise in applications. We can begin this investigation by tabulating the beginning values of the derangement numbers.

n	Derangements of $1\ 2\ 3\ \ldots\ n$	D_n
1	$\varnothing$	0
2	2 1	1
3	2 3 1, 3 1 2	2
4	2 1 4 3, 2 3 4 1, 2 4 1 3, 3 1 4 2, 3 4 1 2, 3 4 2 1, 4 1 2 3, 4 3 1 2, 4 3 2 1	9

So far, there is no obvious pattern for the derangement numbers. However, we can obtain a triangle identity by using the combinatorial meaning of the derangement numbers.

Theorem 2.55 Define $D_0 = 1$. Then the derangement numbers satisfy

$$D_n = (n-1)(D_{n-1} + D_{n-2}), \quad n \geq 2 \tag{2.28}$$

Proof. Let π be any derangement of the set $[n]$, so that $\pi(n) = j$, for some $1 \leq j \leq n-1$. There are then two distinct types of derangements.

Type 1: π Interchanges n and j. Any permutation that interchanges n with j corresponds to a derangement of the remaining $(n-2)$-element set $[n] - \{j, n\}$. Therefore, there are D_{n-2} derangements of type 1.

Type 2: π Does not interchange n and j. Therefore, $\pi(k) = n$ for some $k \neq j$. This means that we can associate π with the permutation π' defined on $[n-1]$ by

$$\pi' = \begin{pmatrix} 1 & 2 & \cdots & k & \cdots & n-1 \\ \pi(1) & \pi(2) & \cdots & j & \cdots & \pi(n-1) \end{pmatrix}.$$

This is a derangement on $[n-1]$ since $\pi'(k) = j \neq k$ and $\pi'(i) = \pi(i) \neq i$ for all $i \in [n-1] - \{k\}$. The association is clearly reversible, so there are D_{n-1} derangements of type 2.

For each j, $1 \leq j \leq n-1$, we have shown there are $D_{n-2} + D_{n-1}$ derangements of $[n]$ with $\pi(n) = j$. Therefore, the total number of derangements of $[n]$ is given by equation (2.28), since there are $n-1$ choices for j. ■

A simple algebraic rearrangement of (2.28) shows that

$$D_n - nD_{n-1} = -[D_{n-1} - (n-1)D_{n-2}] \tag{2.29}$$

Equation (2.29) can be iterated to show that

$$\begin{aligned} D_n - nD_{n-1} &= -[D_{n-1} - (n-1)D_{n-2}] = (-1)^2[D_{n-2} - (n-2)D_{n-3}] \\ &= \cdots = (-1)^{n-2}[D_2 - 2D_1] = (-1)^n \end{aligned}$$

since $D_2 = 1$ and $D_1 = 0$. This calculation gives us a first-order recursion relation for the derangement numbers.

Theorem 2.56

$$D_n = nD_{n-1} + (-1)^n, \quad n \geq 1 \tag{2.30}$$

The recursion relation (2.30) makes it easy to create a table of the derangement numbers (see Table 2.2).

TABLE 2.2 Derangement Numbers D_n, $n = 0, 1, 2, \ldots, 10$

n	0	1	2	3	4	5	6	7	8	9	10
D_n	1	0	1	2	9	44	265	1854	14833	133496	1334961

We can also obtain an explicit formula for D_n using (2.30).

Theorem 2.57

$$D_n = n!\left(1 - \frac{1}{1!} + \frac{1}{2!} - \frac{1}{3!} + \cdots + (-1)^n \frac{1}{n!}\right) \tag{2.31}$$

Proof. Dividing the terms in (2.30) by $n!$, we see that it can be rewritten in the equivalent form

$$\frac{D_n}{n!} - \frac{D_{n-1}}{(n-1)!} = \frac{(-1)^n}{n!} \tag{2.32}$$

Thus

$$\begin{aligned} \frac{D_n}{n!} - 1 &= \frac{D_n}{n!} - \frac{D_0}{0!} = \left(\frac{D_n}{n!} - \frac{D_{n-1}}{(n-1)!}\right) + \left(\frac{D_{n-1}}{(n-1)!} - \frac{D_{n-2}}{(n-2)!}\right) + \cdots \\ &\quad + \left(\frac{D_2}{2!} - \frac{D_1}{1!}\right) + \left(\frac{D_1}{1!} - \frac{D_0}{0!}\right) \\ &= (-1)^n \frac{1}{n!} + (-1)^{n-1} \frac{1}{(n-1)!} + \cdots + (-1)^2 \frac{1}{2!} + (-1)^1 \frac{1}{1!} \end{aligned}$$

which is easily rearranged to become (2.31). ■

Recall from elementary calculus that

$$e^x = 1 + \frac{x}{1!} + \frac{x^2}{2!} + \frac{x^3}{3!} + \cdots + \frac{x^n}{n!} + \cdots$$

so that setting $x = -1$ gives us

$$e^{-1} = 1 - \frac{1}{1!} + \frac{1}{2!} - \frac{1}{3!} + \cdots + \frac{(-1)^n}{n!} + \cdots$$

This is a rapidly converging alternating series, so that from (2.31) we obtain the approximate equation $D_n \approx n!/e$. More precisely, it can be shown that

$$D_n = \left\{ \frac{n!}{e} \right\}, \quad n \geq 1 \tag{2.33}$$

where $\{x\}$ denotes the integer nearest x.

Example 2.58 Suppose that checks are randomly stuffed into six preaddressed envelopes. What is the probability that

(a) no check is placed in the correct envelope?
(b) exactly one check is in the correct envelope?
(c) two or more checks are in the correct envelopes?

Solution

(a) There are 6! ways to stuff the envelopes, and D_6 ways that every check is in an incorrect envelope. Therefore, the probability that no check is in the correct envelope is $D_6/6! = \frac{265}{720} \doteq 0.368$.
(b) There are 6 ways to correctly insert exactly one of the checks into the correct envelope, and D_5 ways to place the remaining checks into incorrect envelopes. The probability of exactly one check placed in the correct envelope is therefore $6D_5/6! = D_5/5! = \frac{44}{120} \doteq 0.367$.
(c) By parts (a) and (b), the probability that more that one check is in the correct envelope is

$$\left(1 - \frac{D_6}{6!}\right) - \frac{D_5}{5!} \doteq 1 - 0.368 - 0.367 = 0.265$$

Rather surprisingly, it is more likely that exactly one check has been placed correctly compared with more than one correctly stuffed envelope. ■

PROBLEMS

2.6.1. There are eight people at a dinner party, including four men and four women. In how many ways can they be seated at a circular table if

(a) there are no restrictions?

(b) no two men can be seated side by side?

(c) Mr. and Mrs. Smith insist on sitting next to each other?

(d) George and Alice insist on not sitting next to each another?

2.6.2. King Arthur, together with 10 knights and 10 ladies, are to be seated at the round table. How many ways can this be done if no two knights and no two ladies sit side by side?

2.6.3. There are five adults and eight children at a party. In how many ways can they be seated at a circular table if at least one child is seated between any two adults?

2.6.4. In a carnival ride, five arms rotate about a pivot, and a revolving circular set of four chairs is at the end of each arm. In how many ways can groups of people of these sizes be seated:

(a) a group of 20?

(b) a group of 17?

(c) a group of 25, with 5 people unable to be seated?

2.6.5. **(a)** In how many ways can eight people be seated at a square table, with two persons on each side of the table?

(b) In how many ways can 12 people be seated at a square table, with three persons on each side of the table?

2.6.6. How many permutations of [15]

(a) have no number in its natural position?

(b) leave the even numbers in even positions?

(c) leave the even numbers in even positions but not in their natural positions?

2.6.7. How many permutations π: $[n] \to [n]$ satisfy $\pi(1) \neq 1$?

2.6.8. Prove that D_n has the opposite parity of its index n.

2.6.9. Players A and B each shuffle their own deck of nine cards numbered 1, ..., 9. Each player then shows a card. If the two cards match, player A wins the game, but if the two cards are different then the game continues with each player revealing another card. Player A wins if two matching cards are revealed on any turn, and player B wins the game if no match occurs after all 9 turns have been taken.

(a) Would you prefer to be player A or B?

(b) Would one of the players benefit if either 8 or 10 cards were used?

2.6.10. Let d_k be the number of permutations that fix k elements of $[5]$ and derange the remaining elements. Determine d_k for $k = 0, 1, \dots, 5$, and calculate the sum $d_0 + d_1 + \cdots + d_5$.

2.6.11. **(a)** How many permutations π: $[n] \to [n]$ derange k elements and leave the remaining elements fixed?

(b) Use combinatorial reasoning to proved the identity $n! = \sum_{k=0}^{n} \binom{n}{k} D_k$.

2.6.12. The 3! permutations of 1 2 3 are $\underline{1}\,\underline{2}\,\underline{3}$, $\underline{1}\,3\,2$, $2\,3\,1$, $2\,1\,\underline{3}$, $3\,1\,2$, $3\,\underline{2}\,1$, where the underlines indicate the fixed points. In particular, there are six fixed points in the 3! permutations, so the average number of fixed points is $6/3! = 1$. Show that the average number of fixed points of the 4! permutations of 1 2 3 4 is also 1.

2.6.13. Use the identity of Problem 2.6.11(b) to show that

$$1 = \frac{1}{n!} \sum_{k=0}^{n} k \binom{n}{k} D_{n-k}$$

for all n. Thus, the mean μ (average) number of fixed points of the $n!$ permutations of $[n]$ is 1 (see Problem 2.6.12).

2.6.14. Given any set of numbers $\{x_1, x_2, \dots, x_n\}$, the *standard deviation* σ, given by $\sigma^2 = \frac{(x_1-\mu)^2+(x_2-\mu)^2+\cdots+(x_n-\mu)^2}{n}$, measures the dispersion of the data about the mean $\mu = \frac{x_1+x_2+\cdots+x_n}{n}$. Problem 2.6.13 shows that $\mu = 1$ for all n when the data are the number of fixed points of the permutations of $[n]$. It can be shown (see Problem 3.5.17) that $\sigma = 1$ for all $n \geq 1$. Verify that $\sigma = 1$ for these two cases.

(a) $n = 3$

(b) $n = 4$

2.6.15. One morning, n monkeys lined up so that monkey 2 scratches the back of monkey 1 at the front of the line, monkey 3 scratches the back of monkey 2, and so on. That afternoon, the n monkeys want to line up again, but in one of the ways that no monkey's back will be scratched by the same monkey as in the morning. Let a_n denote the number of realignments of the n monkeys. For example, $a_1 = 1$ and $a_2 = 1$.

(a) Compile lists of the allowed realignments to determine that $a_3 = 3$ and $a_4 = 11$. For example, 3142 and 4231 are two allowable realignments if 1234 is the original order.

(b) Explain why there are $(n-1)\,a_{n-1}$ ways to line up n monkeys, even if monkey n were to decide to leave the line.

(c) Explain why there are $(n-2)\,a_{n-2}$ ways to line up n monkeys, where monkey n cannot leave the newly formed line without creating a forbidden realignment.

(d) Summing parts (b) and (c) shows that $a_n = (n-1)\,a_{n-1} + (n-2)\,a_{n-2}$. Compare this to identity (2.27) to prove that $a_n = D_{n+1}/n$.

2.7 SUMMARY AND ADDITIONAL PROBLEMS

In this chapter, we have encountered several important combinatorial terms, notations, and formulas. These are summarized in Table 2.3.

It was also seen how combinatorial models are used to derive counting formulas and prove identities. These models included:

- *Tiling*—counting the ways in which a rectangular board can be tiled with tiles of a given shape and color
- *Selecting a committee*—counting the ways to select committees, subcommittees, committee chairs, and so on
- *Block walking*—counting the number of paths between two points of a square grid or other system of edges
- *Flagpole arrangements*—counting the number of ways to erect a flagpole and supporting guywires on a row of blocks under various conditions

PROBLEMS

2.7.1. The factorials $n!$ grow very rapidly with n. Indeed, $15! = 1{,}307{,}674{,}368{,}000$ is already more than a trillion. In 1730, Abraham DeMoivre (1667–1754) gave the approximate value $n! \approx (n/e)^n \sqrt{2\pi n}$ that is more often called *Stirling's formula* (James Stirling, 1692–1770).

(a) Use Stirling's formula with a calculator to obtain an estimate of 15!.

(b) Show that a better approximation to 15! is given by Gosper's formula $n! \approx \left(\frac{n}{e}\right)^n \sqrt{\left(2n + \frac{1}{3}\right)\pi}$.

2.7.2. **(a)** Explain why 28! ends with six zeros, since

$$\left\lfloor \frac{28}{5} \right\rfloor + \left\lfloor \frac{28}{5^2} \right\rfloor = 5 + 1 = 6$$

Here $\lfloor x \rfloor$ is the floor function.

(b) How many zeros are at the end of 1001!?

2.7.3. The nth *Catalan number* is given by $C_n = [1/(n+1)]\binom{2n}{n}$. Use Stirling's formula from Problem 2.7.1 to derive the approximate formula $C_n \approx 4^n/(\sqrt{\pi}n^{3/2})$.

TABLE 2.3 Summary of Combinatorial Terms, Notations, Formulas, and Meanings in the Chapter

Combinatorial term	Symbol/Formula	Combinatorial Meaning as the Number of
r-Permutation	$P(n,r)$ or $(n)_r$	Ordered lists of r objects taken from a set of n distinct elements
Binomial coefficient	$\binom{n}{r}$ or $C(n,r)$ "n choose r"	r-Subsets (combinations) of an n set Injective (one-to-one) distributions of n distinct objects to r distinct recipients Solutions in integers $x_k \in \{0,1\}$ of $x_1 + x_2 + \cdots + x_r = n$ One-to-one functions $f : [r] \to [n]$
Multinomial coefficient	$\binom{n}{n_1, n_2, \ldots, n_r}$, $n_1 + n_2 + \cdots + n_r = n$	Permutations of multiset $\{n_1 \cdot a_1, n_2 \cdot a_2, \ldots, n_r \cdot a_r\}$ Distributions of n distinct objects to r distinct recipients so recipient k receives n_k objects
Multichoose coefficient	$\left(\!\binom{n}{m}\!\right)$	Multisubsets (combinations with repetition) with n objects from the multiset $\{\infty \cdot a_1, \infty \cdot a_2, \ldots, \infty \cdot a_n\}$ Distributions of m identical objects to n distinct recipients Nondecreasing functions $f : [m] \to [n]$ Solutions in nonnegative integers of $x_1 + x_2 + \cdots + x_n = m$
Distribution number	$T(m,n)$	Distributions of m distinct objects to n distinct recipients
Circular permutation of an n-set	$(n-1)!$	Ways to seat n people around a circular table
r-Circular permutation of an n-set	$\frac{P(n,r)}{r}$	Ways to seat r people around a circular table, selected from a set of n people
Derangement number	D_n	Permutations of a set of n distinct objects with no element mapped to itself (i.e., no fixed points)
Combinatorial Fibonacci numbers	f_n	Ways to tile a $1 \times n$ board with squares and dominoes
Catalan numbers	$C_n = \frac{1}{n+1}\binom{2n}{n}$	Paths in Pascal's triangle from $A(0,0)$ to $B(2n,n)$ that never pass through a point on east side of vertical line from A to B (additional interpretations of the Catalan numbers will be found in Chapter 6)

2.7.4. **(a)** Prove algebraically that the Catalan numbers $C_n = [1/(n+1)]\binom{2n}{n}$ satisfy the recursion relation $C_n = [(4n-2)/(n+1)]C_{n-1}$, $n \geq 1$.

(b) Use the recursion relation of part (a) to make a table of the Catalan numbers $C_0, C_1, C_2, \ldots, C_{12}$, where $C_0 = 1$.

2.7.5. A *bridge hand* is a set of 13 cards dealt from a 52-card deck. How many bridge hands

(a) contain all four aces?

(b) contain four spades, four hearts; three diamonds, and two clubs?

(c) have a 5-3-3-2 distribution? In other words, the hand contains five cards of one suit, three cards each of two other suits, and two cards from the remaining suit.

2.7.6. **(a)** How many bridge hands (13-card hands dealt from a 52-card deck) have no diamonds?

(b) Explain why $\binom{4}{1}\binom{39}{13}$ does *not* give the number of bridge hands with exactly one missing suit.

(c) Explain why

$$\binom{4}{1}\binom{13}{1}^3\binom{36}{10}$$

does *not* give the number of bridge hands with exactly one missing suit.

2.7.7. **(a)** Explain why the number of paths from P_1 to P_n through the triangular grid shown, always moving in the direction of the arrows, is the nth Fibonacci number F_n:

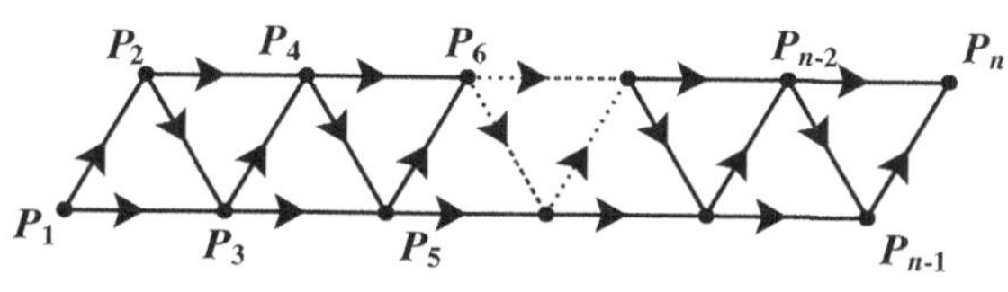

(b) How many paths connect point P_a to point P_b?

(c) Use the triangular grid to prove the identity $F_{a+b} = F_aF_{b+1} + F_{a-1}F_b$.

2.7.8. Explain the following curious pattern in Pascal's triangle. A vertical bar is placed between any two adjacent coefficients of some row of Pascal's Triangle. Why is the ratio of the two binomial coefficients to the left and right of the bar the same as the ratio of the number of terms to the left and to the right of the bar in the row? For example, in row 11 of Pascal's triangle, suppose that the bar is placed between the 330 and the 462: 1 11 55 165 330

| 462 462 330 165 55 11 1. There are five terms to the left and seven to the right of the bar, and $\frac{330}{462} = \frac{5}{7}$.

2.7.9. Prove the identity

$$\binom{n}{p}\binom{n}{q} = \sum_{k\geq 0}\binom{n}{k}\binom{n-k}{p-k}\binom{n-p}{q-k}$$

by counting the number of ways to tile a $2 \times n$ board with p gray tiles in the top row, q gray tiles in the bottom row, and the remaining cells left empty. [*Hint:* Consider the number of columns that are tiled with two gray squares.]

2.7.10. **(a)** Show that $\binom{k-m}{m}$ is the number of paths in a square grid from (0,0) to the point (m,n) on the diagonal d_k with equation $y = k - 2x$.

(b) Give a block walking argument for the identity

$$F_k = \sum_{m=0}^{\lfloor k/2 \rfloor}\binom{k-m}{m}$$

where F_k is the kth Fibonacci number.

2.7.11. Give a combinatorial proof that

$$\binom{n}{2}^2 = \binom{n}{2} + 6\binom{n}{3} + 6\binom{n}{4}$$

2.7.12. Use the tiling model to prove the identity

$$\sum_{k=1}^{n} k^3 = \binom{n+1}{2}^2$$

[*Hint:* Tile a $2 \times (n+1)$ board with two red tiles in the top row and two black tiles in the bottom row.]

2.7.13. Give a combinatorial proof that $\sum_{r=1}^{n} r = \left(\!\binom{n}{2}\!\right)$.

REFERENCE

1. A. T. Benjamin and J. Quinn, *Proofs that Really Count*, Mathematical Association of America, 2003.

3

BINOMIAL SERIES AND GENERATING FUNCTIONS

3.1 INTRODUCTION

In this chapter, we associate a sequence $a_0, a_1, \dots, a_n, \dots$ with a function $f(x)$ called its *generating function*. The association is very simple; the terms in the sequence are the coefficients of a series expansion of $f(x)$ of a specified type. The eminent mathematician George Pólya (1887–1985) informally called f a "bag function," since the terms of the sequence are conveniently packaged by the function. Similarly, in his book on generating functions, Herb Wilf [1] described a generating function as a "clothesline" on which the terms of the sequence are hung up for display. These two descriptions remind us that the association goes in both directions; if we start with the sequence, the generating function bags them up into a convenient package. In the opposite direction, if we start with the generating function, the series expansion displays the sequence of coefficients as if hung on a clothesline.

There are several types of generating functions, but we will focus on these two: the ordinary generating function (OGF) and the exponential generating function (EGF).

Definition 3.1 Given the sequence $a_0, a_1, \dots, a_k, \dots$ its *ordinary generating function* (OGF) is the function $f(x)$ defined by the power series

$$f(x) = a_0 + a_1x + a_2x^2 + \cdots + a_kx^k + \cdots = \sum_{k \geq 0} a_kx^k \tag{3.1}$$

Combinatorial Reasoning: An Introduction to the Art of Counting, First Edition. Duane DeTemple and William Webb.

Perhaps the simplest example of an OGF is the geometric series familiar from elementary calculus, where the geometric series

$$f(x) = \frac{1}{1-x} = 1 + x + x^2 + \cdots + x^n + \cdots = \sum_{n\geq 0} x^n \tag{3.2}$$

shows that $f(x) = 1/(1-x)$ is the OGF of the constant sequence 1,1,1, ...

The second type of generating function that we will consider, the exponential generating function (EGF), has this definition.

Definition 3.2 Given the sequence $a_0, a_1, \ldots, a_k, \ldots$, its *exponential generating function* (EGF) is the function defined by the power series

$$f^{(e)}(x) = a_0 + a_1\frac{x}{1!} + a_2\frac{x^2}{2!} + \cdots = \sum_{k\geq 0} a_k\frac{x^k}{k!} \tag{3.3}$$

For example, the sequence 1, 1, 1, ... has the EGF

$$f^{(e)}(x) = 1 + \frac{x}{1!} + \frac{x^2}{2!} + \cdots = \sum_{k\geq 0}\frac{x^k}{k!} = e^x \tag{3.4}$$

Most sequences of combinatorial numbers (e.g., binomial, multinomial, and multichoose coefficients; derangement numbers, Fibonacci numbers) have surprisingly simple generating functions, as we will see. Of special interest is how properties of generating functions reveal corresponding properties of the sequence of its coefficients. Thus, generating functions give us a new approach to both discovering and proving identities. We will also see how counting problems can be solved with generating functions, both OGFs and EGFs.

3.2 THE BINOMIAL AND MULTINOMIAL THEOREMS

3.2.1 The Binomial Theorem

By expanding the powers $(w + x)^n$ for $n = 0, 1, 2, \ldots$ we obtain the following pattern:

$$\begin{aligned}
(w+x)^0 &= \mathbf{1}\\
(w+x)^1 &= \mathbf{1}w + \mathbf{1}x\\
(w+x)^2 &= \mathbf{1}w^2 + \mathbf{2}wx + \mathbf{1}x^2\\
(w+x)^3 &= \mathbf{1}w^3 + \mathbf{3}w^2x + \mathbf{3}wx^2 + \mathbf{1}x^3\\
(w+x)^4 &= \mathbf{1}w^4 + \mathbf{4}w^3x + \mathbf{6}w^2x^2 + \mathbf{4}wx^3 + \mathbf{1}x^4\\
&\vdots \qquad\qquad \vdots
\end{aligned}$$

The coefficients of the expansions shown in bold are those in Pascal's triangle, and we have the following theorem.

Theorem 3.3 (Binomial Theorem)

$$(w+x)^n = \sum_{k=0}^{n} \binom{n}{k} w^{n-k} x^k \tag{3.5}$$

Proof. First note that $(w+x)^n = (w+x)(w+x)\cdots(w+x)$ is a product of the n binomial factors $w + x$. The coefficient of $w^{n-k}x^k$ is therefore the number of ways to choose the variable x from k of the binomial factors, with the variable w taken from the remaining $n-k$ binomial factors. This is the number of k-combinations from the set of n binomial factors, so there are $\binom{n}{k}$ ways to select the terms that contribute to the coefficient $w^{n-k}x^k$. ■

If we set the variable $w = 1$ in (3.5), we obtain the OGF $f_n(x) = (1+x)^n$ for the binomial coefficients $\binom{n}{0}, \binom{n}{1}, \binom{n}{2}, \ldots, \ldots, \binom{n}{k}, \ldots, \binom{n}{n}, 0, 0, \ldots$ that appear in row n of Pascal's triangle.

Theorem 3.4

$$(1+x)^n = \sum_{k=0}^{n} \binom{n}{k} x^k \tag{3.6}$$

Identities for the binomial coefficients can now be obtained by choosing special values of x in (3.6). For example, the choice of $x = 1$ proves that

$$2^n = \sum_{k=0}^{n} \binom{n}{k}, \quad n \geq 0 \tag{3.7}$$

which is a new derivation of identity (1.23) of Chapter 1.

If we let $x = -1$ in (3.6), then $(1-1)^n = 0^n = 0$ for all $n \geq 1$. Therefore, we obtain, the following new identity.

Theorem 3.5

$$0 = \sum_{k=0}^{n} (-1)^k \binom{n}{k}, \quad n \geq 1 \tag{3.8}$$

In the next examples, we will illustrate some typical applications of the binomial theorem (3.6). We will start by giving a new proof of Pascal's identity.

Theorem 3.6 (Pascal's Identity via the Binomial Theorem)

$$\binom{n}{k} = \binom{n-1}{k} + \binom{n-1}{k-1} \tag{3.9}$$

Proof. The calculation that follows illustrates some important techniques often employed in working with generating functions. Starting with (3.6), we see that

$$\begin{aligned}
(1+x)^n &= (1+x)(1+x)^{n-1} \\
&= (1+x)\sum_{k=0}^{n-1}\binom{n-1}{k}x^k \\
&= \sum_{k=0}^{n-1}\binom{n-1}{k}x^k + x\sum_{k=0}^{n-1}\binom{n-1}{k}x^k \\
&= \sum_{k=0}^{n-1}\binom{n-1}{k}x^k + \sum_{k=0}^{n-1}\binom{n-1}{k}x^{k+1} \qquad (3.10) \\
&= \sum_{k=0}^{n-1}\binom{n-1}{k}x^k + \sum_{k=1}^{n}\binom{n-1}{k-1}x^k \quad \text{(replace } k \text{ with } k-1) \\
&= \sum_{k=0}^{n}\binom{n-1}{k}x^k + \sum_{k=0}^{n}\binom{n-1}{k-1}x^k \quad \left(\text{since } \binom{n-1}{n} = 0 \text{ and } \binom{n-1}{-1} = 0\right)
\end{aligned}$$

The coefficient of x^k in this expansion must be the same as the coefficient of x^k in $(1+x)^n = \sum_{k=0}^{n}\binom{n}{k}x^k$, so we get Pascal's identity. ■

A generating function can often be determined from the recursion relation and initial conditions satisfied by the sequence.[1] For example, to obtain the generating function for the binomial coefficients, let's start with Pascal's identity (3.9). Multiply the terms by x^k and sum over all $k \geq 0$ to get

$$\begin{aligned}
f_n(x) = \sum_{k\geq 0}\binom{n}{k}x^k &= \sum_{k\geq 0}\binom{n-1}{k}x^k + x\sum_{k\geq 0}\binom{n-1}{k-1}x^{k-1} \\
&= f_{n-1}(x) + xf_{n-1}(x) \qquad (3.11)
\end{aligned}$$

[1]Indeed, this application will be discussed in detail in Chapter 5, where we explore how very general recursion relations can be solved with generating functions.

Rearranging (3.11), we see that the generating functions satisfy the recursion relation

$$f_n(x) = (1+x)f_{n-1}(x) \tag{3.12}$$

Since $f_0 = 1$, we can iterate (3.12) to get

$$f_n(x) = (1+x)f_{n-1}(x) = (1+x)^2 f_{n-2}(x) = \cdots = (1+x)^n f_0 = (1+x)^n$$

which is just the OGF we expected on the basis of (3.6).

The next two examples illustrate how additional identities can be derived from the binomial theorem by using algebraic manipulation.

Theorem 3.7

$$\binom{m+n}{r} = \sum_{k=0}^{\min(r,m)} \binom{m}{k}\binom{n}{r-k} \tag{3.13}$$

Proof. First note that the sum on the right side of the identity can just as well be written in the form $\sum_k \binom{m}{k}\binom{n}{r-k}$, where the sum is taken over all integers k. After all, the terms in the sum are 0 when k is negative or when k is larger than either r or m. With this simplification of notation, we have

$$(1+x)^{m+n} = (1+x)^m(1+x)^n = \left(\sum_i \binom{m}{i} x^i\right)\left(\sum_j \binom{n}{j} x^j\right)$$

$$= \sum_i \sum_j \binom{m}{i}\binom{n}{j} x^{i+j} = \sum_k \left(\sum_i \binom{m}{i}\binom{n}{k-i}\right) x^k \quad (\text{let } i+j=k)$$

But we also have the expansion

$$(1+x)^{m+n} = \sum_k \binom{m+n}{k} x^k$$

so that identity (3.13) follows by equating coefficients of x^k. ∎

If the generating function of a sequence is known, then both its derivative and antiderivative give the generating functions of related sequences. The following example illustrates this idea.

Example 3.8 Prove these two identities:

(a)
$$n(n+1)2^{n-2} = \sum_{k=1}^{n} k^2 \binom{n}{k}, \quad n \geq 1 \tag{3.14}$$

(b)
$$0 = \sum_{k=1}^{n} (-1)^k k^2 \binom{n}{k}, \quad n \geq 3 \tag{3.15}$$

Solution. Both of the identities require two differentiations of the binomial formula

$$(1+x)^n = \sum_k \binom{n}{k} x^k$$

The first derivative is

$$n(1+x)^{n-1} = \sum_k k \binom{n}{k} x^{k-1}.$$

To introduce the needed k^2 factor, multiply this identity by x and then take another derivative to get

$$\begin{aligned}\frac{d}{dx}\left(nx(1+x)^{n-1}\right) &= n(1+x)^{n-1} + n(n-1)x(1+x)^{n-2} \\ &= \frac{d}{dx}\sum_k k\binom{n}{k}x^k = \sum_k k^2\binom{n}{k}x^{k-1}\end{aligned}$$

Thus, we have obtained a new polynomial identity:

$$n(nx+1)(1+x)^{n-2} = \sum_k k^2 \binom{n}{k} x^{k-1}$$

Identities (3.14) and (3.15) are obtained by setting $x = 1$ and $x = -1$, respectively. ■

3.2.2 The Multinomial Theorem

The binomial theorem, which gives a formula for the expansion of $(w+x)^n$, will now be generalized to give a formula for the expansion of $\left(x_1 + x_2 + \cdots + x_k\right)^n$. As a beginning example, consider

$$(x+y+z)^5 = (x+y+z)(x+y+z)(x+y+z)(x+y+z)(x+y+z)$$

The expansion is calculated by selecting one of the variables, either x or y or z, from each of the five multinomial factors. For example, if the term y is chosen from the first and third factors, z from the second factor, and x from the last two factors, we get the product $yzyxx$ in the expansion. Of course, this is only one of many ways to contribute to a term in the expansion of the form x^2y^2z. More generally, *every* permutation of the multiset $\{2 \cdot x, 2 \cdot y, 1 \cdot z\}$ gives a product that can be written in the form x^2y^2z. Since the number of permutations is given by the multinomial coefficient $\binom{5}{2,2,1}$, this is the coefficient of $x^2y^2z^1$ in the expansion of $(x + y + z)^5$.

The reasoning just given can be generalized to the expansion of a general k-nomial $(x_1 + x_2 + \cdots + x_k)^n$, and we get the following theorem that explains why $\binom{n}{r_1,r_2,\ldots,r_k}$ is, indeed, a coefficient.

Theorem 3.9 (The Multinomial Theorem)

$$(x_1 + x_2 + \cdots + x_k)^n = \sum \binom{n}{r_1, r_2, \ldots, r_k} x_1^{r_1} x_2^{r_2} \cdots x_k^{r_k} \tag{3.16}$$

where the sum is taken over all k-tuples of nonnegative integers $r_1, r_2, \ldots, r_k$ for which $r_1 + r_2 + \cdots + r_k = n$.

We see that (3.16) gives a generating function for the multinomial coefficients, although it is a polynomial in several variables instead of a single variable.

The following example demonstrates how the multinomial theorem can be used to determine the coefficients of multinomial expansions of a more general form.

Example 3.10 What is the coefficient of xy^2z^3 in the expansion of each of the following expressions:

(a) $(x + y + z)^6$
(b) $(2x - 3y + z)^6$
(c) $(x + y^2 + z)^6$
(d) $(x + y^2 + z)^5$

Solution

(a) $\binom{6}{1,2,3} = 6!/(1!2!3!) = 60$, by a direct application of the multinomial theorem (3.16).

(b) $\binom{6}{1,2,3}(2)(-3)^2 = (18)\,60 = 1080$, since the single term x is always accompanied by a factor of 2 and the two y terms are each accompanied by the factor –3.

(c) 0, since the terms in the expansion have the form $x^r y^{2s} z^t$, $r+s+t=6$ and xy^2z^3 corresponds to $r+s+t = 1+1+3 = 5 \neq 6$.

(d) $\binom{5}{1,1,3} = 20$, as seen in solution (c) above. ■

PROBLEM

3.2.1. Evaluate each of these sums:

(a) $\sum_{n=0}^{8} \binom{8}{n} 9^n$ **(b)** $\sum_{n=0}^{8} (-1)^n \binom{8}{n} 11^n$

3.2.2. Use the binomial theorem to expand:

(a) $(1+4x)^3$ **(b)** $(2a+b)^4$

3.2.3. Determine the coefficient of w^3x^4 in the expansion of each of these binomials:

(a) $(w+2x)^7$ **(b)** $(2w-x^2)^7$ **(c)** $(2w-x^2)^5$

3.2.4. Show how the binomial theorem can simplify these polynomials.

(a) $p(t) = t - 3t^2 + 3t^3 - t^4$

(b) $q(y) = 2y + 16y^2 + 48y^3 + 64y^4 + 32y^5$

3.2.5. Prove algebraically that

(a) $\binom{n}{r} = \binom{n-2}{r-2} + 2\binom{n-2}{r-1} + \binom{n-2}{r}$

(b) $\binom{n}{r} = \binom{n-3}{r-3} + 3\binom{n-3}{r-2} + 3\binom{n-3}{r-1} + \binom{n-3}{r}$

3.2.6. **(a)** Verify that

$$\frac{(1+x)^{n+1}-1}{n+1} = \sum_{r=0}^{n} \binom{n}{r} \frac{x^{r+1}}{r+1}$$

(b) Evaluate

$$\sum_{r=0}^{n} \binom{n}{r} \frac{1}{r+1}$$

(c) Evaluate

$$\sum_{r=0}^{n} \binom{n}{r} \frac{(-1)^r}{r+1}$$

3.2.7. Give a combinatorial proof of the identity

$$0 = \sum_{k=0}^{n} (-1)^k \binom{n}{k}, \quad n \geq 1$$

[*Hint*: Pair subsets of $[n]$ that differ only by the inclusion or exclusion of the element 1.]

3.2.8. **(a)** Prove that if the polynomial $p(x) = a_n x^n + a_{n-1}x^{n-1} + \cdots + a_0$ is 0 for all values of x, then every coefficient $a_n, a_{n-1}, \ldots, a_0$ must be 0.

(b) Assuming that the two polynomials $p(x) = a_n x^n + a_{n-1}x^{n-1} + \cdots + a_0, a_n \neq 0$ and $q(x) = b_m x^m + b_{m-1}x^{m-1} + \cdots + b_0, b_m \neq 0$ satisfy $p(x) = q(x)$ for all x, prove that $m = n$ and $a_r = b_r$ for all $r = 0, 1, \ldots, n$.

3.2.9. **(a)** Use the binomial theorem to prove that $\sum_{r=1}^{n} r\binom{n}{r} = n2^{n-1}$.

(b) Show from part (a) that the average number of elements in the subsets of an n-element set is $n/2$.

(c) Give a combinatorial reason for the result of part (b). [*Hint*: How can one pair up the subsets of the n-set?]

3.2.10. **(a)** Use the binomial identity (3.5) to prove that

$$n(n-1)(w+x)^{n-2} = \sum_{r=1}^{n-1} \binom{n}{r} (n-r) w^{n-r-1} r x^{r-1}$$

(b) Prove that for all $n \geq 2$, $n(n-1)2^{n-2} = \sum_{r=1}^{n-1} (n-r)\binom{n}{r}$.

(c) Give a committee selection proof of the identity given in part (b).

3.2.11. Prove that for all $n \geq 3$, $0 = \sum_{r=1}^{n-1} (-1)^r (n-r) r \binom{n}{r}$.

3.2.12. Let $\sigma_0(n) = n + 1$ and $\sigma_k(n) = 1^k + 2^k + 3^k + \cdots + n^k, k \geq 1$. For example, we know that $\sigma_1(n) = 1 + 2 + \cdots + n = \frac{1}{2}n(n+1)$.

(a) Prove that

$$\sigma_k(1+n) = \sum_{r=0}^{k} \binom{k}{r} \sigma_r(n)$$

[*Hint*: Consider the sum $1 + \sum_{j=1}^{n} (1+j)^k$.]

(b) Rearrange the result of part (a) to obtain the recursion formula

$$\sigma_k(n) = \frac{1}{k+1}\left[(1+n)^{k+1} - (1+n) - \sum_{r=1}^{k-1} \binom{k+1}{r} \sigma_r(n)\right]$$

[*Hint*: Use the formula of part (a), but with $k+1$ replaced with k.]

(c) Use part (b) to show that
$\sigma_2(n) = 1^2 + 2^2 + \cdots + n^2 = [n(n+1)(2n+1)]/6.$

3.2.13. Expand these multinomials:

(a) $(r - 2s + t)^3$ **(b)** $(w + x + 2y - 3z)^2$

3.2.14. Determine the coefficient of $w^2x^4y^3$ in the expansion of each of these trinomials:

(a) $(w + 2x - y)^9$ **(b)** $(3w - x^2 + y)^7$ **(c)** $(3w - x^2 + y)^9$

3.2.15. What is the coefficient of x^2y^4z in

(a) $(x + y + z)^7$ **(b)** $(2x + 3y - z)^7$ **(c)** $(x + 2y + z + 1)^7$

3.2.16. Prove that

$$k^n = \sum_{\substack{r_1, r_2, \ldots, r_k \geq 0 \\ r_1 + r_2 + \cdots + r_k = n}} \binom{n}{r_1, r_2, \ldots, r_k}$$

3.2.17. Prove that

$$\binom{n}{k} k^{n-k} = \frac{1}{k!} \sum_{\substack{r_1, r_2, \ldots, r_k \geq 0 \\ r_1 + r_2 + \cdots + r_k = n}} r_1 r_2 \cdots r_k \binom{n}{r_1, r_2, \ldots, r_k}$$

3.3 NEWTON'S BINOMIAL SERIES

The generating functions for the binomial and multinomial coefficients derived in Section 3.2 are polynomials in one or several variables. However, a generating function is usually an infinite series. For combinatorial purposes, power series can be treated quite informally. In particular, we assume that they converge in some interval about $x = 0$ and we do not need to carefully determine the interval of convergence.

3.3.1 Generating Function for the Multichoose Coefficients

In the previous section, we showed how to determine the generating function for the binomial coefficients by starting with Pascal's identity. Therefore it seems hopeful that the generating function $f_n(x) = \sum_{k \geq 0} \left(\binom{n}{k}\right) x^k$ for the multichoose coefficients $\left(\binom{n}{k}\right) = \binom{k+n-1}{k}$ can be determined from the triangle identity

$$\left(\binom{n}{k}\right) = \left(\binom{n-1}{k}\right) + \left(\binom{n}{k-1}\right), \quad n, k \geq 1 \tag{3.17}$$

that was proved in Chapter 2. If we multiply (3.17) by x^k and sum over all nonnegative integers k, we see that

$$f_n(x) = \sum_{k\geq 0}\left(\!\binom{n}{k}\!\right)x^k = 1 + \sum_{k\geq 1}\left(\!\binom{n}{k}\!\right)x^k$$
$$= 1 + \sum_{k\geq 1}\left(\!\binom{n-1}{k}\!\right)x^k + x\sum_{k\geq 1}\left(\!\binom{n}{k-1}\!\right)x^{k-1} = f_{n-1}(x) + xf_n(x)$$

which gives us this recursion relation for the generating function $f_n(x)$:

$$f_n(x) = \frac{1}{(1-x)}f_{n-1}(x) \tag{3.18}$$

Using iteration, we then see that

$$f_n(x) = \frac{1}{(1-x)}f_{n-1}(x) = \frac{1}{(1-x)^2}f_{n-2}(x) = \cdots = \frac{1}{(1-x)^n}f_0(x) \tag{3.19}$$

Since

$$f_0(x) = \sum_{k\geq 0}\left(\!\binom{0}{k}\!\right)x^k = 1 + \sum_{k\geq 1}\binom{k+0-1}{k}x^k = 1$$

the following theorem has been proved.

Theorem 3.11

$$\frac{1}{(1-x)^n} = \sum_{k=0}^{\infty}\left(\!\binom{n}{k}\!\right)x^k, \quad n \geq 1 \tag{3.20}$$

or, equivalently

$$\frac{1}{(1-x)^n} = \sum_{k=0}^{\infty}\binom{k+n-1}{n-1}x^k, \quad n \geq 1 \tag{3.21}$$

Setting $n = 1$ in (3.21) and recalling that $\left(\!\binom{1}{0}\!\right) = \binom{0}{0} = 1$ we get $1/(1-x) = \sum_{k=0}^{\infty} x^k$, which is just the geometric series seen earlier in equation (3.2). If the variable x is replaced with αx, we obtain this result.

Theorem 3.12 The ordinary generating function of the geometric sequence $c, c\alpha, c\alpha^2, c\alpha^3, \ldots$ is

$$\frac{c}{1-\alpha x} = c \sum_{k=0}^{\infty} \alpha^k x^k \tag{3.22}$$

In the next section, the OGF (ordinary generating function) will be used both to solve a variety of counting problems and to derive combinatorial identities. The type of calculation demonstrated in the next two examples will be very useful. The following notation is convenient when doing the computations.

> **NOTATION**
>
> $[x^n] f(x) =$ the coefficient of x^n in the series expansion of $f(x)$.

We can show with short calculations that

$$[x^n]\left(x^j f(x)\right) = [x^{n-j}] f(x) \quad \text{and} \quad \left[\frac{x^n}{c}\right](f(x)) = c\,[x^n] f(x)$$

Example 3.13 For each rational function below, determine the coefficient of x^{10} in the series expansion. Give your answers as expressions involving binomial coefficients.

(a) $\dfrac{x^2}{(1-x)^7}$

(b) $\dfrac{x^3+5x^2}{(1-x^2)^4}$

(c) $\dfrac{(1-x^3)^4}{(1-x)^6}$

Solution

(a) $[x^{10}]\dfrac{x^2}{(1-x)^7} = [x^8]\dfrac{1}{(1-x)^7} = [x^8]\displaystyle\sum_{k\geq 0}\binom{k+6}{6}x^k = \binom{8+6}{6} = \binom{14}{6}$

(b)
$$\begin{aligned}
[x^{10}]\frac{x^3+5x^2}{(1-x^2)^4} &= [x^7]\frac{1}{(1-x^2)^4} + [x^8]\frac{5}{(1-x^2)^4} \\
&= [x^7]\sum_{k\geq 0}\binom{k+3}{3}x^{2k} + 5\,[x^8]\sum_{k\geq 0}\binom{k+3}{3}x^{2k} \\
&= 0 + 5\binom{4+3}{3} = 5\binom{7}{3}
\end{aligned}$$

(c) $$[x^{10}]\,\frac{(1-x^3)^4}{(1-x)^6} = [x^{10}]\,(1-x^3)^4\frac{1}{(1-x)^6}$$

$$= [x^{10}]\sum_{r\geq 0}(-1)^r\binom{4}{r}x^{3r}\sum_{k\geq 0}\binom{k+5}{5}x^k$$

$$= \binom{4}{0}\binom{10+5}{5} - \binom{4}{1}\binom{7+5}{5} + \binom{4}{2}\binom{4+5}{5} - \binom{4}{3}\binom{1+5}{5}$$

$$= \binom{15}{5} - 4\binom{12}{5} + 6\binom{9}{5} - 4\binom{6}{5}$$ ■

3.3.2 Generalized Binomial Coefficients and Newton's Binomial Series

If x is replaced with $-x$ in (3.21), we see that

$$(1+x)^{-n} = \sum_{k\geq 0}(-1)^k\binom{k+n-1}{n-1}x^k \tag{3.23}$$

When this formula is compared with the binomial series

$$(1+x)^n = \sum_{k\geq 0}\binom{n}{k}x^k$$

it suggests that we should extend the definition of the binomial coefficients by setting

$$\binom{-n}{k} = (-1)^k\binom{k+n-1}{k} \tag{3.24}$$

To allow the upper symbol of binomial coefficients to be a negative integer, or, better yet, an arbitrary real or complex number α, recall that $\binom{n}{k} = (n)_k/k!$, where $(n)_k$ is the Pochhammer symbol defined by

$$(n)_k = n\,(n-1)\,(n-2)\cdots(n-k+1) \tag{3.25}$$

Since $(n)_k$ is a polynomial in the variable n, it is defined for all real or even complex values of n. Therefore, we can extend the binomial coefficients with this definition.

Definition 3.14 For any nonnegative integer k and any real or complex number α, the *generalized binomial coefficient* is given by

$$\binom{\alpha}{k} = \frac{(\alpha)_k}{k!} = \frac{\alpha\,(\alpha-1)\,(\alpha-2)\cdots(\alpha-k+1)}{k!},\quad \binom{\alpha}{0} = 1 \tag{3.26}$$

Choosing $\alpha = -n$ in (3.26), we get

$$\binom{-n}{k} = \frac{(-n)_k}{k!} = \frac{(-n)(-n-1)\cdots(-n-k+1)}{k!}$$
$$= \frac{(-1)^k (n)(n+1)\cdots(n+k-1)}{k!}$$
$$= (-1)^k \binom{n+k-1}{k}$$

This calculation confirms that (3.24) holds for the generalized binomial coefficient $\binom{-n}{k}$.

Many, indeed most, of the properties of the binomial coefficients we have encountered so far continue to hold for the generalized binomial coefficients. The two identities in the next theorem are representative.

Theorem 3.15 For any real or complex number α and any integer $k \geq 1$,

$$\binom{\alpha}{k} = \binom{\alpha-1}{k} + \binom{\alpha-1}{k-1} \tag{3.27}$$

and

$$k\binom{\alpha}{k} = \alpha\binom{\alpha-1}{k-1} \tag{3.28}$$

Proof. Both of the identities follow by a simple algebraic computation. However, an even better proof is to note that all of the terms are polynomials in the variable α. Since we have already shown that the identities hold when α is any positive integer, the polynomial equations hold for infinitely many values of α, and therefore the equations hold for all values of the real or complex variable α. ■

If we let $f_\alpha(x) = \sum_{k\geq 0}\binom{\alpha}{k}x^k$ be the OGF of the generalized binomial coefficients, then identity (3.27) gives us the relation

$$f_\alpha(x) = (1+x)f_{\alpha-1}(x) \tag{3.29}$$

which follows by exactly the same calculation used in the previous section to derive (3.12). However, α is not necessarily a positive integer, so iteration of (3.29) is not an option. Instead, if we take the derivative of the OGF and use identity (3.28), we find that

$$f'_\alpha(x) = \sum_{k\geq 1} k\binom{\alpha}{k}x^{k-1} = \sum_{k\geq 1} \alpha\binom{\alpha-1}{k-1}x^{k-1} = \alpha f_{\alpha-1}(x) \tag{3.30}$$

Now set $g_\alpha(x) = (1+x)^{-\alpha} f_\alpha(x)$ and take its derivative. Using both (3.29) and (3.30), we see that

$$\begin{aligned} g'_\alpha(x) &= (1+x)^{-\alpha} f'_\alpha(x) - \alpha(1+x)^{-\alpha-1} f_\alpha(x) \\ &= (1+x)^{-\alpha-1}\left[\alpha(1+x)f_{\alpha-1}(x) - \alpha f_\alpha(x)\right] \\ &= 0 \end{aligned}$$

Thus, $g_\alpha(x)$ is a constant that can be evaluated by setting $x = 0$. Since $f_\alpha(0)\binom{\alpha}{0} = 1$, we see that $g_\alpha(0) = 1^{-\alpha} f_\alpha(0) = 1$ and therefore $g_\alpha(x) = (1+x)^{-\alpha} f_\alpha(x) = 1$, or, equivalently, $f_\alpha(x) = (1+x)^\alpha$. This proves the following theorem, made famous by Isaac Newton (1642–1727).

Theorem 3.16 (Newton's Binomial Series) For real or complex x and α,

$$(1+x)^\alpha = \sum_{k=0}^{\infty} \binom{\alpha}{k} x^k \tag{3.31}$$

where $\binom{\alpha}{k}$ are the generalized binomial coefficients given by (3.26).

One useful application of the binomial series is the expansion of square roots, where either $\alpha = \frac{1}{2}$ or $\alpha = -\frac{1}{2}$. We first calculate

$$\begin{aligned} \binom{-\frac{1}{2}}{k} &= \frac{\left(-\frac{1}{2}\right)\left(-\frac{1}{2}-1\right)\left(-\frac{1}{2}-2\right)\cdots\left(-\frac{1}{2}-k+1\right)}{k!} \\ &= \frac{(-1)^k\left(\frac{1}{2}\right)\left(\frac{3}{2}\right)\left(\frac{5}{2}\right)\cdots[(2k-1)/2]}{k!} \\ &= \frac{(-1)^k\, 1\cdot 3\cdot 5\cdot\cdots\cdot(2k-1)}{2^k k!} \\ &= (-1)^k\frac{1\cdot 2\cdot 3\cdot 4\cdot 5\cdot\cdots\cdot(2k-1)\cdot(2k)}{2^k k! 2^k k!} \\ &= \frac{(-1)^k}{2^{2k}}\binom{2k}{k} \end{aligned} \tag{3.32}$$

The coefficient corresponding to $\alpha = \frac{1}{2}$ can be calculated in the same way, but it is simpler to use (3.28); when $k \geq 1$, we obtain

$$\binom{\frac{1}{2}}{k} = \frac{1}{2k}\binom{-\frac{1}{2}}{k-1} = \frac{1}{2k}(-1)^{k-1}\frac{\binom{2k-2}{k-1}}{2^{2k-2}} = (-1)^{k-1}\frac{\binom{2k-2}{k-1}}{k2^{2k-1}} \tag{3.33}$$

The binomial series (3.31) in the respective cases $\alpha = -\frac{1}{2}$ and $\alpha = \frac{1}{2}$ give us the expansions

$$\frac{1}{\sqrt{1+x}} = \sum_{k=0}^{\infty} \frac{(-1)^k}{2^{2k}}\binom{2k}{k}x^k \tag{3.34}$$

and

$$\sqrt{1+x} = 1 + \sum_{k=1}^{\infty} \frac{(-1)^{k-1}}{k2^{2k-1}}\binom{2k-2}{k-1}x^k \tag{3.35}$$

PROBLEM

3.3.1. Determine the generating functions of these sequences.

(a) $a_n = 2^n$, $n = 0, 1, 2, \ldots$

(b) $b_n = 3(-4)^n$, $n = 0, 1, 2, \ldots$

(c) $c_n = 2^n - 3(-4)^n$, $n = 0, 1, 2, \ldots$

3.3.2. Determine the nth term a_n of the sequence with OGF $3/[(1-x)(1+2x)]$. [*Hint*: First rewrite the rational function in its partial fraction decomposition.]

3.3.3. What sequences are generated by these OGFs?

(a) $f(x) = \dfrac{1}{1-3x}$ **(b)** $g(x) = \dfrac{1}{2x-2}$ **(c)** $h(x) = \dfrac{1+x}{2-8x+6x^2}$

[*Hint*: Parts (a) and (b) will be helpful for part (c) once $h(x)$ is written in partial fraction form.]

3.3.4. Determine the OGF of the sequence $0, \frac{1}{1}, \frac{1}{2}, \frac{1}{3}, \ldots, \frac{1}{n}, \ldots$.

3.3.5. **(a)** Use differentiation to show that

$$\frac{x}{(1-x)^2} = \sum_{k=0}^{\infty} kx^k$$

assuming that you already know the geometric series expansion given by (3.2).

(b) What sequence has the OGF $x^3/(1-5x)^2$?

(c) What is the OGF of the sequence $0, 6, 24, 72, \ldots, 3n2^n, \ldots$?

3.3.6. Let

$$f(x) = \frac{5x - 4x^2}{(1+x)(1-2x)^2}$$

(a) Verify that

$$f(x) = \frac{1}{(1-2x)^2} - \frac{1}{1+x}$$

(b) What sequence a_k is generated by the OGF $f(x)$?

3.3.7. The *Fibonacci Sequence OGF* is obtained as follows. Suppose that $x^2 = 1 + x$ has the roots φ and $\hat{\varphi}$, so that $x^2 - x - 1 = (x - \varphi)(x - \hat{\varphi})$.

(a) Show that $\varphi + \hat{\varphi} = 1$ and $\varphi\hat{\varphi} = -1$.
[*Hint:* Equate coefficients of x.]

(b) Show that

$$\frac{x}{1-x-x^2} = \frac{1}{\varphi - \hat{\varphi}}\left(\frac{1}{1-\varphi x} - \frac{1}{1-\hat{\varphi}x}\right)$$

(c) Show that

$$\frac{x}{1-x-x^2} = \sum_{n=0}^{\infty} \widehat{F}_n x^n, \quad \text{where} \quad \widehat{F}_n = \frac{\varphi^n - \hat{\varphi}^n}{\varphi - \hat{\varphi}}$$

(d) Verify that $\widehat{F_n} = F_n$ is the nth Fibonacci number, by showing that

$$\widehat{F}_0 = 0, \widehat{F}_1 = 1 \quad \text{and} \quad \widehat{F}_n = \widehat{F}_{n-1} + \widehat{F}_{n-2}$$

[*Hint*: $\varphi^2 = \varphi + 1$ and $\hat{\varphi}^2 = \hat{\varphi} + 1$.]

(e) Show that

$$\widehat{F}_n = F_n = \frac{1}{\sqrt{5}}\left(\left(\frac{1+\sqrt{5}}{2}\right)^n - \left(\frac{1-\sqrt{5}}{2}\right)^n\right)$$

where F_n is the nth Fibonacci number.
[*Note*: We have shown that the OGF of the Fibonacci sequence is $x/(1-x-x^2) = \sum_{n=0}^{\infty} F_n x^n$. The formula for the Fibonacci numbers given in part (e) is known as *Binet's formula*.]

(f) Show that $F_n = \{\varphi^n/\sqrt{5}\}$, $n \geq 1$, where $\{x\}$ is the nearest integer to the x function.
[*Hint*: $|\hat{\varphi}| = (\sqrt{5}-1)/2 = 0.618\ldots < 1$ and $\sqrt{5} > 2$.]

3.3.8. The *Lucas sequence* is defined by $L_0 = 2, L_1 = 1$, and $L_n = L_{n-1} + L_{n-2}$, $n \geq 2$. Given $\varphi = (1+\sqrt{5})/2$ and $\hat{\varphi} = (1-\sqrt{5})/2$ as the roots of $x^2 = 1 + x$ as seen in Problem 3.3.7, show that

(a) $L_n = \varphi^n + \hat{\varphi}^n$ (Binet's formula for the Lucas numbers)

(b) $\frac{2-x}{1-x-x^2} = \sum_{n=0}^{\infty} L_n x^n$ (the OGF of the Lucas numbers)

(c) $L_n = \{\varphi^n\}$, $n \geq 2$, where $\{x\}$ is the nearest integer to x function.
[*Hint*: $|\hat{\varphi}| = (\sqrt{5}-1)/2 = 0.618\ldots < 0.7$.]

(d) use a calculator and part (c) to evaluate L_{10} and L_{11}.

3.3.9. Use differentiation to give an induction proof of $(1-x)^{-n} = \sum_{k\geq 0} \binom{k+n-1}{n-1} x^k$, assuming that you already know the geometric series $(1-x)^{-1} = \sum_{k\geq 0} x^k$, which is the base case of the induction.

3.3.10. Use the *Cauchy product formula* $(\sum_{r\geq 0} a_r x^r)(\sum_{s\geq 0} b_s x^s) = \sum_{n\geq 0}(\sum_{r=0}^{n} a_r b_{n-r})x^n$ and a hockey stick identity to give an induction proof of formula

$$(1-x)^{-n} = \sum_{k\geq 0} \binom{k+n-1}{n-1} x^k$$

assuming that you have proved the geometric series $(1-x)^{-1} = \sum_{k\geq 0} x^k$, which is the base case of the induction.

3.3.11. Prove that

(a) $(-1)_n = (-1)^n\, n!$ **(b)** $(-\alpha)_n = (-1)^n\,(\alpha - n + 1)_n$

3.3.12. Evaluate these generalized binomial coefficients:

(a) $\binom{-3}{4}$ **(b)** $\binom{\frac{5}{2}}{3}$ **(c)** $\binom{\sqrt{2}}{2}$

3.3.13. Evaluate $\binom{\frac{1}{3}}{n}$.

3.3.14. Prove the identities of Theorem 3.15 by algebra.

3.3.15. Use four terms of the binomial series to derive the approximation $1/\sqrt{1.1} \doteq 0.9534375$. How many digits beyond the decimal point are correct?

3.3.16. Note that $\sqrt{2} = \sqrt{\frac{49}{25} \cdot \frac{50}{49}} = \frac{7}{5} \cdot \frac{1}{\sqrt{\frac{49}{50}}} = \frac{7}{5} \cdot \left(1 - \frac{1}{50}\right)^{-1/2}$. Use the first three terms of the binomial series to derive an approximate value of $\sqrt{2}$. How many digits beyond the decimal point are correct?

3.3.17. Prove that

$$\frac{1}{\sqrt{1-4x}} = \sum_{k\geq 0} \binom{2k}{k} x^k$$

3.3.18. Suppose that a bent coin has the probability x of landing heads and probability $1 - x$ of landing tails.

(a) Explain carefully why the probability that it requires $k + n$ flips of the coin to first obtain n tails is given by $\binom{k+n-1}{n-1} x^k (1-x)^n$.

(b) Use part (a) to give a probabilistic proof of

$$(1-x)^{-n} = \sum_{k\geq 0} \binom{k+n-1}{n-1} x^k$$

assuming that n tails is certain to occur eventually after sufficiently many flips.

3.4 ORDINARY GENERATING FUNCTIONS

Some sequences $a_0, a_1, \ldots, a_k, \ldots$ and their corresponding ordinary generating functions $f(x) = \sum_{k\geq 0} a_k x^k$ are shown in the following table:

Sequence	Ordinary Generating Function
Constant: $c, c, c, \ldots$	$\frac{c}{1-x} = \sum_{k\geq 0} cx^k$
Geometric: $c, c\alpha, c\alpha^2, \ldots, c\alpha^k, \ldots$	$\frac{c}{1-\alpha x} = \sum_{k\geq 0} c\alpha^k x^k$
Binomial coefficients:	
$\binom{n}{0}, \binom{n}{1}, \ldots, \binom{n}{k}, \ldots, \binom{n}{n}, 0, 0, \ldots$	$(1+x)^n = \sum_{k\geq 0} \binom{n}{k} x^k$
Multichoose coefficients, $n \geq 1$:	
$1, \left(\!\binom{n}{1}\!\right), \left(\!\binom{n}{2}\!\right), \ldots, \left(\!\binom{n}{k}\!\right), \ldots$	$\frac{1}{(1-x)^n} = \sum_{k\geq 0} \left(\!\binom{n}{k}\!\right) x^k$
or equivalently:	
$1, \binom{1+n-1}{n-1}, \binom{2+n-1}{n-1}, \ldots, \binom{k+n-1}{n-1}, \ldots$	$= \sum_{k\geq 0} \binom{k+n-1}{n-1} x^k$

In this section, we will obtain the OGFs of other sequences, show that the product of two or more OGFs can be interpreted as another OGF, and then illustrate how OGFs are used in counting. The sequences and OGFs in the preceding table will be helpful for both purposes. In later chapters, we will see how OGFs are used to solve recursion relations and prove other combinatorial identities.

3.4.1 Deriving Ordinary Generating Functions

Often, a new generating function of interest can be obtained by operations performed on an OGF that has already been calculated. For example, since we already know that $1/(1-x) = 1 + x + x^2 + \cdots$, a multiplication by x^{k+1} shows that $x^{k+1}/(1-x) = x^{k+1} + x^{k+2} + \cdots$. Subtracting one OGF from the other, we see that

$$1 + x + x^2 + \cdots + x^k = \frac{1 - x^{k+1}}{1 - x} \tag{3.36}$$

Similarly, we see that

$$x^n \frac{1}{(1-x)^{n+1}} = x^n \sum_{j \geq 0} \binom{j+n}{n} x^j = \sum_{j \geq 0} \binom{j+n}{n} x^{j+n} = \sum_{k \geq n} \binom{k}{n} x^k$$

which gives the following result.

Theorem 3.17 For all $n \geq 0$,

$$\frac{x^n}{(1-x)^{n+1}} = \sum_{k \geq n} \binom{k}{n} x^k \tag{3.37}$$

We can also use simple algebra to derive the OGF of the Fibonacci sequence:

$$f_F(x) = \sum_{k=0}^{\infty} F_k x^k = 0 + x + x^2 + 2x^3 + 3x^4 + 5x^5 + \cdots$$

Theorem 3.18 The ordinary generating function of the Fibonacci numbers is

$$f_F(x) = \frac{x}{1 - x - x^2} \tag{3.38}$$

Proof. Since $F_{k+2} = F_{k+1} + F_k$, we see that $\sum_{k=0}^{\infty} F_{k+2} x^{k+2} = x \sum_{k=0}^{\infty} F_{k+1} x^{k+1} + x^2 \sum_{k=0}^{\infty} F_k x^k$; that is, $f_F(x) - x = x f_F(x) + x^2 f_F(x)$, which gives (3.38) by solving for $f_F(x)$. ■

Similarly, a simple algebraic observation can be used to derive the OGF of the sequence of square numbers.

Theorem 3.19 The OGF of the sequence of squares $0,\ 1^2, 2^2, \ldots, k^2, \ldots$ is

$$\sum_{k\geq 0} k^2 x^k = \frac{x\,(1+x)}{(1-x)^3} \tag{3.39}$$

Proof. Note that

$$k^2 = k\,(k-1) + k = 2\binom{k}{2} + \binom{k}{1} \tag{3.40}$$

Then, using (3.37), we see that

$$\begin{aligned}\sum_{k\geq 0} k^2 x^k &= 2\sum_{k\geq 2}\binom{k}{2}x^k + \sum_{k\geq 1}\binom{k}{1}x^k \\ &= 2\frac{x^2}{(1-x)^3} + \frac{x}{(1-x)^2} = \frac{2x^2 + x - x^2}{(1-x)^3} = \frac{x\,(1+x)}{(1-x)^3}\end{aligned}$$

■

We have just derived the OGF of the sequence of square numbers, but the next example shows that the same method of calculation gives us the OGF of any of polygonal number. The polygonal numbers $p_k^{(r)}$ were first encountered in Section 1.4, where we showed that

$$p_k^{(r)} = k + (r-2)\,t_{k-1},\ k \geq 1 \tag{3.41}$$

where $t_{k-1} = \binom{k}{2}$ is the $(k-1)^{\text{st}}$ triangular number.

Example 3.20 Show that the OGF of the r-gonal numbers $p_k^{(r)}$ is

$$f^{(r)}\,(x) = \frac{x + (r-3)\,x^2}{(1-x)^3}. \tag{3.42}$$

Solution. In view of (3.41), we see from (3.37) that

$$\begin{aligned}f^{(r)}\,(x) &= \sum_{k\geq 1} p_k^{(r)} x^k = \sum_{k\geq 1}\binom{k}{1}x^k + (r-2)\sum_{k\geq 2}\binom{k}{2}x^k \\ &= \frac{x}{(1-x)^2} + (r-2)\,\frac{x^2}{(1-x)^3} = \frac{x + (r-3)\,x^2}{(1-x)^3}.\end{aligned}$$

As a check, we note that (3.42) agrees with the (3.39) when $r = 4$. ■

3.4.2 Products of Ordinary Generating Functions

The Cauchy product of two power series, given by $(\sum_{r\geq 0} a_r x^r)(\sum_{s\geq 0} b_s x^s) = \sum_{n\geq 0}(\sum_{r=0}^{n} a_r b_{n-r})x^n$, is another power series, so it follows that the product of ordinary generating functions is another ordinary generating function. Thus, we obtain the following result.

Theorem 3.21 Let the sequences $a_0, a_1, \ldots, a_r, \ldots$ and $b_0, b_1, \ldots, b_s, \ldots$ have the respective OGFs

$$f_A(x) = a_0 + a_1 x + a_2 x^2 + \cdots = \sum_{r\geq 0} a_r x^r \quad \text{and}$$

$$f_B(x) = b_0 + b_1 x + b_2 x^2 + \cdots = \sum_{s\geq 0} b_s x^s$$

Then their Cauchy product

$$h(x) = f_A(x) f_B(x) = a_0 b_0 + \left(a_0 b_1 + a_1 b_0\right) x + \left(a_0 b_2 + a_1 b_1 + a_2 b_0\right) x^2 + \cdots = \sum_{k\geq 0} h_k x^k \tag{3.43}$$

is the OGF of the sequence

$$h_k = a_0 b_k + a_1 b_{k-1} + \cdots + a_k b_0 = \sum_{r=0}^{k} a_r b_{k-r} \tag{3.44}$$

The sequence with the terms $h_k = a_0 b_k + a_1 b_{k-1} + \cdots + a_k b_0$ is called the *convolution product* of the sequences $a_0, a_1, \ldots, a_r, \ldots$ and $b_0, b_1, \ldots, b_s, \ldots$. Convolution products arise frequently in combinatorial formulas.

As an example of Theorem 3.21, suppose that $f_A(x) = (1+x)^\alpha = \sum_{r\geq 0} \binom{\alpha}{r} x^r$ and $f_B(x) = (1+x)^\beta = \sum_{s\geq 0} \binom{\beta}{s} x^s$. Then $\sum_{r=0}^{k} \binom{\alpha}{r}\binom{\beta}{k-r}$ is the kth coefficient of $f_A(x) f_B(x) = (1+x)^\alpha (1+x)^\beta = (1+x)^{\alpha+\beta}$. But we also know that this coefficient is $\binom{\alpha+\beta}{k}$. This gives us

$$\binom{\alpha+\beta}{k} = \sum_{r=0}^{k} \binom{\alpha}{r}\binom{\beta}{k-r} \tag{3.45}$$

which extends identity (3.13) to the generalized binomial coefficients.

It is especially interesting to set $f_B(x) = 1 + x + x^2 + \cdots = 1/(1-x)$ in Theorem 3.21, since then $b_s = 1$ for all s. This special case gives us the following very useful result.

Theorem 3.22 Let $f(x) = a_0 + a_1x + a_2x^2 + \cdots = \sum_{k\geq 0} a_kx^k$ be the OGF for the sequence $a_0, a_1, \ldots, a_k, \ldots$. Then the OGF for the sequence of partial sums $a_0, a_0 + a_1, a_0 + a_1 + a_2, \ldots, a_0 + a_1 + \cdots + a_k, \ldots$ is $[1/(1-x)]f_A(x)$; that is

$$\frac{1}{1-x}f(x) = a_0 + \left(a_0 + a_1\right)x + \left(a_0 + a_1 + a_2\right)x^2 + \cdots + \left(a_0 + a_1 + \cdots + a_k\right)x^k + \cdots \tag{3.46}$$

Example 3.23 Prove that

$$1^1 + 2^2 + \cdots + k^2 = \frac{k(k+1)(2k+1)}{6} \tag{3.47}$$

Solution. We have already shown in (3.39) that $\sum_{k\geq 0} k^2x^k = x(1+x)/(1-x)^3$. Now let $s_k = 1^1 + 2^2 + \cdots + k^2$ denote the partial sum sequence, so that it follows from Theorem 3.22 that

$$\sum_{k\geq 0} s_kx^k = \frac{1}{(1-x)}\frac{x(1+x)}{(1-x)^3}$$

Thus

$$s_k = 1^1 + 2^2 + \cdots + k^2 = \left[x^k\right]\frac{x+x^2}{(1-x)^4} = \left[x^k\right]\left(x+x^2\right)\sum_{m\geq 0}\binom{m+3}{3}x^m$$
$$= \binom{k+2}{3} + \binom{k+1}{3} = \frac{k(k+1)(2k+1)}{6}. \qquad ■$$

3.4.3 Counting with Ordinary Generating Functions

Suppose that we are given a counting problem whose solution we seek is denoted by h_k. The problem is solved if we can construct the generating function $h(x)$ for the sequence h_k and then calculate its kth coefficient, $h_k = \left[x^k\right]h(x)$.

Often, the number h_k is the result of selecting k objects from a set or multiset, where the objects can be of various types, say, type A and type B. The k objects might be a set of tiles where the tiles can be chosen from types that depend on shape or color. Or we may be choosing pieces of fruit to pack in a basket, where different types of fruit are available, or we may contributing k dollars to different charities. Specifically, suppose that there are a_r ways to choose r objects of type A and b_s ways to choose the remaining s objects of type B, where $k = r + s$. Then the total number of ways to select k objects is

$$h_k = \sum_{\substack{r,s\geq 0\\ r+s=k}} a_r b_s = \sum_{r=0}^{k} a_r b_{k-r}$$

But h_k is the kth coefficient of the Cauchy product of the series $f_A(x) = \sum_{r\geq 0} a_r x^r$ and $f_B(x) = \sum_{s\geq 0} b_s x^s$. Thus, the generating function $h(x)$ for the number of selections from two types of objects is

$$h(x) = f_A(x) f_B(x)$$

This formula shows us that we can count by the algebraic procedure of multiplying two OGFs and then determining the relevant coefficient of the product.

The following examples demonstrate this method of solution.

Example 3.24 Isabel found a sheet of ten 4¢ stamps and another sheet of seven 6¢ stamps. In how many ways can she use a combination of these stamps to put exactly 46¢ of postage on a letter?

Solution. The possible amounts of postage using the 4¢ stamps are 0, 4, 8, 12,. . ., 40, which gives the OGF $f_A(x) = 1 + x^4 + x^8 + \cdots + x^{40}$. Similarly, the sheet of 6¢ stamps corresponds to the OGF $f_B(x) = 1 + x^6 + x^{12} + \cdots + x^{42}$. Using a computer algebra system (CAS), we can show that $[x^{46}] f_A(x) f_B(x) = 4$. For example, in *Maple*, we use these commands:

```
fA:=sum(x^(4*j),j=0..10);
fB:=sum(x^(6*k),k=0..7);
coeff(fA*fB,x^46);
```

The computer calculation shows that there are 4 ways to place 46¢ of postage on Isabel's letter. ■

Example 3.25 The garden shop sells both square and round concrete blocks. The round blocks are \$2 each, and the store has an ample supply of red and gray round blocks. The square blocks are \$3 each, and the store has just four square blocks in stock. In how many ways can \$15 be spent on blocks?

Solution. Our goal is to count the number of ways to spend \$15 on blocks. Suppose that r dollars are spent on round blocks and s dollars are spent on square blocks. If i

round blocks are purchased, this will cost $r = 2i$ dollars, and there are $\left(\binom{2}{i}\right) = i + 1$ ways to choose the i blocks from the two colors available, red and gray. This gives us the following OGF for the number of ways to spend r dollars on round blocks:

$$f_R(x) = 1 + 2x^2 + 3x^4 + 4x^6 + \cdots \tag{3.48}$$

Next we will purchase square blocks. If j square blocks are purchased, this will add $s = 3j$ dollars to the total purchase price. Since only four square blocks are available, the OGF for the ways to spend s dollars on square blocks is

$$f_S(x) = 1 + x^3 + x^6 + x^9 + x^{12} \tag{3.49}$$

Now multiply the two generating functions to obtain

$$\begin{aligned} h(x) &= f_R(x) f_S(x) \\ &= \left(1 + 2x^2 + 3x^4 + 4x^6 + \cdots\right)\left(1 + x^3 + x^6 + x^9 + x^{12}\right) \\ &= 1 + 2x^2 + x^3 + 2x^5 + 5x^6 + \cdots + 11x^{15} + \cdots \end{aligned} \tag{3.50}$$

We know that $[x^k]\, h(x)$ gives the number of ways to spend k dollars on blocks. For example, there is $1 = [x^0]\, h(x)$ way to spend \$0 (do nothing!) and $2 = [x^2]\, h(x)$ ways to spend \$2 (buy either a red or a gray round block). There are $\left[x^6\right] h(x) = 5$ ways to spend \$6 buying blocks, namely, SS, 3R, 2R+1G, 1R+2G, 3G, where S denotes a square block and R and G are, respectively, a red and a gray round block. Finally, we see that all \$15 can be spent in $\left[x^{15}\right] h(x) = 11$ ways: S+6R, S+6G, S+5R+G, S+4R+2G, S+3R+3G, S+2R+4G, S+R+5G, 3S+3R, 3S+2R+G, 3S+R+2G, 3S+3G. ■

The following theorem summarizes how we use a product of two OGFs to count the number of ways to make selections where two types of objects can be selected.

Theorem 3.26 Suppose that k objects are to be selected, where r objects of type A can be selected in a_r ways and the remaining $s = k - r$ objects of type B can be selected in b_s ways. Let $f_A(x) = \sum_{r \ge 0} a_r x^r$ and $f_B(x) = \sum_{s \ge 0} b_s x^s$ be the respective ordinary generating functions for the sequences $a_0, a_1, \ldots, a_r, \ldots$ and $b_0, b_1, \ldots, b_s, \ldots$. Then the ordinary generating function for the number of ways, h_k, to select k objects is

$$h(x) = f_A(x) f_B(x) \tag{3.51}$$

In particular, we obtain

$$h_k = \left[x^k\right] f_A(x) f_B(x) = \sum_{r=0}^{k} a_r b_{k-r} \tag{3.52}$$

The following application of Theorem 3.26 demonstrates some computational techniques that are frequently needed to calculate the coefficients of the products of OGFs.

Example 3.27 The Math Club has k dollars to spend on posters from a company that offers large posters for \$2 each and small posters for \$1 each that come in two different colors. In how many ways can they submit their order for posters, where they want at least one of the large posters?

Solution. The possible amounts spent on large posters is 2, 4, 6, ..., which corresponds to the OGF

$$f_A(x) = x^2 + x^4 + x^6 + \cdots = \frac{x^2}{1-x^2}$$

There are also $\left(\!\binom{2}{s}\!\right) = s + 1$ ways to spend s dollars on the small posters, choosing them from the two colors available. This gives the OGF $f_B(x) = 1 + 2x + 3x^2 + 4x^3 + \cdots = 1/(1-x)^2$.

The number of ways to submit an order for posters that cost k dollars is therefore

$$h_k = \left[x^k\right] f_A(x) f_B(x) = \left[x^k\right] \frac{x^2}{\left(1-x^2\right)} \frac{1}{(1-x)^2} = \left[x^k\right] \frac{x^2}{(1+x)(1-x)^3}$$

With some rather tedious algebra (or using a CAS), we obtain the partial fraction decomposition

$$\frac{x^2}{(1+x)(1-x)^3} = \frac{1}{8(1+x)} + \frac{1}{8(1-x)} - \frac{3}{4(1-x)^2} + \frac{1}{2(1-x)^3}$$

Each summand in the partial fraction decomposition can be expanded by a power series, so our answer is given by

$$\begin{aligned} h_k &= \left[x^k\right] \left(\frac{1}{8} \sum_k (-1)^k x^k + \frac{1}{8} \sum_k x^k - \frac{3}{4} \sum_k (k+1) x^k + \frac{1}{2} \sum_k \frac{(k+2)(k+1)}{2} x^k \right) \\ &= \frac{(-1)^k + 1}{8} - \frac{3(k+1)}{4} + \frac{(k+2)(k+1)}{4} = \frac{(-1)^k - 1 + 2k^2}{8} \\ &= \begin{cases} m^2, & \text{if } k = 2m \\ m(m+1), & \text{if } k = 2m+1 \end{cases} \end{aligned}$$

Our algebraic calculations can be checked by noting that $h_0 = h_1 = 0, h_2 = 1, h_3 = 2$, and $h_4 = 4$. For example, here are the $h_4 = 4$ purchases that cost \$4:

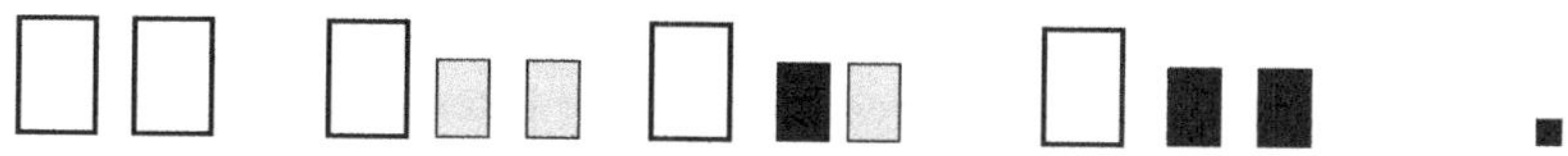

COMPUTER ALGEBRA NOTES

Some of the lengthy algebraic manipulation that accompanies the use of generating functions can be minimized or even avoided if *Maple*, *Mathematica*, *MatLab*, or some other CAS is available. For example, in *Maple*, the partial fraction decomposition required in Example 3.27 is executed with the command

```
convert(x^2/((1+x)*(1-x)^3),parfrac,x);
```

where the final semicolon is important in *Maple* syntax.

Similarly, Taylor series expansions up to a specified number of terms can be automated with a CAS. For example, the *Maple* command

```
series(x^2/((1+x)*(1-x)^3),x=0,8);
```

returns the series

$$x^2 + 3x^3 + 4x^4 + 6x^5 + 9x^6 + 12x^7 + O(x^8).$$

The last term $O(x^8)$ (read as "big oh" of x^8) indicates where the explicit calculation of coefficients has ended. A CAS can also calculate a single coefficient of interest, so the *Maple* command

```
coeff(series(x^2/((1+x)*(1-x)^3),x=0,21),x^20);
```

returns 100, the value of h_{20} in Example 3.27.

Many of these calculations can be carried out by WolframAlpha (http://www.wolframalpha.com/), which is freely available on the Internet.

Theorem 3.26 easily extends to products of any number of OGFs. For example, suppose that k objects are to be selected, where r objects of type A can be selected in a_r ways, s objects of type B can be selected in b_s ways, and so on. Then the number of ways to select k objects is

$$[x^k] f_A(x) f_B(x) \cdots f_D(x) = \sum_{\substack{r \geq 0, s \geq 0, \ldots, u \geq 0 \\ r+s+\cdots+u=k}} a_r b_s \cdots d_u \tag{3.53}$$

where $f_A(x), f_B(x), \ldots, f_D(x)$ are the OGFs for the ways to select objects of the different types.

The next examples each illustrate how (3.53) is used.

Example 3.28 A fruit bowl contains five apples, five bananas, and three cantaloupes. In how many ways can seven pieces of fruit be taken from the bowl on a picnic if there are to be an odd number of apples, an even number of bananas, and either one or two cantaloupes?

Solution. Our multistep task is to select seven pieces of fruit for the picnic. The products of the respective OGFs for the numbers of apples, bananas, and cantaloupes that can be taken on the picnic is

$$\begin{aligned} f_A(x) f_B(x) f_C(x) &= (x + x^3 + x^5)(1 + x^2 + x^4)(x + x^2) \\ &= x^2 + x^3 + 2x^4 + 2x^5 + 3x^6 + 3x^7 + 2x^8 + 2x^9 + x^{10} + x^{11} \end{aligned}$$

We see that there are $[x^7] f_A(x) f_B(x) f_C(x) = 3$ combinations of the fruit selected under the given requirements. ■

Example 3.29 In how many ways can 30 identical tennis balls be distributed to six players so that each player is given at least three balls?

Solution. Each of the six players can be given the number of balls described by the same generating function $f(x) = x^3 + x^4 + x^5 + \cdots = x^3/(1 - x)$. The number of ways to distribute the balls is then described by the generating function $h(x) = [f(x)]^6 = x^{18}/(1 - x)^6$. Therefore, the number of distributions is

$$[x^{30}] \frac{x^{18}}{(1-x)^6} = [x^{12}] \frac{1}{(1-x)^6} = \left(\!\!\binom{6}{12}\!\!\right) = \binom{12+6-1}{12} = \binom{17}{12} = 6188$$

This answer can be expected, since once each of the six players is given the minimum of 3 balls apiece, the remaining 12 balls can be distributed in $\left(\!\binom{6}{12}\!\right)$ ways. ■

We have already noted that counting the number selections and distributions is equivalent to counting the number of solutions in integers of an equation. For instance, Example 3.28 is asking for the number of solutions of the equation $a + b + c = 7, a \in \{1, 3, 5\}, b \in \{0, 2, 4\}, c \in \{1, 2\}$, and Example 3.29 is asking for the number of solutions of $x_1 + x_2 + \cdots + x_6 = 30, x_j \geq 3, j = 1, 2, \ldots, 6$.

The following example illustrates some useful algebraic steps.

Example 3.30 How many solutions in integers exist for the equation $x_1 + x_2 + x_3 + x_4 = 15, 1 \leq x_j \leq 5,\ j = 1, 2, 3, 4$?

Solution. By (3.53), the answer is

$$[x^{15}]\left(x+x^2+\cdots+x^5\right)^4 = [x^{15}]\,x^4\left(1+x+\cdots+x^4\right)^4 = [x^{15}]\,x^4\left(\frac{1-x^5}{1-x}\right)^4$$
$$= [x^{11}]\,\frac{\left(1-x^5\right)^4}{(1-x)^4}$$

But $\left(1-x^5\right)^4 = 1-\binom{4}{1}x^5+\binom{4}{2}x^{10}-\binom{4}{3}x^{15}+\binom{4}{4}x^{20}$ and $1/(1-x)^4 = \sum_{k\geq 0}\binom{k+3}{3}x^k$, so

$$[x^{11}]\left(1-x^5\right)^4\frac{1}{(1-x)^4} = [x^{11}]\left(1-\binom{4}{1}x^5+\binom{4}{2}x^{10}-\binom{4}{3}x^{15}+\binom{4}{4}x^{20}\right)$$
$$\times\sum_{k\geq 0}\binom{k+3}{3}x^k$$
$$= \binom{11+3}{3}-\binom{4}{1}\binom{6+3}{3}+\binom{4}{2}\binom{1+3}{3} = 52 \quad ■$$

Example 3.31 A coin is flipped 20 times, and has landed heads 13 times and tails the other 7 times. What is the probability that 5 successive heads, HHHHH, never occurred?

Solution. There are $\binom{20}{7} = 77{,}520$ ways for seven tails to occur among the 20 flips of the coin. Now we need to count the number of ways that no string of five consecutive heads occurs. Let x_i be the number of heads either before, between, or after the 7 tails, as shown in this diagram:

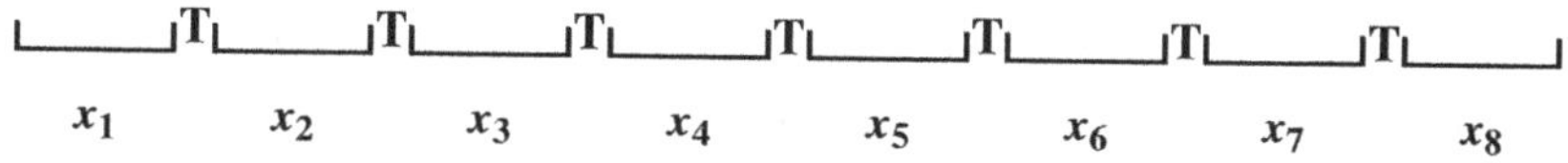

We then have the equation $x_1+x_2+\cdots+x_8 = 13,\ 0\leq x_i\leq 4$, which has the number of solutions given by

$$[x^{13}]\left(1+x+x^2+x^3+x^4\right)^8 = [x^{13}]\left(\frac{1-x^5}{1-x}\right)^8$$
$$= [x^{13}]\left(1-\binom{8}{1}x^5+\binom{8}{2}x^{10}+\cdots\right)\sum_{k\geq 0}\binom{k+7}{7}x^k$$
$$= \binom{13+7}{7}-\binom{8}{1}\binom{8+7}{7}+\binom{8}{2}\binom{3+7}{7} = 29{,}400$$

The probability is therefore $\frac{29{,}400}{77{,}520} \doteq 0.38$. Or course, this also means that there was a probability of about 0.62 that five or more consecutive heads occurred at least once. ■

The next example makes an important point: look for ways to simplify the OGF. This may turn a complicated problem and computation into a simple calculation.

Example 3.32 Kerry is coloring eggs so that an odd number are dyed blue, either one or two are dyed red, and any number are dyed green. How many ways can Kerry color a set of k identical eggs?

Solution. We must calculate $\left[x^k\right] h(x)$, where $h(x) = (x + x^3 + x^5 + \cdots)\left(x + x^2\right)(1 + x + x^2 + \cdots)$ is the generating function given by the product of OGFs corresponding to the ways to color eggs blue, red, or green. But then

$$h(x) = \frac{x}{1 - x^2} x (1 + x) \frac{1}{1 - x} = \frac{x^2}{(1 - x)^2} = \sum_{k \geq 0} \binom{k+1}{1} x^{k+2}$$

We now see easily that $\left[x^k\right] h(x) = \binom{k-1}{1} = k - 1$ for $k \geq 1$. ■

PROBLEM

3.4.1. **(a)** Show that the OGF of the sequence of squares $1, 2^2, 3^2, \ldots, n^2, \ldots$ is $\sum_{k=1}^{\infty} k^2 x^k = (xD)^2\, 1/(1 - x) = (x + x^2)(1 - x)^3$, where $D = d/dx$.

(b) Show that the OGF for the sequence of cubes $1, 2^3, 3^3, \ldots, k^3, \ldots$ is

$$\sum_{k=0}^{\infty} k^3 x^k = (xD)^3 \frac{1}{1 - x} = \frac{x\left(x^2 + 4x + 1\right)}{(1 - x)^4}$$

3.4.2. **(a)** Use the OGF given in Problem 3.4.1(b) to derive the formula $1^3 + 2^3 + \cdots + n^3 = \binom{n+1}{2}^2$.

(b) The following diagram illustrates why $1^3 + 2^3 + 3^3 = (1 + 2 + 3)^2$. Note that 2^3 is represented by two 2×2 squares, one of which is broken into two halves, and 3^3 is represented by three 3×3 squares, and so on.

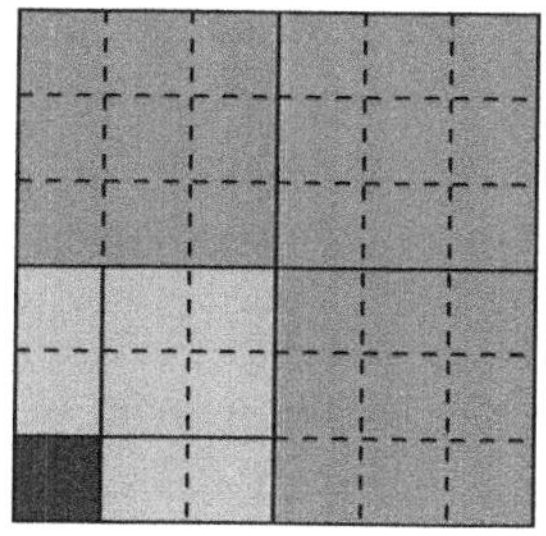

Extend the diagram to illustrate why $1^3 + 2^3 + 3^3 + 4^3 + 5^3 = (1 + 2 + 3 + 4 + 5)^2$.

3.4.3. **(a)** Verify that $k^3 - (k - 1)^3 = 3k^2 - 3k + 1$.

(b) Sum the result of part (a) over all $k = 1, 2, \ldots, n$ and obtain a new derivation of the identity $1^2 + 2^2 + \cdots + n^2 = [n(n + 1)(2n + 1)]/6$.

3.4.4. **(a)** Verify that $k^4 - (k - 1)^4 = 4k^3 - 6k^2 + 4k - 1$.

(b) Sum the result of part (a) over all $k = 1, 2, \ldots, n$ and derive the identity $1^3 + 2^3 + \cdots + n^3 = (1 + 2 + \cdots + n)^2$.

3.4.5. A *centered polygonal number* is the number of dots in a pattern of polygons that surround a central dot and continue to increase the number of dots along each edge of the subsequent polygons. For example, the sequence of centered hexagonal numbers begins $c_0^{(7)} = 1, c_1^{(7)} = 8, c_2^{(7)} = 22$, and $c_3^{(7)} = 43$.

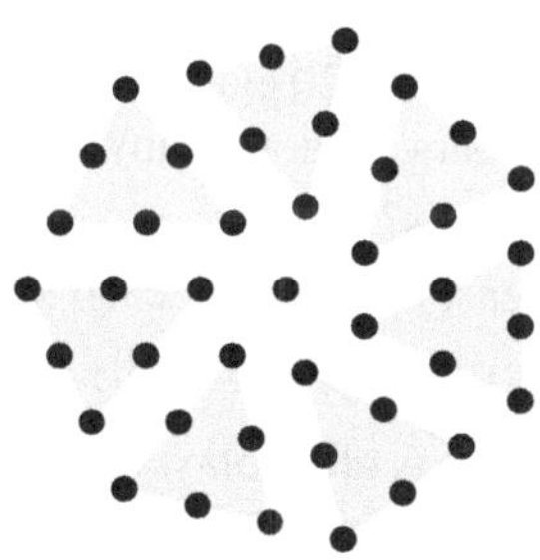

The shading shows that $c_3^{(7)} = 1 + 7t_3 = 1 + 7[(4 \cdot 3)/2]) = 43$, where $t_j = \frac{1}{2}(j + 1)j$ is the jth triangular number. More generally, the kth centered r-gonal number $c_k^{(r)}$ is given by $c_k^{(r)} = 1 + rt_k = 1 + r\binom{k+1}{2}, k \geq 0$.

(a) Show that the generating function for the sequence of centered r-gonal numbers is

$$g^{(r)}(x) = \frac{1 + (r - 2)x + x^2}{(1 - x)^3}$$

(b) Use the OGF obtained in (a) to verify that $c_4^{(10)} = 101$.

3.4.6. Create a combinatorial problem of the type stated in each part below that is answered by $[x^4](1+x)^3(1+x+x^2)^2 = 18$.

(a) A selection problem from a multiset

(b) A distribution problem of identical objects into distinct boxes

(c) The number of solutions in integers of some equation

3.4.7. Find and simplify the OGF for the number of solutions in integers of $x_1 + x_2 + x_3 = n,\ x_1 \geq 0,\ x_2 \geq 3,\ 4 \geq x_1 \geq 1$.

3.4.8. Find and simplify the OGF for the number of solutions in integers of $x_1 + x_2 + x_3 + x_4 + x_5 = n,\ x_i \geq 0,\ x_1$ even, and x_2 odd

3.4.9. **(a)** Determine the form of an OGF for the number of solutions of the equation $a_1 + a_2 + a_3 = n$, where a_1 is a positive square, a_2 is a positive cube, and a_3 is a prime number. Don't attempt to simplify your expression.

(b) Find the number of solutions when $n = 12$.

(c) Use a computer (CAS) to find the number of solutions when $n = 46$.

3.4.10. How many ways can n be written as an ordered sum of a prime number, an odd number, a positive square number, and a multiple of 3 when?

(a) $n = 15$? **(b)** $n = 24$?

3.4.11. Obtain, as a product of series, the generating functions for the number of solutions in integers of the equation $2x_1 + 4x_2 + 5x_3 = n, n \geq 0$, where one of these conditions holds:

(a) x_1, x_2, x_3 are nonnegative integers **(b)** $x_1 \geq 3, x_2 \geq 2, x_3 \geq 1$

3.4.12. Late in the day, a bakery has just 20 glazed, 7 chocolate, and 5 jelly doughnuts left, as well as 15 maple bars. In how many ways can you buy a dozen items

(a) with no other restrictions?

(b) if you must buy at least one item of each kind?

(c) if you must buy an even number of jelly doughnuts?

3.4.13. The Chinese game mahjong is often played with 144 tiles. There are 8 distinct tiles called *flowers* and *seasons*, and 4 copies each of 34 other kinds of tiles. A *hand* consists of 13 tiles.

(a) Write a generating function for the number of mahjong hands and then determine the number of hands using a CAS.

(b) How many hands do not contain either a flower or season? Give an answer in terms of binomial coefficients, and determine the numerical value with a CAS.

(c) Points are awarded for having either a pung—three of a kind—or a kong—four of a kind. What is the probability that a hand has at least one kong?

(d) How many hands contain neither a pung nor a kong?

3.4.14. You have large sheets of 23¢, 37¢, and 44¢, stamps. Can they be used to make exactly \$3.33 in postage? If so, in how many ways can the stamps be used? Use a computer algebra system (CAS) to find your answer. Find all, if any, of the ways to make the exact postage with the stamps.

3.4.15. Pólya [2] asked for the number of ways to make change of an American dollar using coins—pennies, nickels, dimes, quarters, and half-dollars.

(a) Give the OGF $h(x)$ for which the answer to Pólya's question is given by the expression $[x^{100}]h(x)$.

(b) If you have access to *Maple*, *Mathematica*, *MatLab*, or another CAS, show that there are 292 ways to make change for a dollar bill in coins of smaller denominations. For example, in *Maple*, use the commands

```
coeff(series(((1-x^5)*(1-x^10)*(1-x^25)*(1-
x^50)*(1-x))^(-1), (x=0,101)),x^100);
```

3.4.16. **(a)** Verify that the generating function for the number T_n of noncongruent triangles with integer sides and perimeter n is $x^3/[(1-x^2)(1-x^3)(1-x^4)]$.

[*Hint*: Suppose that the sides of the triangle are a, b, and c, where $1 \le a \le b \le c$ and $a + b > c$. Then obtain an equation in the integer variables $x_1 = b - a, x_2 = c - b, x_3 = a + b - c$].

(b) Use a computer algebra system to continue this table of values of T_n for $n \le 20$.

n	3	4	5	6	7	8	9	10	11
T_n	1	0	1	1	2	1	3	2	4

(c) Determine the four noncongruent triangles of integer sides with perimeter 11.

(d) Use the OGF found in part (a) to prove that $T_{2m} = T_{2m-3}$ for $m \ge 3$.

3.4.17. Give a new derivation of the closed form of the generating function

$$f_n(x) = \sum_{k\ge 0} \left(\!\binom{n}{k}\!\right) x^k$$

by using this model: let $f_n(x)$ be the generating function for the number of ways to buy k cones at an ice cream parlor that offers n flavors. Now justify these steps.

(a) If only vanilla is available, then $f_1(x) = 1/(1-x)$.

(b) If vanilla and n other flavors are available, then $f_{n+1}(x) = f_n(x) f_1(x)$.

(c) $f_n(x) = \sum_{k\ge 0} \left(\!\binom{n}{k}\!\right) x^k = \frac{1}{(1-x)^n}$

3.4.18. Starting with the OGF

$$f(x) = \frac{1}{(1-x)^{n+1}} = \sum_{j\geq 0} \binom{j+n}{j} x^j$$

prove the hockey stick identity:

$$\binom{n+r+1}{r} = \sum_{k=0}^{r} \binom{k+n}{k}$$

3.4.19. Starting with the OGF

$$g(x) = \frac{1}{(1-x)^{r+1}} = \sum_{k\geq 0} \binom{k+r}{r} x^k$$

prove the hockey stick identity:

$$\binom{n+1}{r+1} = \binom{r}{r} + \binom{r+1}{r} + \cdots + \binom{n}{r}$$

3.4.20. Mary was planting a row of n tulips bulbs. She had two bags of bulbs. Bag A contained red, yellow, and white bulbs, and bag B contained pink and purple bulbs. She decided to first use bulbs from bag A, and then finish the row with bulbs from bag B, although it wasn't essential that any bulbs from either bag were used.

(a) In how many ways can the bulbs have been planted if the row began with r bulbs from bag A?

(b) In how many ways could the row have been planted?

3.4.21. In Example 3.32, suppose that Kerry has six eggs to color, an odd number of which are to be dyed blue, either one or two to be dyed red, and the rest dyed green. What are his coloring choices?

3.4.22. **(a)** Create a problem that is solved by $a_n = [x^n]f(x)$, where $f(x) = (1 + x^2 + x^4)\left(x + x^3 + x^5 + \cdots\right)\left(1 + x^6 + x^{12} + x^{18} + \cdots\right)$.

(b) Determine a formula for a_n.

(c) Give a list of the a_{15} ways that answer the problem posed in part (a).

3.4.23. A $1 \times n$ board will be tiled with red or green unit squares at the left, then by 1×2 dominoes, and finished with unit squares that include exactly one black tile and the rest white tiles. One possible tiling of a board of length 10 is the RRGDDWBW tiling shown here:

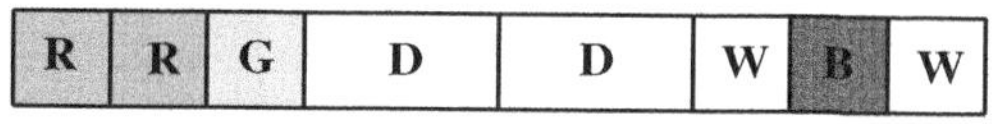

(a) In how many ways, h_n, can the board of length n be tiled?

(b) Verify your formula in part (a) by listing the 12 tilings of a board of length 3.

3.4.24. **(Sicherman Dice)** The OGF of a standard 6-faced die is $f(x) = x^1 + x^2 + \cdots + x^6$.

(a) Verify that the OGF of a roll of two standard dice corresponding to the sum of spots on the dice is $f^2(x) = x^2 + 2x^2 + 3x^4 + 4x^5 + 5x^6 + 6x^7 + 5x^8 + 4x^9 + 3x^{10} + 2x^{11} + x^{12}$.

(b) Verify the factorization $f(x) = x(1 + x)(1 + x + x^2)(1 - x + x^2)$:

(c) Show that $g(x) = x(1 + x)(1 + x + x^2)$ and $h(x) = x(1 + x)(1 + x + x^2)(1 - x + x^2)^2$ are the respective OGFs for these dice, known as the *Sicherman dice*.

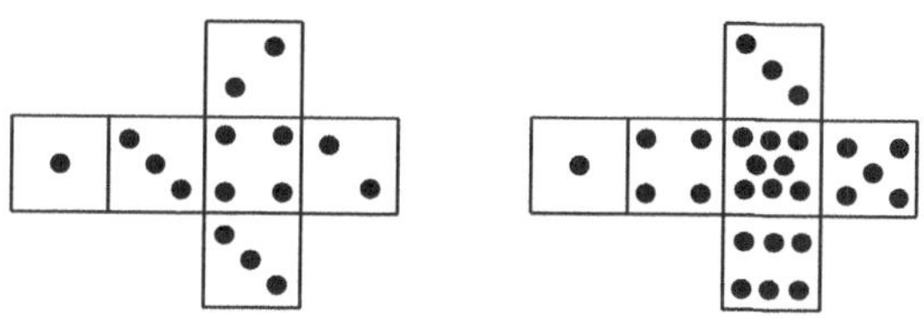

(d) Why is the frequency of outcomes of the Sicherman pair of dice the same as that of a pair of standard dice? (Sicherman dice were discovered by George Sicherman and were first described by Martin Gardner [3]. The numbers on the Sicherman dice can be arranged so the sums on opposite faces are 5 on one die and 9 on the other, as seen in the figure above.)

3.5 EXPONENTIAL GENERATING FUNCTIONS

The discussion of exponential generating functions (EGFs) that follows will parallel the development of OGFs in the preceding section. We will first enlarge our library of EGFs, next show why the product of EGFs is another EGF, and then investigate how EGFs are used to solve counting problems. Unlike OGFs, which are used primarily to count problems about selections and combinations, EGFs are used primarily to count arrangements and permutations where order is a consideration.

3.5.1 Deriving Exponential Generating Functions

So far we have calculated just the EGF for the constant sequence 1, 1, 1, … , namely

$$e^x = 1 + \frac{x}{1!} + \frac{x^2}{2!} + \frac{x^3}{3!} + \cdots + \frac{x^k}{k!} + \cdots \tag{3.54}$$

If x is replaced with αx, we would find that the EGF of the geometric sequence $1, \alpha, \alpha^2, \ldots, \alpha^k, \ldots$ is

$$e^{\alpha x} = \sum_{k\geq 0} \frac{(\alpha x)^k}{k!} = \sum_{k\geq 0} \alpha^k \frac{x^k}{k!} \tag{3.55}$$

In particular, if we set $\alpha = -1$ in (3.55), it follows that the EGF for the sequence 1, −1, 1, −1,… is

$$e^{-x} = 1 - \frac{x}{1!} + \frac{x^2}{2!} - \frac{x^3}{3!} + \cdots + (-1)^k \frac{x^k}{k!} + \cdots \tag{3.56}$$

If (3.54) is added to (3.56), and the sum is divided by 2, we see that the sequence 1,0,1,0,1,… has the EGF given by the hyperbolic cosine:

$$\cosh(x) = \frac{e^x + e^{-x}}{2} = 1 + \frac{x^2}{2!} + \frac{x^4}{4!} + \frac{x^6}{6!} + \cdots + \frac{x^{2k}}{(2k)!} + \cdots \tag{3.57}$$

Likewise, if we either differentiate (3.57) or subtract (3.56) from (3.54) and divide by 2, we see that the EGF for the sequence 0,1,0,1,… is the hyperbolic sine function:

$$\sinh(x) = \frac{e^x - e^{-x}}{2} = x + \frac{x^3}{3!} + \frac{x^5}{5!} + \frac{x^7}{7!} + \cdots + \frac{x^{2k+1}}{(2k+1)!} + \cdots \tag{3.58}$$

Example 3.33 For an arbitrary real or complex variable α, show that the sequence of Pochhammer symbols $(\alpha)_k = \alpha(\alpha-1)(\alpha-2)\cdots(\alpha-k+1)$ has the EGF

$$(1+x)^\alpha = \sum_{k\geq 0} (\alpha)_k \frac{x^k}{k!} \tag{3.59}$$

Solution. Recall that $\binom{\alpha}{k} = (\alpha)_k/k!$. Thus $\sum_{k\geq 0} (\alpha)_k (x^k/k!) = \sum_{k\geq 0} \binom{\alpha}{k} x^k = (1+x)^\alpha$,where the last step of the calculation is a consequence of Newton's binomial series. ■

When $\alpha = m$, a positive integer, we see that $(1+x)^m$ is the EGF for the permutations $(m)_n = m(m-1)\cdots(m-n+1) = P(m,n)$, whereas $(1+x)^m$ is the OGF for the combinations $C(m,n)$. This comparison shows a general distinction between ordinary and exponential generating functions; for problems involving combinations (unordered selections), use OGFs, and for problems involving permutations (ordered arrangements), use EGFs.

3.5.2 Products of Exponential Generating Functions

In Section 3.4, we showed that the product of any number of OGFs is another OGF. We will now see that a product of EGFs can also be viewed as an EGF. We begin with the simplest case in which two EGFs are multiplied by one another.

Theorem 3.34 Given two exponential generating functions $f^{(e)}(x) = \sum_{r=0}^{\infty} a_r \frac{x^r}{r!}$ and $g^{(e)}(x) = \sum_{s=0}^{\infty} b_s \frac{x^s}{s!}$ their product is the EGF given by

$$h^{(e)}(x) = f^{(e)}(x)\, g^{(e)}(x) = \sum_{k\geq 0} h_k \frac{x^k}{k!} \tag{3.60}$$

where

$$h_k = \left[\frac{x^k}{k!}\right] h^{(e)}(x) = \sum_{r=0}^{k} \binom{k}{r} a_r b_{k-r} \tag{3.61}$$

Proof.

$$\begin{aligned} h^{(e)}(x) &= f^{(e)}(x)\, g^{(e)}(x) = \left(\sum_{r=0}^{\infty} a_r \frac{x^r}{r!}\right)\left(\sum_{s=0}^{\infty} b_s \frac{x^s}{s!}\right) \\ &= \sum_{r=0}^{\infty}\sum_{s=0}^{\infty} a_r b_s \frac{x^{r+s}}{r!s!} = \sum_{k=0}^{\infty}\left(\sum_{r=0}^{k} a_r b_{k-r} \frac{k!}{r!\,(k-r)!}\right)\frac{x^k}{k!} \qquad (k = r+s) \\ &= \sum_{k=0}^{\infty} h_k \frac{x^k}{k!}, \quad \text{where} \quad h_k = \sum_{r=0}^{k} \binom{k}{r} a_r b_{k-r} \end{aligned}$$

■

As an application of Theorem 3.34, we will prove a property of the Pochhammer symbols $(\alpha)_r = \alpha(\alpha-1)(\alpha-2)\cdots(\alpha-r+1)$.

Theorem 3.35 For arbitrary real or complex numbers α and β,

$$(\alpha+\beta)_n = \sum_{r=0}^{n} \binom{n}{r} (\alpha)_r (\beta)_{n-r} \tag{3.62}$$

Proof. From (3.59), we have $f_\alpha^{(e)}(x) = (1+x)^\alpha = \sum_{r\geq 0} (\alpha)_r (x^r/r!)$ and $f_\beta^{(e)}(x) = (1+x)^\beta = \sum_{s\geq 0} (\beta)_s (x^s/s!)$. Thus

$$(\alpha+\beta)_n = \left[\frac{x^n}{n!}\right](1+x)^{\alpha+\beta} = \left[\frac{x^n}{n!}\right](1+x)^\alpha (1+x)^\beta$$
$$= \left[\frac{x^n}{n!}\right] f_\alpha^{(e)}(x) f_\beta^{(e)}(x) = \sum_{r=0}^{n} \binom{n}{r} (\alpha)_r (\beta)_{n-r}$$

■

In Chapter 2, we showed that the derangement number D_k is given by

$$\frac{D_k}{k!} = 1 - \frac{x}{1!} + \frac{x^2}{2!} - \frac{x^3}{3!} + \cdots + (-1)^k \frac{x^k}{k!} \tag{3.63}$$

This formula, together with Theorem 3.34, allows us to derive the EGF for the derangement numbers.

Theorem 3.36 The exponential generating function $D(x)$ for the derangement numbers is

$$D(x) = \sum_{k\geq 0} D_k \frac{x^k}{k!} = \frac{e^{-x}}{1-x} \tag{3.64}$$

Proof. The right side of (3.63) is the nth partial sum $1 - (x/1!) + (x^2/2!) - (x^3/3!) + \cdots + (-1)^k (x^k/k!)$ of the power series of the EGF $e^{-x} = \sum_{k=0}^{\infty} (-1)^k (x^k/k!)$. Since the power series with coefficients that are partial sums of another power series is obtained by dividing the series by $1-x$, we see that $D(x) = e^{-x}/(1-x)$. ■

We now generalize Theorem 3.34 to show how to view the product of any number of EGFs as yet another EGF.

Theorem 3.37 Given any number of EGFs $f_A^{(e)}(x) = \sum_{r\geq 0} a_r (x^r/r!)$, $f_B^{(e)}(x) = \sum_{s\geq 0} b_s (x^s/s!), \ldots, f_D^{(e)}(x) = \sum_{u\geq 0} d_u (x^u/u!)$, their product is the EGF given by

$$h^{(e)}(x) = f_A^{(e)} f_B^{(e)} \cdots f_D^{(e)} = \sum_{k\geq 0} h_k \frac{x^k}{k!} \tag{3.65}$$

where the coefficients are given by

$$h_k = \sum_{\substack{r,s,\ldots,u\geq 0\\ r+s+\cdots u=k}} \binom{k}{r,s,\ldots,u} a_r b_s \cdots d_u \tag{3.66}$$

Proof. Recall that the multinomial coefficient is given by

$$\binom{k}{r, s, \ldots, u} = \frac{k!}{r!s! \cdots u!}, \qquad r + s + \cdots + u = k$$

Therefore,

$$\begin{aligned} f_A^{(e)}(x) f_B^{(e)}(x) \cdots f_D^{(e)}(x) &= \left(\sum_{r \geq 0} a_r \frac{x^r}{r!} \right) \left(\sum_{s \geq 0} b_s \frac{x^s}{s!} \right) \cdots \left(\sum_{u \geq 0} d_u \frac{x^u}{u!} \right) \\ &= \sum_{r,s,\ldots,u \geq 0} a_r b_s \cdots d_u \frac{x^{r+s+\cdots+u}}{r!s! \cdots u!} \\ &= \sum_{k \geq 0} \left(\sum_{\substack{r,s,\ldots,u \geq 0 \\ r+s+\cdots+u=k}} a_r b_s \cdots d_u \frac{k!}{r!s! \cdots u!} \right) \frac{x^k}{k!} = \sum_{k \geq 0} h_k \frac{x^k}{k!} \end{aligned}$$

■

Theorem 3.35 can be used to extend identity (3.62).

Theorem 3.38 For any finite sequence of real or complex numbers $\alpha, \beta, \ldots, \delta$, we have

$$(\alpha + \beta + \cdots + \delta)_n = \sum_{\substack{r \geq 0, s \geq 0, \ldots, u \geq 0 \\ r+s+\cdots+u=n}} \binom{n}{r, s, \ldots, u} (\alpha)_r (\beta)_s \cdots (\delta)_u \qquad (3.67)$$

3.5.3 Counting with Exponential Generating Functions

In Section 3.4, we saw that OGFs could be used to count *selections* of a total of n objects, where there are various types of objects available. If there are a_r ways to choose r objects of type A, and b_s ways to choose another s objects of type B, and so on, then the number of ways to select an unordered combination of n objects is given by $[x^n] f_A(x) f_B(x) \cdots$, where $f_A(x), f_B(x), \ldots$ are the OGFs of the sequences a_r, $b_s, \ldots$. It is important to note that *combinations* are being counted with no attention given to order. For example, if we are selecting pieces of fruit for a picnic, what was important is the amount of fruit selected of each available type. The fruit selected is not put into any order.

Now suppose that we are counting *arrangements* of k objects, where again the objects can be chosen from various types that are available. Here we must first choose a subset of r positions in the overall arrangement that can be filled with objects of type A in a_r ways, and next we choose another subset of s of the remaining positions that can be filled with objects of a second type B in b_s ways, and so on. We will now show that the number of permutations is given by $[x^k/k!]f_A^{(e)}(x)f_B^{(e)}(x)\cdots$, where $f_A^{(e)}(x), f_B^{(e)}(x), \ldots$ are of the EGFs of the sequences $a_r, b_s, \ldots$.

To begin as simply as possible, suppose that there are only two types of objects available, A and B. We then have this theorem.

Theorem 3.39 Suppose that an arrangement is formed with a sequence of k objects, where two types of objects are available, A or B. Any subset of r positions in the sequence can be filled in a_r ways with objects of type A and any subset of s position can be filled in b_s ways with objects of type B. Let $f_A^{(e)}(x)$ and $f_B^{(e)}(x)$ be the respective EGFs for the sequences a_r and b_s. Then the EGF for the number of arrangements, h_k, of size k is

$$h^{(e)}(x) = f_A^{(e)}(x)f_B^{(e)}(x) \tag{3.68}$$

That is, the number of arrangements of size k is given by

$$h_k = \left[\frac{x^k}{k!}\right] f_A^{(e)}(x)f_B^{(e)}(x) \tag{3.69}$$

Proof. For $r = 0, 1, \ldots, k$, choose, in $\binom{k}{r}$ ways, r positions of the sequence to be filled with objects of type A, and then fill these positions in a_r ways. This leaves b_s ways to fill the remaining $s = k - r$ positions with objects of type B. Summing over all r shows that there are $h_k = \sum_{r=0}^{k} \binom{k}{r} a_r b_{k-r}$ ways to complete the arrangement with all k positions filled. But $\sum_{r=0}^{k} \binom{k}{r} a_r b_{k-r}$ is the coefficient of $x^k/k!$ in the product of the two EGFs $f_A^{(e)}(x)$ and $f_B^{(e)}(x)$ by Theorem 3.34. ■

The following example allows you to compare the use of OGFs and EGFs.

Example 3.40 Mischa is making a garden walkway from 10 stepping stones set in a row. She wants to have an even number of round stepping stones in both red and gray, and any number of square stones, where square stones come only in gray.

(a) In how many ways can she order the 10 stones to be used for the walkway?
(b) How many different walkways are possible?

Solution

(a) The stones being purchased form a combination, so OGFs are used to obtain the answer. If $2r$ round stones are ordered, and these are chosen from the two available colors, there are $a_{2r} = \left(\!\binom{2}{2r}\!\right) = 2r + 1$ ways to purchase the round stones. This corresponds to the OGF

$$f_A(x) = \sum_{r=0}^{\infty} (2r+1)x^{2r} = \frac{d}{dx}\sum_{r=0}^{\infty} x^{2r+1} = \frac{d}{dx}\frac{x}{1-x^2} = \frac{1+x^2}{\left(1-x^2\right)^2}$$

Since 10 is even, there must also be an even number of square stones ordered. The OGF for the even number of square stones is $f_B(x) = \sum_{s=0}^{\infty} x^{2s} = 1/(1-x^2)$. Therefore, 10 stones can be ordered in

$$\begin{aligned}\left[x^{10}\right] f_A(x) f_B(x) &= \left[x^{10}\right] \frac{1+x^2}{\left(1-x^2\right)^3} = \left[x^{10}\right](1+x^2)\sum_{n=0}^{\infty}\binom{n+2}{2}x^{2n} \\ &= \binom{7}{2} + \binom{6}{2} = 36\end{aligned}$$

ways.

(b) We now wish to count arrangements, so EGFs are used. If $2r$ round stones are used, there are 2^{2r} ways to lay the stones with a choice of red or gray in each position. This gives us the EGF for the round stones:

$$f_A^{(e)}(x) = \sum_{r=0}^{\infty} 2^{2r}\frac{x^{2r}}{(2r)!} = \sum_{r=0}^{\infty}\frac{(2x)^{2r}}{(2r)!} = \cosh 2x = \frac{e^{2x}+e^{-2x}}{2}$$

Similarly, the EGF for the even number of square stones is $f_B^{(e)}(x) = \cosh x = (e^x + e^{-x})/2$, so the number of 10-stone paths is

$$\begin{aligned}\left[\frac{x^{10}}{10!}\right] f_A^{(e)}(x) f_B^{(e)}(x) &= \left[\frac{x^{10}}{10!}\right]\left(\frac{e^{2x}+e^{-2x}}{2}\right)\left(\frac{e^{x}+e^{-x}}{2}\right) \\ &= \left[\frac{x^{10}}{10!}\right]\frac{e^{3x}+e^{x}+e^{-x}+e^{-3x}}{4} = \frac{3^{10}+1+1+3^{10}}{4} \\ &= \frac{3^{10}+1}{2}\end{aligned}$$

It shouldn't be surprising that the number of walkways far exceeds the number of purchase combinations. Unless all ten stones are identical, there are many arrangements of a combination that includes a variety of stones. ■

Example 3.41 The 10 members of the Combinatorics Club have lined up for their yearbook photo. They want to display at least one of their banners, which are identical and all say "You can count on us!" Each banner is held by two club members, one at each end, and no banner overlaps another. Those club members not holding an end of a banner will each hold either a white or a black pennant. In how many ways can banners and pennants be distributed to the n club members in the line? The figure below shows one way to distribute banners and pennants to the club members:

Solution. The banners are distributed to a positive even number of club members, so the EGF is

$$f_A^{(e)}(x) = \frac{x^2}{2!} + \frac{x^4}{4!} + \frac{x^6}{6!} + \cdots = \cosh x - 1 = \frac{e^x + e^{-x}}{2} - 1$$

If s club members are selected to hold a pennant, they can hold white and black pennants in 2^s ways. Therefore, the EGF for the arrangement of pennants is

$$f_B^{(e)}(x) = 1 + 2\frac{x}{1!} + 2^2\frac{x^2}{2!} + \cdots = e^{2x}$$

Under the given constraints, the number of distributions of at least one banner and pennants to the 10 club members in the line is

$$\begin{aligned} h_{10} &= \left[\frac{x^{10}}{10!}\right] f_A^{(e)}(x) f_B^{(e)}(x) = \left[\frac{x^{10}}{10!}\right] \left(\frac{e^x + e^{-x}}{2} - 1\right) e^{2x} \\ &= \left[\frac{x^{10}}{10!}\right] \left(\frac{1}{2}e^{3x} + \frac{1}{2}e^{x} - e^{2x}\right) = \frac{1}{2}\left(3^{10}\right) + \frac{1}{2} - 2^{10} = \frac{1}{2}\left(3^{10} + 1\right) - 2^{10} = 28{,}501 \end{aligned}$$

■

In the next example, recall that if $f(x) = a_0 + a_1x + a_2x^2 + \cdots,$ then $[1/(1-x)]f(x) = \sum_{n=0}^{\infty} s_n x^n$, where s_n is the partial sum $s_n = a_0 + a_1 + a_2 + \cdots + a_n$.

Example 3.42 In how many ways can six distinct places at a table be set with dishes taken from a stacks of identical yellow, red, and blue plates, and a fourth stack of handpainted plates that are all different? Only plates from the top of the stacks can be used:

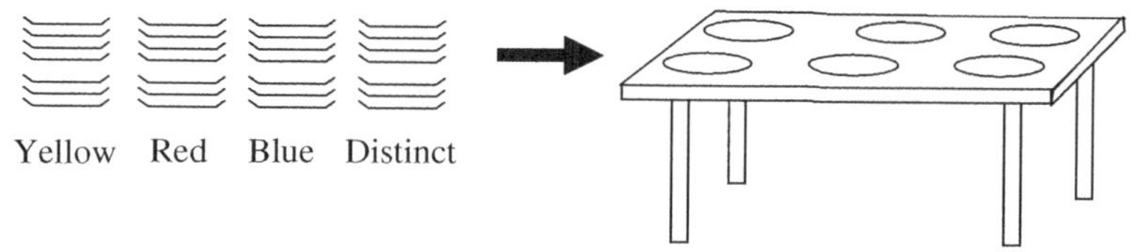

Solution. Two types of plates can be arranged:

Type A—*plain*: Choose r place settings that each get a yellow, red, or blue dish.

Type B—*handpainted*: Take the top s plates from the stack of distinct plates and place them at a subset of s seats of the table.

There are three choices of color at r selected seats, so there are 3^r ways to set these place settings. The corresponding EGF for the number of ways to distribute plain plates is $f_A^{(e)}(x) = \sum_{r\geq 0} 3^r \frac{x^r}{r!} = e^{3x}$. There are $s!$ ways to place s distinct handpainted dishes at s seats at the table, so the EGF corresponding to the number of ways to distribute s handpainted dishes from the top of the stack is

$$f_B^{(e)}(x) = \sum_{s\geq 0} s!\frac{x^s}{s!} = \sum_{s\geq 0} x^s = \frac{1}{1-x}$$

The number of table settings is therefore

$$\begin{aligned} h_6 &= \left[\frac{x^6}{6!}\right] f_A^{(e)}(x) f_B^{(e)}(x) = \left[\frac{x^6}{6!}\right]\left(e^{3x}\frac{1}{1-x}\right) \\ &= 6!\left(1 + \frac{3}{1!} + \frac{3^2}{2!} + \cdots + \frac{3^6}{6!}\right) = 13{,}977 \end{aligned}$$ ■

Theorem 3.37 allows us to consider any number of types of objects, and we have this extension of Theorem 3.39.

Theorem 3.43 Let $f_A^{(e)}(x), f_B^{(e)}(x), \ldots, f_D^{(e)}(x)$ be the respective EGFs for the sequences $a_r, b_s, \ldots, d_u$, where a subset of r positions can be filled in a_r ways with objects of type A, a subset of s positions can be filled in b_s ways with objects of type B, and so on. Then the number of permutations of k objects of types A, B, and so forth is

$$h_k = \left[\frac{x^k}{k!}\right] f_A^{(e)}(x) f_B^{(e)}(x) \cdots f_D^{(e)}(x) \tag{3.70}$$

Proof. Suppose that a permutation of k objects contains r objects of type A, s objects of type B,..., and u objects of type D, where $k = r + s + \cdots + u$. There are $\binom{k}{r,s,\ldots,u}$ permutations of the multiset $\{r \cdot A, s \cdot B, \ldots, u \cdot D\}$ that count the number of placements of the objects of each type. If there are a_r, b_s, ..., d_u ways to fill the designated positions with objects of respective types A, B, ... D, there are then $\binom{n}{r,s,\ldots,u} a_r b_s \cdots d_u$ permutations with r, s, ..., u objects of the respective types A, B, ..., D. Summing over all r, s, ..., u gives the total number of permutations as

$$h_k = \sum_{\substack{r\geq 0, s\geq 0,\ldots,u\geq 0 \\ r+s+\cdots+u=k}} \binom{k}{r, s, \ldots, u} a_r b_s \cdots d_u$$

By Theorem 3.37, h_k is given by (3.70). ■

Example 3.44 How many six-letter codewords (i.e., words or nonword letter combinations used as codes) can be formed under the given conditions.

(a) Any of the 26 letters A, B, ..., Z, can be used with no further restrictions.

(b) Only the letters A, B, C, and D are allowed, and each letter must appear at least once.

(c) Only the letters A, B and C are allowed, and each letter must appear an even number of times.

Solution

(a) Each of the 26 letters has the same EGF, namely $f_A^{(e)}(x) = f_B^{(e)}(x) = \cdots = f_Z^{(e)}(x) = e^x$. Therefore, the number of six-letter codewords is

$$\left[\frac{x^6}{6!}\right] f_A^{(e)}(x) f_B^{(e)}(x) \cdots f_Z^{(e)}(x) = \left[\frac{x^6}{6!}\right] \left(e^x\right)^{26} = \left[\frac{x^6}{6!}\right] e^{26x} = 26^6$$

This answer should not be a surpise!

(b) The EGFs are $f_A^{(e)}(x) = f_B^{(e)}(x) = f_C^{(e)}(x) = f_D^{(e)}(x) = e^x - 1$, so the number of codewords in which each of the four letters appears at least once is

$$\left[\frac{x^6}{6!}\right] f_A^{(e)}(x) f_B^{(e)}(x) f_C^{(e)}(x) f_D^{(e)}(x) = \left[\frac{x^6}{6!}\right] \left(e^x - 1\right)^4$$

$$= \left[\frac{x^6}{6!}\right] \left(e^{4x} - 4e^{3x} + 6e^{2x} - 4e^x + 1\right)$$

$$= 4^6 - 4 \cdot 3^6 + 6 \cdot 2^6 - 4 = 1560$$

(c) The EGFs are $f_A^{(e)}(x) = f_B^{(e)}(x) = f_C^{(e)}(x) = \cosh x = [(e^x + e^{-x})/2]$, so the number of codewords in which each of the three letters appears an even number of times is

$$\left[\frac{x^6}{6!}\right](\cosh x)^3 = \left[\frac{x^6}{6!}\right]\left(\frac{e^x + e^{-x}}{2}\right)^3 = \left[\frac{x^6}{6!}\right]\frac{e^{3x} + 3e^x + 3e^{-x} + e^{-3x}}{8}$$
$$= \frac{3^6 + 3 + 3 + 3^6}{8} = 183 \qquad \blacksquare$$

3.5.4 Comparison between Exponential and Ordinary Generating Functions

The exponential generating function counts ordered arrangements, since the terms in the sum

$$h_n = \sum_{\substack{r,s,\ldots,u \geq 0 \\ r+s+\cdots u=n}} \binom{n}{r, s, \ldots, u} a_r b_s \cdots d_u$$

first count the *permutations* $\binom{n}{r,s,\ldots,u}$ of the multiset $\{r \cdot \mathrm{A}, s \cdot \mathrm{B}, \ldots, u \cdot D\}$ and then takes into account that r objects of type A can be placed in a_r ways, that s objects of type B can be placed in b_s ways, and so on.

In the following example, order is important, so it is solved with EGFs.

Example 3.45 Determine the number of passwords of length 10 that can be formed for which each password satisfies all of these three conditions:

A—includes any number of the letters X, Y, and Z

B—includes an odd number of the symbol #

C—includes an even number of the symbol $

Solution. The EGFs are $f_A^{(e)}(x) = \sum_{r\geq 0} 3^r(x^r/r!) = e^{3x}$, $f_B^{(e)}(x) = \sum_{s\geq 0} [x^{2s+1}/(2s+1)!] = (e^x - e^{-x})/2$, and $f_C^{(e)}(x) = \sum_{t\geq 0} [x^{2t}/(2t)!] = (e^x + e^{-x})/2$. The number of passwords of length 10 is therefore

$$h_{10} = \left[\frac{x^{10}}{10!}\right] f_A^{(e)}(x) f_B^{(e)}(x) f_C^{(e)}(x)$$
$$= \left[\frac{x^{10}}{10!}\right] e^{3x}\left(\frac{e^x + e^{-x}}{2}\right)\left(\frac{e^x - e^{-x}}{2}\right)$$
$$= \left[\frac{x^{10}}{10!}\right] \frac{e^{3x}\left(e^{2x} - e^{-2x}\right)}{4} = \left[\frac{x^{10}}{10!}\right] \frac{e^{5x} - e^x}{4} = \frac{5^{10} - 1}{4} = 2{,}441{,}406 \qquad \blacksquare$$

Example 3.45 should now be compared with the following combination problem that is solved with ordinary generating functions.

Example 3.46 Determine the number of ways to pack a picnic basket with 10 items that meets all three of these conditions.

A—includes any number of the carrots, celery sticks, or cucumbers

B—includes an odd number of oranges

C—includes an even number of bagels

Solution. The order in which the items are placed in the basket is of no importance, so we use ordinary generating functions. The OGFs for the vegetables, oranges, and bagels are

$$f_A(x) = \sum_{r\geq 0} \left(\!\!\binom{3}{r}\!\!\right) x^r = \frac{1}{(1-x)^3}, \quad f_B(x) = \sum_{s\geq 0} x^{2s+1} = \frac{x}{1-x^2},$$

$$f_C(x) = \sum_{t\geq 0} x^{2t} = \frac{1}{1-x^2}$$

Therefore 10 items can be packed in $h_{10}(x) = \left[x^{10}\right] f_A(x) f_B(x) f_C(x)$ ways.

Expanding the product of the generating function in a partial fraction decomposition, we obtain

$$\begin{aligned} f_A(x) f_B(x) f_C(x) &= \frac{x}{(1-x)^3\left(1-x^2\right)^2} = \frac{x}{(1-x)^5(1+x)^2} \\ &= \frac{1}{4(1-x)^5} - \frac{1}{16(1-x)^3} - \frac{1}{16(1-x)^2} - \frac{3}{64(1-x)} \\ &\quad - \frac{1}{32(1+x)^2} - \frac{3}{64(1+x)} \\ &= \frac{1}{4}\sum_{k\geq 0} \left(\!\!\binom{5}{k}\!\!\right) x^k - \frac{1}{16}\sum_{k\geq 0} \left(\!\!\binom{3}{n}\!\!\right) x^k - \frac{1}{16}\sum_{k\geq 0} (k+1)x^k \\ &\quad - \frac{3}{64}\sum_{k\geq 0} x^n - \frac{1}{32}\sum_{k\geq 0} (-1)^k (k+1) x^k - \frac{3}{64}\sum_{k\geq 0} (-1)^k x^k \end{aligned}$$

Thus $h_{10} = \frac{1}{4}\left(\!\!\binom{5}{10}\!\!\right) - \frac{1}{16}\left(\!\!\binom{3}{10}\!\!\right) - \frac{1}{16}(11) - \frac{3}{64} - \frac{1}{32}(11) - \frac{3}{64} = 245.$ ■

PROBLEMS

3.5.1. Let $h(x) = \sum_{n\geq 0} h_n x^n$ be a power series. Verify that $[x^n/n!]f(x) = n![x^n]f(x)$.

3.5.2. Give the sequence that corresponds to each of these EGFs:

(a) $f^{(e)}(x) = 3e^{2x}$ **(b)** $g^{(e)}(x) = e^{3x} + 5e^{-x}$

3.5.3. Give the EGF that corresponds to each of these sequences:

(a) $1,\ 0,\ -1,\ 0,\ 1,\ 0,\ -1,\ 0, \ldots$ **(b)** $0,\ 1,\ 0,\ -1,\ 0,\ -1,\ 0, \ldots$

[*Hint*: Both EGFs are a common trigonometric function.]

3.5.4. The complex number $\omega = (1 + i\sqrt{3})/2$ is the primitive root of unity that satisfies $\omega^3 = 1$.

(a) Show that $\omega^2 + \omega + 1 = 0$. [*Hint*: Factor the polynomial $\omega^3 - 1$.]

(b) Show that $\omega^{2n} + \omega^n + 1 = \begin{cases} 3, & \text{when } n = 0, 3, 6, \ldots \\ 0, & \text{otherwise} \end{cases}$

(c) Find a closed form of the EGF

$$\sum_{n=0}^{\infty} \frac{x^{3n}}{(3n)!} = 1 + \frac{x^3}{3!} + \frac{x^6}{6!} + \ldots$$

3.5.5. Let $f^{(e)}(x) = \sum_{n\geq 0} a_n(x^n/n!)$ be the EGF of the sequence, $a_0,\ a_1,\ a_2, \ldots,$ $a_n, \ldots$. Verify that

(a) $f^{(e)\prime}(x) = \sum_{n\geq 0} a_{n+1}(x^n/n!)$ is the EGF of the sequence $a_1,\ a_2, \ldots,$ $a_{n+1}, \ldots$.

(b) $xf^{(e)}(x)$ is the EGF of the sequence $0,\ a_0,\ 2a_1,\ 3a_2, \ldots, na_{n-1}, \ldots$.

(c) $\left[\frac{x^n}{n!}\right] x^r f^{(e)}(x) = \begin{cases} 0, & 0 \leq n < r \\ (n)_r a_{n-r}, & r \geq n \end{cases}$

3.5.6. In how many ways can a 1×5 board be tiled with red and blue square tiles, where there must be an odd number of red tiles?

3.5.7. Let h_n denote the number of ways that a $1 \times n$ rectangle can be tiled with red, blue, green, and yellow squares, where there must be an even number of red tiles, an even number of blue tiles, and at least two green tiles.

(a) Find a formula for h_n.

(b) Use the formula to determine h_3 and then list the tilings of the 1×3 board.

3.5.8. Consider these two problems:

- how many 10-bit binary strings have an even number of 1s?
- how many ways can 1×10 rectangular board be tiled with evenly many red tiles and any number of blue tiles?

(a) why are the problems equivalent to one another?

(b) solve the problems using the EGF method.

3.5.9. Find the EGF that can be used to find the number of ways to distribute 10 distinct candy bars to four children, where the oldest child must receive either, 2 3, 5, or 8 candy bars.

3.5.10. How many n-digit decimal strings include at least one 5 and at least one 7?

3.5.11. The *Bell number* $B(n)$ denotes the number of ways to partition a set of n distinct objects into any number of nonempty subsets, with $B(0) = 1$. Clearly $B(1) = 1$, since {a} itself is the only nonempty subset of itself; moreover $B(2) = 2$, since $\{a, b\}$ can be partitioned in two ways: $\{a, b\}$ itself or the two singleton sets $\{a\}$ and $\{b\}$.

(a) Verify that $B(3) = 5$ and $B(4) = 15$.

(b) Use combinatorial reasoning to derive the recursion formula

$$B(n+1) = \sum_{r=0}^{n} \binom{n}{r} B(r)$$

(c) Let $B(x) = \sum_{n \geq 0} B(n)\,(x^n/n!)$ be the EGF of the Bell numbers. Use the result of Problem 3.5.5 to show that $B'(x) = e^x B(x)$.

(d) Solve the differential equation obtained in part (c) to show that $B(x) = e^{(e^x - 1)}$.

3.5.12. Verify that each of these sequences has the EGF given:

(a) $a_k = k \leftrightarrow f^{(e)}(x) = xe^x$

(b) $b_k = k^2 \leftrightarrow g^{(e)}(x) = x(x+1)e^x$

(c) $c_k = (k-1)^2 \leftrightarrow h^{(e)}(x) = (x^2 - x + 1)e^x$

3.5.13. The product of the first n even positive integers is the *double* factorial $(2n)!! = (2n) \cdot (2n-2) \cdot \cdots \cdot 4 \cdot 2$. We also define $0!! = 1$.

(a) Verify that $(2n)!! = 2^n n!$.

(b) Show that the EGF of the double factorial sequence of even positive integers is

$$\frac{1}{1-2x} = \sum_{n=0}^{\infty} (2n)!! \frac{x^n}{n!}$$

3.5.14. The product of the first n odd positive integers is the double factorial $(2n-1)!! = (2n-1) \cdot (2n-3) \cdot \cdots \cdot 3 \cdot 1$. We also define $(-1)!! = 1$.

(a) Verify that $(2n-1)!! = (2n)!/2^n n!$.

(b) Show that the EGF of the double factorial sequence of odds, namely 1,1,3·1,5·3·1,7·5·3·, . . . , is

$$\frac{1}{\sqrt{1-2x}} = \sum_{n\geq 0} (2n-1)!!\frac{x^n}{n!}$$

[*Hint*: See Problem 3.3.17].

3.5.15. Use combinatorial reasoning to prove that $P(j+k,n) = \sum_{r=0}^{n} \binom{n}{r} P(j,r)P(k,n-r)$, where j and k are nonnegative integers.

3.5.16. The $n!$ permutations of n distinct elements can be partitioned into subsets according to the number of elements left fixed by the permutation. Since there are $\binom{n}{k} D_k$ permutations that derange exactly k elements and leave the remaining $n-k$ elements fixed, where $k = 0,1,\ldots, n$, it follows that $n! = \sum_{k=0}^{n} \binom{n}{k} D_k$.

(a) Use Theorem 3.37 and the formula just given to give a new proof that $D(x) = \sum_{n\geq 0} D_n(x_n/n!) = e^{-x}/(1-x)$ is the exponential generating function (EGF) of the sequence D_n of the derangement numbers.

(b) Use the EGF $D(x) = e^{-x}/(1-x)$ to show that

$$D_n = n!\left(1 - \frac{1}{1!} + \frac{1}{2!} - \frac{1}{3!} + \cdots + (-1)^n\frac{1}{n!}\right)$$

(c) Use $(1-x)D(x) = e^{-x}$ to show that $D_n - nD_{n-1} = (-1)^n$, $n \geq 1$.

(d) Use the result of part (c) to prove that $D_{n+1} = n(D_n + D_{n-1})$.

3.5.17. The $n!$ permutations of an n-set contain $\binom{n}{k} D_{n-k}$ permutations that have exactly k fixed points, since there are $\binom{n}{k}$ ways to choose a subset of k elements that are fixed and D_{n-k} ways to derange the remaining $n-k$ elements. The average number μ_n of fixed points of the $n!$ permutations is then given by

$$\mu_n = \frac{1}{n!}\sum_{k=0}^{n} k\binom{n}{k}D_{n-k}$$

and the standard deviation σ_n of the number of fixed points is given by

$$\sigma_n^2 = \frac{1}{n!}\sum_{k=0}^{n} (k-\mu_n)^2 \binom{n}{k} D_{n-k}$$

(a) prove that the average number of fixed points is $\mu_n = 1$, all $n \geq 1$.

(b) prove that the standard deviation of the fixed points of the permutations of n-set is $\sigma_1 = 0$ and $\sigma_n = 1$ for all $n \geq 2$.
[*Hint*: Use Theorem 3.34 and the results of Problem 3.5.12.]

3.5.18. How many 8-letter words (or nonword letter sequences) can be formed by permutations of 1, 2, or 3 As; 2, 3, or 4 Bs; and 0, 2, or 4 Cs?

3.5.19. In how many ways can a 10-letter word be formed from the letters A, B, C, D, and E where each letter must appear at least once?

3.5.20. In how many ways can a 6-letter word be formed from the letters A, B, and C, so that each letter appears an even number of times in the word?

3.5.21. In how many ways can n pieces of fruit be lined up on a shelf, where there must be an even number of both apples and kiwi fruits and an odd number of both bananas and oranges?

3.5.22. Suppose that you wish to know in how many ways can eight different toys be given to five children under these conditions:

Child A is given 2, 4, 6, . . . toys
Child B is given 1, 3, 5, . . . toys
Child C is given at least one toy
Child D is given at least two toys
Child E is given any number of toys, or possibly none at all

(a) express your answer as the coefficient of a product of EGFs.
(b) use a computer algebra system to get an exact numerical answer.

3.5.23. Let c_n be the number of ways to seat n people around any number of indistinguishable circular tables, with $c_0 = 1$. For example, $c_1 = 1$ and $c_2 = 2$.
(a) Verify that $c_3 = 6$.
(b) Verify the recursion formula

$$c_{n+1} = \sum_{r=0}^{n} \binom{n}{r} c_r (n-r)!$$

(c) Let $C(x)$ be the EGF of the sequence $c_0, c_1, c_2, \ldots c_n, \ldots$. Use the result of Problem 3.5.5 and Theorem 3.34 to show that $C'(x) = C(x)/(1-x)$.
(d) Use the result of part (c) to find $C(x)$ and give a general formula for c_n.
(e) You may be surprised at your answer for part (c). Use combinatorial reasoning to give a more direct derivation by calculating how many ways can person 1 be seated; person 2 can be seated, once person 1 has been seated; . . . ; person $r + 1$ can be seated, once persons 1 through r have been seated.

3.5.24. Describe the h_3 passwords described in Example 3.45 that follow the stated conditions.

3.5.25. A group of adventurers decides to split into two groups. Group A will ride the two-leg zip line, where the first leg is ridden in the order of decreasing age and the second leg is crossed so that no one is in the same position as on the first leg. The second group B will take a swim in Lago Crocodilo.

(a) In how many ways can the n adventurers pursue the activities?

(b) Describe the ways for $n = 3$ adventurers to pursue the activities.

3.5.26. Given unlimited supplies of red and white square tiles.

(a) In how many ways can 10 tiles be selected?

(b) In how many ways can a $1 \times n$ board be tiled?

(c) Is this statement true?

(Number of ways to select 10 tiles)
$\times$ (number of ways to permute the 10 selected tiles)
= (number of tilings of a 1×10 board)

Explain carefully why or why not.

3.5.27. An unlimited supply of red and white tiles is available.

(a) In how many ways can 10 tiles be selected if there must be an odd number of each color?

(b) In how many ways can a 1×10 board be tiled with an odd number of red and an odd number of white tiles?

3.5.28. Recall from Definition 2.34 (in Chapter 2) that the distribution number $T(m, n)$ is the number of ways to distribute m distinct objects onto n distinct recipients in such a way that each recipient is given at least one object. Let $f_a^{(e)}(x) = \sum_{m\geq a}^{\infty} T(m, a)\,(x^m/m!)$.

(a) Verify that $f_1^{(e)}(x) = e^x - 1$.

(b) It can be shown that $T(m, a+b) = \sum_{k=0}^{m} \binom{m}{k} T(k,a)T(m-k,b)$ (see Problem 2.5.14). Use this identity to show that $f_{a+b}^{(e)}(x) = f_a^{(e)}(x)\,f_b^{(e)}(x)$.

(c) Prove that $f_a^{(e)}(x) = \sum_{m\geq a}^{\infty} T(m, a)\,(x^m/m!) = (e^x - 1)^a$.

3.6 SUMMARY AND ADDITIONAL PROBLEMS

A *generating function* is a series of a specified form whose coefficients are associated with a sequence of constants or functions $a_0, a_1, a_2, \ldots$. Combinations can be counted with ordinary generating functions, and permutations with the exponential generating functions. The following tables summarize some useful ordinary and exponential generating functions (OGFs and EGFs).

Sequence	OGF, $f(x) = \sum_{k=0}^{\infty} a_k x^k$
Constant sequence, 1,1, ...	$\frac{1}{1-x} = \sum_{k=0}^{\infty} x^k$
Geometric sequence, $1, \alpha, \alpha^2, \alpha^3, \ldots$	$\frac{1}{1-\alpha x} = \sum_{k=0}^{\infty} \alpha^k x^k$
Binomial coefficients	$(1+x)^n = \sum_{k=0}^{n} \binom{n}{k} x^k$
Multinomial coefficients	$(x_1 + x_2 + \cdots + x_j)^n = \sum_{\substack{r_1 \geq 0, r_2 \geq 0, \ldots, r_j \geq 0 \\ r_1 + r_2 + \cdots + r_j = n}} \binom{n}{r_1, r_2, \ldots, r_j} x_1^{r_1} x_2^{r_2} \cdots x_j^{r_j}$
Multichoose coefficients	$\frac{1}{(1-x)^n} = \sum_{k=0}^{\infty} \left(\binom{n}{k}\right) x^k,\ n \geq 1$
Generalized binomial coefficients, $\binom{\alpha}{r}$	$(1+x)^\alpha = \sum_{k=0}^{\infty} \binom{\alpha}{k} x^k$
Fibonacci numbers, $F_0 = 0, F_1 = 1, F_2 = 1, F_3 = 2, F_4 = 3, F_5 = 5, \ldots$	$\frac{x}{1-x-x^2} = \sum_{k=0}^{\infty} F_k x^k$

Sequence	EGF, $f^{(e)}(x) = \sum_{k=0}^{\infty} a_k \frac{x^k}{k!}$
Constant sequence, 1,1, ...	$e^x = \sum_{k=0}^{\infty} \frac{x^k}{k!}$
Geometric sequence $1, \alpha, \alpha^2, \alpha^3, \ldots$	$e^{\alpha x} = \sum_{k=0}^{\infty} \alpha^k \frac{x^k}{k!}$
$1, 0, 1, 0, \ldots$	$\cosh x = \frac{e^x + e^{-x}}{2} = \sum_{k=0}^{\infty} \frac{x^{2k}}{(2k)!}$
$0, 1, 0, 1, \ldots$	$\sinh x = \frac{e^x - e^{-x}}{2} = \sum_{k=0}^{\infty} \frac{x^{2k+1}}{(2k+1)!}$
Pochhammer symbols, $(\alpha)_k$	$(1+x)^\alpha = \sum_{k=0}^{\infty} (\alpha)_k \frac{x^k}{k!}$
Derangement numbers, D_k	$D(x) = \frac{e^{-x}}{1-x} = \sum_{k=0}^{\infty} D_k \frac{x^k}{k!}$
Distribution numbers, $T(k,n)$	$(e^x - 1)^n = \sum_{k \geq n}^{\infty} T(k,n) \frac{x^k}{k!}$ (see Problem Set 3.5.28)

PROBLEMS

3.6.1. Use the geometric series $1/(1-x) = \sum_{n=0}^{\infty} x^n$ to calculate

(a) $[x^n]\dfrac{1}{1-5x}$ **(b)** $\left[x^n\right]\dfrac{1}{1-x^3}$ **(c)** $\left[x^n\right]\dfrac{1}{1+x}$ **(d)** $\left[x^{50}\right]\dfrac{x^7}{1+2x}$

3.6.2. Use the generating function $e^x = \sum_{k\geq 0}\frac{x^k}{k!}$ to prove that $e^x e^y = e^{x+y}$.

3.6.3. Let $\varphi = (1+\sqrt{5})/2$ and $\hat{\varphi} = (1-\sqrt{5})/2$ be the two roots of the polynomial equation $x^2 = x+1$. Then $\varphi^2 = \varphi + 1$ so that $\varphi^n = \varphi^{n-1} + \varphi^{n-2}$ and it is seen that $a_n = \varphi^n$ is a sequence that satisfies the Fibonacci recursion relation $a_n = a_{n-1} + a_{n-2}$, $n \geq 2$, with the initial values $a_0 = 1, a_1 = \varphi$. Similarly, the sequence $b_n = \hat{\varphi}^n$ satisfies $b_0 = 1, b_1 = \hat{\varphi}, b_n = b_{n-1} + b_{n-2}$, $n \geq 2$. Now show that any generalized Fibonacci sequence $c_0, c_1, c_n = c_{n-1} + c_{n-2}$, $n \geq 2$, can be written in the form $c_n = r\varphi^n + s\hat{\varphi}^n$ if the constants r and s are chosen suitably.

3.6.4. The OGFs of the Fibonacci numbers $F_0 = 0, F_1 = 1, F_2 = 1, F_3 = 2, F_4 = 3, F_5 = 5, \ldots$ and the Lucas numbers $L_0 = 2, L_1 = 1, L_2 = 3, L_3 = 4, L_4 = 7, L_5 = 11, \ldots$ are, respectively, $f_F(x) = x/(1-x-x^2)$ and $f_L(x) = (2-x)(1-x-x^2)$.

(a) Show that $[x(1+x^2)]/(1-x-x^2) = [(2-x)x]/(1-x-x^2) - x$.

(b) Explain carefully why $F_{n+2} + F_n = L_{n+1}$, $n \geq 0$, using part (a).

3.6.5. Let $f_F(x)$ and $f_L(x)$ be the OGFs of the Fibonacci and Lucas numbers (see Problem 3.6.4).

(a) Show that $[(2-x)\left(1+x^2\right)]/(1-x-x^2) - 2 - x = [5x^2/(1-x-x^2)]$

(b) Explain carefully why $L_{n+2} + L_n = 5F_{n+1}$, $n \geq 0$, using part (a).

3.6.6. Use the OGF $f_F(x) = x/(1-x-x^2)$ of the Fibonacci numbers 0,1,1,2, 3,5,... to show that $F_0 + F_1 + \cdots + F_n = F_{n+2} - 1$, $n \geq 0$. [*Hint*: First simplify $x/(1-x-x^2) - x/(1-x)$.]

3.6.7. **(OGF of the Perrin Sequence)** Let ρ, σ, and τ be the roots of the cubic polynomial equation $x^3 = x+1$, so that $x^3 - x - 1 = (x-\rho)(x-\sigma)(x-\tau)$. Derive the OGF of the Perrin sequence $p_0 = 3, p_1 = 0, p_2 = 2, p_n = p_{n-2} + p_{n-3}$, $n \geq 3$ by carrying out the following steps.

(a) Compare coefficients of the equation $x^3 - x - 1 = (x-\rho)(x-\sigma)(x-\tau)$ to show that $\rho + \sigma + \tau = 0, \rho\sigma + \sigma\tau + \rho\tau = -1, \rho\sigma\tau = 1$.

(b) Use part (a) to show that $\rho^2 + \sigma^2 + \tau^2 = 2$.

(c) Let $\hat{p}_n = \rho^n + \sigma^n + \tau^n$. Show that $\hat{p}_n = p_n$, for all $n \geq 0$.

(d) Show that

$$\sum_{n=0} p_n x^n = \frac{3-x^2}{1-x^2-x^3}$$

3.6.8. **(a)** Use the OGF of the multichoose coefficients to derive the identity

$$\left(\!\!\binom{n+1}{r}\!\!\right) = \left(\!\!\binom{n}{0}\!\!\right) + \left(\!\!\binom{n}{1}\!\!\right) + \cdots + \left(\!\!\binom{n}{r}\!\!\right)$$

(b) What identity is obtained when the multichoose coefficients in part (a) are written in terms of binomial coefficients?

3.6.9. Ana discovered six boxes of stepping stones in her basement. Each box contains five identical stones of the same color, and no two boxes contain stones of the same color. In how many ways can she make a path with

(a) 30 stones? **(b)** 29 stones? **(c)** 15 stones?

(Evaluate numerically if a CAS is available.)

REFERENCES

1. H.S. Wilf, *Generating Functionology*, 2nd ed., Academic Press, San Diego, **1994**.
2. G. Pólya, On Picture Writing, *The American Mathematical Monthly*, Vol. 63, No. 10 (Dec., 1956), pp. 689–697.
3. M. Gardner, Mathematical games. *Sci. Am. 238*(2):19–32.

4

ALTERNATING SUMS, INCLUSION-EXCLUSION PRINCIPLE, ROOK POLYNOMIALS, AND FIBONACCI NIM

4.1 INTRODUCTION

A sum of numbers that alternate between positive and negative is known as an *alternating sum*, which can be expressed as follows:

$$\sum_{k=0}^{n} (-1)^k a_k, \quad a_k \geq 0 \tag{4.1}$$

Often the unsigned sum $\sum_{k=0}^{n} a_k$ can be viewed as a count of the number of objects in some set S. If a pairing of some of these objects can be described in which one object is positive and the other negative, then these pairs contribute a 0 to the alternating sum (4.1). This means that only the exceptional objects that have not been paired must be counted to compute the alternating sum. We will see how combinatorial reasoning can sometimes be used to evaluate an alternating sum, with a method that involves three considerations:

D—a description of the objects that are counted by the unsigned sum $\sum_{k=0}^{n} a_k$

I—a sign-reversing involution that describes how at least some of the objects with opposite sign can be paired off

E—an accounting of the exceptional objects that are not in any oppositely signed pair.

Combinatorial Reasoning: An Introduction to the Art of Counting, First Edition. Duane DeTemple and William Webb.

Combinatorialist Arthur Benjamin calls this the DIE method of summation,[1] and this will be the topic of Section 4.2.

The DIE method will be applied to several alternating sums, including a sum that is essential for our proof of the principle of inclusion-exclusion (PIE), the topic discussed in Section 4.3. PIE is an alternating sum that gives the number of elements in a union of m arbitrary sets $A_1 \cup A_2 \cup \cdots \cup A_m$. The PIE formulas for the cases of $m = 2$ and $m = 3$ are

$$\begin{aligned} |A_1 \cup A_2| &= (|A_1| + |A_2|) - (|A_1 \cap A_2|) \\ |A_1 \cup A_2 \cup A_3| &= (|A_1| + |A_2| + |A_3|) - (|A_1 \cap A_2| + |A_1 \cap A_3| + |A_2 \cap A_3|) \\ &\quad + (|A_1 \cap A_2 \cap A_3|). \end{aligned} \tag{4.2}$$

Section 4.4 covers rook polynomials. The name stems from the rook in chess, which can move any distance along a row or column of a chessboard to attack an enemy piece. Therefore, if several rooks are placed on a chessboard, no rook can attack another if at most one rook is placed in any row or column of the board. While it's easy to see that there are $n!$ ways to place n nonattacking rooks on an $n \times n$ board, the problem becomes more interesting if there is some set of squares that are not allowed to be occupied by a rook. We will see how both the inclusion–exclusion principle and generating functions help solve this combinatorial problem of restricted permutations.

This chapter concludes with an optional excursion into Zeckendorf representations of the nonnegative integers, with an unexpected application to a two-person game called *Fibonacci Nim*.

4.2 EVALUATING ALTERNATING SUMS WITH THE DIE METHOD

Suppose that we have been asked to evaluate the alternating sum

$$\sum_{k=0}^{3} (-1)^k \binom{4}{k} = \binom{4}{0} - \binom{4}{1} + \binom{4}{2} - \binom{4}{3} \tag{4.3}$$

The binomial coefficients have a familiar combinatorial description. Namely, $\binom{4}{k}$ is the number of k-element subsets of the set $\{1, 2, 3, 4\}$, where $0 \leq k \leq 3$. Therefore the unsigned sum $\sum_{k=0}^{3} \binom{4}{k}$ counts the number of proper subsets of $\{1, 2, 3, 4\}$. In the alternating sum, the factor $(-1)^k$ tells us that a subset is counted positively when it has an even number of elements, and a subset is counted negatively when it has an odd number of elements. The following table lists all of the 15 proper subsets of $\{1, 2, 3, 4\}$, where the subsets are separated into two columns according to their sign.

[1]Benjamin and Quinn [1] attribute the DIE method to the American mathematician and journalist Fabian Franklin, 1853–1939.

The four-element subset $\{1, 2, 3, 4\}$ is not included here since it is the entire set and is not a proper subset:

Positive; k Even	Negative; k Odd
$\varnothing$	$\{1\}$
$\{1, 2\}$	$\{2\}$
$\{1, 3\}$	$\{3\}$
$\{1, 4\}$	$\{4\}$
$\{2, 3\}$	$\{1, 2, 3\}$
$\{2, 4\}$	$\{1, 2, 4\}$
$\{3, 4\}$	$\{1, 3, 4\}$
	$\{2, 3, 4\}$

Each row except the last contains a positive subset that has been paired with a negative subset. The rule for the pairing is easy to see; any subset containing element 1 is associated with the subset obtained when element 1 is removed, and any subset without element 1 is paired to the subset obtained by inserting element 1 into the set. Such a back-and-forth function is known as an *involution*, and since this involution pairs a set containing an even number of elements with a set containing an odd number of elements, we see that it is a sign reversing involution. There is an exceptional subset $\{2, 3, 4\}$ in the last row that is not paired, since if element 1 were inserted into the set, this would result in a four-element set that is not being counted in the sum given in (4.3).

The sum (4.3) is now easy to compute. The paired sets combine to add 0 to the sum and the one exceptional set $\{2, 3, 4\}$ has a negative sign, so it contributes a -1 to the sum. In other words, we see that $\sum_{k=0}^{3} (-1)^k \binom{4}{k} = -1$.

For the related sum, $\sum_{k=0}^{4} (-1)^k \binom{4}{k}$, the corresponding unsigned sum $\sum_{k=0}^{4} \binom{4}{k}$ counts all 16 subsets of $\{1, 2, 3, 4\}$, and each subset occurs in exactly one pair of oppositely signed subsets. Since there are no exceptional unpaired subsets, it follows that

$$\sum_{k=0}^{4} (-1)^k \binom{4}{k} = 0 \tag{4.4}$$

The combinatorial approach to obtain the sum of an alternating series can be described in the three steps of the DIE method.

THE DIE METHOD TO SUM $\sum_{k=0}^{n} (-1)^k a_k$

D (description)—describe how the unsigned sum $\sum_{k=0}^{n} a_k$ can be viewed as a count of the elements from some set *S*.

I (involution that reverses sign)—find a pairing of at least some of the elements in the set *S*, where one of the elements in the pair is counted negatively and the other is counted positively.

E (exceptions)—determine how many objects in set *S* were not paired by the involution, and determine the sign of these objects.

The sum of the series is the algebraic sum of the exceptional elements.

4.2.1 Using the DIE Method

The following theorem generalizes the result of equation (4.4), showing that any set has equally many subsets with an even number of elements as subsets with an odd number of elements. An earlier derivation of this result was given as a consequence of the binomial theorem in Chapter 3.

Theorem 4.1 For any $n \geq 1$,

$$\sum_{k=0}^{n} (-1)^k \binom{n}{k} = 0 \tag{4.5}$$

Solution.

D. $\sum_{k=0}^{n} \binom{n}{k}$ is the number of subsets of $\{1, 2, \ldots, n\}$. In the alternating sum, a set is counted positively when it contains an even number of elements, and is counted negatively when it has an odd number of elements.

I. Pair a subset containing element 1 with the subset obtained by deleting element 1, and pair a subset not containing element 1 with the subset obtained by inserting element 1. This involution pairs sets of opposite parity, so it is a sign reversing involution.

E. Every subset is paired to one of the opposite sign, so there are no exceptional subsets.

Since there are no exceptional subsets, the alternating sum is 0 and we obtain identity (4.5). ■

Example 4.2 Prove the identity

$$\sum_{k=1}^{n} (-1)^k k \binom{n}{k} = \begin{cases} -1, & n = 1 \\ 0, & n > 1 \end{cases} \tag{4.6}$$

Solution. The identity is trivial if $n = 1$, so consider the cases for which $n > 1$.

D. The unsigned sum $\sum_{k=1}^{n} k \binom{n}{k}$ counts the number of ways to choose a committee and a chair from a group of n members. The signed sum counts a committee with an even number of members positively, and a committee with an odd number of members negatively.

I. Since there are at least two members, suppose that two of these members are persons A and B. If A is not the chair, either add A to the committee or remove A from the committee. If A is the chair, either add or remove person B. In either way, the number of committee members is changed by one member, so this is a sign reversing involution.

E. There are no exceptions, so the signed sum is 0. ■

Example 4.3 Prove the identity

$$\sum_j (-1)^j \binom{n}{j} \binom{jk}{n} = (-1)^n k^n, \quad n, k \geq 0 \tag{4.7}$$

Solution. The identity is obvious for either n or k zero, so assume that both are positive numbers. We now apply the DIE method:

D. Imagine that there are n closed egg cartons, numbered 1 through n, and each carton has k distinct cups that can each hold one egg. Some of cartons are opened and then n identical eggs are placed into the cups of the opened cartons. We can do this in $\sum_j \binom{n}{j}\binom{jk}{n}$ ways, where j is the number of cartons that are opened to reveal jk cups that are selected to hold the n eggs.

I. The number of ways to distribute the eggs is counted positively if an even number of cartons are opened and negatively otherwise. Any distribution of eggs can be associated with two sets of empty cartons, those with an open lid and those with a closed lid. If the lid of the empty carton with lowest number is either closed or opened, reverse it to give a distribution with the opposite sign.

E. The only exceptional distributions are those where no carton is left without an egg. Therefore, all n cartons are open, giving the distribution sign $(-1)^n$, and each carton has an egg in one of the k positions. That is, there are k^n ways to put one egg into one of the k cups of each of the n cartons.

The sum of the exceptional cases is $(-1)^n k^n$, which proves the identity. ■

For the next theorem, it is convenient to express the answer in terms of the Kronecker delta function (Leopold Kronecker, 1823–1891).

Definition 4.4 The *Kronecker delta* function is given by

$$\delta_{n,m} = \begin{cases} 1, & \text{if } n = m \\ 0, & \text{if } n \neq m \end{cases} \tag{4.8}$$

Theorem 4.5 For all $m, n \geq 0$,

$$\sum_{k \geq 0} (-1)^k \binom{n}{k} \binom{k}{m} = (-1)^n \delta_{n,m} \tag{4.9}$$

Solution. Both sides of (4.9) are 0 when $n < m$, so we are left with the cases for which $n \geq m$. Imagine a club of n members, assigned the numbers $1, 2, \ldots, n$.

D. $\sum_{k=m}^{n} \binom{n}{k}\binom{k}{m}$ is the number of ways to choose a committee of k members from n club members, together with a subcommittee of m members chosen from the k committee members.

I. The alternating sum counts committees with k members with the sign $(-1)^k$. Let A be the set of club members not on the committee, and B be the set of people on the committee but not on the subcommittee. Let $p \in A \cup B$ be the person with the lowest number. Moving person p from A to B or vice versa gives a new committee/subcommittee selection with either one more committee member or one fewer. Therefore, this is a sign reversing involution.

E. The only committee/subcommittee selection that cannot be paired occurs when $A = B = \emptyset$. But this means all n club members are selected for both the committee and the subcommittee. Thus, there is one exceptional case of sign $(-1)^n$ when $k = n = m$, and no exceptional cases when $n \neq m$. The sum is then given by $(-1)^n \delta_{n,m}$. ■

Example 4.6 Recall that the derangement number D_n gives the number of permutations of an n-set with no fixed point (see Section 2.6). Use the DIE method to give a new derivation of the identity

$$D_n = n! \sum_{k=0}^{n} \frac{(-1)^k}{k!} \tag{4.10}$$

Solution. First rewrite the sum by setting $k = n - j$, so that it takes the form

$$n! \sum_{k=0}^{n} \frac{(-1)^k}{k!} = \sum_{j=0}^{n} \frac{(-1)^{n-j} n!}{(n-j)!} = (-1)^n \sum_{j=0}^{n} (-1)^j P(n,j)$$

where $P(n,j)$ is a j-permutation of $[n]$. The alternating sum $\sum_{j=0}^{n} (-1)^j P(n,j)$ will now be evaluated using the DIE method:

D. The unsigned sum $\sum_{j=0}^{n} P(n,j)$ gives the number of permutations of the subsets of $[n]$.

I. In the alternating sum, a j-permutation is given the sign of $(-1)^j$. Let π be a j-permutation of $[n]$. Now consider these two sets: the set A of fixed points of π and the set B of values of $[n]$ not in the permutation. Let $r \in A \cup B$ be the smallest element of the two sets. If $r \in A$ is a fixed point, remove it from π to obtain a $(j-1)$-permutation π'. If $r \in B$ is a point not in the permutation, insert it into the rth position of π to form a $(j+1)$-permutation with r as its first fixed point.

E. The exceptional permutations are those with no fixed points and no omitted points; that is, $A = B = \emptyset$. These are the n-permutations of $[n]$ with no fixed points.

There are D_n exceptional permutations, namely, the derangements, each with sign $(-1)^n$, so we see that $\sum_{j=0}^{n}(-1)^j P(n,j) = (-1)^n D_n$.This gives the desired formula. ■

Example 4.7 Recall (from Section 2.5) that the distribution number $T(m,n)$ gives the number of ways that m distinct objects can be distributed onto n distinct recipients so that each recipient is given at least one of the objects. Use the DIE method to derive the formula

$$T(m,n) = (-1)^n \sum_{k=0}^{n} (-1)^k \binom{n}{k} k^m, \quad m,n \geq 1 \tag{4.11}$$

Solution. We will use the DIE method to evaluate the alternating sum $\sum_{k=0}^{n}(-1)^k \binom{n}{k} k^m$.

D. Suppose that m distinct toys, labeled 1 through m, are to be distributed into n distinct toy boxes that are labeled 1 through n by first opening the lids of k of the boxes and then putting the toys into some of the opened boxes. Some of the opened boxes can remain empty. There are $\binom{n}{k}$ ways to choose the k boxes to be opened and k^m ways to place m distinct objects into the opened boxes. The unsigned sum $\sum_{k=0}^{n}\binom{n}{k} k^m$ therefore counts all of the ways to open some of the boxes and distribute the toys into some of the opened boxes.

I. In the alternating sum, a distribution is given the sign $(-1)^k$ if k boxes are opened. For any distribution, there are two sets of boxes for which no object has been assigned: the set A of boxes that were opened but have been left empty and the set B of $n-k$ boxes that were left closed. Let $r \in A \cup B$ be the smallest label of a box left empty by the distribution. This distribution can be paired with the distribution in which the lid to box r is opened or closed to transfer the box from A to B or vice versa. The new distribution has one more or one fewer boxes opened, so it has the opposite sign.

E. The only unpaired distributions are those for which all n boxes are open, $B = \emptyset$, and the distribution leaves no toy box empty, $A = \emptyset$.

There are $T(m,n)$ exceptional distributions: the distributions for which the m toys are distributed to the n toy boxes, all opened, and no box is left empty. Each exceptional distributions has the sign $(-1)^n$. Therefore, the DIE principle gives us the sum $\sum_{k=1}^{n}(-1)^k\binom{n}{k}k^m = (-1)^n T(m,n)$, which is equivalent to the desired formula. ■

4.2.2 Applications of the DIE Method to Tiling Problems

Recall that the number of unrestricted tilings of a $1 \times n$ board with squares and dominoes is given by the combinatorial Fibonacci number f_n.

Example 4.8 Use the DIE method to derive the identity

$$\sum_{k=1}^{n}(-1)^k f_k = (-1)^n f_{n-1}, \quad n \geq 1 \tag{4.12}$$

where $f_k = F_{k+1}$ denotes the kth combinatorial Fibonacci number.

Solution.

D. The unsigned sum $\sum_{k=1}^{n} f_k$ counts the number of ways to tile n boards of length k, $1 \leq k \leq n$, with squares and dominoes.

I. In the alternating sum, the tilings of a board of length k are given the sign of $(-1)^k$. Consider a $1 \times k$ board tiled with squares and dominoes. If the rightmost tile is a domino, that domino can be replaced by a square to pair it with a tiling of a $1 \times (k-1)$ board. If the rightmost tile is a square and $k < n$, that square can be replaced by a domino, pairing it with a tiling of a $1 \times (k+1)$ board.

E. The exceptional tilings are those of length n that end in a square, since the rightmost square cannot be replaced by a domino without exceeding the maximum length n of the boards being tiled.

There are f_{n-1} exceptional tilings, each with sign $(-1)^n$, so we obtain the desired formula. ■

If a $1 \times n$ board is tiled with k dominoes, then there are $n - 2k$ squares in the tiling. There are then $n - k$ tiles in all that can be in any of $\binom{n-k}{k}$ orders. In the following example, $\binom{n-k}{k}$ is viewed as the number of tilings of an n-board with k dominoes.

Example 4.9 For $n > 0$, let

$$s_n = \sum_{k\geq 0}(-1)^k\binom{n-k}{k} \tag{4.13}$$

If $n \equiv r(\text{mod } 6)$,[2] use the DIE method to prove that

$$s_n = \begin{cases} 1, & \text{if } r = 0 \text{ or } r = 1 \\ 0, & \text{if } r = 2 \text{ or } r = 5 \\ -1, & \text{if } r = 3 \text{ or } r = 4 \end{cases} \tag{4.14}$$

Solution.

D. Since $\binom{n-k}{k}$ is the number of tilings of a $1 \times n$ board that each use exactly k dominoes, the unsigned sum $\sum_{k\geq 0} \binom{n-k}{k}$ is the total number of ways, namely, f_n, to tile a board of length n.

I. In the alternating sum, those tilings with an even number of dominoes are counted positively, and tilings with an odd number of dominoes are counted negatively. The following diagram shows that $s_1 = 1, s_2 = 0, s_3 = -1$, in agreement with (4.14).

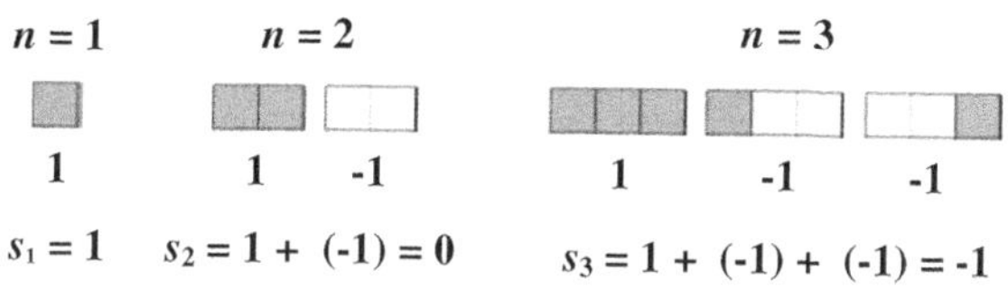

Any tiling with k dominoes that ends with two squares at its right end can be paired with the tiling with $k + 1$ dominoes that is obtained by replacing the two squares with a domino. In the opposite direction, any tiling ending with a domino can be paired with the tiling with one fewer domino that is obtained by replacing the rightmost domino with two squares:

E. Since s_1, s_2, s_3 have already been determined, assume that $n \geq 4$. The unpaired tilings contributing to the sum s_n are those ending with a domino followed by a square. Any such tiling corresponds to a tiling of a board of length $n - 3$ with one fewer domino; therefore, the sum of the exceptional tilings is $-s_{n-3}$.

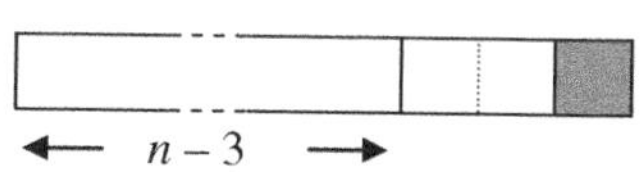

[2]We say that "n is congruent to r modulo m" and write "$n \equiv r(\text{mod } m)$" when r is the remainder when n is divided by m; that is, $n \equiv r(\text{mod } m)$ if and only if $n = qm + r$ for some quotient q and remainder r, $0 \leq r < m$.

We have now shown that $s_n = -s_{n-3}$ for all $n \geq 4$. Thus $s_4 = -s_1 = -1, s_5 = -s_2 = 0$, $s_6 = -s_3 = 1$, which shows that (4.14) holds for $n = 1, 2, \ldots, 6$. Since $s_n = s_{n-6}$ for $n > 6$, it follows that (4.14) holds for all positive n. ∎

PROBLEMS

4.2.1. An *involution* is a bijection $f : X \to X$ for which $f(f(x)) = x$ for all $x \in X$. Show that each of the following functions is an involution:

(a) $f(x) = -x, X = \mathbb{R}$

(b) $g(x) = \frac{1}{x}, X = \mathbb{R} - \{0\}$

(c) $h(x) = -\frac{1}{x}, X = \mathbb{R} - \{0\}$

(d) $k(x) = a + \frac{1}{x-a}, X = \mathbb{R} - \{a\}, a \in \mathbb{R}$

4.2.2. Let I_n be the number of involutions of $[n]$. Show that

(a) $I_1 = 1, I_2 = 2, I_3 = 4, I_4 = 10$

(b) $I_n = n! \sum_{k=0}^{\lfloor n/2 \rfloor} \frac{1}{(n-2k)!2^k k!}$

(c) $I_n = I_{n-1} + (n-1)I_{n-2}, n \geq 3$

4.2.3. Prove that

$$\sum_{k=0}^{m} (-1)^k \binom{n}{k} = (-1)^m \binom{n-1}{m}$$

for all $m \geq 0$ and $n > 0$.

4.2.4. Prove that

$$\sum_{k=m}^{n} (-1)^k \binom{n}{k} = (-1)^m \binom{n-1}{m-1}$$

4.2.5. **(a)** Why is

$$\sum_{k=0}^{n} (-1)^k \binom{n}{k} k^m = 0$$

for all $m < n$?

(b) Why is

$$\sum_{k=0}^{n} (-1)^k \binom{n}{k} k^n = (-1)^n n!$$

for all $n > 0$?

4.2.6. Prove that

$$\sum_{k=0}^{m} (-1)^k \binom{n}{k} \binom{n-k}{n-m} = 0$$

[*Hint*: A $1 \times n$ board is tiled with m square tiles that can be either red or blue, and the remaining $n - m$ cells of the board are covered with white tiles.]

4.2.7. Prove that

$$\sum_{k=0}^{m} (-1)^k \binom{m}{k} \binom{n-k}{r-k} = \binom{n-m}{r}, \quad m + r \leq n$$

[*Hint*: Consider tilings of a $1 \times n$ board with red, blue, and white square tiles, where a total of r tiles are red or blue and the red tiles are allowed only on a subset of the first m cells of the board.]

4.2.8. Prove that

$$\sum_{k=0}^{n} (-1)^k \binom{n+1}{k} \binom{n+m-k}{m} = \binom{m-1}{n}, \quad n < m$$

[*Hint*: Consider tilings of a $1 \times (n+m)$ board with red, blue, and white square tiles, where these conditions are satisfied: (1) there are a total of n red and blue tiles together with m white tiles, and (2) the red tiles are allowed only on some subset of the first $n + 1$ cells of the board.]

4.2.9. Show that $(-1)^n \sum_{k=0}^{n} (-1)^k \binom{n}{k} k^{n+1} = n(n+1)!/2$ by

(a) using Example 4.7

(b) direct application of the DIE method

4.2.10. Show that identity (4.11) can also be written

$$T(m,n) = \sum_{j=0}^{n} (-1)^j \binom{n}{j} (n-j)^m$$

4.2.11. Prove that

$$\sum_{k=m}^{n-1} (-1)^k \binom{n}{k+1} \binom{k}{m} = (-1)^m$$

for all integers $m, n \geq 0$.

[*Hint*: Consider the committee–subcommittee selection model in a club of n members that are numbered $1, 2, \ldots, n$. Let the committee be chaired by the committee member with the highest number. Consider the set A of the club members not on the committee with numbers lower than the chair, and the set B of committee members not on the subcommittee and not the chair.]

4.2.12. Prove that

$$\sum_{k=0}^{m}(-1)^k\binom{m}{k}\left(\!\binom{k}{n}\!\right)=(-1)^m\left(\!\binom{m}{n-m}\!\right)$$

for all integers $m, n \geq 0$.

[*Hint*: Consider the number of ways to order n doughnuts from a bakery offering m types of doughnuts that have been numbered 1 through m.]

4.2.13. Use equation (4.13) to explain why

$$\sum_{k\geq 0}(-1)^{n-k}\binom{n-k}{k}=\begin{cases}1, & \text{if } n\equiv 0 \pmod 3\\ 0, & \text{if } n\equiv 2 \pmod 3\\ -1, & \text{if } n\equiv 1 \pmod 3\end{cases}$$

4.2.14. Complete the DIE method to prove that

$$\sum_{k\geq 0}(-1)^k\binom{n-k}{k}2^{n-2k}=n+1$$

for all $n \geq 0$, given that the *describe* step is;

D. $\binom{n-k}{k}2^{n-2k}$ is the number of ways to tile a $1\times n$ board with k dominoes and $n-2k$ squares that are either red or blue.

4.2.15. In Theorem 4.5, we showed that

$$\sum_{k\geq 0}(-1)^k\binom{n}{k}\binom{k}{m}=(-1)^n\delta_{n,m}$$

Use this identity to prove that $a_n=\sum_{i\geq 0}(-1)^i\binom{n}{i}b_i$ if and only if $b_n=\sum_{j\geq 0}(-1)^j\binom{n}{j}a_j$.

4.2.16. **(a)** In Example 4.7, we showed that

$$T(m,n)=(-1)^n\sum_{k=0}^{n}(-1)^k\binom{n}{k}k^m$$

Use this and the result of Problem 4.2.15 to show that $n^m=\sum_{k\geq 0}\binom{n}{k}T(m,k)$.

(b) Explain why it follows from part (a) that $x^m=\sum_{k\geq 0}(x)_kT(m,k)/k!$, where $(x)_k=x(x-1)(x-2)\cdots(x-k+1)$.

4.2.17. The $n!$ permutations of $[n]$ can be partitioned according to the exact number k of elements that are deranged, leaving the remaining $n - k$ elements fixed. Therefore, $n! = \sum_{k \geq 0} \binom{n}{k} D_k$. Use this and the result shown in Problem 4.2.15 to obtain formula (4.10) for D_n.

4.3 THE PRINCIPLE OF INCLUSION–EXCLUSION (PIE)

There are many subsets of interest in an ordinary deck D of 52 playing cards. For example, the hearts form a 13-element set H. Similarly, the jack, queen, and king in each of the four suits form the 12-element set F known as the *face cards*. Now let's determine the number of cards in a deck that are either a heart or a face card; that is, let us evaluate $|H \cup F|$. It may be tempting to say 25, since $13 + 12 = 25$. But this is incorrect since the jack, queen, and king of hearts have been counted twice, one time as a heart and a second time as a face card. To compensate for the double counting of the three cards in the set intersection $H \cap F$, we must *exclude* the number of cards in the intersection $|H \cap F|$ from the number of cards that has been *included* twice in the sum $|H| + |F|$. This gives us the formula

$$|H \cup F| = |H| + |F| - |H \cap F|$$

that correctly tells us that there are $|H \cup F| = 13 + 12 - 3 = 22$ cards in the deck that are either a heart, a face card, or both.

If we had been asked to determine the number of cards in the deck D that are neither a heart nor a face card, we would express this as $|\overline{H} \cap \overline{F}| = |D| - |H \cup F| = |D| - (|H| + |F|) + |H \cap F|$. Therefore, there are $|\overline{H} \cap \overline{F}| = 52 - 22 = 30$ cards in the deck that are neither a heart nor a face card.

The correct counts are given by alternating sums that include some terms and exclude others, a balancing act that precisely compensates for overcounting and undercounting along the way. The formulas given above for $|H \cup F|$ and $|\overline{H} \cap \overline{F}|$ are the two simplest cases of counting by the inclusion–exclusion principle (PIE).

The reasoning applies equally well to any two subsets A_1 and A_2 of a set U, and we have the following theorem.

Theorem 4.10 Let $A_1 \subseteq U$ and $A_2 \subseteq U$. Then

$$|A_1 \cup A_2| = |A_1| + |A_2| - |A_1 \cap A_2| \tag{4.15}$$

$$|\overline{A_1} \cap \overline{A_2}| = |U| - (|A_1| + |A_2|) + |A_1 \cap A_2|. \tag{4.16}$$

Although the formulas in Theorem 4.10 are the simplest cases of PIE, they are surprisingly useful for solving problems that would be difficult by other means.

Example 4.11 The 30 members at the class reunion will be divided into three groups of 10 to ride in vans to the picnic. Unfortunately, persons a and b do not want to ride together, nor do persons x and y. In how many ways can the three groups of 10 be formed? The order of the groups or the order of the people within any group is not important.

Solution. Let U be the set of all ways the 30 people can be split into groups of 10, including the ways that a and b are in the same group and x and y are placed in the same group. Therefore

$$|U| = \frac{1}{3!}\binom{30}{10, 10, 10}$$

where the division by 3! is because it is unimportant in which order the three groups of equal size are formed. Now let A be the set of groupings in which a and b are in the same group of 10. Thus, set A is formed by selecting a group of 8 and two other groups of 10, and then including a and b with the group of 8. We see that

$$|A| = \frac{1}{2!}\binom{28}{8, 10, 10}$$

where the division by 2! is needed because it is unimportant which of the two groups of 10 is chosen first. Similarly, if B is the set of groupings in which x and y are in the same group, then

$$|B| = \frac{1}{2!}\binom{28}{8, 10, 10}$$

The groupings in $A \cap B$ are of two types.

- The pairs ab and xy are in different groups. There are $\binom{26}{8,8,10}$ ways to form a first group of 8 to which ab are added and a second group of 8 to which xy are added. None of a, b, x, or y belong to the remaining group of 10.
- All four of a, b, x, and y belong to the same group. There are $(1/2!)\binom{26}{6,10,10}$ ways to form a group of 6 to which a, b, x, and y are added, together with two groups of 10. Again there is a division by 2! since it makes no difference in which order the two groups of 10 are selected.

We now see that

$$|A \cap B| = \binom{26}{8, 8, 10} + \frac{1}{2!}\binom{26}{6, 10, 10}$$

By formula (4.16), the number of ways to create the three groups of 10 people with a and b in different groups and x and y in different subgroups is given by

$$|\overline{A} \cap \overline{B}| = \frac{1}{6}\binom{30}{10, 10, 10} - \binom{28}{8, 10, 10} + \binom{26}{8, 8, 10} + \frac{1}{2}\binom{26}{6, 10, 10} \qquad \blacksquare$$

4.3.1 PIE for Three Subsets

Suppose that there are three subsets of interest. For example, let the universal set be the set of the first 25 positive integers, $U = [25]$, and let E, P, and F be, respectively, the subsets of even integers, prime numbers, and Fibonacci numbers in U. The Venn diagram in Figure 4.1 illustrates how these sets are related.

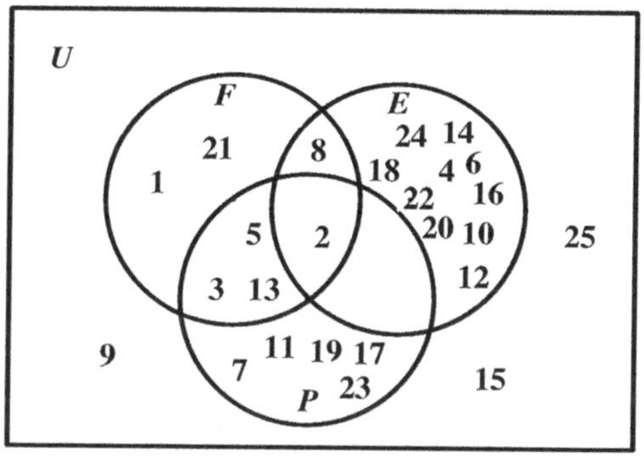

FIGURE 4.1 Venn diagram for the sets of even numbers E, prime numbers P, and Fibonacci numbers F in the set $U = [25]$.

Suppose that we wish to count how many positive integers less than 25 are either even, prime, or a Fibonacci number; that is, we want to determine $|E \cup F \cup P|$. We can start by including all of them in the sum $|E| + |P| + |F| = 12 + 9 + 7 = 28$. But each number in U that is in two of the subsets has been counted twice, not once, so 28 over counts the desired sum by $|E \cap F| + |F \cap P| + |E \cap P| = 2 + 4 + 1 = 7$. Unfortunately, $(|E| + |P| + |F|) - (|E \cap F| + |F \cap P| + |E \cap P|) = 28 - 7 = 21$ is also incorrect, since the number $2 \in E \cap F \cap P$ was included 3 times in $|E| + |P| + |F|$ and then excluded 3 times in $|E \cap F| + |F \cap P| + |E \cap P|$. To include it in the count, we must include $|E \cap F \cap P| = |\{2\}| = 1$. Finally, we have arrived at the correct answer; $|E \cup F \cup P|$ is expressed as

$$|E \cup F \cup P| = (|E| + |P| + |F|) - (|E \cap F| + |F \cap P| + |E \cap P|) \\ + (|E \cap F \cap P|) = 28 - 7 + 1 = 22$$

If we had wanted to know how many numbers in U were neither even, prime, nor Fibonacci, the answer would be expressed as

$$\begin{aligned}|\overline{E}\cap\overline{P}\cap\overline{F}| &= |U| - |E\cup P\cup F|\\ &= |U| - (|E|+|P|+|F|) + (|E\cap F| + |F\cap P| + |E\cap P|) - (|E\cap F\cap P|)\\ &= 25 - 22 = 3\end{aligned}$$

The reasoning just given works the same for any three subsets A_1, A_2, A_3 of a set U, so we have the following theorem.

Theorem 4.12 Let A_1, A_2, A_3 be three subsets of a set U. Then the number of elements in at least one of $A_1, A_2,$ or A_3 is given by

$$\begin{aligned}|A_1\cup A_2\cup A_3| &= (|A_1| + |A_2| + |A_3|)\\ &\quad - (|A_1\cap A_2| + |A_2\cap A_3| + |A_1\cap A_3|) + (|A_1\cap A_2\cap A_3|)\end{aligned} \tag{4.17}$$

and the number of elements of U not in any of the sets A_1, A_2, and A_3 is given by

$$\begin{aligned}|\overline{A_1}\cap\overline{A_2}\cap\overline{A_3}| &= |U| - |A_1\cup A_2\cup A_3|\\ &= |U| - (|A_1| + |A_2| + |A_3|)\\ &\quad + (|A_1\cap A_2| + |A_2\cap A_3| + |A_1\cap A_3|) - (|A_1\cap A_2\cap A_3|)\end{aligned} \tag{4.18}$$

Frequently, the number of elements in a set or intersection of sets depends on knowing how many of the numbers in $[n]$ are multiples of some positive integer k. Equivalently, we need to know the largest positive integer j for which $jk \leq n$. This is the integer j given by the floor function[3] $j = \lfloor n/k \rfloor$.

Example 4.13 How many positive integers no larger than 1000 are

(a) divisible by either 4, 5, or 6?
(b) not divisible by any of 4, 5, or 6?

Solution.

(a) Let $U = [1000]$, and let A_1, A_2, and A_3, be, respectively, the subsets of U that are the multiples of 4, 5, and 6. Since $\left\lfloor \frac{1000}{4} \right\rfloor = 250$, $\left\lfloor \frac{1000}{5} \right\rfloor = 200$, and $\left\lfloor \frac{1000}{6} \right\rfloor = \lfloor 166.6\ldots \rfloor = 166$, we see that $|A_1| = 250$, $|A_2| = 200$, and $|A_3| = 166$. Any multiple of both 4 and 5 is a multiple of 20, and $\left\lfloor \frac{1000}{20} \right\rfloor = 50$,

[3]The *floor function* $\lfloor x \rfloor$ is the largest integer n for which $n \leq x$.

so $|A_1 \cap A_2| = 50$. Similarly, any multiple of both 5 and 6 is a multiple of 30, and since $\left\lfloor \frac{1000}{30} \right\rfloor = \lfloor 33.33\ldots \rfloor = 33$, we get $|A_2 \cap A_3| = 33$. It might be tempting to claim that a multiple of both 4 and 6 is a necessarily a multiple of 24, but this overlooks the fact that both 4 and 6 are multiples of 2. The correct observation is that the multiples of both 4 and 6 are the multiples of 12, which is the *least* common multiple of 4 and 6. Thus $|A_1 \cap A_3| = 83$ since $\left\lfloor \frac{1000}{12} \right\rfloor = \lfloor 83.33\ldots \rfloor = 83$. Similarly, the least common multiple of 4, 5, and 6 is 60 and $\left\lfloor \frac{1000}{60} \right\rfloor = \lfloor 16.6\ldots \rfloor = 16$. Therefore, $|A_1 \cap A_2 \cap A_3| = 16$. We have now calculated all of the values needed to apply the PIE formula: By (4.17), there are

$$|A_1 \cup A_2 \cup A_3| = (250 + 200 + 166) - (50 + 33 + 83) + 16 = 466$$

positive integers no larger than 1000 that are divisible by at least one of 4, 5, or 6.

(b) There are $|\overline{A_1} \cap \overline{A_2} \cap \overline{A_3}| = |U| - |A_1 \cup A_2 \cup A_3| = 1000 - 466 = 534$ positive integers less than 1000 that are not divisible by any of 4, 5, or 6. ■

Example 4.14 How many permutations of the 11 letters in MISSISSIPPI do not contain any of the blocks IIII, SSSS, or PP?

Solution. Let U be the set of all permutations of the 11 letters of the multiset, so $|U| = \binom{11}{1,4,4,2} = 34{,}650$. If A_1, A_2, and A_3 are the respective subsets of permutations that contain the blocks IIII, SSSS, and PP, then we see that $|A_1| = |A_2| = \binom{8}{1,1,4,2} = 840$ and $|A_3| = \binom{10}{1,4,4,1} = 6300$. Similarly, $|A_1 \cap A_2| = \binom{5}{1,1,1,2} = 60$, $|A_1 \cap A_3| = |A_2 \cap A_3| = \binom{7}{1,1,4,1} = 210$, and $|A_1 \cap A_2 \cap A_3| = \binom{4}{1,1,1,1} = 24$. We can now apply formula (4.18) to see that there are $34{,}650 - (840 + 840 + 6300) + (60 + 210 + 210) - 24 = 27{,}126$ permutations of the 11 letters in which no consecutive block IIII, SSSS, or PP appears. ■

4.3.2 PIE in General

Theorem 4.15 (PIE) Let $A_1, A_2, \ldots, A_n$ be subsets of the finite set U. Then

(a)
$$\begin{aligned} |A_1 \cup A_2 \cup \cdots \cup A_n| = & \sum_{1 \le i \le n} |A_i| - \sum_{1 \le i < j \le n} |A_i \cap A_j| \\ & + \sum_{1 \le i < j < k \le n} |A_i \cap A_j \cap A_k| \\ & + \cdots + (-1)^{n+1} |A_1 \cap A_2 \cap \cdots \cap A_n| \end{aligned} \tag{4.19}$$

(b)
$$|\overline{A_1} \cap \overline{A_2} \cap \cdots \cap \overline{A_n}| = |U| - \sum_{1\le i\le n} |A_i| + \sum_{1\le i<j\le n} |A_i \cap A_j| - \sum_{1\le i<j<k\le n} |A_i \cap A_j \cap A_k| + \cdots + (-1)^n |A_1 \cap A_2 \cap \cdots \cap A_n|. \tag{4.20}$$

Proof.

(a) We must show that any element $x \in A_1 \cup A_2 \cup \cdots \cup A_n$ is counted exactly once by the right side of (4.19). Renumbering the sets if necessary, suppose that $x \in A_1 \cap A_2 \cap \cdots \cap A_m$ but $x \notin A_j$ for any $j > m$. Then we see that x is counted m times by the first term $\sum_{1\le i\le n} |A_i|$ of the PIE formula, since x is an element of m of the n sets that appear in the sum. Also, x is counted $\binom{m}{2}$ times by the second term $\sum_{1\le i<j\le n} |A_i \cap A_j|$, once for every pair of the sets taken from $A_1, A_2, \ldots, A_m$. Similarly, x is counted $\binom{m}{3}$ times by $\sum_{1\le i<j<k\le n} |A_i \cap A_j \cap A_k|$, once for every triple of sets from $A_1, A_2, \ldots, A_m$. The last term in (4.19) that counts x a positive number of times is the sum of m-fold intersections, and here just the one intersection $|A_1 \cap A_2 \cap \cdots \cap A_m|$ counts x once. No intersection of more than m sets contains x, so the remaining terms in the PIE formula do not count x. Altogether we see that x is counted by the right side of (4.19) a total of

$$t = \binom{m}{1} - \binom{m}{2} + \binom{m}{3} - \binom{m}{4} + \cdots + (-1)^{m+1}\binom{m}{m}$$

times. Using either (3.8) or (4.5), we have $0 = \sum_{k=0}^{m} (-1)^k \binom{m}{k} = 1 - t$. Thus, x is counted $t = 1$ time by the PIE formula, as needed.

(b) This is a consequence of $|\overline{A_1} \cap \overline{A_2} \cap \cdots \cap \overline{A_n}| = |U| - |A_1 \cup A_2 \cup \cdots \cup A_n|$ and (4.19). ■

Example 4.16 Late in the day, the bakery's stock is down to five loaves of rye, six of sourdough, seven of raisin, and eight of multigrain. In how many ways can 20 loaves be purchased, including at least one loaf of each type?

Solution. Let U be the number of orders that include at least one loaf of each type, but possibly call for more loaves of some type or types than are available. Thus

$$|U| = \left(\!\!\binom{4}{20-1-1-1-1}\!\!\right) = \binom{19}{16} = 969$$

Now let A_1 be the orders that cannot be filled since they include six or more loaves of rye. Then

$$|A_1| = \left(\!\!\binom{4}{20-6-1-1-1}\!\!\right) = \binom{14}{11} = 364$$

Similarly, let A_2, A_3, and A_4 be the orders that, respectively, include seven or more rye loaves, eight or more raisin loaves, and nine or more multigrain loaves. Then

$$|A_2| = \left(\!\!\binom{4}{20-1-7-1-1}\!\!\right) = \binom{13}{10} = 286$$

$$|A_3| = \left(\!\!\binom{4}{20-1-1-8-1}\!\!\right) = \binom{12}{9} = 220$$

$$|A_2| = \left(\!\!\binom{4}{20-1-1-1-9}\!\!\right) = \binom{11}{8} = 165$$

The intersection $A_1 \cap A_2$ contains orders with at least 6 rye and 7 sourdough, and at least one of the other two types as well. Thus

$$|A_1 \cap A_2| = \left(\!\!\binom{4}{20-6-7-1-1}\!\!\right) = \binom{8}{5} = 56$$

The same type of calculation tells us

$$|A_1 \cap A_3| = \binom{7}{4} = 35, |A_1 \cap A_4| = \binom{6}{3} = 20, |A_2 \cap A_3| = \binom{6}{3} = 20,$$

$$|A_2 \cap A_4| = \binom{5}{2} = 10, |A_3 \cap A_4| = \binom{4}{1} = 4$$

The triple intersection $A_1 \cap A_2 \cap A_3$ is empty, since $6 + 7 + 8 + 1 > 20$. Indeed, all of the triple intersections are empty, and so we can now use PIE to find that there are

$$\begin{aligned}|\overline{A_1} \cap \overline{A_2} \cap \overline{A_3} \cap \overline{A_4}| &= 969 - (364 + 286 + 220 + 165)\\ &\quad + (56 + 35 + 20 + 20 + 10 + 4) = 79\end{aligned}$$

ways to buy the 20 loaves of bread from the limited stock on hand, with at least one loaf of each type. ■

Note that the selection problem in Example 4.16 can be restated as a distribution problem or as a count of the number of solutions of an equation in integers:

An Equivalent Distribution Problem. In how many ways can \$20 be given to Ryan, Sue, Ray, and Gary so each is given at least \$1, Ryan is given no more than \$5, Sue no more than \$6, Ray no more than \$7, and Gary no more than \$8?

An Equivalent Integer Solutions Problem. How many solutions in integers exist for the equation

$$x_1 + x_2 + x_3 + x_4 = 20,\ \ 1 \le x_1 \le 5, 1 \le x_2 \le 6, 1 \le x_3 \le 7, 1 \le x_4 \le 8?$$

This means that Example 4.16 can also be solved with ordinary generating functions. Indeed

$$[x^{20}](x + x^2 + \cdots + x^5)(x + x^2 + \cdots + x^6)(x + x^2 + \cdots + x^7)(x + x^2 + \cdots + x^8)$$

$$= [x^{20}]x^4 \frac{1-x^5}{1-x}\frac{1-x^6}{1-x}\frac{1-x^7}{1-x}\frac{1-x^8}{1-x} = [x^{16}]\frac{(1-x^5)(1-x^6)(1-x^7)(1-x^8)}{(1-x)^4}$$

$$= [x^{16}](1 - x^5 - x^6 - x^7 - x^8 + x^{11} + x^{12} + 2x^{13}$$

$$+ x^{14} + x^{15} + O(x^{16}))\sum_{k=0}^{\infty}\binom{3+k}{3}x^k$$

$$= \binom{19}{3} - \binom{14}{3} - \binom{13}{3} - \binom{12}{3} - \binom{11}{3} + \binom{8}{3} + \binom{7}{3}$$

$$+ 2\binom{6}{3} + \binom{5}{3} + \binom{4}{3}$$

$$= 969 - 364 - 286 - 220 - 165 + 56 + 35 + 40 + 10 + 4 = 79$$

The reader might note that the same binomial coefficients appear in both the PIE and OGF solutions.

Example 4.17 Use PIE to prove that the derangement number D_n is given by

$$D_n = n!\left(1 - \frac{1}{1!} + \frac{1}{2!} - \frac{1}{3!} + \cdots + (-1)^n\frac{1}{n!}\right)$$

Solution. Let U be the set of all of the permutations of $[n]$, so $|U| = n!$. For any $k \in [n]$, let $A_k = \{\pi \in U | \pi(k) = k\}$ be the set of permutations that fix the element k and perhaps other elements as well. Thus, for $1 \le i < j < k < \cdots \le n$, we have

$$|A_k| = (n-1)!, |A_j \cap A_k| = (n-2)!,$$
$$|A_i \cap A_j \cap A_k| = (n-3)!, \ldots, |A_1 \cap A_2 \cap \cdots \cap A_n| = 1$$

Using (4.20), we see that

$$
\begin{aligned}
D_n &= |\overline{A_1} \cap \overline{A_2} \cap \cdots \cap \overline{A_n}| \\
&= n! - \sum_{1\le k\le n}^{n} (n-1)! + \sum_{1\le j<k\le n}^{n} (n-2)! - \sum_{1\le i<j<k\le n}^{n} (n-2)! + \cdots + (-1)^n \\
&= n! - n(n-1)! + \binom{n}{2}(n-2)! - \binom{n}{3}(n-3)! + \cdots + (-1)^n \\
&= n!\left(1 - \frac{1}{1!} + \frac{1}{2!} - \frac{1}{3!} + \cdots + \frac{(-1)^n}{n!}\right)
\end{aligned}
$$

■

Example 4.18 Determine the number of permutations q_n of $1\ 2\ 3\ \cdots\ n$ that do not take any adjacent pair $(j, j+1)$ to an adjacent pair in the same order; that is, $\pi(j+1) \neq \pi(j) + 1$. There is one permissible permutation of 1 2, namely 2 1 so $q_2 = 1$. There are three permissible permutations of 1 2 3 namely 1 3 2, 2 1 3, 3 2 1 so $q_3 = 3$. The $q_4 = 11$ permissible permutations of 1 2 3 4 are 1 3 2 4, 1 4 3 2, 2 1 4 3, 2 4 3 1, 2 4 1 3, 3 1 4 2, 3 2 1 4, 3 2 4 1, 4 1 3 2, 4 2 1 3, and 4 3 2 1. So far, we have $q_1 = 1$, $q_2 = 1$, $q_3 = 3$, and $q_4 = 11$.

Solution. Let U be the set of all permutations of $[n]$, so $|U| = n!$. For any $k \in [n-1]$, let $A_k = \{\pi \in U | \pi(k+1) = \pi(k) + 1\}$. Thus, $|A_k| = (n-1)!$, since A_k is the set of permutations of the $n-1$ elements of the sequence $1\ 2 \ldots (k, k+1) \ldots n$. For $i < j$, $|A_i \cap A_j| = (n-2)!$. This is clearly the case if $i + 1 < j$, since then $A_i \cap A_j$ is the set of permutations of $(n-2)$ member sequence $1\ 2 \ldots (j, j+1) \ldots (k, k+1) \ldots n$. But $A_i \cap A_{i+1}$ is the set of permutations of the $(n-2)$ element sequence $1\ 2 \ldots (i, i+1, i+2) \ldots n$, so there are still $(n-2)!$ permutations. More generally, any intersection of r distinct sets $A_1, A_2, \ldots, A_{n-1}$ contains $(n-r)!$ permutations.

The PIE formula (4.20) now gives

$$
\begin{aligned}
q_n &= |\overline{A_1} \cap \overline{A_2} \cap \cdots \cap \overline{A_{n-1}}| = \sum_{k=0}^{n-1} (-1)^k \binom{n-1}{k} (n-k)! \\
&= \sum_{k=0}^{n-1} (-1)^k \frac{(n-1)!}{(n-1-k)!k!}(n-k)! = \sum_{k=0}^{n} (-1)^k \frac{(n-1)!}{k!}(n-k) \\
&= n! \sum_{k=0}^{n} (-1)^k \frac{1}{k!} + (n-1)! \sum_{k=1}^{n} (-1)^{k-1} \frac{1}{(k-1)!} \\
&= D_n + D_{n-1}
\end{aligned}
$$

Thus, each q_n is a sum of two successive derangement numbers. Alternatively, the identity proved in Theorem 2.55 shows that $q_n = (D_{n+1}/n)$. ■

4.3.3 An Application of PIE to Number Theory

For our final application of PIE, we need two concepts from number theory.

Definition 4.19 Two integers m and n are said to be *relatively prime* if no positive integer other than 1 divides both m and n.

To check whether two integers m and n are relatively prime, it is sufficient to see if they have any common prime factors. For example, the prime factors of $75 = 3^1 \cdot 5^2$ are 3 and 5 and the prime factors of $56 = 2^3 \cdot 7^1$ are 2 and 7. Since there is no prime number that divides both 75 and 56, it follows that 75 and 56 are relatively prime. On the other hand, $56 = 2^3 \cdot 7^1$ and $91 = 7 \cdot 13$ share the prime factor 7, so 56 and 91 are not relatively prime.

Definition 4.20 The *Euler phi function* (Leonard Euler, 1707–1783), $\varphi(n)$, is the number of integers m, $1 \leq m \leq n$ that are relatively prime to the positive integer n.

For example, the four integers 1, 3, 5, and 7 are relatively prime to 8, so $\varphi(8) = 4$. The six integers 1, 2, 4, 5, 7, and 8 are relatively prime to 9, so $\varphi(9) = 6$. If p is a prime number, it is divisible only by itself and 1, so $\varphi(p) = p - 1$. For example, $1, 2, \ldots, 12$ are all relatively prime to the prime number 13, and $\varphi(13) = 12$.

Here is a short table of values of the phi function $\varphi(n)$ for $1 \leq n \leq 10$:

n	1	2	3	4	5	6	7	8	9	10
$\varphi(n)$	1	1	2	2	4	2	6	4	6	4

To calculate $\varphi(n)$, it may be easiest to determine the numbers that are *not* relatively prime to n. These are the numbers that have a prime divisor in common with n. Therefore, if p is any prime number that divides n, then the multiples of p are not relatively prime to n. For example, consider $n = 12 = 2^2 \cdot 3$, which has the two prime divisors $p_1 = 2$ and $p_2 = 3$. The sets of multiples of these two primes in [12] are $A_1 = \{2,4,6,8,10,12\}$ and $A_2 = \{3,6,9,12\}$, and these two sets must contain all of the integers $m \in [12]$ *not* relatively prime to 12. Since $|A_1 \cup A_2| = |A_1| + |A_2| - |A_1 \cap A_2| = 6 + 4 - 2 = 8$, this means $\varphi(12) = 12 - 8 = 4$.

The PIE formula allows us to generalize this calculation of $\varphi(n)$ to any n that is represented as a product of powers of its prime factors.

Theorem 4.21 Let $n = p_1^{e_1} p_2^{e_2} \cdots p_m^{e_m}$ be the representation of the positive integer n as a product of powers of the prime numbers $p_1, p_2, \ldots, p_m$ taken to the respective powers $e_1, e_2, \ldots, e_m$. Then

$$\varphi(n) = n\left(1 - \frac{1}{p_1}\right)\left(1 - \frac{1}{p_2}\right) \cdots \left(1 - \frac{1}{p_m}\right) \tag{4.21}$$

Proof. Let $U = [n]$ and let A_j be the subset of multiples of the prime p_j in U. Since n itself is a multiple of p_j, we have $|A_j| = \lfloor n/p_j \rfloor = n/p_j$. For $i \neq j$, $\gcd(p_i, p_j) = 1$ so $A_i \cap A_j$ is the subset of multiples of $p_i p_j$ and $|A_i \cap A_j| = n/(p_i p_j)$. Similarly, $A_i \cap A_j \cap A_k$ is the subset of multiples of $p_i p_j p_k$ and $|A_i \cap A_j \cap A_k| = n/(p_i p_j p_k)$. The pattern should now be clear, and we can apply the PIE formula (4.20) to find that

$$\varphi(n) = |\overline{A_1} \cap \overline{A_2} \cap \cdots \cap \overline{A_m}| = n - \sum_{1 \leq i \leq m} \frac{n}{p_i} + \sum_{1 \leq i < j \leq m} \frac{n}{p_i p_j}$$

$$- \sum_{1 \leq i < j < k \leq m} \frac{n}{p_i p_j p_k} + \cdots + \frac{n(-1)^m}{p_1 p_2 \cdots p_m}$$

$$= n \left(1 - \sum_{1 \leq i \leq m} \frac{1}{p_i} + \sum_{1 \leq i < j \leq m} \frac{1}{p_i p_j} - \sum_{1 \leq i < j < k \leq m} \frac{1}{p_i p_j p_k} + \cdots + \frac{(-1)^m}{p_1 p_2 \cdots p_m} \right)$$

$$= n \left(1 - \frac{1}{p_1} \right) \left(1 - \frac{1}{p_2} \right) \cdots \left(1 - \frac{1}{p_m} \right)$$

■

As an example of (4.21), let $n = 60 = 2^2 3^1 5^1$, so $p_1 = 2, p_2 = 3$, and $p_3 = 5$. Then $\varphi(60) = 60 \left(1 - \frac{1}{2}\right)\left(1 - \frac{1}{3}\right)\left(1 - \frac{1}{5}\right) = 60 \left(\frac{1}{2}\right)\left(\frac{2}{3}\right)\left(\frac{4}{5}\right) = 16$.

The 16 integers relatively prime to 60 are 1, 7, 11, 13, 17, 19, 23, 29, 31, 37, 41, 43, 47, 49, 53, and 59.

PROBLEMS

4.3.1. How many of the first one million positive integers are

(a) either a square or a cube?

(b) neither a square nor a cube?

4.3.2. How many permutations of $\{0, 1, 2, \ldots, 9\}$ do not begin with either 0 or 1 and do not end with either 8 or 9?

4.3.3. How many of the first 1000 positive integers are divisible by either 6, 7, or 8?

4.3.4. How many of the first million positive integers are either a fourth, fifth, or sixth power?

4.3.5. How many permutations of the multiset $S = \{3 \cdot X, 4 \cdot Y, 2 \cdot Z\}$ contain none of the blocks *XXX*, *YYYY*, *ZZ*?

4.3.6. In a group of 25 people, 15 can speak Spanish, 20 can speak English, and 3 do not speak either language. How many speak both languages? Only Spanish? Only English?

4.3.7. In a poll of 100 voters, asking them if they had a favorable opinion about candidates A, B, and C, 40 liked candidate A, 47 liked candidate B, 53 liked candidate C, 7 liked both A and B, 28 liked B and C, and 13 liked A and C. Another three liked none of the candidates. How many of those polled liked all of the candidates?

4.3.8. Can the following diagram serve as a Venn diagram for the four subsets A, B, C, and D of a universal set U? Explain carefully why or why not.

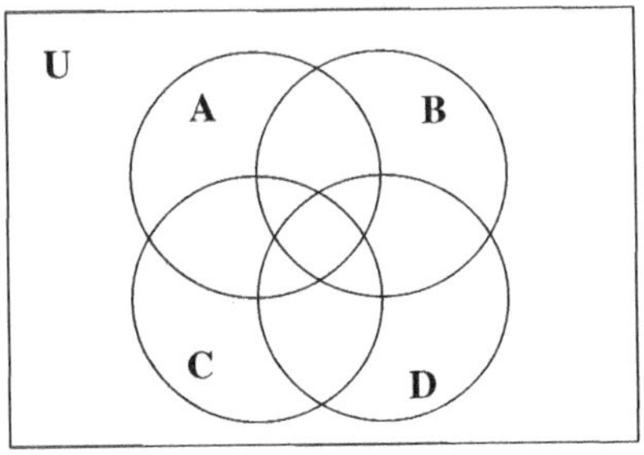

4.3.9. How many permutations of the letters in WHATSUPDOC do not contain any of the substrings WHAT, UP, or DOC?

4.3.10. How many passwords of length n using the symbols A, B, C, # contain at least one symbol of each type?

4.3.11. **(a)** Use PIE to calculate the number of ways that a 5-card hand can be dealt from a 52-card deck with at least one card of every suit.

(b) Check your answer to part (a) by using a more direct approach.

4.3.12. Show that there are 9 solutions in integers of the equation $y_1 + y_2 + y_3 = 11$, where $1 \leq y_1 \leq 3, 0 \leq y_2 \leq 5$, and $-2 \leq y_3 \leq 6$.

4.3.13. **(a)** Use PIE to show that there are six solutions in integers of the equation $t_1 + t_2 + t_3 = 16$, where $2 \leq t_1 \leq 6, -1 \leq t_2 \leq 4$, and $0 \leq y_3 \leq 8$.

(b) List the six solutions.

4.3.14. **(a)** In how many ways can four couples be seated at a circular table so that no couple is seated side by side?

(b) What is the probability that at least one couple is seated together in a random seating?

4.3.15. What theorem is proved when PIE is applied to the n sets $A_1 = A_2 = \cdots = A_n = \{1\}$?

4.3.16. Determine theses values of the Euler phi function $\varphi(n)$:

(a) $\varphi(21)$ **(b)** $\varphi(1200)$

4.3.17. The Euler–Fermat theorem states that if n and a have no common prime divisor, then $a^{\varphi(n)} - 1$ is divisible by n. Verify this theorem for

(a) $n = 24$ and $a = 5$ **(b)** $n = 17$ and $a = 2$

4.4 ROOK POLYNOMIALS

Suppose that four people, call them A, B, C, and D, are to be assigned to perform jobs 1, 2, 3, and 4. If each person can do each job, there are four ways to assign a job to person A, then three ways to assign a job to person B, and so on. This means the jobs can be assigned in $4! = 24$ ways.

Any assignment of jobs can be visualized as an arrangement of nonattacking rooks on a chess board, as shown in Figure 4.2. For example, the diagram shows that job 1 has been assigned to person B, job 2 to person A, and so on. Since a rook is free to move to any other square in its row or column, a nonattacking placement of the rooks means that no person is assigned to two jobs, and no job has been assigned to two people.

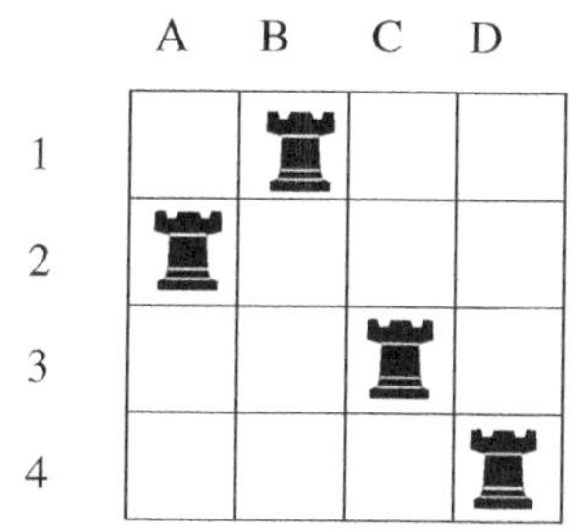

FIGURE 4.2 A job assignment can be visualized as an arrangement of nonattacking rooks on a chessboard.

> From now on it is always assumed the rooks must be in nonattacking arrangements, with no two rooks in the same row or column.

Many variations of rook placement problems can be considered. For example, given a board with five rows and eight columns, in how many ways can a red rook, a white rook, and a black rook be put on the board? The red rook can be placed anywhere, so there are 5×8 choices. This removes both a row and column, leaving 4×7 choices for the white rook. Similarly, this leaves 3×6 choices for the black rook, and altogether the three distinct rooks can be placed in $(5 \times 8)(4 \times 7)(3 \times 6) = 20{,}160$ ways. Another way to obtain this answer is to check that there are $5 \times 4 \times 3$ 3–permutations of the five rows that have, respectively, the red, white, and black rook, and similarly there are $8 \times 7 \times 6$ 3-permutations of the columns, which shows that there are $(5 \times 4 \times 3)(8 \times 7 \times 6)$ placements of the three distinct rooks. If the rooks are identical, there are 3! ways to permute the positions chosen for the three distinct rooks, so there are $20{,}160/3! = 3360$ ways to place three identical rooks on the board.

So far we have assumed that each person is qualified to perform any of the available jobs, but more often there will be some jobs that cannot be assigned to certain individuals. This will be indicated by shading some of the squares of the diagram to identify forbidden assignments. In the example in Figure 4.3, there are just two arrangements of rooks (i.e., job assignments) with the constraint that no rook can be placed in a shaded square.

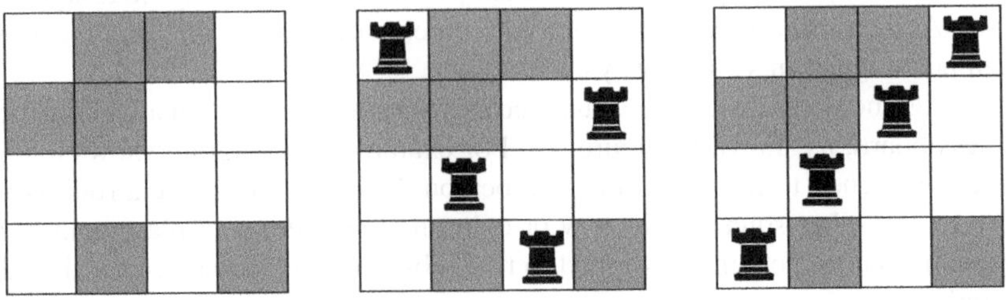

FIGURE 4.3 A board with forbidden squares and the two rook arrangements that avoid these squares.

As the number of people and jobs increase, the job assignment problem becomes more difficult. It may not even be clear whether any job assignment is possible. In the branch of combinatorics known as *graph theory*, this is called a *matching problem* and there are algorithms to find a matching of jobs and people if one exists. In enumerative combinatorics, we want to calculate the number of job assignments. We will see that both inclusion–exclusion (PIE) and ordinary generating functions (OGF) provide answers.

4.4.1 PIE Applications

A simple example will illustrate how PIE is applied. Suppose that we wish to place four rooks on this board with two forbidden squares:

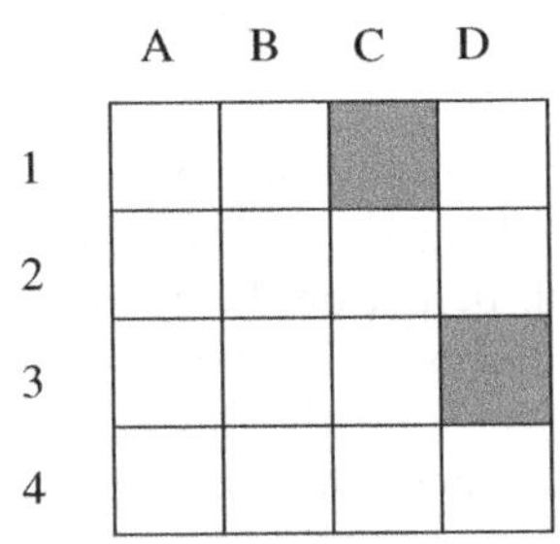

If the forbidden squares are disregarded, there are 4! ways to place the four rooks. We must exclude those arrangements with a rook in a forbidden square. For example, there are 3! arrangements with a rook in the forbidden square 1C, and 3! arrangements with a rook in 3D. These 3! + 3! arrangements with a rook in 1C or 3D both include the 2! arrangements with rooks in both 1C and 3D. Therefore, by PIE the correct count of arrangements with no rook in a forbidden square is $4! - (3! + 3!) + 2! = 14$.

To apply PIE in the general case where n rooks are to be placed on an $n \times n$ board with forbidden squares, we introduce the following sets:

A_i— the set of assignments with a rook in a forbidden square of row i and the remaining $n-1$ rooks on either forbidden or allowed squares

A_{ij}— the set of assignments of two rooks in forbidden squares of rows i and j, where $i < j$ and the remaining $n-2$ rooks on either forbidden or allowed squares

A_{ijk}— the set of assignments of three rooks in forbidden squares of each of the rows i,j and k, $i < j < k$, and the remaining $n-3$ rooks on either forbidden or allowed squares

...

$\boldsymbol{B}$— the board of only forbidden squares

We also introduce these enumeration values:

$r_k(\boldsymbol{B})$ = number of ways to place $k \geq 1$ rooks (nonattacking, of course) on $\boldsymbol{B}$

$r_0(\boldsymbol{B}) = 1$

$S_1 = \sum_i |A_i|$

$S_2 = \sum_{i,j} |A_{ij}|$

...

$S_k = \sum_{i_1,i_2,\ldots,i_k} \left|A_{i_1 i_2 \cdots i_k}\right|$

...

We see that $S_k = r_k(\boldsymbol{B})(n-k)!$ since S_k is the number of ways to place k rooks on forbidden squares times the number of ways to place the remaining $n-k$ rooks in the squares of the unused $n-k$ rows and $n-k$ columns.

The allowable rook placements (or job assignments) are those belonging to none of the sets A_i. These can be enumerated by PIE to give us this theorem.

Theorem 4.22 The number of allowable rook placements or job assignments is given by

$$N = \sum_{k=0}^{n} (-1)^k r_k(\boldsymbol{B})(n-k)! \tag{4.22}$$

Example 4.23 Suppose that the n squares along the main diagonal of an $n \times n$ chessboard are forbidden. Then any placement of n rooks corresponds to a permutation of $[n]$ with no fixed points if a rook in row i and column j indicates that the permutation takes i to j; that is, the number of rook placements is the derangement number D_n. Now use formula (4.22) to derive a formula for D_n.

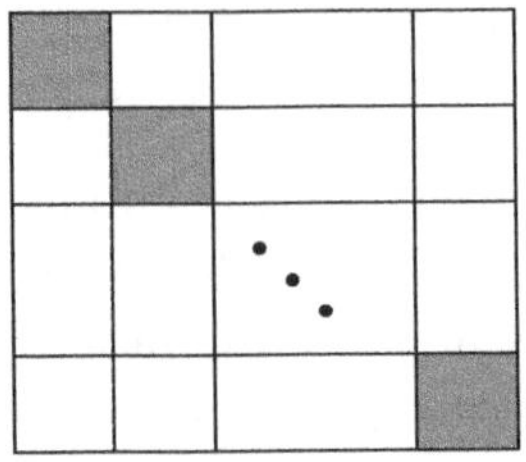

Solution. Since the forbidden squares are in distinct rows and columns, we see that $r_k(\boldsymbol{B}) = \binom{n}{k}$. Using formula (4.22), we can express the number of derangements of $[n]$ a is

$$D_n = \sum_{k\geq 0} (-1)^k \binom{n}{k} (n-k)! = \sum_{k\geq 0} (-1)^k \frac{n!}{k!} = n! \left(1 - \frac{1}{1!} + \frac{1}{2!} - \cdots + (-1)^n \frac{1}{n!}\right)$$

■

Fortunately, in Example 4.23 the values of $r_k(\boldsymbol{B})$ were easily determined. Usually more effort is required, as seen in the next example.

Example 4.24 Calculate the number of rook placements on this chessboard:

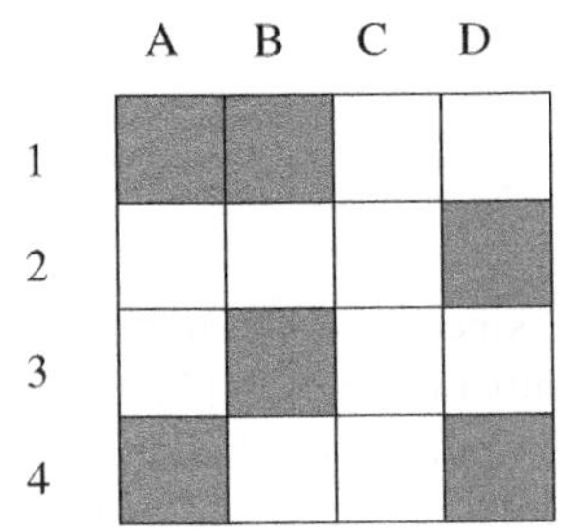

Solution. Although it is not a necessary step, it is often useful to permute rows and columns so that the forbidden squares are as adjacent to one another as possible. The following diagram shows the permuted board on the left and its corresponding board $\boldsymbol{B}$ on the right:

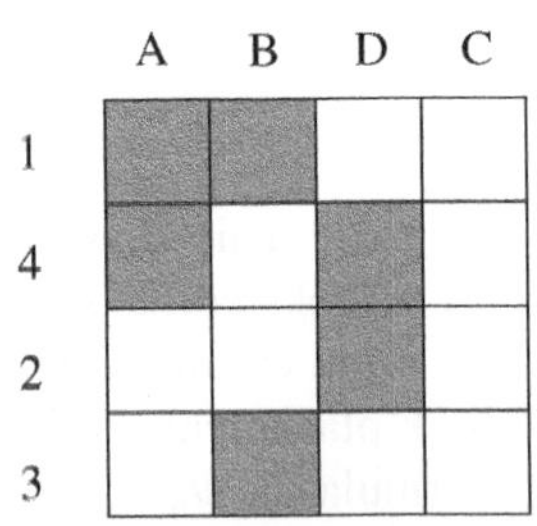

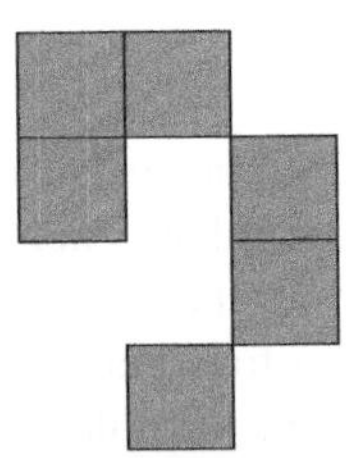

To apply formula(4.22), we obtain these counts of $r_k(\mathcal{B})$ by direct observation.

$r_1(\mathcal{B}) = 6$, since a rook can be placed in any of the six squares of $\mathcal{B}$
$r_2(\mathcal{B}) = 10$, since we have these ten arrangements with two rooks:

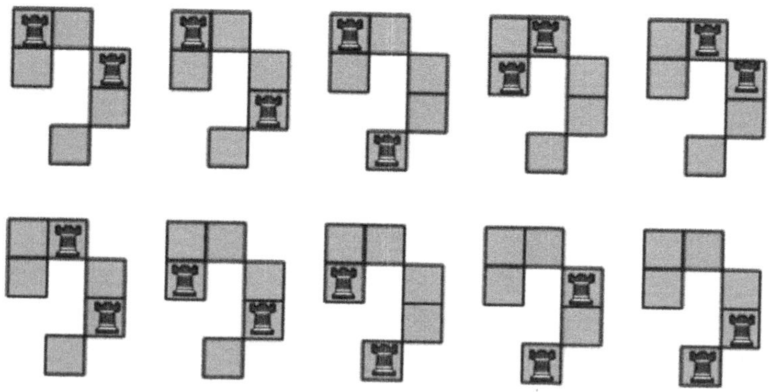

$r_3(\mathcal{B}) = 4$, since we have these four arrangements of three nonattacking rooks:

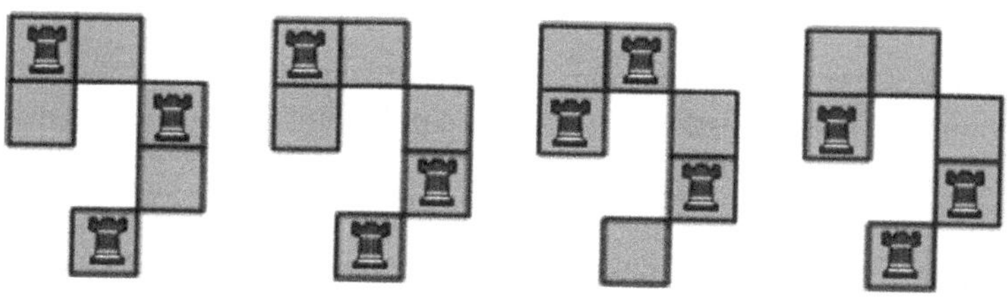

$r_4(\mathcal{B}) = 0$, since there are only three columns

Therefore, using equation (4.22), the number of rook placements is $N = 1 \cdot 4! - 6 \cdot 3! + 10 \cdot 2! - 4 \cdot 1! + 0 \cdot 0! = 24 - 36 + 20 - 4 + 0 = 4$. ■

Most often, the difficult part of applying Theorem 4.22 is calculating the values of $r_k(\mathcal{B})$. A large n and a complicated board $\mathcal{B}$ presents a severe challenge. Fortunately, we still have another trick up our sleeve—ordinary generating functions.

4.4.2 Using Ordinary Generating Functions

Two boards, $\mathcal{B}_1$ and $\mathcal{B}_2$ are said to be *disjoint* if they have no rows or columns in common. Therefore, the placement of the rooks on board $\mathcal{B}_1$ does not affect the placement of rooks on board $\mathcal{B}_2$. Sometimes this decomposition requires a permutation of rows and columns, as shown in Figure 4.4.

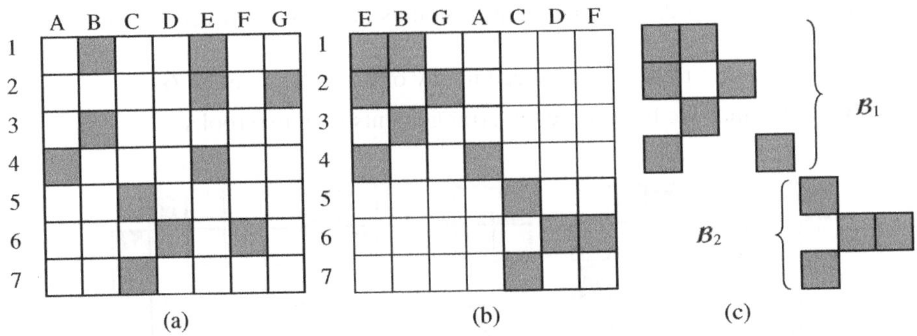

FIGURE 4.4 A permutation of the columns of the board in panel (a) decomposes $\mathcal{B}$ into two disjoint boards, $\mathcal{B}_1$ and $\mathcal{B}_2$: (a) original board; (b) permuted board; (c) decomposition into two disjoint boards.

This observation gives us the following theorem.

Theorem 4.25 Suppose that board $\mathcal{B}$ can be decomposed into the two disjoint boards $\mathcal{B}_1$ and $\mathcal{B}_2$. Then the number of ways to place k rooks on $\mathcal{B}$ is

$$r_k(\mathcal{B}) = r_0(\mathcal{B}_1)r_k(\mathcal{B}_2) + r_1(\mathcal{B}_1)r_{k-1}(\mathcal{B}_2) + \cdots + r_k(\mathcal{B}_1)r_0(\mathcal{B}_2) \tag{4.23}$$

We see that (4.23) is a convolution, the type of sum obtained as the product of two polynomials. This suggests that we define a polynomial that is associated with a board $\mathcal{B}$.

Definition 4.26 The *rook polynomial* $R(x, \mathcal{B})$ of a board $\mathcal{B}$ of forbidden squares is the ordinary generating function of the sequence $r_0(\mathcal{B}), r_1(\mathcal{B}), r_2(\mathcal{B}), \ldots$:

$$R(x, \mathcal{B}) = \sum_{j\geq 0} r_j(\mathcal{B})x^j \tag{4.24}$$

Rook polynomials give us a succinct way to express the result of Theorem 4.25:

$$R(x, \mathcal{B}) = R(x, \mathcal{B}_1)R(x, \mathcal{B}_2) \tag{4.25}$$

If the board decomposes into more than two disjoint boards, say, $\mathcal{B}_1, \mathcal{B}_2, \mathcal{B}_3, \ldots$, then the extension given in the following theorem is clear.

Theorem 4.27 If a board $\mathcal{B}$ can be partitioned into j pairwise disjoint boards $\mathcal{B}_1, \mathcal{B}_2, \ldots \mathcal{B}_j$, then

$$R(x, \mathcal{B}) = R(x, \mathcal{B}_1)R(x, \mathcal{B}_2) \cdots R(x, \mathcal{B}_j) \tag{4.26}$$

Example 4.28 In how many ways can nonattacking rooks be placed on this 5×5 chessboard?

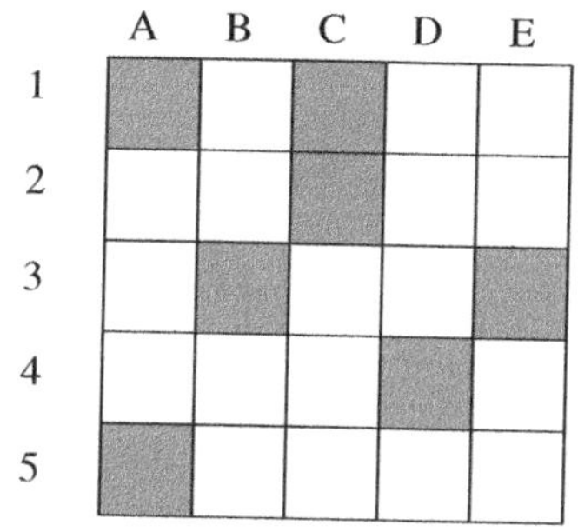

Solution. Permutations of the rows and columns show that $\boldsymbol{B}$ can be decomposed into three disjoint boards:

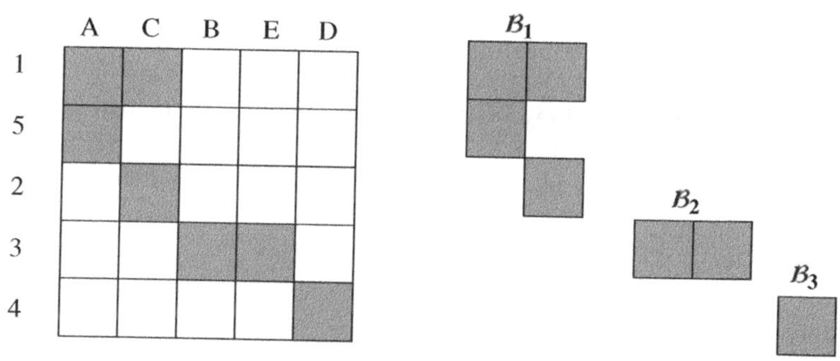

It is easy to obtain the rook polynomials for the three small boards:

$$R(x, \boldsymbol{B}_1) = 1 + 4x + 3x^2, R(x, \boldsymbol{B}_2) = 1 + 2x, R(x, \boldsymbol{B}_3) = 1 + x.$$

Therefore

$$\begin{aligned} R(x, \boldsymbol{B}) &= (1 + 4x + 3x^2)(1 + 2x)(1 + x) \\ &= 1 + 7x + 17x^2 + 17x^3 + 6x^4 \end{aligned}$$

and so we see that there are $N = 1 \cdot 5! - 7 \cdot 4! + 17 \cdot 3! - 17 \cdot 2! + 6 \cdot 1! = 26$ ways to place nonattacking rooks on the chessboard that avoid the forbidden squares. ∎

Example 4.29 In how many ways can six rooks be placed on this chessboard?

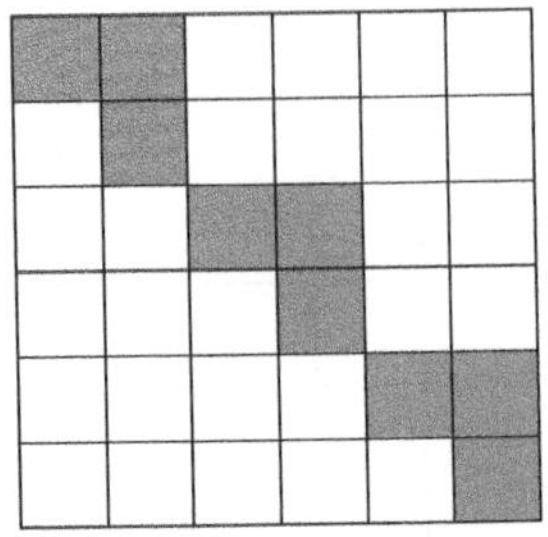

Solution. The board $\mathcal{B}$ of forbidden squares decomposes into three identical boards each with the rook polynomial $1 + 3x + x^2$. Therefore $R(x, \mathcal{B}) = (1 + 3x + x^2)^3 = 1 + 9x + 33x^2 + 63x^3 + 66x^4 + 36x^5 + 8x^6$, which shows that there are $N = 6! - 9 \cdot 5! + 33 \cdot 4! - 63 \cdot 3! + 66 \cdot 2! - 36 \cdot 1! + 8 \cdot 0! = 158$ rook placements on the board. ■

4.4.3 Another Computational Strategy

Suppose that we want to count the number of ways $r_k(\mathcal{B})$ in which k rooks can be placed on a board $\mathcal{B}$. If $\mathcal{S}$ is any square of $\mathcal{B}$, let $\mathcal{B} - \mathcal{S}$ be the board $\mathcal{B}$ from which square $\mathcal{S}$ has been removed and $\mathcal{B}_{\mathcal{S}}$ be the board $\mathcal{B}$ from which the row and column containing $\mathcal{S}$ have been removed.

Since we either do or do not place a rook on square $\mathcal{S}$ it follows that

$$r_k(\mathcal{B}) = r_k(\mathcal{B} - \mathcal{S}) + r_{k-1}(\mathcal{B}_{\mathcal{S}}) \tag{4.27}$$

and therefore

$$R(x, \mathcal{B}) = R(x, \mathcal{B} - \mathcal{S}) + xR(x, \mathcal{B}_{\mathcal{S}}) \tag{4.28}$$

These recursion formulas are valid for any choice of the square $\mathcal{S}$, but the best strategy is to choose a square whose removal leaves boards decomposed into disjoint boards.

Example 4.30 Determine the rook polynomial of this board:

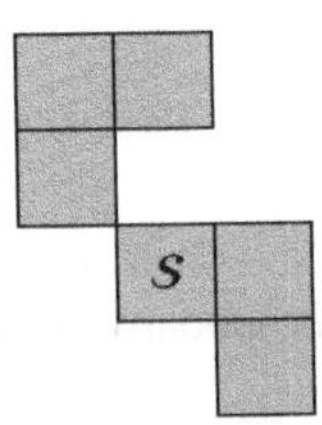

Solution. Let $\mathcal{S}$ be the square of the board of forbidden squares as shown above. Then

$R(x, \mathcal{B}) = R(x,$  $)R(x,$  $) + xR(x,$  $)R(x,$  $)$

$$= (1 + 3x + x^2)(1 + 2x) + x(1 + 2x)(1 + x)$$
$$= 1 + 6x + 10x^2 + 4x^3$$

■

Example 4.31 In how many ways can the five children, A, B, C, D, E, be given five family heirlooms from their parents' estate, where the shaded squares indicate those items that are not of interest to one of the heirs. See diagram:

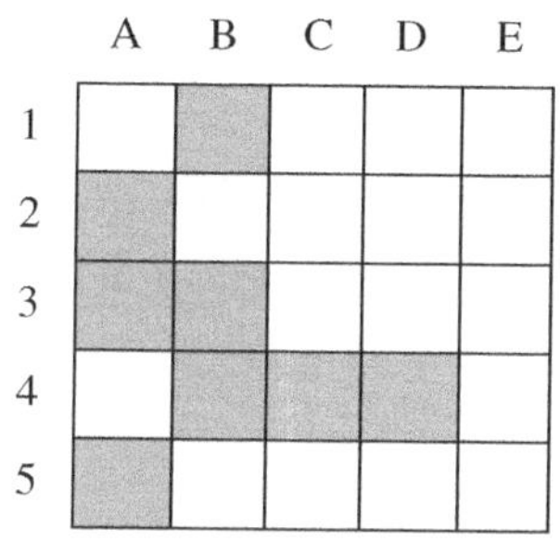

Solution. If $\mathcal{S}$ is square 4B, we see that $\mathcal{B} - \mathcal{S}$ decomposes into two boards, a board $\mathcal{B}_1$ with five squares and a 1×2 board $\mathcal{B}_2$:

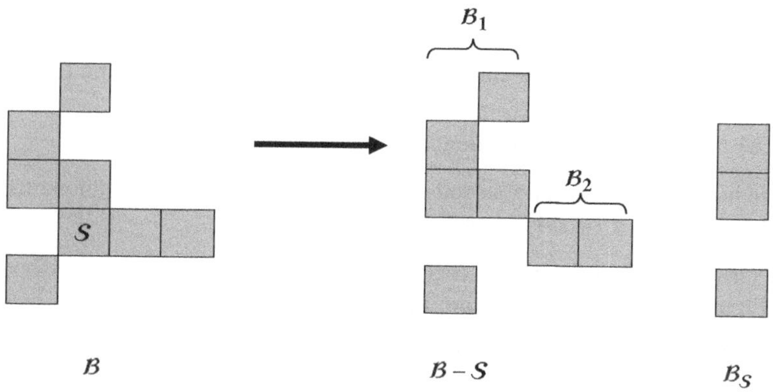

Therefore

$$\begin{aligned} R(x, \mathcal{B}) &= R(x, \mathcal{B} - \mathcal{S}) + xR(x, \mathcal{B}_S) \\ &= R(x, \mathcal{B}_1)R(x, \mathcal{B}_2) + xR(x, \mathcal{B}_S) \\ &= (1 + 5x + 5x^2)(1 + 2x) + x(1 + 3x) \\ &= 1 + 8x + 18x^2 + 10x^3 \end{aligned} \tag{4.29}$$

As a check, suppose that square S' is removed from B_1:

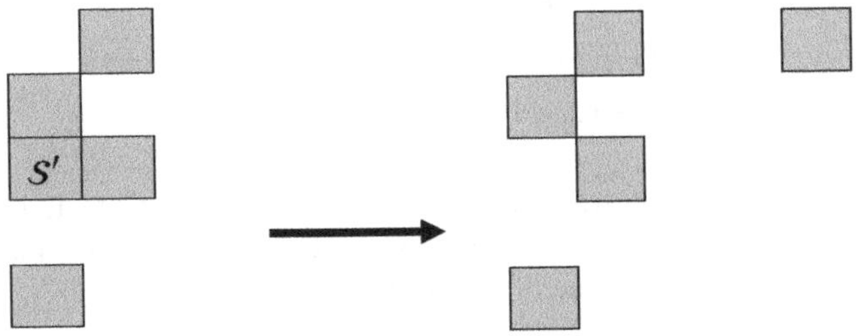

This shows that

$$\begin{aligned} R(x, B_1) &= R(x, B_1 - S') + xR(x, B_{1S'}) \\ &= (1 + 4x + 4x^2) + x(1 + x) \\ &= 1 + 5x + 5x^2 \end{aligned}$$

and confirms our calculation in (4.29).

Finally, using (4.29), we see that the five heirlooms can be distributed in $5! - 8 \cdot 4! + 18 \cdot 3! - 10 \cdot 2! = 16$ ways. ■

PROBLEMS

4.4.1. **(a)** In how many ways can four nonattacking distinct rooks be placed on a 7×9 board?

(b) What if the rooks are identical?

4.4.2. In how many ways can four red rooks and three green rooks, all nonattacking, be placed on a 10×11 board?

4.4.3. Verify that there are two rook placements for the 4×4 chessboard in Figure 4.3, using rook polynomials.

4.4.4. If a board B of forbidden squares has a rook polynomial of the form $R(x, B) = 1 + kx + tx^2 + \cdots$, what is the maximum value of t?

4.4.5. Compute the rook polynomial $R(x, B)$ for each of these boards:

(a) a $1 \times n$ rectangle

(b) a $2 \times n$ rectangle

(c) a $3 \times n$ rectangle

4.4.6. Obtain the rook polynomial for the board in Example 4.24 using the recursion formula $R(x, B) = R(x, B - S) + xR(x, B_S)$.

4.4.7. How many ways are there to place six rooks on this chessboard?

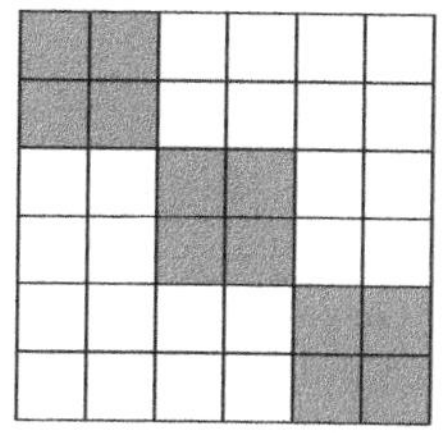

4.4.8. How many ways are there to place six rooks on this chessboard?

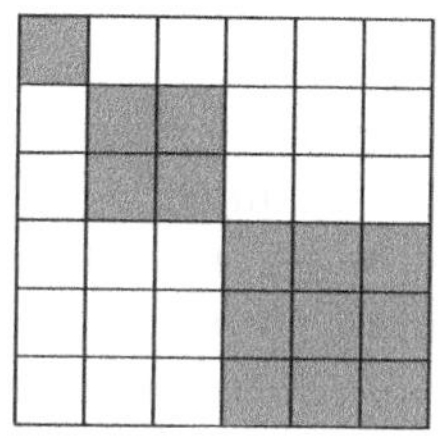

4.4.9. **(a)** Compute the rook polynomial $R(x, \boldsymbol{B})$ for this board by direct counting:

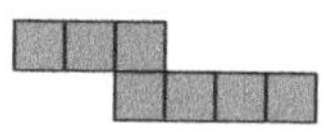

(b) Answer part (a) using equation (4.28), where $\boldsymbol{S}$ is the leftmost square of the lower row.

4.4.10. Use equation (4.28) to obtain the rook polynomials of these boards where the square $\boldsymbol{S}$ has been identified:

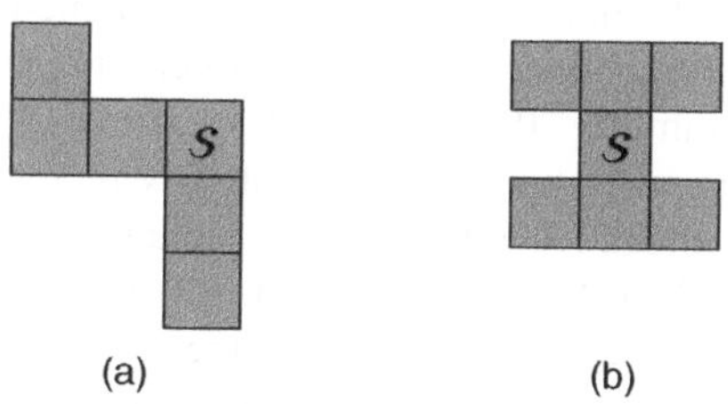

4.4.11. Obtain the rook polynomial of this board.

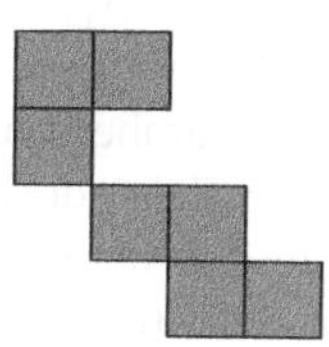

4.5 (*OPTIONAL*) ZECKENDORF REPRESENTATIONS AND FIBONACCI NIM

We are most accustomed to writing an integer as a *positional base ten* numeral. Base ten numeration is convenient, especially for numerical computation, but other representations can also be useful. In this section, we will see how *Zeckendorf's theorem* gives a unique way to represent any nonnegative integers as a sum of Fibonacci numbers, no two of which are consecutive. For example, $20 = 13 + 5 + 2$.

Quite unexpectedly, Zeckendorf's theorem gives a winning strategy of a simple two-person game called *Fibonacci Nim.*

4.5.1 Zeckendorf's Theorem

There are many ways to represent a positive integer n. The most common representation is as a base ten numeral. For example, $n = 75{,}304$ has the expanded form $n = 75{,}304_{\text{ten}} = 7 \cdot 10^4 + 5 \cdot 10^3 + 3 \cdot 10^2 + 0 \cdot 10 + 4$. As a base ten numeral, any positive integer n is a unique sum of powers of 10, where the coefficients of the powers are the ten digits 0, 1, 2, ..., 9. In some applications, bases other than ten may be advantageous. For example, in base two there are only two digits, 0 and 1, and an integer in base two is represented by a binary string that expresses the integer as a sum of powers of 2. A so-called *greedy algorithm* can be used to determine the binary string. For example, suppose that we wish to determine the binary numeral for $n = 75{,}304$. Since $2^{16} = 65{,}536 < n < 2^{17} = 131{,}072$ and $75{,}304 - 65{,}536 = 9768$, we see that $75{,}304 = 1 \cdot 2^{16} + 9768$. Similarly, $2^{13} = 8192 < 9768 < 2^{14} = 16{,}384$ and $9768 - 8192 = 1576$, so $75{,}304 = 1 \cdot 2^{16} + 1 \cdot 2^{13} + 1576$. Continuing in this way, the greedy algorithm finally gives us $75{,}304_{\text{ten}} = 2^{16} + 2^{13} + 2^{10} + 2^9 + 2^5 + 2^3 = 10010011000101000_{\text{two}}$, where the binary string at the right is the base two numeral. Of course, any integer $b \geq 2$ can serve as the base; just apply the greedy algorithm to the increasing sequence $1 = b^0, b^1, b^2, b^3, \ldots$ and use a set of b symbols $0, 1, 2, \ldots, b - 1$ as the digits. In computer science, it is sometimes convenient to use the hexadecimal system (base sixteen) with the sixteen digits 0, 1, 2, 3, 4, 5, 6, 7, 8, 9, A, B, C, D, E, F; for example, $2061_{\text{ten}} = 8 \cdot 16^2 + 0 \cdot 16 + 13 = 80\text{D}_{\text{sixteen}}$.

The greedy algorithm can be applied to any strictly increasing sequence to give a representation of a nonnegative integer. This includes the sequence of distinct Fibonacci numbers 1, 2, 3, 5, 8, 13, 21, 34, 55, 89, 144, 233, 377, ... that begin with $F_2 = 1$. As an example, if $n = 120$, then the largest Fibonacci number not exceeding 120 is $F_{11} = 89$, so $120 = F_{11} + 31$. The largest Fibonacci number in the remainder 31 is $F_8 = 21$, so $120 = F_{11} + F_8 + 10$. Finally, $10 = 8 + 2 = F_6 + F_3$, where the algorithms ends the first time that the remainder is a Fibonacci number. Altogether, we see $120 = 89 + 21 + 8 + 2 = F_{11} + F_8 + F_6 + F_3$, which is a sum of distinct Fibonacci numbers. The sum has no two consecutive Fibnonacci numbers since their sum would be a Fibonacci number already selected by the algorithm. This sum can also be expressed as the Zekendorf numeral 10010100100_Z, where the digits reading from right to left denote whether to include the Fibonacci numbers $F_2, F_3, F_4, \ldots$ as an addend of the sum. Only the binary digits 0 and 1 are needed, since if $2F_j \le n$, then $F_{j+1} = F_j + F_{j-1} < 2F_j \le n$, which shows that the greedy algorithm would have placed the larger number F_{j+1} in the sum. Moreover, there are never two consecutive 1s in a Zeckendorf numeral, since this would correspond to a sum of two consecutive Fibonacci numbers, which is not allowed in a Zeckendorf representation.

Zeckendorf's theorem,[4] which follows, guarantees that the greedy algorithm is successful. That is, it gives both the existence and the uniqueness of the representation of any positive integer as a sum of nonconsecutive Fibonacci numbers.

Theorem 4.32 Every positive integer n has a unique representation as a sum of the distinct Fibonacci numbers $1, 2, 3, 5, 8, \ldots$, where no two consecutive Fibonacci numbers appear in the sum; that is

$$n = F_{a_1} + F_{a_2} + \cdots + F_{a_r} \tag{4.30}$$

where $a_1 \ge a_2 + 2 \ge a_3 + 2 \ge \cdots \ge a_r + 2$ and $a_r \ge 2$. Equivalently, every nonnegative integer n has a unique representation as a binary string of 0s and 1s, where no two 1s are consecutive.

Proof. The greedy algorithm gives us at least one representation of any positive integer m as a sum of nonconsecutive Fibonacci numbers F_n, $n \ge 2$. To show that the representation is unique, suppose that some integer m has two different representations. Any terms that appear in both sums can be deleted, so we can assume that the largest Fibonacci number in either sum, say, F_k, appears in just one of the two representations. Now use the identities $F_{2j-2} + \cdots + F_4 + F_2 = F_{2j-1} - 1$ and $F_{2j-1} + \cdots + F_5 + F_3 = F_{2j} - 1$, both easily proved by mathematical induction (see Problem 4.5.6), to see that the second representation is no larger than $F_{k-1} + F_{k-3} + F_{k-5} + \cdots = F_k - 1 < F_k$, and therefore it cannot equal the sum containing F_k. An alternate proof of uniqueness is to show (or recall Problem 1.5.11) that there are F_n binary strings of length $n-2$, just the number needed to represent the nonnegative integers 0, 1, 2, ..., $F_n - 1$. ■

Example 4.33 Determine the Zeckendorf representation and Zeckendorf numeral of $n = 1270$.

Solution. It is helpful to make a list of Fibonacci numbers, beginning with $F_2 = 1$:

F_2	F_3	F_4	F_5	F_6	F_7	F_8	F_9	F_{10}	F_{11}	F_{12}	F_{13}	F_{14}	F_{15}	F_{16}	F_{17}
1	2	3	5	8	13	21	34	55	89	144	233	377	610	987	1597

[4]Edouard Zeckendorf (1901–1983) was a Belgian physician, army officer, and mathematician.

Now apply the greedy algorithm to see that

$$n = 1270 = 987 + 283 = F_{16} + 233 + 50 = = F_{16} + F_{13} + F_9 + 16$$
$$= F_{16} + F_{13} + F_9 + 13 + 3 = F_{16} + F_{13} + F_9 + F_7 + F_4$$

The corresponding 15-digit Zeckendorf numeral is $n = 1270 = 100{,}100{,}010{,}100{,}100_Z$.

■

4.5.2 Fibonacci Nim

There are many variations on the two-player game Nim but they share the common feature that coins are removed in turn from one or more piles of coins, where certain rules restrict how coins are removed, and the person who takes the last coin (or forces the opponent to do so) is the winner. Fibonacci Nim is played as follows.

> There is an initial pile of n coins, from which player A begins play by removing some but not all of the coins. Player B must then remove at least one coin from the pile, but no more that twice the number removed on the previous turn by player A. Player A then must remove at least one coin from the pile, but no more than twice the number removed from the pile by player B on the previous turn. Play continues in the same way, and the player removing the last coin is the winner.

To understand Fibonacci Nim better, suppose that the two players are Ana and Ben, and Ana has first turn to remove coins from a pile of 20 coins. Ben is confident that he will win the game, since he realizes that he should never remove a third or more of the coins. If he did so, Ana will win on the next move since she can remove the other two thirds of the pile. Ana, who is an enthusiastic math major who has learned about the Zeckendorf representation, has a different strategy; at each turn, she will represent the number of coins in the pile as a sum of nonconsecutive Fibonacci numbers and will then remove the number of coins given by the smallest Fibonacci number in the Zeckendorf sum.

Here's how their game proceeds, where each player's "position" is given by (n,b), where n is the number of coins currently in the pile and b is the upper bound on the number of coins that can be removed by the player at that turn:

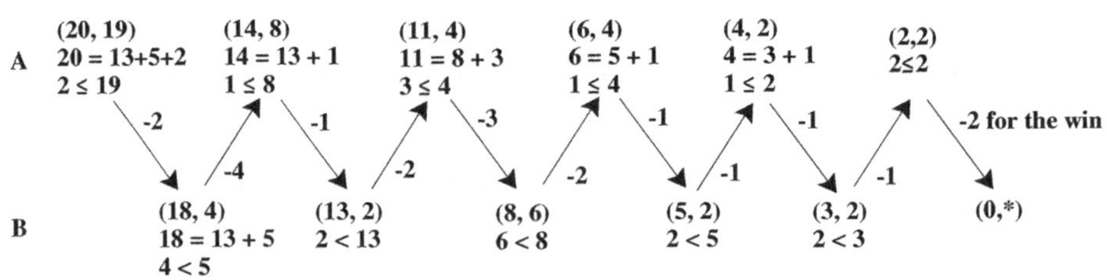

We see that Ana's strategy has worked beautifully. At each move, the smallest Fibonacci number in the Zeckendorf representation was never more than the bound on the number of coins that Ana could remove. Ben, on the other hand, didn't realize that at each of his moves he could only remove fewer coins than even the smallest Fibonacci term of the Zeckendorf representation. Since he always left a positive number of coins, it was inevitable that Ana would eventually win the game.

Let's now see how Ana's strategy works. We assume that Ana starts with a position (n,b), and writes n in its Zeckendorf representation $n = F_{a_1} + F_{a_2} + \cdots + F_{a_r}$ as a decreasing sum of nonconsecutive Fibonacci numbers. We say that (n,b) is a *winning position* if $b \geq F_{a_r}$, and a *losing position* if $b < F_{a_r}$. In the example above, Ana starts with the position (20,19); since $20 = 13 + 5 + 2$ and $2 \leq 19$, Ana has started with a winning position. After Ana removed two coins, Ben is left with the position (18,4); since $18 = 13 + 5$ and $5 > 4$, this is a losing position. Ben next decided to remove four coins, and this left Ana with a winning position once again. If the game can be shown to proceed this way—with Ana always having a winning position for any of Ben's allowed moves, and Ben always having a losing position when Ana removes the number of coins given by the smallest Fibonacci number in the Zeckendorf representation—then it is clear that Ana wins the game since the number of coins decreases at each turn, and so eventually Ana has a position (F_a, b) and wins the game by removing the last F_a coins.

To see why Ana's strategy works, we must prove the following two claims:

Claim 1. If $(F_{a_1} + F_{a_2} + \cdots + F_{a_{r-1}} + F_{a_r}, b)$ is a winning position, then, taking F_{a_r} coins produces $(F_{a_1} + F_{a_2} + \cdots + F_{a_{r-1}}, 2F_{a_r})$, which is a losing position.

Proof. Since $b \geq F_{a_r}$, it is allowable to take away F_{a_r} coins and give the opponent the position $(F_{a_1} + F_{a_2} + \cdots + F_{a_{r-1}}, 2F_{a_r})$. Since $2F_{a_r} < F_{a_r} + F_{a_r+1} = F_{a_r+2} \leq F_{a_{r-1}}$, this is a losing position. ■

Claim 2. If (n,b) is a losing position, any allowable move leaves a winning position.

Proof. Let $n = F_{a_1} + F_{a_2} + \cdots + F_{a_r}$ and $b < F_{a_r}$, so that (n,b) is a losing position. If k coins are removed, where $1 \leq k \leq b$, this leaves the position $(n - k, 2k)$. Now consider the Zeckendorf representations $F_{a_r} - k = F_{b_1} + F_{b_2} + \cdots + F_{b_s}$ and $k = F_{c_1} + F_{c_2} + \cdots + F_{c_t}$, so that $F_{a_r} = F_{b_1} + F_{b_2} + \cdots + F_{b_s} + F_{c_1} + F_{c_2} + \cdots + F_{c_t}$. Since a Zeckendorf representation is unique, the sum on the right side cannot be a Zeckendorf sum, so $c_1 \geq b_s - 1$. Since $k \geq F_{c_1}$ we see that $2k \geq 2F_{c_1} \geq 2F_{b_s-1} \geq F_{b_s}$. But this means that $(n - k, 2k) = (F_{a_1} + F_{a_2} + \cdots + F_{a_{t-1}} + F_{b_1} + F_{b_2} + \cdots + F_{b_s}, 2k)$ is a winning position. ■

The player making the first move has the position $(n, n - 1)$. If n is not a Fibonacci number, then its Zeckendorf representation is $n = F_{a_1} + F_{a_2} + \cdots + F_{a_r}$, where $r > 1$. Therefore, $n - 1 \geq F_{a_r}$, which shows that $(n, n - 1)$ is a winning position. However, if $n = F_a$, then $(F_a, F_a - 1)$ is a losing position since $b = F_a - 1 < F_a$.

We have now unraveled the mysteries of Fibonacci Nim.

Theorem 4.34 In Fibonacci Nim with an initial pile of n coins, the first player to move will necessarily win the game if n is not a Fibonacci number, and at every move the smallest Fibonacci number in the Zeckendorf representation of the current number of coins is the number of coins removed. If n is a Fibonacci number, the second player will necessarily win by following that strategy.

PROBLEMS

4.5.1. Give the Zeckendorf representation of

(a) 27 **(b)** 37 **(c)** 86

4.5.2. Give the base ten numerals for these Zeckendorf numerals.

(a) $1{,}001{,}010{,}001_Z$ **(b)** $10{,}101{,}001{,}010_Z$

4.5.3. Let $m = 101{,}010_Z$ and $n = 1{,}010{,}101_Z$. Without determining the base 10 numerals of m and n, explain how to find the Zeckendorf numerals of

(a) $m + 1$ **(b)** $n + 1$ **(c)** $m + n$

4.5.4. Halley's comet is expected to return to the inner solar system in the year 2061. Give the Zeckendorf representation and numeral for 2061.

4.5.5. The stacked divisions by 5 shown below gives the sequence of remainders 2, 3, and 4:

$$\begin{array}{rl} 0 & R = 4 \\ 5\overline{)4} & R = 3 \\ 5\overline{)23} & R = 2 \\ 5\overline{)117} & \end{array}$$

(a) Explain why $117_{\text{ten}} = 4 \times 5^2 + 3 \times 5 + 2 = 432_{\text{five}}$.

For parts (b) and (c), use the stacked division method to determine the numerals that represent.

(b) 1164_{ten} in base twelve, using the digits 0, 1, … , 9, T, E, where T is ten and E is 11.

(c) 959_{ten} in base sixteen, with the digits 0,1, … , 9,A,B,C,D,E,F.

4.5.6. Prove these two identities for the Fibonacci numbers:

$$F_2 + F_4 + \cdots + F_{2k} = F_{2k+1} - 1 \text{ and } F_3 + F_5 + \cdots + F_{2k+1} = F_{2k+2} - 1, k \geq 1$$

4.5.7. Recall that there are F_{n+2} binary sequences of length n for which no consecutive 1s appear (see Problem 1.5.11). Use this result to prove the uniqueness of the Zeckendorf representation.

4.5.8. Let u_n be an increasing sequence of positive integers with $u_0 = 1$; that is, $1 = u_0 < u_1 < u_2 < u_3 < \cdots$. Prove that given any positive integer m, the greedy algorithm gives a unique representation of the form $m = \sum_{k \geq 0} d_k u_k$, where $d_k \geq 0$ are the "digits."

4.5.9. A greedy pirate and a smart pirate have discovered a treasure chest filled with gold bars weighing 25, 17, and 13 lb. It is a long walk back through a crocodile-infested swamp to the galleon, so the pirates know that a 60-lb load is the most each can carry. The greedy pirate put two of the heaviest bars in his pack. What did the smart pirate do?

4.5.10. The diagram at the left below shows cities $A,B,C,\ldots,G,H$. The numbers along the edges between some pairs of the cities are the costs, in millions of dollars, to build a communication line between the two cities. The problem is to connect all of the cities at a minimum total cost. One possible communication system is shown at the right, whose total cost is \$22 million using the darker edges. Notice that it would be wasteful to use edges that formed a cycle. For example, cities C,D,E, and F can communicate with one another if just the three connections CF,DF, and EF are built, and any additional connection such as CD would create the unnecessary cycle CDF.

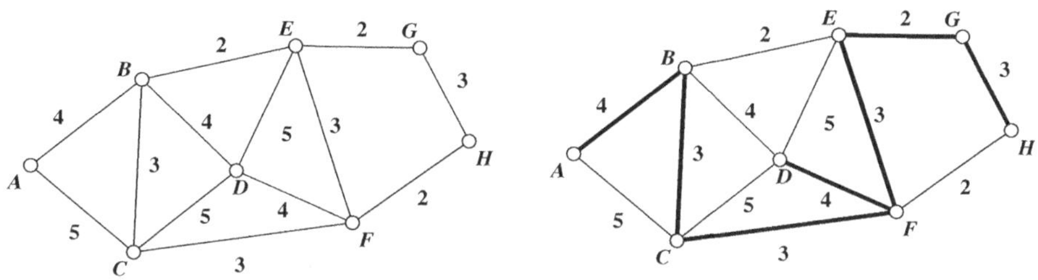

$4 + 3 + 3 + 4 + 3 + 2 + 3 = 22$

(a) Attempt to design a communication system less costly than \$22 million.

(b) The least costly system is one constructed by a greedy algorithm due to Joseph Kruskal [2]. It starts by selecting a least expensive link between two cities, then adds the next link of least cost, and so on, but at no time adds a link that creates a cycle. Apply Kruskal's algorithm to find a communication system whose total cost is just \$20 million.

4.5.11. Suppose that $(F_{a_1} + F_{a_2} + \cdots + F_{a_{j-1}} + F_{a_j} + \cdots + F_{a_{r-1}} + F_{a_r}, b)$ is a winning Fibonacci Nim position. Show that if the "tail" of the Zeckendorf sum $k = F_{a_j} + \cdots + F_{a_{r-1}} + F_{a_r}$ satisfies $k \le b$ and $2k < F_{a_{j-1}}$, then k coins can be safely taken from the pile of coins.

4.6 SUMMARY AND ADDITIONAL PROBLEMS

This chapter has addressed the following topics by means of combinatorial reasoning:

1. *Summation of Alternating Series.* An alternating series is viewed as counting some set of objects, some negatively and some positively. The DIE method of summation has three steps: (D) describe the set of objects that the unsigned series counts, (I) show that there is a sign reversing *i*nvolution that pairs off

many of the objects into opposite-sign objects that then have a zero contribution to the alternating sum, and determine the (E) exceptional objects that have not been paired. The sum of the alternating series is then just the sum of the exceptional objects.

2. *Inclusion–Exclusion Principle.* Many combinatorial problems require us to know the number of members of some union of sets $A_1 \cup A_2 \cup \cdots \cup A_n$. Simply adding the number of members of the sets is incorrect, since elements that belong to two of the sets are doubly counted. Removing these double counts helps, but overcompensates for elements that belong to three of the sets. The PIE formula is an alternating series that correctly counts the number of members of $A_1 \cup A_2 \cup \cdots \cup A_n$.
3. *Rook Polynomials.* The job assignment enumeration problem was modeled as a rook placement problem in which the number of ways to place a given number of rooks on a chessboard was to be determined. No two rooks could be in the same row or column of the board, nor were rooks allowed in some set of forbidden squares of the board. PIE gave rise to an enumeration formula that, in turn, suggested an approach through ordinary generating functions known as *rook polynomials*.
4. *Zeckendorf Representations and Fibonacci Nim.* To win at Fibonacci Nim the number of coins is represented by its Zeckendorf sum of nonconsecutive Fibonacci numbers, and the player who can remove the number of coins given by the smallest Fibonacci number in the Zeckendorf numeral is unbeatable.

PROBLEMS

4.6.1. Prove the identity

$$\sum_{k\geq 0} (-1)^k \binom{m}{k} \binom{n+k}{r} = (-1)^m \binom{n}{m-r}$$

[*Hint*: Start with two bags, the first with red balls numbered 1 through m and a second with green balls numbered 1 through n].

4.6.2. Let $s_n = \sum_{k\geq 0} 2k \binom{n}{2k}$ and $t_n = \sum_{k\geq 0} (2k+1) \binom{n}{2k+1}$.

(a) Use the DIE method to prove that $s_n = t_n$ for all $n \geq 0$.

(b) Evaluate s_n and t_n.

4.6.3. Let $s_n = \sum_{k=1}^{n} (-1)^{k+1} k(n+1-k)$. Use the interpretation that $k(n+1-k)$ gives the number of ways to place three square tiles on a $1 \times (n+2)$ board, with the middle tile on cell $k+1$. Arrangements with the middle tile on an even cell are counted positively. Now show that

(a) $s_{2m} = 0$, using the DIE method

(b) $s_{2m-1} = m$

4.6.4. Prove that the combinatorial Fibonacci numbers satisfy the identity $\sum_{k=1}^{n}(-1)^k k f_k = (-1)^n(nf_{n-1} + f_{n-3}) + 1$ by continuing with the DIE method started as follows. In step D (discription), assume that kf_k is the number of tilings of a $1 \times k$ board by squares and dominoes, together with a penny placed on one of the k cells on the board. (See example given by Benjamin and Quinn [1]). [*Hint*: Suppose that a tiled board with a penny in place is paired with the board in which the rightmost tile is switched from a square to a domino or vice versa and the penny is still in the same cell. Determine the exceptions and calculate their contributions to the alternating sum, where the result of Example 4.8 may be helpful.]

4.6.5. The PIE formula can be written conveniently as $|A_1 \cup A_2 \cup \cdots \cup A_n| = S_1 - S_2 + \cdots + (-1)^{n+1}S_n$, where

$$S_1 = \sum_{1\le i\le n} |A_i|,\ S_2 = \sum_{1\le i<j\le n} |A_i \cap A_j|,\ S_3 = \sum_{1\le i<j<k\le n} |A_i \cap A_j \cap A_k|, \ldots,$$

$$S_n = |A_1 \cap A_2 \cap \cdots \cap A_n|.$$

Using this notation, show that

(a) $N_1 = S_1 - 2S_2 + 3S_3 - 4S_4 + \cdots + (-1)^{n-1}nS_n$ is the number of elements in $A_1 \cup A_2 \cup \cdots \cup A_n$ that belong to exactly one of the sets $A_1, \ldots, A_n$.

(b) $N_2 = S_2 - \binom{3}{2}S_3 + \binom{4}{2}S_4 - \binom{5}{2}S_5 + \cdots + (-1)^{n-2}\binom{n}{2}S_n$ is the number of elements in $A_1 \cup A_2 \cup \cdots \cup A_n$ that belong to exactly two of the sets $A_1, \ldots, A_n$.

4.6.6. Generalize the formulas in Problem 4.6.5 by proving this result: the number of elements that belong to exactly m of the sets $A_1, \ldots, A_n$ is

$$N_m = S_m - \binom{m+1}{m}S_{m+1} + \cdots + (-1)^{k-m}\binom{k}{m}S_k + \cdots + (-1)^{n-m}\binom{n}{m}S_n$$

4.6.7. The number of elements that belong to exactly one of the four sets A_1, A_2, A_3, A_4 is given (see Problem 4.6.5(a)) by $N_1 = S_1 - 2S_2 + 3S_3 - 4S_4$ where

$$S_1 = \sum_{1\le i\le 4} |A_i|,\ S_2 = \sum_{1\le i<j\le 4} |A_i \cap A_j|,\ S_3 = \sum_{1\le i<j<k\le 4} |A_i \cap A_j \cap A_k|,$$

$$S_4 = |A_1 \cap A_2 \cap A_3 \cap A_4|$$

(a) use the formula for N_1 to count the number of permutations of four elements with exactly one fixed point.

(b) verify that the answer obtained in part (a) is $4D_3$.

4.6.8. **(a)** Prove that there are $D_n + D_{n-1}$ rook placements on an $n \times n$ chessboard with $n - 1$ disjoint forbidden squares.

(b) Use the result of part (a) to prove the identity $D_{n+1} = n(D_n + D_{n-1})$.

4.6.9. Determine $r_2(\boldsymbol{B})$ and $r_3(\boldsymbol{B})$ for each of these boards $\boldsymbol{B}$ of forbidden squares.

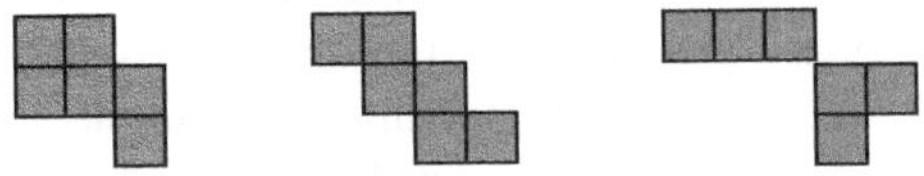

(a) (b) (c)

4.6.10. For each of the three chessboards shown, first compute $r_i(\boldsymbol{B})$, $i = 0, 1, 2, 3, 4$ and then calculate the number of ways to place four rooks on each board.

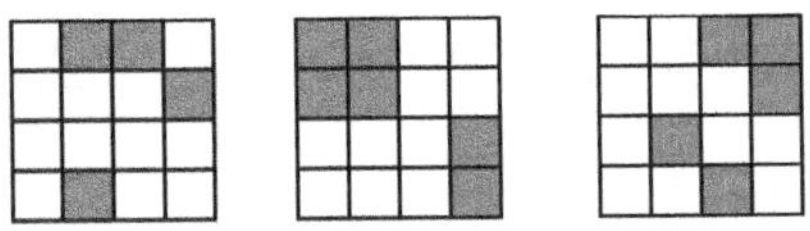

(a) (b) (c)

4.6.11. Show that two identical nonattacking rooks can be placed on the "staircase" board of height n shown in $2n^2 - 3n + 1$ ways.

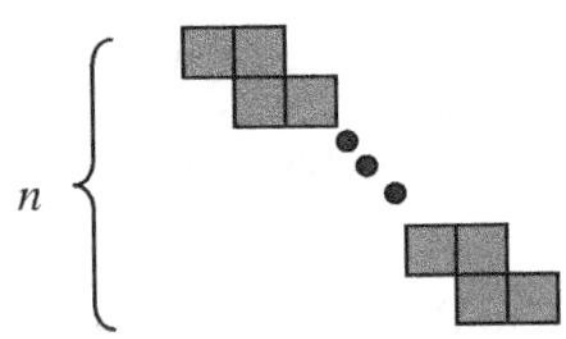

4.6.12. Calculate the rook polynomial $R(x, \boldsymbol{B})$ for the following chessboard and use it to compute the number of ways to place five rooks on the board.

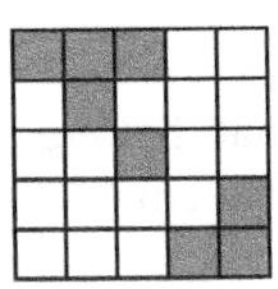

REFERENCES

1. A. Benjamin and J. Quinn, An alternate approach to alternating sums: A method to die for, *Coll. Math. J.* **39**:191–201 (2008).
2. J. Kruskal, On the shortest spanning subtree of a graph and the traveling salesman problem, *Proc. Am. Math. Soc.* 7(1):48–50 (Feb. 1956).

5

RECURRENCE RELATIONS

5.1 INTRODUCTION

Many—indeed, most—of the sequences $h_0, h_1, h_2, \dots, h_n, \dots$ encountered in earlier chapters satisfy a *recurrence relation*. In other words, if initial values of the sequence $h_0, h_1, h_2, \dots, h_{k-1}$ either are given to us or can be determined, then every additional term of the sequence is given by a function $h_n = f(h_0, h_1, h_2, \dots, h_{n-1})$, $n \geq k$, which depends only on earlier terms of the sequence.

A simple example is the recurrence relation $h_n = \alpha h_{n-1}$, $n \geq 1$, where an initial condition $h_0 = A$ is to be satisfied. Since each new term depends on just the preceding term, this is a recurrence relation of *first order*. Iteration shows that $h_n = \alpha h_{n-1} = \alpha^2 h_{n-2} = \dots = \alpha^n h_0$, so the solution is the geometric sequence $h_n = A\alpha^n$, $n \geq 0$.

In Section 5.2, we will investigate the Fibonacci recurrence relation $h_n = h_{n-1} + h_{n-2}$, $n \geq 2$. Each additional term of the sequence is determined by the previous two terms, so this is an example of a *second-order* recurrence relation. The initial conditions $h_0 = 0$ and $h_1 = 1$ correspond to the Fibonacci number sequence 0,1,1,2,3,5,8,... , and the initial conditions $h_0 = 2$ and $h_1 = 1$ correspond to the Lucas sequence 2, 1, 3, 4, 7, 11, ...

In Section 5.3, general second-order recurrence relations of the form $h_n = a_1 h_{n-1} + a_2 h_{n-2}$, $n \geq 2$ are considered, where a_1 and a_2 are constants. This is known as a *second-order linear homogeneous recurrence relation with constant coefficients*. Mathematical induction shows that the solution is uniquely specified when two initial conditions $h_0 = A, h_1 = B$ are satisfied.

Combinatorial Reasoning: An Introduction to the Art of Counting, First Edition. Duane DeTemple and William Webb.

In Section 5.4, we extend our methods to solve recurrence relations of arbitrary order k that have the form

$$h_n = a_1 h_{n-1} + a_2 h_{n-2} + \cdots + a_k h_{n-k} \tag{5.1}$$

where $a_1, a_2, \ldots, a_k$ are constants. A recurrence relation with the form of (5.1) is described as a *linear homogeneous recurrence relation of order k with constant coefficients.*

Section 5.5 considers *nonhomogeneous,* or *inhomogeneous,* recurrence relations of the form

$$h_n = a_1 h_{n-1} + a_2 h_{n-2} + \cdots + a_k h_{n-k} + q_n \tag{5.2}$$

where q_n is a given sequence. If q_n satisfies a homogeneous linear recurrence relation, say, of order j, then (5.2) can be shown to satisfy a related homogeneous recurrence of order $j + k$ for which we already have a method of solution.

In Section 5.6, we will see that generating functions provide a way to solve many recurrence relations. The generating function approach applies to linear equations of the type (5.1) and (5.2), and also gives a method that may solve more general types of recurrence relations that do not have either of these forms. This includes recurrences with variable coefficients, additional parameters, and certain nonlinear recurrences.

5.2 THE FIBONACCI RECURRENCE RELATION

A combinatorialist will often solve an enumeration problem with recursive reasoning, which is described by these steps:

1. Introduce notation for which h_n represents the answer that is to be found for the nth case.
2. Determine a few of the beginning values, say, h_1, h_2, and h_3, by direct counting.
3. Use combinatorial reasoning to obtain a recurrence relation that expresses h_n as a function of previous values of the sequence.
4. Solve the recurrence relation, using initial values of the sequence to determine the unique solution.

Some examples of recursive reasoning are given in the following three examples.

Example 5.1 In how many ways can n people seated in a row of n seats at the theater be reseated after intermission so that each person returns either to the same seat or to a seat that is adjacent to the one occupied before intermission?

Solution. Let h_n denote the number of seatings following intermission. Therefore, $h_1 = 1$, since a single person just returns to the one seat in the row. Also, $h_2 = 2$, since now two people either return to their original seats or else switch seats. To obtain a recurrence relation, let $n \geq 3$ and note that there are two disjoint types of reseatings of the n people depending on whether person 1 returns to the same seat or persons 1 and 2 switch seats. Since there are h_{n-1} seatings in which person 1 returns to seat 1, and there are h_{n-2} seatings in which persons 1 and 2 switch seats, we get the Fibonacci recurrence $h_n = h_{n-1} + h_{n-2}$. Together with the initial conditions $h_1 = 1, h_2 = 2$, we conclude that $h_n = f_n = F_{n+1}$. ■

The answer to Example 5.1 should not be a surprise, since any person retaining the same seat can be represented by a square and a pair of people who trade seats can be represented by a domino. Therefore, the number of reseatings of n people is the same as the number of tilings of n-boards with squares and dominoes.

Example 5.2 How many binary sequences (sequences of the symbols 0 and 1) of length n have the property that no two 1s are adjacent?

Solution. Let h_n denote the number of binary sequences of length n with no "11" block. For example, $h_1 = 2$ since there are two sequences of length 1, namely, 0 and 1. Also, $h_2 = 3$, since 00, 01, and 10 are the three permissible sequences of length 2. Now consider the sequences of length $n \geq 3$. Of these h_n sequences, there are h_{n-1} that begin with a 0, since we can place a 0 at the beginning of any admissible sequence of length $n - 1$. A sequence that begins with a 1 must then begin with a 10 to avoid having a block 11, so there are h_{n-2} sequences that begin with a 1. We see that $h_n = h_{n-1} + h_{n-2}$, which once again is the Fibonacci recurrence but now with the initial conditions $h_1 = 2$ and $h_2 = 3$. Comparing this with the Fibonacci number sequence, we see that $h_n = F_{n+2}$. ■

Example 5.3 How many subsets of $[n]$ contain no pair of consecutive integers?

Solution. Let h_n denote the number of subsets of $[n] = \{1, 2, \ldots, n\}$ with no two consecutive elements. Thus, $h_1 = 2$, since this counts the two subsets $\oslash$ and $\{1\}$. Similarly, $h_2 = 3$ counts the three subsets $\oslash$, $\{1\}$, and $\{2\}$. Those subsets not containing the element n are the h_{n-1} subsets of $[n - 1]$ with no consecutive pair. Any subset that contains n cannot also contain the element $n - 1$, so these correspond to the h_{n-2} subsets of $[n - 2]$ with no consecutive pairs to which element n is appended. Altogether, $h_n = h_{n-1} + h_{n-2}$, with the initial conditions $h_1 = 2$ and $h_2 = 3$, and therefore $h_n = F_{n+2}$. ■

The identity of the following theorem is a consequence of Example 5.3.

Theorem 5.4 There are $\binom{n+1-k}{k}$ subsets of $[n]$ with exactly k elements and with no two consecutive integers. The total number of subsets of $[n]$ with no two consecutive integers is F_{n+2}. Therefore,

$$\sum_{k \geq 0} \binom{n+1-k}{k} = F_{n+2} \tag{5.3}$$

Proof. Given the k-element subset $S = \{s_1, s_2, \ldots, s_k\} \subseteq [n]$ with no pair of consecutive elements, define the subset $T = \{t_1, t_2, \ldots, t_k\}$ by $t_1 = s_1 - 0, t_2 = s_2 - 1, t_3 = s_3 - 2, \ldots, t_k = s_k - (k-1)$. Then $\{t_1, t_2, \ldots, t_k\}$ is a k-element subset of $[n-(k-1)]$ since $1 \le s_1 = t_1 < t_2 < \cdots < t_{k-1} < t_k = s_k - (k-1) \le n - (k-1)$. Moreover, the mapping is bijective, since it is inverted by letting $s_1 = t_1 + 0, s_2 = t_2 + 1, s_3 = t_3 + 2, \ldots, s_k = t_k + (k-1)$. Since there are $\binom{n+1-k}{k}$ subsets of $[n+1-k]$ with k-elements, this is the number of k-element subsets of $[n]$ with no pair of consecutive integers. In view of Example 5.3, each side of (5.3) counts the total number subsets of $[n]$ with no pair of consecutive integers in any of the subsets. ∎

We were fortunate in Examples 5.1, 5.2, and 5.3 to not only obtain the Fibonacci recurrence but also to have initial conditions that were two successive Fibonacci numbers. This allowed us to answer the problems in the form of the Fibonacci sequence. But what if we had to contend with more general initial conditions, say, $h_1 = 1$, and $h_2 = 5$? Evidently, we need a better understanding of how to solve the Fibonacci recurrence.

5.2.1 The *E* Operator and Its Application to First-Order Recurrences

Our approach to understanding and solving recurrence relations will be facilitated by introducing the E operator.

Definition 5.5 Given any expression $f(n)$ that depends on the integer variable n, the *successor operator* (or *shift operator*) E is defined by

$$Ef(n) = f(n+1) \tag{5.4}$$

Therefore, the successor operator E simply replaces n with $n+1$.

The following theorem lists some useful algebraic properties of the E operator. In particular, we see that E is a linear operator.

Theorem 5.6 Let α, c, c_1, and c_2 be constants, and $P(x)$ and $Q(x)$ be polynomials. Then

$$\begin{aligned} E\alpha^n &= \alpha^{n+1}, \quad E^k h_n = h_{n+k}, \quad E^{j+k} h_n = E^j(E^k h_n) = h_{n+j+k} \\ E(a_n b_n) &= (Ea_n)(Eb_n) = a_{n+1} b_{n+1} \\ E(cx_n) &= cx_{n+1} = cE(x_n) \\ E(c_1 x_n + c_2 y_n) &= c_1 x_{n+1} + c_2 y_{n+1} = c_1 E(x_n) + c_2 E(y_n) \\ P(E)Q(E)h_n &= Q(E)P(E)h_n \end{aligned} \tag{5.5}$$

Proof. Each property given in (5.5) except the last one follows by inspection. To verify the commutative property, suppose that $P(x) = \sum_{j=0}^{r} a_j x^j$ and $Q(x) = \sum_{k=0}^{s} b_k x^k$.

Then

$$P(E)Q(E)\,h_n = \left(\sum_{j=0}^{r} a_j E^j\right)\left(\sum_{k=0}^{s} b_k E^k\right) h_n = \left(\sum_{j=0}^{r} a_j E^j\right)\sum_{k=0}^{s} b_k h_{n+k}$$

$$= \sum_{j=0}^{r}\sum_{k=0}^{s} a_j b_k h_{n+k+j} = \sum_{k=0}^{s}\sum_{j=0}^{r} b_k a_j h_{n+k+j} = Q(E)\,P(E)h_n$$ ■

The E operator gives us a new way to write recurrence relations. For example, the first-order linear homogeneous recurrence relation $h_{n+1} = \alpha h_n$, where α is a constant, can be written

$$h_{n+1} - \alpha h_n = Eh_n - \alpha h_n = (E - \alpha)\,h_n = 0$$

or, equivalently, as

$$C(E)\,h_n = (E - \alpha)\,h_n = 0$$

The linear polynomial $C(x) = x - \alpha$ is known as the *characteristic polynomial* of the first-order recurrence relation $h_{n+1} = \alpha h_n$, and we see that the sequence h_n is *annihilated* (i.e., mapped to 0) by the operator $C(E) = E - \alpha$. The equation $C(x) = x - \alpha = 0$ is known as the *characteristic equation*, and its root α is called an *eigenvalue*. The eigenvalue is of importance since the following calculation shows that the geometric sequence $h_n = c\alpha^n$ is annihilated by the operator $C(E) = E - \alpha$ for any constant c:

$$C(E)\,c\alpha^n = (E - \alpha)\,c\alpha^n = cE\alpha^n - \alpha c\alpha^n = c\alpha^{n+1} - c\alpha^{n+1} = 0$$

If an initial condition $h_0 = A$ is prescribed, then $A = h_0 = c\alpha^0 = c$, and we see that $h_n = A\alpha^n$ is the unique solution of the recurrence relation $h_{n+1} = \alpha h_n$ that satisfies the initial condition $h_0 = A$.

Example 5.7 (Compound Interest) An initial deposit of \$100 is placed in a bank that offers an annual percentage yield (APY) of $i = 3\%$. If the interest is paid annually and is added to the account, interest in subsequent years is paid on the interest earned in previous years, resulting in what is known as *compound interest*.

(a) How much is on deposit after n years of receiving 3% annual compound interest?

(b) How many years are required to have at least \$200 in the account?

Solution

(a) Let P_n denote the principal after n years, where $P_0 = 100$ dollars is the initial amount deposited at the beginning of the first year. The principal P_{n+1} after

$n + 1$ years is the sum of the principal P_n at the beginning of the year plus the interest $0.03\,P_n$. Thus $P_{n+1} = P_n + 0.03P_n = (1.03)P_n$. Therefore, $P_{n+1} - (1.03)\,P_n = 0,\ P_0 = 100$. The root of the characteristic polynomial $C(x) = x - (1.03)$ is $x = 1.03$, so the recurrence relation is solved by $P_n = c\,(1.03)^n$. The constant c is determined from the initial condition $100 = P_0 = c\,(1.03)^0 = c$, so at the end of n years the amount on deposit is $P_n = 100\,(1.03)^n$.

(b) The account is doubled in the first year N at which $200 \le 100\,(1.03)^N$. Thus

$$N = \left\lceil \frac{\log 2}{\log(1.03)} \right\rceil = \lceil 23.449... \rceil = 24$$

where the ceiling function is used to round up to the year in which the interest is paid. This result illustrates the *rule of 72* (see Problem 5.2.11), which says that at an interest rate of i, it will take about $72/i$ years to double the principal. For $i = 3\%$, we see that $\frac{72}{3} = 24$ agrees with our exact computation. ■

5.2.2 The *E* Operator and the Fibonacci Recurrence

The Fibonacci recurrence $h_{n+2} - h_{n+1} - h_n = 0$, when expressed with the E operator, is

$$(E^2 - E - 1)h_n = 0 \tag{5.6}$$

Therefore, its characteristic polynomial is

$$C(x) = x^2 - x - 1 \tag{5.7}$$

The roots of the characteristic equation $C(x) = x^2 - x - 1 = 0$, found by the quadratic formula, are

$$\varphi = \frac{1+\sqrt{5}}{2}, \qquad \hat{\varphi} = \frac{1-\sqrt{5}}{2} \tag{5.8}$$

where φ is a very interesting number known as the *golden mean,* or the *golden ratio.*[1] The characteristic polynomial $C(x)$ can now be written as a product of linear factors, $C(x) = (x - \varphi)\,(x - \hat{\varphi})$, so the Fibonacci recurrence becomes

$$(E - \varphi)\,(E - \hat{\varphi})\,h_n = 0 \tag{5.9}$$

[1]There is an extensive popular literature on the role of the golden ratio in art, architecture, biological form, aesthetics, and so on. However, many of the claims made are dubious at best. An interesting discussion is found in Underwood Dudley's book [1].

Since the geometric sequences $c_1\varphi^n$ and $c_2\hat{\varphi}^n$ are annihilated by $E - \varphi$ and $E - \hat{\varphi}$, both of these sequences solve (5.9). Moreover, by the linearity property observed earlier in Theorem 5.6, the linear sum of powers

$$h_n = c_1\varphi^n + c_2\hat{\varphi}^n \tag{5.10}$$

also solves the Fibonacci recurrence for any constants c_1 and c_2.

All that remains is to ensure that if h_0 and h_1 are prescribed initial conditions, say, $h_0 = A$ and $h_1 = B$, then we can solve uniquely for c_1 and c_2. In other words, we must solve the two linear equations $A = h_0 = c_1 + c_2$ and $B = h_1 = c_1\varphi + c_2\hat{\varphi}$. In matrix form, this pair of linear equations becomes

$$\begin{bmatrix} 1 & 1 \\ \hat{\varphi} & \varphi \end{bmatrix} \begin{bmatrix} c_2 \\ c_1 \end{bmatrix} = \begin{bmatrix} A \\ B \end{bmatrix} \tag{5.11}$$

The determinant of the coefficients matrix is

$$\det \begin{bmatrix} 1 & 1 \\ \hat{\varphi} & \varphi \end{bmatrix} = \varphi - \hat{\varphi} = \sqrt{5} \neq 0$$

This means that the matrix

$$\begin{bmatrix} 1 & 1 \\ \hat{\varphi} & \varphi \end{bmatrix}$$

is nonsingular and therefore (5.11) is uniquely solvable for c_1 and c_2 for any initial conditions $h_0 = A$ and $h_1 = B$.

We have now obtained a general theorem for the Fibonacci recurrence.

Theorem 5.8 Every solution of the Fibonacci recurrence relation $h_n = h_{n-1} + h_{n-2}$ has the form $h_n = c_1\varphi^n + c_2\hat{\varphi}^n$, where $\varphi = (1 + \sqrt{5})/2$, $\hat{\varphi} = (1 - \sqrt{5})/2$ and c_1 and c_2 are arbitrary constants. If the sequence is given prescribed initial values A and B, then c_1 and c_2 are the unique solution of (5.11).

5.2.3 The Binet Formulas

The Fibonacci sequence is the unique solution of the Fibonacci recurrence with the initial conditions $F_0 = 0$ and $F_1 = 1$. In other words, $0 = F_0 = c_1 + c_2$ and $1 = F_1 = c_1\varphi + c_2\hat{\varphi}$, so that $c_2 = -c_1$ and $1 = c_1(\varphi - \hat{\varphi}) = c_1\sqrt{5}$, from which it follows that $F_n = (\varphi^n - \hat{\varphi}^n)/\sqrt{5}$. Similarly, to obtain the Lucas sequence $L_0 = 2, L_1 = 1, L_{n+2} = L_{n+1} + L_n$, we use the initial conditions $2 = L_0 = c_1 + c_2$ and $1 = L_1 = c_1\varphi + c_2\hat{\varphi} = (c_1 + c_2)/2 + ((c_1 - c_2)/2)\sqrt{5}$ to see that $c_1 = c_2 = 1$ and therefore $L_n = \varphi^n + \hat{\varphi}^n$.

These formulas for the Fibonacci and Lucas numbers are most often called the *Binet* formulas (Jacques Binet, 1786–1856), although they were derived earlier by Abraham de Moivre (1667–1754).

Theorem 5.9 **(The Binet Formulas)** Let $\varphi = (1+\sqrt{5})/2$, $\hat{\varphi} = (1-\sqrt{5})/2$. Then the Fibonacci and Lucas sequences are given by

$$F_n = \frac{\varphi^n - \hat{\varphi}^n}{\sqrt{5}} \tag{5.12}$$

$$L_n = \varphi^n + \hat{\varphi}^n \tag{5.13}$$

It follows from this theorem that any solution of the Fibonacci recurrence can just as well be written in the form $h_n = c_1F_n + c_2L_n$ with constants c_1 and c_2 that are uniquely determined from the initial conditions imposed. For example, if $h_0 = A$ and $h_1 = B$, then $h_n = [B - (A/2)]F_n + (A/2)L_n$.

Alternatively, any solution of the Fibonacci recurrence relation can be written in the form $h_n = c_1F_n + c_2F_{n+1}$, which makes it convenient to represent a solution in terms of the well-known Fibonacci numbers. This observation is used to prove the following theorem.

Theorem 5.10 $\varphi^n = F_{n+1} - \hat{\varphi}F_n$ for all $n \geq 0$

Proof. Since φ^n satisfies the Fibonacci recurrence, we know that there are constants c_1 and c_2 for which $\varphi^n = c_1F_{n+1} + c_2F_n$. Letting $n = 0$, we get the equation $1 = c_1F_1 + c_2F_0 = c_1 + 0$, so $c_1 = 1$. Letting $n = 1$, we get the equation $\varphi = F_2 + c_2F_1 = 1 + c_2$, so $c_2 = \varphi - 1 = -\hat{\varphi}$. ■

Alternate Proof. The following proof illustrates a very productive way of reasoning. Each term φ^n, F_{n+1}, and F_n is annihilated by the Fibonacci operator $C(E) = E^2 - E - 1$, so the sequence $h_n = \varphi^n - F_{n+1} + \hat{\varphi}F_n$ is annihilated as well. But $h_0 = \varphi^0 - F_1 + \hat{\varphi}F_0 = 1 - 1 + 0 = 0$ and $h_1 = \varphi^1 - F_2 + \hat{\varphi}F_1 = \varphi - 1 + \hat{\varphi} = 0$ so we conclude, since $h_{n+2} = h_{n+1} + h_n$, $h_0 = h_1 = 0$, that $h_n = 0$ for all $n \geq 0$. This is equivalent to the formula needed. ■

The ordinary generating function of the Fibonacci numbers can be derived easily from the Binet formula by first noticing that $\varphi + \hat{\varphi} = 1$, $\varphi - \hat{\varphi} = \sqrt{5}$, and $\varphi\hat{\varphi} = -1$. The OGF of the Fibonacci numbers is therefore

$$f_F(x) = \sum_{n=0}^{\infty} F_n x^n = \frac{1}{\sqrt{5}} \sum_{n=0}^{\infty} (\varphi^n - \hat{\varphi}^n)x^n = \frac{1}{\sqrt{5}} \left(\frac{1}{1-\varphi x} - \frac{1}{1-\hat{\varphi}x} \right)$$

$$= \frac{1}{\sqrt{5}} \frac{(\varphi - \hat{\varphi})x}{1 - (\varphi + \hat{\varphi})x + \varphi\hat{\varphi}x^2} = \frac{x}{1 - x - x^2} \tag{5.14}$$

Additional connections between recurrence relations and generating functions will be discussed in Section 5.6.

The Binet formulas (5.12) and (5.13) can often be used to derive identities by simple algebraic calculations. For example, if the two formulas are multiplied, we see that

$$F_n L_n = \frac{\varphi^{2n} - \hat{\varphi}^{2n}}{\sqrt{5}} = F_{2n} \tag{5.15}$$

Sometimes a Fibonacci number identity is most easily proved by taking advantage of the recurrence relation itself to give a proof by an algebraic calculation or by mathematical induction. The following example illustrates these ideas.

Theorem 5.11 The Fibonacci numbers satisfy the following identities:

(a)
$$F_1 + F_2 + \cdots + F_n = F_{n+2} - 1 \tag{5.16}$$

(b)
$$\sum_{k \geq 0} \binom{n}{k} F_k = F_{2n} \tag{5.17}$$

(c)
$$\sum_{k \geq 0} \binom{n}{k} F_{k+1} = F_{2n+1} \tag{5.18}$$

Proof.

(a) Since a telescoping sum is easy to evaluate, we see that

$$\begin{aligned} F_1 + F_2 + F_3 + \cdots + F_n &= (F_3 - F_2) + (F_4 - F_3) \\ &\quad + (F_5 - F_4) + \cdots + (F_{n+2} - F_{n+1}) \\ &= -F_2 + F_{n+2} = F_{n+2} - 1. \end{aligned}$$

(b) and (c) We use the principle of mathematical induction, proving both identities at the same time. For $n = 0$, the respective identities hold since $0 = 0$ and $1 = 1$. Now assume that *both* (5.17) and (5.18) hold some $n \geq 0$. Then, using Pascal's identity and the Fibonacci recurrence, we see that

$$\begin{aligned} \sum_{k \geq 0} \binom{n+1}{k} F_k &= \sum_{k \geq 0} \left[\binom{n}{k} + \binom{n}{k-1} \right] F_k = \sum_{k \geq 0} \binom{n}{k} F_k + \sum_{k \geq 1} \binom{n}{k-1} F_k \\ &= F_{2n} + F_{2n+1} = F_{2n+2} \end{aligned}$$

$$\begin{aligned} \sum_{k \geq 0} \binom{n+1}{k} F_{k+1} &= \sum_{k \geq 0} \left[\binom{n}{k} + \binom{n}{k-1} \right] F_{k+1} \\ &= \sum_{k \geq 0} \binom{n}{k} F_{k+1} + \sum_{k \geq 1} \binom{n}{k-1} (F_k + F_{k-1}) \\ &= F_{2n+1} + F_{2n+1} + F_{2n} = F_{2n+1} + F_{2n+2} = F_{2n+3} \end{aligned}$$

This calculation shows that both identities hold for $n + 1$ if it is assumed both identities hold for some $n \geq 0$. By the principle of mathematical induction, the identities hold for all $n \geq 0$. ■

PROBLEMS

5.2.1. Why should you not be surprised in Example 5.2 to discover that the number of binary sequences of length n with no consecutive 1s is given by the combinatorial Fibonacci number f_{n+1}, the number of tilings of a $1 \times (n+1)$ board with squares and dominoes?

5.2.2. Find the recurrence relation for the number of ways h_n to ascend a flight of n stairs if steps are taken either one or two stairs at a time.

5.2.3. In the block walking diagram below, there are two choices at each intersection—move to the next horizontal row below by turning to the southeast, or move directly south to the intersection two rows below:

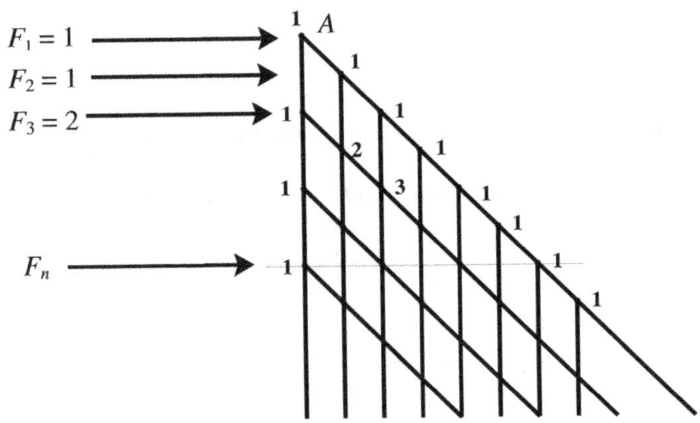

Let F_n denote the total number of paths that reach row n that start at point A in row 1. [*Hint*: F_n also gives the number of paths that move from a starting point in row r to a point in row $r + n - 1$.]

(a) Is the notation used in the diagram appropriate?

(b) Give a block walking proof of the identity $F_{m+n} = F_m F_{n+1} + F_{m-1} F_n$.

(c) Use the identity from part (b) to prove that F_m divides F_{km} for all $k \geq 0$ and $m \geq 1$.

5.2.4. Modify the block walking diagram in Problem 5.2.3 so that the number of paths that pass through a point of row n is given by the Lucas number L_n, $n \geq 0$. [*Hint*: Since $L_0 = 2$, there must be two starting points in row 0.]

5.2.5. In Fibonacci's influential book *Liber Abaci,* he proposed this problem: "A certain man put a pair of rabbits in a place surrounded by a wall. If the rabbits can breed during January, how many pairs of rabbits can be produced in a year, if it is supposed that every month each pair produces a new pair which can breed from the second month?" Find a recurrence relation and then solve Fibonacci's rabbit problem.

5.2.6. Prove that the solution of $h_{n+2} = h_{n+1} + h_n$, with the initial conditions $h_0 = A$, $h_1 = B$ is $h_n = AF_{n-1} + BF_n$, where $F_{-1} = 1$.

5.2.7. Prove the identity $F_0 + F_2 + F_4 + \cdots + F_{2m} = F_{2m+1} - 1$, $m \geq 0$ by

(a) mathematical induction

(b) summing $F_{2n+1} - F_{2n-1} = F_{2n}$

5.2.8. Prove the identity $F_1 + F_3 + F_5 + \cdots + F_{2m-1} = F_{2m}$, $m \geq 1$ by

(a) mathematical induction.

(b) summing $F_{2n+2} - F_{2n} = F_{2n+1}$.

5.2.9. Determine a simple expression involving one Lucas number for each of these sums of Lucas numbers.

(a) $L_0 + L_1 + L_2 + \cdots + L_m$

(b) $L_0 + L_2 + L_4 + \cdots + L_{2m}$

(c) $L_1 + L_3 + L_5 + \cdots + L_{2m-1}$

5.2.10. Generalize identities (5.17) and (5.18) by proving that

$$\sum_{k \geq 0} \binom{n}{k} F_{k+r} = F_{2n+r}, \quad r \geq 0$$

5.2.11. If i is the annual percentage yield, then the principal is doubled in $N = \lceil f(i) \rceil$ years, where $2 = [1 + (i/100)]^{f(i)}$ (see Example 5.7). Graph both $y = f(x)$ and $y = 72/x$ to see how accurately the rule of 72 estimates doubling times of an investment growing at the annual interest rate $x = i$. [*Hint*: Use the logarithm to solve for the function $f(i)$.]

5.2.12. Rewrite the Fibonacci recurrence as $F_{n-1} = F_{n+1} - F_n$ to define F_{-k} for $k \geq 1$. For example, $F_{-1} = F_1 - F_0 = 1 - 0 = 1 = F_1$ and $F_{-2} = F_0 - F_{-1} = 0 - 1 = -1 = -F_2$. Prove that $F_{-k} = (-1)^{k+1} F_k$.

5.2.13. Revise Problem 5.2.12 to define the Lucas numbers L_{-k} and prove a formula that expresses how L_{-k} is related to L_k for all $k \geq 0$.

5.2.14. Use mathematical induction to prove these identities:

(a) $F_{n-1} + F_{n+1} = L_n$, $n \geq 1$ **(b)** $L_{n-1} + L_{n+1} = 5F_n$, $n \geq 1$

5.2.15. Prove these identities:

(a) $\varphi^n = \varphi F_n + F_{n-1}$ **(b)** $\hat{\varphi}^n = \hat{\varphi} F_n + F_{n-1}$

[*Hint*: Apply the operator $C(E) = E^2 - E - 1$ to the sequences $h_n = \varphi^n - \varphi F_n - F_{n-1}$ and $g_n = \hat{\varphi}^n - \hat{\varphi} F_n - F_{n-1}$.]

5.2.16. Let $h_n = a\varphi^n + b\hat{\varphi}^n$ be a general solution of the Fibonacci recurrence. Use the Binet formulas to find the uniquely determined constants c, d, γ, and δ, depending on a and b, for which

(a) $h_n = cF_n + dL_n$ **(b)** $h_n = \gamma F_{n+1} + \delta L_n$

5.2.17. Use the Binet formulas to prove these identities:

(a) $F_{n+1} = \dfrac{F_n + L_n}{2}$ **(b)** $L_{n+1} = \dfrac{5F_n + L_n}{2}$

5.2.18. Give new proofs of the identities of Problem 5.2.17 by applying the operator $C(E) = E^2 - E - 1$ to the sequences $h_n = 2F_{n+1} - F_n - L_n$ and $g_n = 2L_{n+1} - 5F_n - L_n$.

5.2.19. Use the Binet formula to prove these identities:

(a) $F_n^2 + F_{n+1}^2 = F_{2n+1}$ **(b)** $L_n^2 + L_{n+1}^2 = 5F_{2n+1}$

(c) $F_{n+1}^2 - F_n^2 = \dfrac{F_{2n+1} + 4(-1)^n}{5}$ **(d)** $L_n^2 - 5F_n^2 = 4(-1)^n$

5.2.20. Prove that $\lim\limits_{n\to\infty} (F_{n+1}/F_n) = \varphi$.

5.2.21. Prove that

(a) $\begin{bmatrix} F_{n+1} & F_n \\ F_n & F_{n-1} \end{bmatrix} = \begin{bmatrix} 1 & 1 \\ 1 & 0 \end{bmatrix}^n$ **(b)** $F_{n+1}F_{n-1} - F_n^2 = (-1)^n$

5.3 SECOND-ORDER RECURRENCE RELATIONS

In this section we will investigate the general homogeneous linear second-order recurrence relation with constant coefficients given by

$$h_{n+2} = a_1 h_{n+1} + a_2 h_n,\ n \geq 0 \tag{5.19}$$

By mathematical induction, the sequence h_n is uniquely determined when (5.19) is accompanied by two *initial conditions*

$$h_0 = A \quad \text{and} \quad h_1 = B \tag{5.20}$$

Fortunately, the recurrence relation (5.19) and initial conditions (5.20) can be solved in much the same way that was successful in the last section to solve the Fibonacci recurrence.

As before, the E operator makes it easy to understand the algebraic considerations that arise. Note first that (5.19) can be written

$$C(E)h_n = (E^2 - a_1 E - a_2)h_n = 0 \tag{5.21}$$

where

$$C(x) = x^2 - a_1 x - a_2 \tag{5.22}$$

is the *characteristic polynomial* of the associated homogeneous recurrence relation $h_{n+2} = a_1 h_{n+1} + a_2 h_n$. The roots α and β of the characteristic equation $C(x) = 0$ are known as the *eigenvalues* of the problem and allow us to write the characteristic polynomial in the form $C(x) = (x - \alpha)(x - \beta)$. When $\alpha \neq \beta$, each of the geometric sequences α^n and β^n solves the homogeneous recurrence. For example, we see that $C(E)\alpha^n = (E - \beta)(E - \alpha)\alpha^n = (E - \beta)(\alpha^{n+1} - \alpha\alpha^n) = (E - \beta)0 = 0$, with a similar calculation showing $C(E)\beta^n = (E - \alpha)(E - \beta)\beta^n = 0$. By linearity, $C(E)(c_1\alpha^n + c_2\beta^n) = c_1 C(E)\alpha^n + c_2 C(E)\beta^n = 0$, which tells us that the *power sum* $c_1\alpha^n + c_2\beta^n$ solves the homogeneous recurrence for any constants c_1 and c_2. We will see later that if $C(x) = (x - \alpha)^2$, so the eigenvalue α has multiplicity 2, then it is the *generalized power sum* $c_0\alpha^n + c_1 n\alpha^n$ that solves the recurrence relation for any constants c_0 and c_1.

5.3.1 Solving Homogeneous Second-Order Linear Recurrence Relations

Example 5.12 In how many ways can a $1 \times n$ board (i.e., an *n-board*) be tiled with squares and dominoes, where the squares are all red and dominoes are blue or green?

Solution. Let h_n denote the number of tilings of an n-board. Then $h_1 = 1$, since there is just one tiling of a 1-board. A 2-board can be tiled with two red squares in one way and a domino in two ways, so $h_2 = 3$. An n-board can end with a red square added at the right end of any of the boards of length $n - 1$, or it can end with either a blue or green domino added to the right end of any of the tilings of boards of length $n - 2$. This gives us the second-order homogeneous linear recurrence relation

$$h_n = h_{n-1} + 2h_{n-2} \tag{5.23}$$

Using the initial conditions $h_1 = 1$ and $h_2 = 3$, the recurrence relation remains valid for all $n \geq 0$ if we define $h_0 = 1$. The characteristic polynomial corresponding to (5.23) is $x^2 - x - 2 = (x - 2)(x + 1)$, so there are two distinct eigenvalues, $\alpha = 2$ and $\beta = -1$. This gives us the general solution as a power sum, $h_n = c_1 2^n + c_2(-1)^n$. The initial conditions $h_0 = 1$ and $h_1 = 1$ will be satisfied if $1 = h_0 = c_1 + c_2$ and $1 = h_1 = 2c_1 + (-1)c_2$. Adding these equations shows that $2 = 3c_1$. Therefore, $c_1 = \frac{2}{3}$ and $c_2 = 1 - c_1 = \frac{1}{3}$, and we have our final solution $h_n = [2^{n+1} + (-1)^n]/3$. ■

Example 5.13 Solve the recurrence relation $h_{n+2} = 3h_{n+1} - h_n$ with the initial conditions $h_0 = 0$ and $h_1 = 1$.

Solution. The characteristic polynomial is $C(x) = x^2 - 3x + 1$, which has the two roots $\alpha = (3 + \sqrt{5})/2 = 1 + \varphi = \varphi^2$ and $\beta = (3 - \sqrt{5})/2 = 1 + \hat{\varphi} = \hat{\varphi}^2$, where $\varphi = (1 + \sqrt{5})/2$ and $\hat{\varphi} = (1 - \sqrt{5})/2$ are the eigenvalues of the Fibonacci recurrence. Thus, the recurrence relation is solved by the linear combination $h_n = c_1\varphi^{2n} + c_2\hat{\varphi}^{2n}$. The first initial condition $0 = h_0 = c_1 + c_2$ tells us that $c_2 = -c_1$, and the second initial condition shows that $1 = c_1(\varphi^2 - \hat{\varphi}^2) = c_1\sqrt{5}F_2 = c_1\sqrt{5}$. Therefore, $h_n = (\varphi^{2n} - \hat{\varphi}^{2n})/\sqrt{5} = F_{2n}$ using Binet's formula. ■

Example 5.14 In the block walking diagram shown below, consider the paths starting at point A in row 1 and moving downward to lower rows. There are three choices at each intersection: (1) turn left or turn right to reach the next horizontal row downward, or (2) proceed straight downward to move two rows lower. Determine P_n, the number of paths that arrive at row n starting at point A:

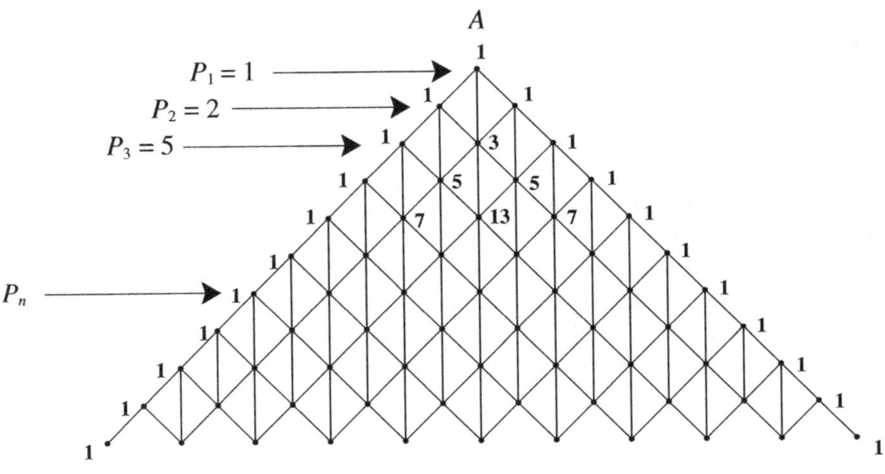

Solution. The diagram shows that $P_1 = 1$ and $P_2 = 2$. To obtain the recurrence relation, note that the P_{n+2} paths that reach row $n + 2$ from point A can be partitioned into three subsets determined by the first block walked when leaving point A. There are P_{n+1} paths that begin with a right turn, another P_{n+1} paths that begin with a left turn, and P_n paths that begin straight downward skipping over row 2. Thus, we obtain the homogeneous second order recurrence relation

$$P_{n+2} = 2P_{n+1} + P_n \tag{5.24}$$

The recurrence continues to hold for $n = 0$ if we define P_0 so that $2 = 2 \cdot 1 + P_0$. Thus, we define $P_0 = 0$. In terms of the E operator, the recurrence relation has the form

$(E^2 - 2E - 1)P_n = 0$, so the characteristic polynomial is $C(x) = x^2 - 2x - 1$. By the quadratic formula, there are two distinct eigenvalues, $\alpha = 1 + \sqrt{2}$ and $\beta = 1 - \sqrt{2}$ of the characteristic equation. The general solution of the homogeneous recurrence is therefore $P_n = c_1\alpha^n + c_2\beta^n$. To determine the constants c_1 and c_2, we must satisfy the initial conditions $0 = P_0 = c_1 + c_2$ and $1 = P_1 = c_1\alpha + c_2\beta = (c_1 + c_2) + \sqrt{2}(c_1 - c_2)$. Solving this system, we find that $c_1 = -c_2 = 1/2\sqrt{2}$. The solution of the recurrence relation $P_{n+2} = 2P_{n+1} + P_n$ with the initial conditions $P_0 = 0$ and $P_1 = 1$ is therefore

$$P_n = \frac{(1 + \sqrt{2})^n - (1 - \sqrt{2})^n}{2\sqrt{2}} \tag{5.25}$$

■

The numbers given by (5.25) are known as the *Pell numbers*,[2] which begin 0,1,2,5,12,29,70,.... They were known in ancient times, since they arise in the search for rational number approximations to the irrational number $\sqrt{2}$ (see Problems 5.3.21 and 5.3.24). The eigenvalue $1 + \sqrt{2}$ that appears in equation (5.25) is sometimes known as the *silver ratio*.

5.3.2 A Coupled First-Order System of Linear Recurrence Relations

The next example shows that a coupled system of two first-order linear recurrence relations can be replaced with a single linear recurrence relation of order 2.

Example 5.15 In how many ways can a $3 \times 2n$ board be tiled with horizontally or vertically arranged dominoes? Three tilings of a 3×12 board are shown below:

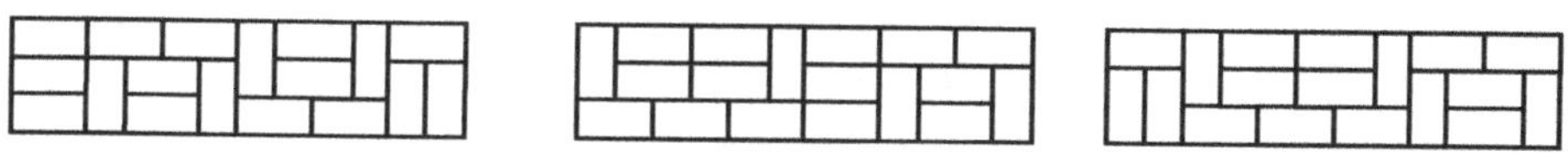

Solution. Let u_n denote the number of tilings of a board of length $2n$. It is easy to verify that a 3×2 board has three tilings, so $u_1 = 3$. To obtain a recurrence, consider a board of length $2n + 2$ and note that any tiling covers the cells in the leftmost column in three distinct ways: with three horizontal dominos, with a vertical domino above a horizontal domino, or with a vertical domino below a horizontal domino. These three possibilities to cover the left column of a board are shown in the tilings shown above. If the dominoes covering the left column are removed, the remainder of the board is

[2]The Pell numbers arise as integer solutions to *Pell's equation* $x^2 - 2y^2 = 1$. The name was given by Leonard Euler (1707–1783), who mistakenly thought that the equation had been first investigated by the seventeenth-century English mathematician John Pell (1611–1685). In fact, Pell contributed very little to the theory of Pell's equations.

either a rectangular board of length $2n$ or else a pruned board of length $2n + 1$ with a cell deleted at the bottom or top of the leftmost column. These three possibilities are shown by these diagrams.

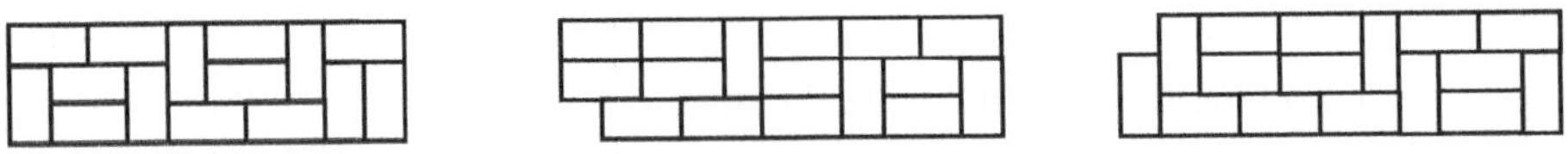

Let v_n denote the number of tilings of a pruned board of length $2n + 1$ with an upper or lower corner removed from the leftmost column. Considering the three ways in which the left column of a rectangular board of length $2n+ 2$ can be covered, we obtain the recurrence

$$u_{n+1} = u_n + 2v_n \tag{5.26}$$

Since $v_0 = 1$, this equation suggests that we define $u_0 = 1$ since $u_1 = 3$. Now consider how to cover the leftmost column of a pruned board. If a pruned board of length $2n + 3$ starts with a pair of horizontal dominoes, it necessarily has an additional horizontal domino that covers the remaining cell of the second column. This leaves a pruned board of length $2n + 1$. Similarly, if a pruned board of length $2n + 3$ begins with a vertical domino, it leaves a rectangular $3 \times (2n + 2)$ board. Thus, we have another first-order linear recurrence relation

$$v_{n+1} = v_n + u_{n+1} \tag{5.27}$$

which shows that $v_1 = 4$. In terms of the E operator, the coupled recurrences (5.26) and (5.27) become

$$\begin{aligned} (E-1)\,u_n &= 2v_n \\ (E-1)\,v_n &= Eu_n \end{aligned} \tag{5.28}$$

We now see that $(E-1)(E-1)\,u_n = (E-1)\,2v_n = 2\,(E-1)\,v_n = 2Eu_n$, which shows that the operator

$$C\,(E) = (E-1)^2 - 2E = E^2 - 4E + 1 \tag{5.29}$$

annihilates u_n. But we also see that $(E-1)\,(E-1)\,v_n = (E-1)\,Eu_n = E\,(E-1)\,u_n = 2Ev_n$, so $C(E)$ also annihilates v_n. Thus, *both* u_n and v_n satisfy the same second-order recurrence relation

$$w_{n+2} = 4w_{n+1} - w_n \tag{5.30}$$

The corresponding characteristic equation is $C(x) = x^2 - 4x + 1$, which has the eigenvalues $\alpha = 2 + \sqrt{3}$ and $\beta = 2 - \sqrt{3}$. It is now straightforward to determine the constants needed to satisfy the desired initial conditions. The conditions $u_0 = 1$ and $u_1 = 3$ give us

$$u_n = \left(\frac{\sqrt{3}+1}{2\sqrt{3}}\right)\left(2+\sqrt{3}\right)^n + \left(\frac{\sqrt{3}-1}{2\sqrt{3}}\right)\left(2-\sqrt{3}\right)^n$$

Similarly, the conditions $v_0 = 1$ and $v_1 = 4$ give us

$$v_n = \frac{\left(2+\sqrt{3}\right)^{n+1} - \left(2-\sqrt{3}\right)^{n+1}}{2\sqrt{3}}$$

■

5.3.3 Repeated Eigenvalues

We were fortunate in each of the Examples 5.12–5.15 that there were two distinct eigenvalues. But what if the characteristic equation has the form $C(x) = (x - \alpha)^2$ for which α is a repeated (i.e., double) root? We are left with just one solution of the homogeneous equation, $c_1\alpha^n$. We expect that a second linearly independent solution will be needed to ensure that two initial conditions can be satisfied. The following theorem shows us how to obtain the general solution, which is unique when two initial conditions $h_0 = A$ and $h_1 = B$ are prescribed.

Theorem 5.16 The second-order homogeneous recurrence relation $(E - \alpha)^2 h_n = 0$, $\alpha \neq 0$ is solved by each of the two sequences α^n and $n\alpha^n$ and therefore by any sequence $h_n = c_0\alpha^n + c_1 n\alpha^n$. In particular, $h_n = A\alpha^n + [(B/\alpha) - A]n\alpha^n$ is the unique solution of the recurrence relation that satisfies the initial conditions $h_0 = A$ and $h_1 = B$.

Proof. The geometric sequence α^n solves the recurrence relation since $(E - \alpha)^2 \alpha^n = (E - \alpha)(E - \alpha)\alpha^n = (E - \alpha)0 = 0$. The sequence $n\alpha^n$ also solves the homogeneous recurrence since

$$\begin{aligned}(E - \alpha)^2 n\alpha^n &= (E - \alpha)(E - a)n\alpha^n \\ &= (E - \alpha)[(n+1)\alpha^{n+1} - n\alpha^{n+1}] = (E - \alpha)\alpha^{n+1} = 0\end{aligned} \tag{5.31}$$

The solution of the recurrence given by the linear sum $h_n = c_0\alpha^n + c_1 n\alpha^n$ will satisfy the initial conditions if the constants c_0 and c_1 solve the equations $h_0 = c_0 = A$ and $h_1 = c_0\alpha + c_1\alpha = B$. Thus, $c_0 = A$ and $c_1 = (B/\alpha) - A$. ■

Example 5.17 In the game of chess, a king can attack any piece in an adjacent square, either horizontally, vertically, or diagonally. The following diagram shows one of the ways that five kings can be positioned on a 2×10 board so that no king can attack another king:

K		K					K		
				K					K

Let h_n denote the total number of ways to position n nonattacking kings on a $2 \times 2n$ board.

(a) Show that $h_n = 4h_{n-1} - 4h_{n-2}$, $n \geq 2$, where $h_1 = 4$ and $h_2 = 12$.
(b) Solve for h_n.

Solution.

(a) A king can be placed in any of the four cells of the 2×2 board, so $h_1 = 4$. For a 2×4 board, suppose that a king is placed in one of the four cells of the 2×2 board at the left and the second king is placed in the 2×2 board at the right. These 16 placements include the four ways to a king place in columns 2 and 3 of the board, so there are $16 - 4 = 12$ nonattacking arrangements of the two kings. More generally, given a $2 \times 2n$ board, there must be one king in the 2×2 subboard at the left and this king can be placed in 4 ways. If $n - 1$ additional kings are placed on the remaining board of length $2n - 2$, this can be done in h_{n-1} ways; however, the $4h_{n-2}$ arrangements with kings in columns 2 and 3 must be deleted. This gives the recurrence relation $h_n = 4h_{n-1} - 4h_{n-2}$.
(b) The characteristic equation is $C(x) = x^2 - 4x + 4 = (x - 2)^2$, which shows that $\alpha = 2$ is a double eigenvalue. The general solution is then $h_n = c_0 2^n + c_1 n 2^n$. The constants must be chosen so that $h_1 = 2c_0 + 2c_1 = 4$ and $h_2 = 4c_0 + 8c_1 = 12$. These equations show that $c_0 = c_1 = 1$, and therefore there are $h_n = (n + 1)\, 2^n$ ways to place n nonattacking kings on a $2 \times 2n$ board. ■

5.3.4 Complex Eigenvalues

Since the characteristic polynomial of a second-order linear recurrence relation is a quadratic equation, it can be anticipated that the eigenvalues will often be irrational numbers such as the golden or silver ratios that we have already encountered for the Fibonacci and Pell recurrences. However, it is also possible for the eigenvalues to be complex numbers. If the recurrence relation $h_{n+2} = a_1 h_{n+1} + a_2 h_n$ has real coefficients a_1 and a_2, then the characteristic equation has either real roots or a pair of complex conjugate roots α and $\bar{\alpha}$, where $\bar{z} = x - iy$ is the *complex conjugate* of the complex number $z = x + iy$ and $x = \mathrm{Re}\,(z) = \frac{1}{2}\,(z + \bar{z})$ is the *real part* of z.

The example that follows illustrates how to use the arithmetic of complex numbers to solve a recurrence relation with complex eigenvalues.

Example 5.18 Solve the second-order recurrence relation $h_{n+2} = 4h_{n+1} - 5h_n$, where $h_0 = 1$ and $h_1 = 2$.

Solution. The recurrence can be written as $C(E)h_n = (E^2 - 4E + 5)h_n = 0$, so the characteristic equation is $C(x) = x^2 - 4x + 5 = 0$. This equation can be written in the algebraically equivalent form $(x - 2)^2 = -1$, which makes it clear that the eigenvalues are the distinct complex conjugate numbers $\alpha = 2 + i$ and $\bar{\alpha} = 2 - i$. Therefore the general solution of the recurrence relation is $h_n = c_1(2 + i)^n + c_2(2 - i)^n$, where the constants c_1 and c_2 may be complex numbers. Letting $n = 0$ and $n = 1$, we get $1 = h_0 = c_1 + c_2$ and $2 = h_1 = c_1(2 + i) + c_2(2 - i) = 2(c_1 + c_2) + i(c_1 - c_2) = 2 + i(c_1 - c_2)$. Then $c_1 = c_2 = \frac{1}{2}$ and $h_n = \frac{1}{2}(2 + i)^n + \frac{1}{2}(2 - i)^n = \mathrm{Re}(2 + i)^n$. ■

5.3.5 A Connection with Linear Algebra

The reader has probably noticed that the terminology used in our discussion of linear recurrence relations—eigenvalue and linear sum—is much like that used in linear algebra. To explain why this is so, note that the homogeneous linear recurrence relation $h_{n+2} = a_1 h_{n+1} + a_2 h_n$ can be written as the equivalent matrix equation

$$\begin{bmatrix} h_{n+1} \\ h_{n+2} \end{bmatrix} = \begin{bmatrix} 0 & 1 \\ a_2 & a_1 \end{bmatrix} \begin{bmatrix} h_n \\ h_{n+1} \end{bmatrix} = M \begin{bmatrix} h_n \\ h_{n+1} \end{bmatrix}$$

where

$$M = \begin{bmatrix} 0 & 1 \\ a_2 & a_1 \end{bmatrix}$$

The characteristic polynomial of M is given by the determinant

$$\begin{aligned} \det(xI - M) &= \det \begin{bmatrix} x & -1 \\ -a_2 & x - a_1 \end{bmatrix} \\ &= x(x - a_1) - a_2 = x^2 - a_1 x - a_2 = C(x) \end{aligned}$$

Thus, the homogeneous recurrence relation $h_{n+2} = a_1 h_{n+1} + a_2 h_n$ *and* the matrix M have the same characteristic polynomial. Thus, M is known as the *companion matrix* of the polynomial $C(x) = x^2 - a_1 x - a_2$. We now see that the roots of $C(x) = 0$ are, indeed, eigenvalues—they are the eigenvalues of the companion matrix M.

Additional connections of linear recurrence relations with linear algebra will be explored in Chapter 7.

PROBLEMS

5.3.1. Let $h_n = 2h_{n-1} + 3h_{n-2}$, $n \geq 2$.

(a) What are the corresponding eigenvalues?

(b) What is the general solution (power sum) with unspecified initial conditions?

(c) Solve the recurrence relation with the initial conditions $h_0 = 0$ and $h_1 = 1$.

(d) Solve the recurrence relation with the initial conditions $h_0 = 3$ and $h_1 = 1$.

5.3.2. Let $h_n = 2h_{n-2}$, $n \geq 2$.

(a) What are the corresponding eigenvalues?

(b) What is the general solution (power sum) with unspecified initial conditions?

(c) Solve the recurrence relation with the initial conditions $h_0 = 0$ and $h_1 = 1$.

5.3.3. Let $h_n = h_{n-1} - 3h_{n-2}$, $n \geq 2$.

(a) What are the corresponding eigenvalues?

(b) What is the general solution (power sum) with unspecified initial conditions?

5.3.4. Let $h_n = 8h_{n-1} - 16h_{n-2}$, $n \geq 2$.

(a) What are the corresponding eigenvalues?

(b) What is the general solution (power sum) with unspecified initial conditions?

(c) Solve the recurrence relation with the initial conditions $h_0 = 1$ and $h_1 = 5$.

5.3.5. Let $h_n = 2h_{n-1} - h_{n-2}$, $n \geq 2$.

(a) What are the corresponding eigenvalues?

(b) What is the general solution (power sum) with unspecified initial conditions?

5.3.6. Solve the recurrence relation $h_{n+2} = 6h_{n+1} - 9h_n$, $n \geq 0$, with the initial conditions $h_0 = 1$ and $h_1 = 4$.

5.3.7. Verify that every linear polynomial $c_0 + c_1 n$ is annihilated by $(E - 1)^2$.

5.3.8. Show that the Lucas number L_{2n} solves the recurrence relation $h_{n+2} = 3h_{n+1} - h_n$ with the initial conditions $h_0 = 2$ and $h_1 = 3$.

5.3.9. **(a)** What recurrence relation has the eigenvalues φ^2 and $\hat{\varphi}^2$?

(b) Why can the general solution of the recurrence relation be written as $c_1 F_{2n} + c_2 L_{2n}$?

5.3.10. Big Bob is a daily customer at Ferdinand's Ice Cream Shoppe, where he buys either a strawberry or a mint cone for \$1, or he buys a \$2 milkshake in one of the flavors chocolate, raspberry, or pineapple. In how many ways can Bob make a sequence of purchases that come to \$10?

5.3.11. Let α and β be the eigenvalues of the characteristic polynomial $C(x) = x^2 - a_1x - a_2$. Show that $\alpha + \beta = a_1$ and $\alpha\beta = -a_2$.

5.3.12. What homogeneous recurrence relation, together with initial conditions, is solved by $g_n = [(1 + \sqrt{3})^n - (1 - \sqrt{3})^n]/2\sqrt{3}$?

5.3.13. **(a)** Show that if $g_n = c_1\alpha^n + c_2\beta^n$, where α and β are distinct eigenvalues, then the constants c_1 and c_2 are uniquely determined by the initial conditions $g_0 = A$ and $g_1 = B$.

(b) Suppose that $g_n = c_1\alpha^n + c_2n\alpha^n$ is the general solution of a second-order recurrence relation for which $\alpha \neq 0$ is a repeated eigenvalue. Show that the constants c_0 and c_1 are uniquely determined by the initial conditions $g_0 = A$ and $g_1 = B$.

5.3.14. Prove that any solution of the Pell recurrence relation $h_{n+2} = 2h_{n+1} + h_n$ with the initial conditions $h_0 = A$ and $h_1 = B$ can be solved by uniquely determining coefficients c_1 and c_2 so that $h_n = c_1P_n + c_2P_{n+1}$.

5.3.15. Show that the Pell number P_n gives the number of ways that a $1 \times n$ board can be tiled with squares and dominoes, where the squares are either red or blue and the dominoes are all white.

5.3.16. Prove that the Pell number P_{n+1} is given by

$$P_{n+1} = \sum_{d\geq 0} \binom{n-d}{d} 2^{n-2d}$$

for every $n \geq 0$. [*Hint*: Use the tiling model of Pell numbers given in Problem 5.3.15.]

5.3.17. Prove that

(a) the Pell numbers P_n are given by

$$\begin{bmatrix} P_{n+1} & P_n \\ P_n & P_{n-1} \end{bmatrix} = \begin{bmatrix} 2 & 1 \\ 1 & 0 \end{bmatrix}^n, \quad n \geq 1$$

(b) $P_{n+1}P_{n-1} - P_n^2 = (-1)^n$

5.3.18. Use the recursive relation for the Pell numbers to show that

(a) $P_1 + P_2 + \cdots + P_n = \frac{1}{2}(P_{n+1} + P_n - 1)$ [*Hint*: Sum $P_{k+1} - P_k = P_k + P_{k-1}$ over $k = 1, 2, \ldots, n$]

(b) $P_1 + P_3 + \cdots + P_{2n+1} = \frac{1}{2}P_{2n+2}$

(c) $P_2 + P_4 + \cdots + P_{2n} = \frac{1}{2}(P_{2n+1} - 1)$

5.3.19. **(a)** Use a block walking argument (see Example 5.14) to show that $P_{m+n} = P_m P_{n+1} + P_{m-1} P_n$. [*Hint*: The paths from point A to row $m + n$ are of two types: those that pass through a point of row m and those that move directly downward from row $m - 1$ to row $m + 1$.]

(b) Use part (a) to prove that $P_{2n+2} = P_{n+1} P_{n+2} + P_n P_{n+1}$.

(c) Use the result of Problem 5.3.18(b) to prove that P_{n+1} divides $\sum_{k=0}^{n} P_{2k+1}$.

[*Hint*: What is the parity of $P_{n+2} + P_n$?]

5.3.20. Prove that the Pell number P_m is a divisor of P_{mk} for all $k \geq 0$.
[*Hint*: Use the identity $P_{m+n} = P_m P_{n+1} + P_{m-1} P_n$ of Problem 5.3.19.]

5.3.21. **(a)** Prove that the limit of the ratio of successive Pell numbers is the silver ratio; that is, prove that $\lim_{n\to\infty} (P_{n+1}/P_n) = 1 + \sqrt{2}$.

(b) Prove that $\lim_{n\to\infty} (P_n + P_{n-1})/P_n = \sqrt{2}$

5.3.22. The *half-companion* Pell numbers are defined by the recurrence relation $H_{n+2} = 2H_{n+1} + H_n$, $n \geq 0$ with the initial conditions $H_0 = H_1 = 1$.

(a) Show that $H_n = [(1 + \sqrt{2})^n + (1 - \sqrt{2})^n]/2$ for all $n \geq 0$.

(b) Complete the table below for the Pell and half-companion Pell numbers:

n	0	1	2	3	4	5	6	7	8	9	10
H_n	1	1	3	7	17						
P_n	0	1	2	5	12						

(c) Prove that $H_n = (P_{n+1} + P_{n-1})/2$ for all $n \geq 1$.

(d) Prove that $P_n = \frac{1}{4}(H_{n+1} + H_{n-1})$ for all $n \geq 1$.

5.3.23. Prove the following recurrences for the Pell and half-companion Pell numbers defined in Problem 5.3.22.

(a) $H_{n+1} = H_n + 2P_n$ for all $n \geq 0$

(b) $P_{n+1} = H_n + P_n$ for all $n \geq 0$

5.3.24. Since $\sqrt{2}$ is irrational, there are no integers H and P for which $H^2 - 2P^2 = 0$. However, if H and P are positive integers that satisfy the Pell equation $H^2 - 2P^2 = \pm 1$, their H/P ratio will give a rational number approximation of $\sqrt{2}$.

(a) Show that H_1 and P_1 satisfy the Pell equation $H^2 - 2P^2 = -1$ and H_2 and P_2 satisfy the Pell equation $H^2 - 2P^2 = 1$.

(b) Prove that

$$\begin{bmatrix} H_n & 2P_n \\ P_n & H_n \end{bmatrix} = \begin{bmatrix} 1 & 2 \\ 1 & 1 \end{bmatrix}^n,$$

for all $n \geq 0$, where

$$\begin{bmatrix} 1 & 2 \\ 1 & 1 \end{bmatrix}^0 = \begin{bmatrix} 1 & 0 \\ 0 & 1 \end{bmatrix}$$

is the identity matrix. [*Hint*: See the identities of Problem 5.3.23.]

(c) Prove that $H_{n+1}^2 - 2P_{n+1}^2 = -\left(H_n^2 - 2P_n^2\right) = (-1)^{n+1}$.

(d) Complete the following table, showing the convergence of the rational number sequence H_n/P_n to $\sqrt{2}$:

n	0	1	2	3	4	5	6	7	8	9	10
H_n	1	1	3	7	17						
P_n	0	1	2	5	12						
H_n/P_n	—	1	1.5	1.4	1.416667						

5.4 HIGHER-ORDER LINEAR HOMOGENEOUS RECURRENCE RELATIONS

Our goal in this section is to solve recurrence relations of the form

$$h_{n+k} = a_1 h_{n+k-1} + a_2 h_{n+k-2} + \cdots + a_k h_n, \ n \geq 0 \tag{5.32}$$

where $a_1, a_2, \ldots, a_k$ are constants. That is, h_{n+k} is a linear sum of the k immediately preceding terms. It is assumed that the coefficient $a_k \neq 0$, and we say that the recurrence relation has *order k*.

An equation with the form (5.32) is said to be a *homogeneous linear recurrence relation of order k with constant coefficients.* If a solution of the recurrence can be found that satisfies the k initial conditions

$$h_0 = A_0, h_1 = A_1, \ldots, h_{k-1} = A_{k-1} \tag{5.33}$$

then, by mathematical induction, this sequence is the unique solution of (5.32) and (5.33).

Fortunately, the ideas used to solve second order linear recurrences can be generalized to show that there exists a unique solution of (5.32) and (5.33). As before, it is helpful to introduce the E operator. This allows us to view a solution of the recurrence (5.32) as a sequence h_n that is annihilated by the operator

$$C(E) = E^k - a_1E^{k-1} - a_2E^{k-2} - \cdots - a_k \tag{5.34}$$

where

$$C(x) = x^k - a_1x^{k-1} - a_2x^{k-2} - \cdots - a_k \tag{5.35}$$

As before, $C(x)$ is the *characteristic polynomial* of the linear recurrence, but it is now of degree k. If the *characteristic equation* $C(x) = 0$ has roots $\alpha, \beta, \ldots, \lambda$ with the respective multiplicities $r, s, \ldots, w$, we then have the factorization

$$C(x) = (x-\alpha)^r (x-\beta)^s \cdots (x-\lambda)^w . \tag{5.36}$$

The roots $\alpha, \beta, \ldots, \lambda$, which may be complex numbers, are the *eigenvalues* of the linear recurrence. Choosing $x = 0$ in (5.35) and (5.36), we see that $C(0) = -a_k = \alpha^r\beta^s \cdots \lambda^w$; therefore, 0 is not an eigenvalue.

We will begin with the simplest case when the k eigenvalues are distinct, so that $r = s = \cdots = w = 1$. The eigenvalues are then said to be *nonrepeated*, or *simple*, eigenvalues. However, we will see later that there is little additional difficulty in handling the case of repeated eigenvalues.

5.4.1 Nonrepeated Eigenvalues

Suppose that the characteristic polynomial has k simple, or nonrepeated, roots $\alpha, \beta, \ldots, \lambda$. Then the characteristic polynomial has the factorization $C(x) = (x-\alpha)(x-\beta)\cdots(x-\lambda)$. Moreover, since $E-\alpha$ annihilates α^n, $E-\beta$ annihilates β^n, and so on, we see that each of the k geometric sequences $\alpha^n, \beta^n, \ldots, \lambda^n$ is a solution of the recurrence $C(E)h_n = 0$. Since $C(E)$ is a linear operator, the *power sum* (PS) $h_n = c_1\alpha^n + c_2\beta^n + \cdots + c_k\lambda^n$ also solves the recurrence for any choice of the constants $c_1, c_2, \ldots, c_k$.

Example 5.19 Solve the recurrence $h_{n+3} = 7h_{n+1} + 6h_n$ with the initial conditions $h_0 = 0, h_1 = 1, h_2 = 2$.

Solution. The characteristic polynomial of this linear recurrence relation of order $k = 3$ is $C(x) = x^3 - 7x - 6 = (x-3)(x+1)(x+2)$, where the factorization shows that there are three simple eigenvalues, $\alpha = 3$, $\beta = -1$, and $\gamma = -2$. The power sum $h_n = c_13^n + c_2(-1)^n + c_3(-2)^n$ therefore solves the recurrence. To satisfy the initial conditions, the constants c_1, c_2, and c_3 must be chosen so that

$$\begin{aligned} 0 = h_0 &= c_1 + c_2 + c_3 \\ 1 = h_1 &= 3c_1 - c_2 - 2c_3 \\ 2 = h_2 &= 9c_1 + c_2 + 4c_3 \end{aligned}$$

The system is solved (using row operations, for example) to find that $c_1 = \frac{1}{4}, c_2 = -\frac{1}{4}, c_3 = 0$ and therefore $h_n = [3^n - (-1)^n]/4$. Since h_n is a linear sum of just the two eigenvalues $\alpha = 3$ and $\beta = -1$, we see that h_n also satisfies the second order recurrence relation $(E-3)(E+1)h_n = 0$. With different initial conditions, this would not be the case. ■

Example 5.20 In how many ways can a $1 \times n$ board be tiled with 1×1 squares, 1×2 dominoes, and 1×3 trominoes? Three tilings of a 1×10 board are shown here:

Solution. Let h_n denote the number of tilings of a board of length n. It is easy to check that $h_1 = 1, h_2 = 2,$ and $h_3 = 4$. The h_n tilings of a board of length n for $n \geq 4$ can be partitioned into three classes depending on the type of the rightmost tile. There are, respectively, h_{n-1}, h_{n-2}, and h_{n-3}, tilings that end with a square, domino, and tromino. This gives us the third-order linear recurrence relation

$$h_n = h_{n-1} + h_{n-2} + h_{n-3} \tag{5.37}$$

If we define $h_0 = 1$, then (5.37) will hold even for $n = 3$. The graph of the characteristic polynomial $y = C(x) = x^3 - x^2 - x - 1$ is plotted here.

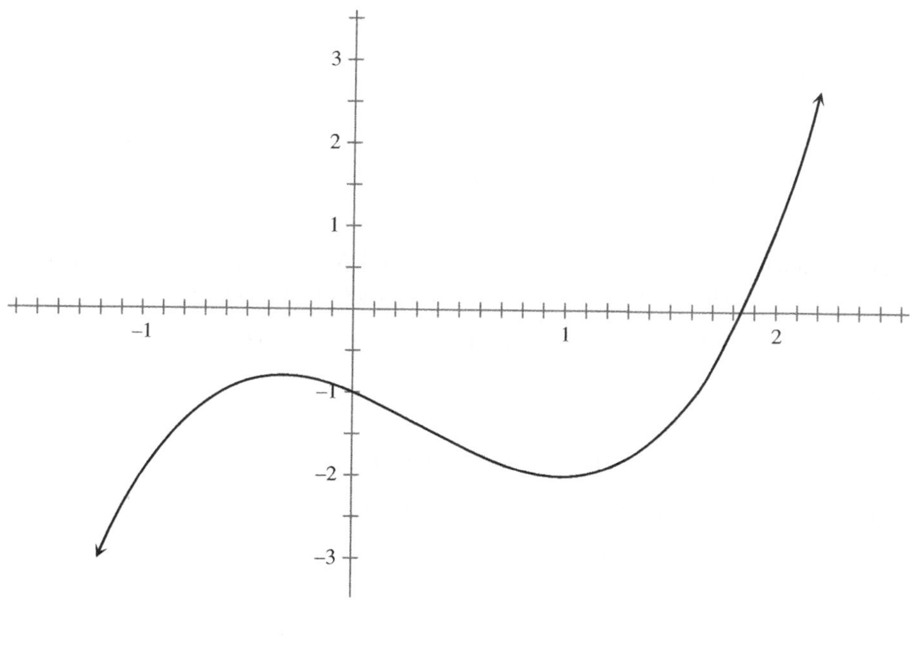

The plot shows that there is a real root, α, with $1 < \alpha < 2$. This is a nonrepeated eigenvalue, since the graph is not tangent to the x axis at α. The other two eigenvalues, β and γ, must then be a pair of complex conjugate numbers, so we now know that the three eigenvalues α, β, and γ are simple. A numerical calculation would show that the real eigenvalue is about 1.89 and the conjugate pair about $-0.4 \pm 0.6i$. It follows that h_n is a power sum (PS) of the form $h_n = c_1\alpha^n + c_2\beta^n + c_3\gamma^n$. It only remains to show that there exist constants c_1, c_2, and c_3 for which the initial conditions are satisfied. In matrix form, we must solve the equation

$$\begin{bmatrix} h_0 \\ h_1 \\ h_2 \end{bmatrix} = \begin{bmatrix} 1 \\ 1 \\ 2 \end{bmatrix} = \begin{bmatrix} 1 & 1 & 1 \\ \alpha & \beta & \gamma \\ \alpha^2 & \beta^2 & \gamma^2 \end{bmatrix} \begin{bmatrix} c_1 \\ c_2 \\ c_3 \end{bmatrix}$$

The determinant of the coefficient matrix can be obtained by expanding by cofactors of the last column, showing that

$$\begin{aligned} \det \begin{bmatrix} 1 & 1 & 1 \\ \alpha & \beta & \gamma \\ \alpha^2 & \beta^2 & \gamma^2 \end{bmatrix} &= \gamma^2 (\beta - \alpha) - \gamma(\beta^2 - \alpha^2) + (\alpha\beta^2 - \beta\alpha^2) \\ &= (\beta - \alpha)(\gamma^2 - \gamma\beta - \gamma\alpha + \alpha\beta) = (\beta - \alpha)(\gamma - \beta)(\gamma - \alpha). \end{aligned} \tag{5.38}$$

Since the eigenvalues α, β, and γ are distinct, the determinant is nonzero and so there exists a unique solution c_1, c_2, c_3 to satisfy the initial conditions, although we will not calculate them explicitly. ■

If the recurrence (5.37) is extended backward two additional terms, we obtain the *tribonacci* sequence 0,0,1,1,2,4,7,13,24,.... The tribonacci numbers are defined by

$$T_0 = 0, T_1 = 0, T_2 = 1, \; T_n = T_{n-1} + T_{n-2} + T_{n-3}, \quad n \geq 3 \tag{5.39}$$

The solution to the question posed in Example 5.20 can then be written as $h_n = T_{n+2}$. (The tribonacci numbers also have an interesting connection with the "four number game," as explored later in Problem 5.7.15.)

Example 5.21 In a Foxtrot cartoon[3] Jason and Marcus declare that they are "... sick of football favoring the brawny." Their version of football goes like this.

Jason: "Hut 0, Hut 1, Hut 1, Hut 2, Hut 3, Hut 5, Hut 8, Hut 13"

Marcus: "Is it the Fibonacci series?"

Jason: "Correct! Touchdown, Marcus!"

Marcus: "Hut 3, Hut 0, Hut 2, Hut 3, Hut 3, Hut 5"

How should Jason respond to Marcus to score a touchdown?

[3] Bill Amend's cartoon can be viewed at WolframMathWorld: `http://mathworld.wolfram.com/PerrinSequence.html`

Solution. The sequence that begins 3,0,2,3,2,5,... is similar to the Fibonacci sequence, but the sum of two successive terms gives, not the next term, but the term that follows the next term. Thus, the sequence is defined by the linear recurrence relation

$$p_{n+3} = p_{n+1} + p_n, \quad n \geq 0 \tag{5.40}$$

and satisfies the initial conditions

$$p_0 = 3, p_1 = 0, p_2 = 2 \tag{5.41}$$

The roots of the characteristic equation $C(x) = x^3 - x - 1 = 0$ can be solved, say, by turning to a programmable calculator or CAS, to show that there is one real eigenvalue $\alpha \approx 1.32$ and two complex eigenvalues $\beta \approx -0.66 + 0.56i$ and $\gamma \approx -0.66 - 0.56i$. This is also revealed by the graph of $y = C(x)$, which is similar to that shown in Example 5.20. However, the numerical values are of less interest than the algebraic relationships of the eigenvalues. First note that

$$x^3 - x - 1 = (x-\alpha)(x-\beta)(x-\gamma) = x^3 - (\alpha+\beta+\gamma)x^2 + (\alpha\beta+\beta\gamma+\alpha\gamma)x - \alpha\beta\gamma$$

Then, by equating coefficients of equal powers of x, we see that $\alpha + \beta + \gamma = 0, \alpha\beta + \beta\gamma + \alpha\gamma = -1$, and $\alpha\beta\gamma = 1$. Therefore, $\alpha^2 + \beta^2 + \gamma^2 = (\alpha+\beta+\gamma)^2 - 2(\alpha\beta+\beta\gamma+\alpha\gamma) = 0^2 - 2(-1) = 2$. These calculations show that $\alpha^n + \beta^n + \gamma^n$ solves the recurrence and satisfies the initial conditions $\alpha^0 + \beta^0 + \gamma^0 = 3$, $\alpha + \beta + \gamma = 0$, $\alpha^2 + \beta^2 + \gamma^2 = 2$. Therefore, $p_n = \alpha^n + \beta^n + \gamma^n$.

Jason can search the *Online Encyclopedia of Integer Sequences*[4] (see Appendix B) to learn that he will score a touchdown by answering Marcus with the response "3,0,2,3,2,5,5,7,10,12,17,... is the Perrin sequence". We conjecture that this is the only popular comic strip featuring the Perrin numbers. ■

The next example shows that a system of linear recurrence relations, one first order and the other second-order, can be converted to a single third-order linear recurrence relation.

Example 5.22 In how many ways can a $2 \times n$ board be tiled with squares and dominoes? The dominoes can be oriented either vertically or horizontally, as shown in the sample tilings below:

Solution. Let u_n denote the number of tilings of a $2 \times n$ board. It can be checked that $u_1 = 2$ and $u_2 = 7$. The diagram above shows that if the tiles covering the leftmost column are removed, the remaining board is either another rectangular board or else

[4]Search the values 3, 0,2,3,2,5 at `http://oeis.org/`.

a pruned board with a deleted corner cell. Therefore, it will be helpful to let v_n denote the number of tilings of a pruned $2 \times n$ board with a deleted corner cell. Then $v_1 = 1$ and $v_2 = 3$. It is now straightforward to obtain the pair of coupled linear recurrence relations

$$\begin{aligned} u_{n+2} &= 2u_{n+1} + u_n + 2v_{n+1} \\ v_{n+1} &= v_n + u_n \end{aligned} \tag{5.42}$$

Equations (5.42) hold for $n = 0$ if we define $u_0 = 1$ and $v_0 = 0$. We also see that

$$u_3 = 2\,(7) + 2 + 2\,(3) = 22 \quad \text{and} \quad v_3 = 3 + 7 = 10 \tag{5.43}$$

In terms of the E operator, these coupled linear recurrence relations (5.42) can be written

$$\begin{aligned} (E^2 - 2E - 1)u_n &= 2Ev_n \\ u_n &= (E - 1)\,v_n \end{aligned} \tag{5.44}$$

Then $0 = (E^2 - 2E - 1)u_n - 2Ev_n = (E^2 - 2E - 1)(E - 1)v_n - 2Ev_n = (E^3 - 3E^2 - E + 1)v_n$, which shows that v_n is annihilated by the operator $C(E) = E^3 - 3E^2 - E + 1$. Similarly, $C\,(E)\,u_n = C\,(E)\,(E - 1)\,v_n = (E - 1)\,C\,(E)\,v_n = (E - 1)\,0 = 0$, so $C(E)$ also annihilates u_n. Thus, both sequences u_n and v_n satisfy the same third-order recurrence

$$w_{n+3} = 3w_{n+2} + w_{n+1} - w_n \tag{5.45}$$

The graph of $y = C(x) = x^3 - 3x^2 - x + 1$ shows there are three distinct real eigenvalues α, β, and γ:

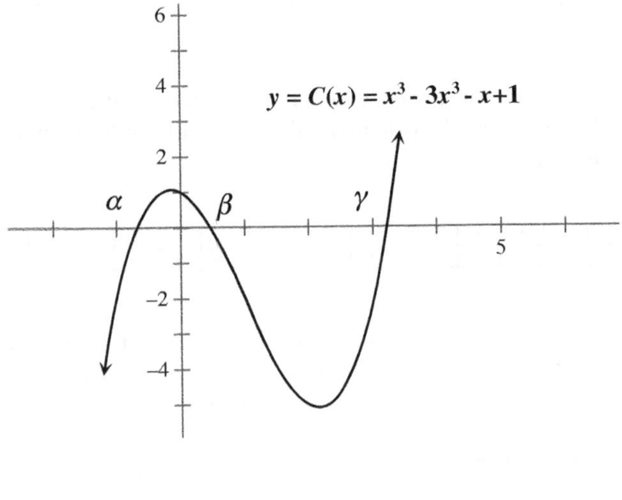

We now know that both u_n and v_n are power sums of these eigenvalues, although with different coefficients determined by their corresponding initial conditions. Since these sequences both solve (5.45) with the initial conditions $u_0 = 1, u_1 = 2, u_2 = 7$ and $v_0 = 0, v_1 = 1, v_2 = 3$, we must solve the two systems of linear equations

$$\begin{bmatrix} 1 & 1 & 1 \\ \alpha & \beta & \gamma \\ \alpha^2 & \beta^2 & \gamma^2 \end{bmatrix} \begin{bmatrix} c_1 \\ c_2 \\ c_3 \end{bmatrix} = \begin{bmatrix} 1 \\ 2 \\ 7 \end{bmatrix} \quad \text{and} \quad \begin{bmatrix} 1 & 1 & 1 \\ \alpha & \beta & \gamma \\ \alpha^2 & \beta^2 & \gamma^2 \end{bmatrix} \begin{bmatrix} \hat{c}_1 \\ \hat{c}_2 \\ \hat{c}_3 \end{bmatrix} = \begin{bmatrix} 0 \\ 1 \\ 3 \end{bmatrix} \tag{5.46}$$

As shown already in (5.38), the coefficient matrix is nonsingular since for distinct eigenvalues we know that its determinant is $(\beta - \alpha)(\gamma - \beta)(\gamma - \alpha) \neq 0$. ■

In each of the four Examples 5.19–5.22, we considered a linear homogeneous recurrence relation with three distinct eigenvalues, α, β, and γ. The recurrence relation was then solved by any power sum $h_n = c_1\alpha^n + c_2\beta^n + c_3\gamma^n$. Moreover, the coefficients c_1, c_2, c_3 for any set of prescribed initial conditions $h_0 = A_1, h_1 = A_2, h_2 = A_3$ are given by the matrix equation

$$\begin{bmatrix} 1 & 1 & 1 \\ \alpha & \beta & \gamma \\ \alpha^2 & \beta^2 & \gamma^2 \end{bmatrix} \begin{bmatrix} c_1 \\ c_2 \\ c_3 \end{bmatrix} = \begin{bmatrix} A_1 \\ A_2 \\ A_3 \end{bmatrix}$$

Since the determinant of the matrix is $(\beta - \alpha)(\gamma - \beta)(\gamma - \alpha) \neq 0$, there always exists a unique solution of this equation for the constants c_1, c_2, c_3.

There is little change needed to solve a linear recurrence of order k whose characteristic equation has k distinct eigenvalues $\alpha_1, \alpha_2, \alpha_3, \ldots, \alpha_k$. As before, the recurrence relation is solved for any power sum $h_n = c_1\alpha_1^n + c_2\alpha_2^n + \cdots + c_k\alpha_k^n$. Therefore, it only remains to show that there exists a unique solution when k initial conditions $h_0 = A_1, h_1 = A_2, \ldots, h_{k-1} = A_k$ are prescribed; that is, we must show that the system of equations $c_1\alpha_1^j + c_2\alpha_2^j + \cdots c_k\alpha_k^j = A_{j+1}, j = 0, 1, \ldots, k-1$ always has a unique solution for the coefficients $c_1, c_2, \ldots, c_k$. This is a consequence of the following theorem, in which a formula for the determinant of the coefficient matrix of the equations is derived. The determinant is seen to be the product of the differences of the eigenvalues, so it is not zero when the k eigenvalues are distinct.

Theorem 5.23 Let

$$V_k(\alpha_1, \alpha_2, \alpha_3, \ldots, \alpha_k) = \det \begin{bmatrix} 1 & 1 & 1 & \cdots & 1 & 1 \\ \alpha_1 & \alpha_2 & \alpha_3 & \cdots & \alpha_{k-1} & \alpha_k \\ \alpha_1^2 & \alpha_2^2 & \alpha_3^2 & \cdots & \alpha_{k-1}^2 & \alpha_k^2 \\ \vdots & \vdots & \vdots & \ddots & \vdots & \vdots \\ \alpha_1^{k-2} & \alpha_2^{k-2} & \alpha_3^{k-2} & \cdots & \alpha_{k-1}^{k-2} & \alpha_k^{k-2} \\ \alpha_1^{k-1} & \alpha_2^{k-1} & \alpha_3^{k-1} & \cdots & \alpha_{k-1}^{k-1} & \alpha_k^{k-1} \end{bmatrix} \tag{5.47}$$

define the *Vandermonde*[5] *determinant* of the k parameters $\alpha_1, \alpha_2, \alpha_3, \dots, \alpha_k$. Then

$$V_k(\alpha_1, \alpha_2, \alpha_3, \dots, \alpha_k) = \prod_{1 \le i < j \le k} (\alpha_j - \alpha_i), k \ge 2 \tag{5.48}$$

Proof. Our proof is by mathematical induction. Since

$$V_2(\alpha_1, \alpha_2) = \det \begin{bmatrix} 1 & 1 \\ \alpha_1 & \alpha_2 \end{bmatrix} = \alpha_2 - \alpha_1,$$

the formula holds for $k = 2$. Now suppose that the expansion (5.48) holds for a Vandermonde determinant of order k, and then consider the Vandermonde determinant $V_{k+1}(\alpha_1, \alpha_2, \alpha_3, \dots, \alpha_k, x)$ of order $k + 1$, viewed as a function of the variable x. By expanding the determinant by cofactors of the last column, we see that $p_k(x) = V_{k+1}(\alpha_1, \alpha_2, \alpha_3, \dots, \alpha_k, x) = cx^k +$ (terms of lower powers of x), so $p_k(x)$ is a polynomial of degree at most k. Whenever $x \in \{\alpha_1, \alpha_2, \alpha_3, \dots, \alpha_k\}$, there are two identical columns of the determinant, so $\alpha_1, \alpha_2, \alpha_3, \dots, \alpha_k$ are the k zeros of the polynomial. This means that $p_k(x)$ has the factorization $p_k(x) = c(x - \alpha_1)(x - \alpha_2) \cdots (x - \alpha_k)$. But c is the cofactor of x^k, which is the Vandermonde determinant $V_k(\alpha_1, \alpha_2, \alpha_3, \dots, \alpha_k)$. By the induction hypothesis, $V_k(\alpha_1, \alpha_2, \alpha_3, \dots, \alpha_k)$ is given by (5.48). Therefore

$$p_k(x) = V_k(\alpha_1, \alpha_2, \alpha_3, \dots, \alpha_k) \prod_{i=1}^{k} (x - \alpha_i) = V_{k+1}(\alpha_1, \alpha_2, \alpha_3, \dots, \alpha_k, x),$$

which completes the proof by mathematical induction. ∎

5.4.2 Repeated Eigenvalues

Theorem 5.16 shows that if α is an eigenvalue of multiplicity 2, then both α^n and $n\alpha^n$ are annihilated by $(E - \alpha)^2$. The following theorem generalizes this result in the case that α is an eigenvalue of any multiplicity r.

Theorem 5.24 Suppose that a homogeneous linear recurrence relation has an eigenvalue α of multiplicity $r \ge 1$. Then the recurrence relation is solved by each of the r sequences

$$\alpha^n, \binom{n}{1}\alpha^{n-1}, \binom{n}{2}\alpha^{n-2}, \dots, \binom{n}{r-1}\alpha^{n-(r-1)}$$

[5] Alexandre-Théophile Vandermonde (1735–1796) was a French musician, mathematician, and chemist. Curiously, the Vandermonde determinant is not explicitly mentioned in his collected papers on the foundations of determinant theory.

Proof. The characteristic polynomial of the linear recurrence contains the factor $(x-\alpha)^r$, so it suffices to show that each of the r sequences in annihilated by the operator $(E-\alpha)^r$. We already know that $(E-\alpha)\,\alpha^n = 0$. For any $j \geq 1$, using Pascal's identity, we get

$$(E-\alpha)\binom{n}{j}\alpha^{n-j} = \binom{n+1}{j}\alpha^{n+1-j} - \alpha\binom{n}{j}\alpha^{n-j} = \binom{n}{j-1}\alpha^{n-(j-1)}$$

and therefore, by induction, we see that

$$(E-\alpha)^j\binom{n}{j}\alpha^{n-j} = \binom{n}{j-j}\alpha^{n-(j-j)} = \alpha^n$$

Finally, we obtain

$$(E-\alpha)^r\binom{n}{j}\alpha^{n-j} = (E-\alpha)^{r-j}\,(E-\alpha)^j\binom{n}{j}\alpha^{n-j} = (E-\alpha)^{r-j}\,\alpha^n = 0$$

for all j, $0 \leq j \leq r-1$. ■

By the linearity of the operator $C(E)$, any linear sum of the sequences $\alpha^n, \binom{n}{1}\alpha^{n-1}, \binom{n}{2}\alpha^n, \ldots, \binom{n}{r-1}\alpha^{n-(r-1)}$ also solves the homogeneous linear recurrence relation if α is an eigenvalue of multiplicity r, and the same is true of linear sums of each of the different eigenvalues. This gives us the following theorem.

Theorem 5.25 Let $C(x) = (x-\alpha_1)^{r_1}(x-\alpha_2)^{r_2}\cdots(x-\alpha_m)^{r_m}$ be the characteristic polynomial of the homogeneous linear recurrence relation $h_{n+k} = a_1 h_{n+k-1} + a_2 h_{n+k-2} + \cdots + a_k h_n$, $n \geq 0$. Then any *generalized power sum* (GPS) of the $r_1 + r_2 + \cdots + r_m = k$ sequences in the set

$$S = \left\{\binom{n}{j}\alpha_i^{n-j} : i = 1, \ldots, m,\ j = 0, 1, \ldots, r_i - 1\right\} \tag{5.49}$$

is a solution of the recurrence relation; that is, for any set of constants c_{ij}, the sequence

$$h_n = \sum_{j=0}^{r_1-1} c_{1j}\binom{n}{j}\alpha_1^{n-j} + \sum_{j=0}^{r_2-1} c_{2j}\binom{n}{j}\alpha_2^{n-j} + \cdots + \sum_{j=0}^{r_m-1} c_{mj}\binom{n}{j}\alpha_m^{n-j} \tag{5.50}$$

solves the recurrence relation.

The word *generalized* reflects the fact that the powers of the eigenvalues are multiplied by polynomials in n. The GPS shown in (5.50) is called a *general solution* of the homogeneous recurrence relation. The general solutions of a recurrence relation can be expressed in alternate ways. For example, instead of using a linear sum of the terms given by the set S shown in (5.49), a general solution can also be written as a linear sum of the sequences in the set $S' = \left\{\binom{n}{j}\alpha_i^n : i = 1, \ldots, m, j = 0, 1, \ldots, r_i - 1\right\}$, or as a linear sum of the sequences in $S'' = \left\{n^j\alpha_i^n : i = 1, \ldots, m, j = 0, 1, \ldots, r_i - 1\right\}$. Often the calculations are simplest when the set S is chosen. In addition, there are some theoretical advantages to using S. For example, the determinant corresponding to the system of linear equations (5.50) is known as the *generalized Vandermonde determinant* $V\left(\alpha_1^{(r_1)}, \alpha_2^{(r_2)}, \ldots, \alpha_m^{(r_m)}\right)$, given by

$$V\left(\alpha_1^{(r_1)}, \alpha_2^{(r_2)}, \ldots, \alpha_m^{(r_m)}\right) = \det \begin{bmatrix} 1 & 0 & \cdots & 1 & 0 & \cdots \\ \alpha_1 & 1 & \cdots & \alpha_2 & 1 & \cdots \\ \alpha_1^2 & 2\alpha_1 & \cdots & \alpha_2^2 & 2\alpha_2 & \cdots \\ \alpha_1^3 & 3\alpha_1^2 & \cdots & \alpha_2^3 & 3\alpha_2^2 & \cdots \\ \vdots & \vdots & \vdots & \vdots & \vdots & \vdots \\ \alpha_1^{k-1} & \binom{k-1}{1}\alpha_1^{k-2} & \cdots & \alpha_2^{k-1} & \binom{k-1}{1}\alpha_2^{k-2} & \cdots \end{bmatrix} \tag{5.51}$$

where $r_1 + r_2 + \cdots + r_m = k$.

Proof of the following theorem is found in Appendix C for the interested reader.

Theorem 5.26 For all $m \geq 2$,

$$V\left(\alpha_1^{(r_1)}, \alpha_2^{(r_2)}, \ldots, \alpha_m^{(r_m)}\right) = \prod_{1 \leq i < j \leq m} (\alpha_j - \alpha_i)^{r_i r_j} \tag{5.52}$$

By combining the previous two theorems, we obtain the following important result.

Theorem 5.27 Every linear homogeneous recurrence relation with constant coefficients of order k subject to any set of k initial conditions has a unique solution given by a generalized power sum of its eigenvalues.

Example 5.28 Find the general solution of the recurrence relation $h_{n+9} = 15h_{n+8} - 78h_{n+7} + 138h_{n+6} + 111h_{n+5} - 561h_{n+4} + 144h_{n+3} + 744h_{n+2} - 240h_{n+1} - 400h_n$, where the characteristic polynomial has the factorization $C(x) = x^9 - 15x^8 + \cdots + 240x + 400 = (x-2)^4 (x+1)^3 (x-5)^2$.

Solution. A general solution of the recurrence relation of order $k = 9$ is a GPS of the nine terms in the set

$$S = \left\{2^n, \binom{n}{1} 2^{n-1}, \binom{n}{2} 2^{n-2}, \binom{n}{3} 2^{n-3}, (-1)^n, \binom{n}{1} (-1)^{n-1}, \binom{n}{2} (-1)^{n-2}, 5^n, \binom{n}{1} 5^{n-1}\right\}$$

In other words

$$h_n = c_0 2^n + c_1 \binom{n}{1} 2^{n-1} + c_2 \binom{n}{2} 2^{n-2} + c_3 \binom{n}{3} 2^{n-3}$$
$$+ d_0 (-1)^n + d_1 \binom{n}{1} (-1)^{n-1} + d_2 \binom{n}{2} (-1)^{n-2} + e_0 5^n + e_1 \binom{n}{1} 5^{n-1}$$

for any constants $\{c_0, c_1, c_2, c_3, d_0, d_1, d_2, e_0, e_1\}$. By Theorem 5.26, there is a unique solution of the recurrence relation for any set of k initial conditions $h_0 = A_0, h_1 = A_1, \ldots, h_{k-1} = A_{k-1}$. Note that the generalized Vandermonde determinant is $V(2^{(4)}, (-1)^{(3)}, 5^{(2)}) = (-1-2)^{12}(5-2)^8(5-(-1))^6 \neq 0$. ■

Example 5.29 Alice has inherited \$1000 and wishes to donate it by making a sequence of contributions that follow this plan. Each month she will write a check for either

- \$200, given to one of six charities serving the homeless
- \$300, given to one of eight environmental organizations
- \$400, given to one of three educational institutions.

She may well donate to the same cause repeatedly. In how many ways, counting order, can she make her \$1000 in donations?

Solution. Let h_n denote the number of ways to donate $n \times 100$ dollars. Then $h_1 = 0, h_2 = 6, h_3 = 8$, and $h_4 = 6 \cdot 6 + 3 = 39$. Partitioning cases according to the amount of the last check that was written, $h_{n+4} = 6h_{n+2} + 8h_{n+1} + 3h_n$. The corresponding characteristic polynomial is therefore $C(x) = x^4 - 6x^2 - 8x - 3 = (x+1)^3(x-3)$. The factorization shows that $\alpha = -1$ is a repeated eigenvalue of multiplicity 3 and $\beta = 3$ is a simple eigenvalue. In view of this, the general solution can be expressed as a GPS of the terms $\left\{(-1)^n, \binom{n}{1}(-1)^{n-1}, \binom{n}{2}(-1)^{n-2}, 3^n\right\}$:

$$h_n = c_0 (-1)^n + c_1 \binom{n}{1} (-1)^{n-1} + c_2 \binom{n}{2} (-1)^{n-2} + d_0 3^n$$

Iteration of the recurrence relation shows that we should define $h_0 = 1$. The initial conditions are satisfied if the constants in the GPS are chosen to satisfy

$$h_0 = c_0 + d_0 = 1, h_1 = -c_0 + c_1 + 3d_0 = 0$$
$$h_2 = c_0 - 2c_1 + c_2 + 9d_0 = 6, h_3 = -c_0 + 3c_1 - 3c_2 + 27d_0 = 8$$

Solving the system shows that $c_0 = \frac{37}{64}, c_1 = -\frac{11}{16}, c_2 = \frac{1}{4}, d_0 = \frac{27}{64}$ so $h_n = \frac{1}{64}\left[37(-1)^n - 44n(-1)^{n-1} + 8n(n-1)(-1)^{n-2} + 27(3)^n\right]$.

In particular, $h_{10} = 24{,}930$. ■

Example 5.30

(a) Find the general solution of the recurrence relation

$$w_{n+4} = 2w_{n+2} - w_n \tag{5.53}$$

(b) Explain why (5.53) has a unique solution given any initial conditions $w_0 = A_0, w_1 = A_1, w_2 = A_2, w_3 = A_3$.

Solution

(a) The characteristic equation is $C(x) = x^4 - 2x^2 + 1 = (x^2-1)^2 = (x-1)^2(x+1)^2$, so the eigenvalues are $\alpha = 1$ and $\beta = -1$, each with multiplicity 2. Therefore, the general solution can be expressed as the GPS $w_n = c_0 + nc_1 + d_0(-1)^n + d_1 n(-1)^{n-1}$ with the constants c_0, c_1, d_0, d_1.

(b) The linear system of equations $w_0 = A_0, w_1 = A_1, w_2 = A_2, w_3 = A_3$ uniquely determine the constants c_0, c_1, d_0, d_1 since the generalized Vandermonde determinant is

$$V(1^{(2)}, (-1)^{(2)}) = ((-1) - 1)^{2\cdot 2} = 2^4 \neq 0. \tag{5.54}$$

■

PROBLEMS

5.4.1. Let $h_n = 2h_{n-1} + 3h_{n-2} - 6h_{n-3}$, $n \geq 3$.
(a) What are the corresponding eigenvalues?
(b) What is the general solution (GPS) with unspecified initial conditions?

5.4.2. Let $h_n = 6h_{n-1} - 3h_{n-2} - 10h_{n-3}$, $n \geq 3$.
(a) What are the corresponding eigenvalues?
(b) What is the general solution (GPS) with unspecified initial conditions?
(c) What is the order of the recurrence relation?
(d) Does the sequence $g_n = 2(-1)^n + 5^n$ satisfy the recurrence relation?
(e) What is the order of the sequence g_n? Give a recurrence of this order satisfied by g_n.

5.4.3. Suppose that a sequence satisfies a linear homogeneous recurrence relation of order 5. Support your answers to these questions:
(a) could it satisfy a recurrence of order 3?
(b) must it satisfy a recurrence of order 7?

5.4.4. Consider the sequence that begins with the five terms 0,6,6,18,30. For each part below, either find a homogeneous linear recurrence with constant coefficients of order k satisfied by the sequence or explain why no such recurrence exists:

(a) $k = 1$ **(b)** $k = 2$ **(c)** $k = 5$

5.4.5. **(a)** Show that the recurrence relation $h_{n+3} = a_1 h_{n+2} + a_2 h_{n+1} + a_n h_n$ can be written as the equivalent matrix equation

$$\begin{bmatrix} h_{n+3} \\ h_{n+2} \\ h_{n+1} \end{bmatrix} = M \begin{bmatrix} h_{n+2} \\ h_{n+1} \\ h_n \end{bmatrix}$$

where

$$M = \begin{bmatrix} a_1 & a_2 & a_3 \\ 1 & 0 & 0 \\ 0 & 1 & 0 \end{bmatrix}$$

is the *companion matrix* of the characteristic polynomial $C(x) = x^3 - a_1 x^2 - a_2 x - a_3$.

(b) Show that $C(x)$ is the characteristic polynomial of M.

5.4.6. Let

$$M = \begin{bmatrix} a_1 & a_2 & a_3 & \cdots & a_k \\ 1 & 0 & 0 & \cdots & 0 \\ 0 & 1 & 0 & \cdots & 0 \\ \vdots & \vdots & \vdots & \ddots & \vdots \\ 0 & 0 & \cdots & 1 & 0 \end{bmatrix}$$

be the companion matrix (see Problem 5.4.5) of the polynomial $C_k(x) = x^k - a_1 x^{k-1} - a_2 x^{k-2} - \cdots - a_k$. Prove that $\det\left(xI_k - M\right) = C_k(x)$ for all $k \geq 1$, where I_k is the $k \times k$ identity matrix.

5.4.7. What GPS (generalized power sum) is annihilated by $(E-2)^2\,(E+1)^3$?

5.4.8. Solve the recurrence relation $h_{n+3} = 2h_{n+2} + 4h_{n+1} - 8h_n$, where $h_0 = 0$, $h_1 = 5, h_2 = 4$.

5.4.9. Solve the recurrence relation $h_{n+4} = 6h_{n+2} - 8h_{n+1} + 3h_n$, where $h_0 = 1$, $h_1 = 0, h_2 = 21, h_3 = 0$.

5.4.10. In Example 5.21 we showed that the Perrin numbers are given by the recurrence relation $p_{n+3} = p_{n+1} + p_n$, $n \geq 0$ with the initial conditions $p_0 = 3, p_1 = 0, p_2 = 2$. Use telescoping sums to show that

(a) $\sum_{n=0}^{m} p_n = p_{m+5} - 2$

(b) $\sum_{n=0}^{m} p_{2n} = p_{2m+3}$

(c) $\sum_{n=0}^{m} p_{2n+1} = p_{2m+4} - 2$

5.4.11. **(a)** Compute the next 12 Perrin numbers of Example 5.21 in the table started below:

n	0	1	**2**	**3**	4	**5**	6	**7**	8	9	10	**11**
p_n	3	0	**2**	**3**	2	**5**	5	**7**	10	12	17	**22**

(b) What property seems to be true when n is a prime number?

(c) Does the converse hold? Investigate the Perrin numbers with an Internet search.

5.4.12. Show that the Perrin numbers of Example 5.21 satisfy

$$\begin{bmatrix} p_n \\ p_{n+1} \\ p_{n+2} \end{bmatrix} = \begin{bmatrix} 0 & 1 & 0 \\ 0 & 0 & 1 \\ 1 & 1 & 0 \end{bmatrix}^n \begin{bmatrix} 3 \\ 0 \\ 2 \end{bmatrix}, \quad n \geq 0$$

5.4.13. The nth *Padovan number* v_n is the number of compositions (ordered sums) of n for which the summands are odd integers ≥ 3. For example, $v_{10} = 3$ since 10 has the three compositions $3 + 7$, $7 + 3$, and $5 + 5$. Note that summands can be repeated and their order matters. Show that the Padovan numbers satisfy the Perrin recursion $v_{n+3} = v_{n+1} + v_n$, $n \geq 0$ investigated in Example 5.21. (It can be shown that

$$v_n = \frac{(\beta-1)(\gamma-1)}{(\beta-\alpha)(\gamma-\alpha)}\alpha^n + \frac{(\alpha-1)(\gamma-1)}{(\alpha-\beta)(\gamma-\beta)}\beta^n + \frac{(\alpha-1)(\beta-1)}{(\alpha-\gamma)(\beta-\gamma)}\gamma^n$$

where α, β, and γ are the eigenvalues of the Perrin recurrence.)

5.4.14. Prove that any polynomial in the variable n of degree d is annihilated by $(E-1)^{d+1}$.

5.4.15. Suppose that the sequences u_n and v_n satisfy the coupled linear recurrences $P(E)u_n + Q(E)v_n = 0$ and $R(E)u_n + S(E)v_n = 0$, where P, Q, R, and S

are polynomials. Prove that if $T(x) = P(x)S(x) - Q(x)R(x)$, then $T(E)\,u_n = T(E)\,v_n = 0$, so u_n and v_n satisfy a common linear recurrence of order k, where k is the degree of the polynomial T.

5.4.16. Let u_n and v_n that satisfy the system of linear recurrence relations

$$u_{n+2} = u_n + v_{n+1} - v_n$$
$$v_{n+2} = 2v_n - u_{n+1} - u_n.$$

(a) Find a linear recurrence relation that is satisfied by both of the sequences u_n and v_n. [*Hint*: Use the result of Problem 5.4.15.]

(b) Why can the system be solved uniquely for any set of initial conditions $(u_0, u_1) = (A_1, A_2)$ and $(v_0, v_1) = (B_1, B_2)$?

5.4.17. Let u_n denote the number of ways to tile a $2 \times n$ board with squares and horizontal dominoes, and let v_n denote the number of ways to tile a $2 \times n$ pruned board that has a missing corner cell.

(a) Show that $u_{n+2} = u_{n+1} + u_n + 2v_{n+1}$ and $v_{n+1} = u_n + v_n$.

(b) Show that both u_n and v_n satisfy the order three recurrence $h_{n+3} = 2h_{n+2} + 2h_{n+1} - h_n$. [*Hint*: Use the result of Problem 5.4.15.]

(c) Explain why F_{n+1}^2 and F_nF_{n+1} satisfy the recurrence of part (b).

5.4.18. Find a third-order recurrence relation satisfied by both F_n^2 and L_n^2.

5.4.19. Find a fourth-order recurrence relation satisfied by F_n^3, L_n^3, $F_n^2L_n$, and $F_nL_n^2$.

5.4.20. Verify these generalized Vandermonde determinant formulas by direct calculation.

(a) $V(\alpha^{(2)}, \beta) = (\beta - \alpha)^2$

(b) $V(\alpha^{(3)}, \beta) = (\beta - \alpha)^3$

(c) $V(\alpha^{(2)}, \beta^{(2)}) = (\beta - \alpha)^4$

[*Hint:* Use column operations.]

5.5 NONHOMOGENEOUS RECURRENCE RELATIONS

A recurrence relation of the form

$$h_{n+k} = a_1h_{n+k-1} + a_2h_{n+k-2} + \cdots + a_kh_n + q_n, \quad n \geq 0 \tag{5.55}$$

is said to be *nonhomogeneous* because a sequence q_n has been added to the right side of the equation. In terms of the E operator, the recurrence can be written as

$$C(E)\,h_n = q_n \tag{5.56}$$

where

$$C(E) = E^k - a_1 E^{k-1} - a_2 E^{k-2} - \cdots - a_k \tag{5.57}$$

Thus, the operator $C(E)$ no longer annihilates h_n but instead maps it onto the sequence q_n. Even so, we will see that it is still helpful to determine those sequences g_n that are annihilated by $C(E)$. The sequence g_n for which $C(E)g_n = 0$ is called the *general solution* of the associated homogeneous recurrence relation. We have seen that the general solution has the form of a generalized power sum (GPS) of the roots (eigenvalues) of the characteristic equation $C(x) = 0$. Now suppose that a solution of the nonhomogeneous recurrence can be found, say, p_n, for which $C(E)p_n = q_n$. Such a solution is called a *particular solution* of the nonhomogenous recurrence relation. The linearity of the $C(E)$ operator then shows that $h_n = g_n + p_n$ solves the nonhomogeneous recurrence, since we have

$$C(E)h_n = C(E)(g_n + p_n) = C(E)g_n + C(E)p_n = 0 + q_n = q_n$$

The sequence $h_n = g_n + p_n$ that combines the general and particular solutions is known as the *complete solution.*

Two remaining questions must be addressed:

1. Can we be assured that we can solve recurrence (5.55) for any given set of initial conditions, $h_j = A_j$, $j = 0, 1, \ldots, k-1$?
2. How can a particular solution p_n be found?

5.5.1 Solving for Initial Conditions

Let's assume that we have found a particular solution p_n of the nonhomogeneous recurrence and have solved the associated homogeneous recurrence to obtain the general solution g_n in the form of a generalized power sum (GPS) of the eigenvalues with the constants $c_1, c_2, \ldots, c_k$. To satisfy the initial conditions, we must ensure that these constants can be determined so that the initial conditions $h_j = g_j + p_j = A_j$, $j = 0, 1, \ldots, k-1$ are satisfied. But this is the same as the system of equations $g_j = A_j - p_j$, $j = 0, 1, \ldots, k-1$, and we already know that the general solution can always be solved uniquely for any set of initial conditions.

We should first determine a particular solution after which we solve for the initial conditions using the complete solution $h_n = g_n + p_n$. In particular, the initial conditions are not applied to the general solution g_n.

Example 5.31 The "Tower of Brahma" puzzle[6] asks for the lowest number of moves to transfer a stack of n disks from one of three pillars to a second pillar. The

[6]The puzzle was posed by E. Lucas in 1883, when he described the puzzle in the form of a legend in which 64 golden disks are to be moved by Brahmin priests. If a disk is moved every second, it would take about 585 billion years to move all 64 disks. Surprisingly, if the game is generalized to four pillars, determining the least number of moves to transfer n disks is still an open question.

disks are all of different diameter, and a disk can either be moved to an empty pillar or placed on top of a disk of larger diameter. A larger disk can never be placed on top of a smaller one (see diagram):

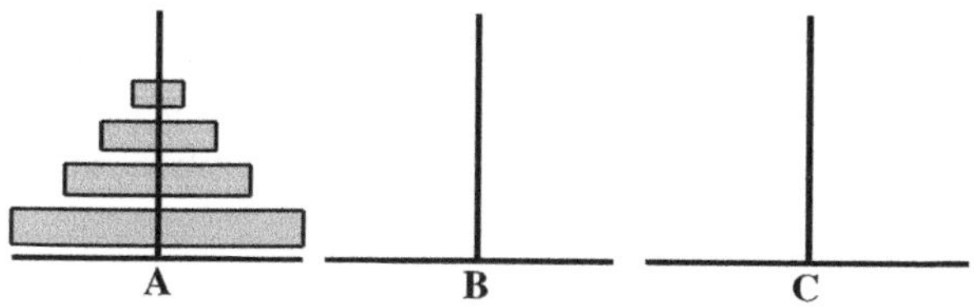

Solution. Let h_n denote the fewest moves to transfer n disks from one pillar to another. Clearly $h_1 = 1$. Also $h_2 = 3$; move the smaller disk from pillar A to pillar B, then the larger disk from A to C, and finally move the smaller disk from B to C. To determine h_3, first move the upper two disks on pillar A to pillar B, which can be done in three moves. Then move the remaining largest disk on pillar A to pillar C in one move, and finally transfer the two disks on pillar B to pillar C in three more moves. Thus $h_3 = 3 + 1 + 3 = 7$. This reasoning can be generalized to determine a recurrence relation for the sequence h_n. Start with $n + 1$ disks on pillar A. Move the upper n disks of pillar A to pillar B in h_n moves, next move the largest disk on pillar A to pillar C in one move, and finally move the n disks on pillar B to pillar C in another h_n moves. Thus, $h_{n+1} = 2h_n + 1$, $n \geq 1$, which is a nonhomogeneous first-order linear recurrence relation. The associated homogeneous recurrence relation is $g_{n+1} = 2g_n$. The eigenvalue is $\alpha = 2$, so the general solution is $g_n = c2^n$ with an unspecified constant c. A particular solution can be found by first noting that the nonhomogeneous term of the recurrence is the constant sequence $q_n = 1$. This suggests that we look for a particular solution that is also some constant sequence, say, $p_n = C$. Inserting this into the nonhomogeneous recurrence $p_{n+1} = 2p_n + 1$, we get $C = 2C + 1$. This shows that $p_n = C = -1$ is a particular solution, so we have now obtained the complete solution $h_n = g_n + p_n = c2^n - 1$. The constant c can be determined by the initial condition $1 = h_1 = c2^1 - 1$, which shows that $c = 1$. This gives us our final solution $h_n = 2^n - 1$. As a check, $h_3 = 2^3 - 1 = 8 - 1 = 7$. ■

5.5.2 Finding a Particular Solution

Suppose, as is most often the case, that the nonhomogeneous term q_n satisfies some homogeneous linear recurrence relation; that is, suppose that there is a polynomial $P(x)$ for which $P(E)\,q_n = 0$. In this case, we see that $P(E)\,C(E)\,p_n = P(E)\,q_n = 0$ and $P(E)\,C(E)\,g_n = P(E)\,0 = 0$. This means that the complete solution $h_n = g_n + p_n$ is annihilated by the operator $P(E)C(E)$, and it follows that h_n is a GPS of the eigenvalues of the characteristic equation $P(x)C(x) = 0$. We already know that g_n is a GPS of the eigenvalues of $C(x) = 0$, so it follows that p_n can be expressed as a GPS of the terms *not* appearing in the linear sum that gives g_n. There is no need for the

GPS that forms p_n to include any of the terms that appear in the general solution of g_n, since these are annihilated by $C(E)$ and need not be a part of p_n.

Example 5.32 The centered octagonal numbers are obtained by starting with a central point, which is then surrounded by concentric octagons that have an increasing number of points along each side. For example, the diagram below shows that the sequence of centered octagonal numbers begins $h_1 = 1, h_2 = 9, h_3 = 25, h_4 = 49, \ldots$. Obtain a recurrence relation for the centered octagonal numbers, and show that they are the odd square numbers.

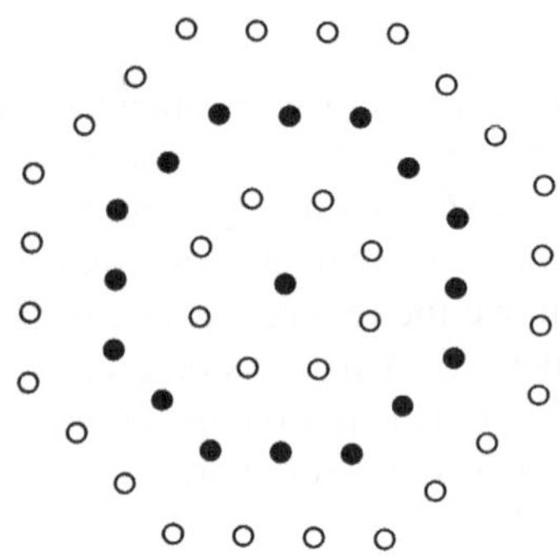

Solution. The diagram makes it clear that $h_{n+1} = h_n + 8n$, a first-order nonhomogeneous linear recurrence relation. The associated homogeneous recurrence is $g_{n+1} = g_n$, with the characteristic equation $C(x) = x - 1$ and the eigenvalue $\alpha = 1$. Thus, the general solution is $g_n = c1^n = c$, where c is a constant. The nonhomogeneous term $q_n = 8n$ is annihilated by the operator $P(E) = (E-1)^2$, since $(E-1)^2\, 8n = 8(E^2 - 2E + 1)n = 8\,(n + 2 - 2\,(n+1) + n) = 0$. Therefore, $P(E)\,C(E) = (x-1)^3$, and it follows that $\alpha = 1$ is an eigenvalue of multiplicity 3. We now know that both g_n and p_n are a GPS of the set $\left\{1, \binom{n}{1}, \binom{n}{2}\right\}$. Since $g_n = c1^n$, we know that the particular solution has the form $p_n = c_1\binom{n}{1} + c_2\binom{n}{2}$. Inserting this equation into the nonhomogeneous recurrence $p_{n+1} = p_n + 8n$, we get $c_1\binom{n+1}{1} + c_2\binom{n+1}{2} = c_1\binom{n}{1} + c_2\binom{n}{2} + 8n$, which simplifies to $c_1\binom{n}{0} + c_2\binom{n}{1} = 8n$, or $c_1 + c_2 n = 8n$. We see that $c_1 = 0$ and $c_2 = 8$ and therefore $p_n = 8\binom{n}{2}$ is a particular solution. The complete solution therefore has the form $h_n = c + 8\binom{n}{2}$. Since $h_1 = 1$, it follows that $c = 1$ and thus $h_n = 1 + 8\binom{n}{2} = 1 + 4n\,(n-1) = (2n-1)^2$, an odd square. ■

It is helpful to know the annihilating operators for some the classes of functions that are often encountered as the nonhomogeneous term q_n. Some of these are shown in the following table; additional factors of powers of n will be needed if an eigenvalue of $P(E)$ repeats an eigenvalue of $C(E)$:

q_n	Operator $P(E)$ that Annihilates q_n	Eigenvalue(s) of $P(E)$	Form of a Particular Solution p_n
Constant	$E-1$	1	Constant
Polynomial of degree d	$(E-1)^{d+1}$	1 with multiplicity $d+1$	Polynomial in n of degree d
Power α^n	$E-\alpha$	α	$c\alpha^n$
Product of a polynomial in n of degree d and α^n	$(E-\alpha)^{d+1}$	α with multiplicity $d+1$	(Polynomial in n of degree d)α^n

For example, if $q_n = 2n^2 + 3^n$, then look for a particular solution of the form $p_n = a + b\binom{n}{1} + c\binom{n}{2} + d3^n$ or $p_n = a + bn + cn^2 + d3^n$. But if $\alpha = 1$ is also an eigenvalue of multiplicity 2 of $C(E)$, then look for a particular solution of the form $p_n = a\binom{n}{2} + b\binom{n}{3} + c\binom{n}{4} + d3^n$ or $p_n = an^2 + bn^3 + cn^4 + d3^n$. The constants $a,b,c,d,\ldots$ in the GPS that form the particular solution are evaluated by inserting p_n into the nonhomogeneous recurrence relation $C(E)p_n = q_n$. Once these constants have been determined, the constants $c_1, c_2, c_3, \ldots$ in the GPS of the complete solution are evaluated by applying the initial conditions $h_j = g_j + p_j = A_j$, $j = 0, 1, \ldots, k-1$.

Example 5.33 Solve the second-order linear nonhomogeneous recurrence relation $h_{n+2} = h_{n+1} + 6h_n + 54n^2$, $n \geq 0$ with the initial conditions $h_0 = 0$, $h_1 = 3$.

Solution. The characteristic polynomial of the homogeneous recurrence $h_{n+2} - h_{n+1} - 6h_n = 0$ is $C(x) = x^2 - x - 6 = (x-3)(x+2)$, so there are two distinct eigenvalues, $\alpha = 3$ and $\beta = -2$. Therefore, the general solution is $g_n = c_1 3^n + c_2(-2)^n$. The nonhomogeneous term is $q(n) = 54n^2$, a quadratic polynomial in the variable n. Therefore, we search for a particular solution of the form $p_n = a + bn + cn^2$. Inserting this expression into the nonhomogeneous recurrence $p_{n+2} = p_{n+1} + 6p_n + 54n^2$, we get $a + b(n+2) + c(n+2)^2 = (a + b(n+1) + c(n+1)^2) + 6(a + bn + cn^2) + 54n^2$, which simplifies to $0 = 6a - b - 3c + 2(3b - 2c)n + 6(c+9)n^2$.

Since the coefficients of the polynomial are 0, we obtain $c = -9, b = c/3 = -3, a = (b + 3c)/6 = (-3 - 27)/6 = -5$, so that our particular solution is $p_n = -5 - 3n - 9n^2$. Adding the particular and general solutions, we get $h_n = c_1 3^n + c_2(-2)^n - 5 - 3n - 9n^2$. The initial conditions will be met if

$$0 = h_0 = c_1 + c_2 - 5$$
$$3 = h_1 = 3c_1 - 2c_2 - 17$$

This linear system of equations is solved by $c_1 = 6$ and $c_2 = -1$, giving us the final solution $h_n = 6 \cdot 3^n - (-2)^n - 5 - 3n - 9n^2$. ■

Example 5.34 Solve the second-order linear nonhomogeneous recurrence relation $h_{n+2} = 4h_{n+1} - 4h_n + 5 \cdot 2^n, n \geq 0$ with the initial conditions $h_0 = 0$, $h_1 = 2$.

Solution. The characteristic polynomial of the homogeneous recurrence $h_{n+2} - 4h_{n+1} + 4h_n = 0$ is $C(x) = x^2 - 4x + 4 = (x-2)^2$, so there is a repeated eigenvalue $\alpha = 2$. The general solution of the homogeneous recurrence is therefore $g_n = c_1 2^n + c_2 n 2^n$. The nonhomogeneous term $5 \cdot 2^n$ is proportional to the power of the eigenvalue of multiplicity 2, and both 2^n and $n2^n$ are annihilated by $C(E) = E^2 - 4E + 4$. Therefore, we seek a particular solution of the nonhomogeneous recurrence that has the form $p_n = an^2 2^n$. We need to determine the constant a such that $C(E)(an^2 2^n) = (E^2 - 4E + 4)(an^2 2^n) = 5 \cdot 2^n$; that is, $a(n+2)^2 2^{n+2} - 4a(n+1)^2 2^{n+1} + 4an^2 2^n = 5 \cdot 2^n$.

This simplifies (or just let $n = 0$) to the equation $8a = 5$, so $p_n = 5n^2 2^{n-3}$. The final step is to determine the constants c_1 and c_2 in $h_n = c_1 2^n + c_2 n 2^n + 5n^2 2^{n-3}$ from the initial conditions $0 = h_0 = c_1$ and $2 = h_1 = c_1 2 + c_2 2 + \frac{5}{4}$. We find that $c_1 = 0$ and $c_2 = 3 \cdot 2^{-3}$ and now we have our final answer: $h_n = 3n2^{n-3} + 5n^2 2^{n-3} = n(3+5n)2^{n-3}$. ∎

The next example reveals an interesting combinatorial property of the totality of all of the tilings of an n-board by squares and dominoes.

Example 5.35 Consider the set of all tilings of a $1 \times n$ board with squares and dominoes. Determine the total number of squares and the total number of dominoes needed to form *all* of the $f_n = F_{n+1}$ tilings.

Solution. Let s_n and d_n denote, respectively, the total number of squares and dominoes in the f_n tilings of the board. The table below shows the values of s_n and d_n for $n = 1,2,3,4$.

n	All Tilings of a $1 \times n$ Board with Squares and Dominoes	Number of Tilings f_n	Number of Squares s_n	Number of Dominoes d_n
1		1	1	0
2		2	2	1
3	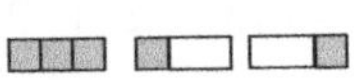	3	5	2
4	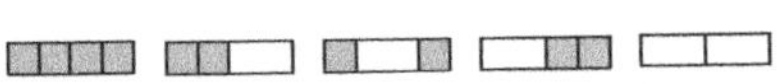	5	10	5

Consider the f_{n+2} tilings of all of the boards of length $n + 2$. There are f_{n+1} of these tilings begining with a square tile, which accounts for $s_{n+1} + f_{n+1}$

squares. There are also s_n squares that are used in the tilings of the $(n+2)$-boards that begin with a domino. Therefore,

$$s_{n+2} = s_{n+1} + s_n + f_{n+1} \tag{5.58}$$

which shows that we should define $s_0 = s_2 - s_1 - f_1 = 2 - 1 - 1 = 0$. The associated homogeneous recurrence $g_{n+2} = g_{n+1} + g_n$ is annihilated by the Fibonacci operator $C(E) = E^2 - E - 1$, so the general solution has the form $g_n = c_1 f_{n+1} + c_2 f_n$. The nonhomogeneous term $q_n = f_{n+1}$ is also annihilated by the Fibonacci operator, so each of the two eigenvalues has multiplicity 2. Therefore, a particular solution can be found with the form $p_n = anf_{n+1} + bnf_n$. Since $p_0 = 0, p_1 = 2a + b, p_2 = 6a + 4b, p_3 = 15a + 9b$, the nonhomogeneous recursion relation (5.58) is satisfied in the cases $n = 0$ and $n = 1$ when $6a + 4b = 2a + b + 0 + 1$ and $15a + 9b = (6a + 4b) + (2a + b) + 2$, which simplifies to $4a + 3b = 1$ and $7a + 4b = 2$. This system is solved to show that $a = \frac{2}{5}$ and $b = -\frac{1}{5}$, giving us the particular solution $p_n = (2nf_{n+1} - nf_n)/5$ for which $p_0 = 0, p_1 = \frac{3}{5}$. Finally, since $s_n = g_n + p_n$, letting $n = 0$ and $n = 1$ gives us $0 = s_0 = g_0 + p_0 = c_1 + c_2 + 0$ and $1 = s_1 = g_1 + p_1 = 2c_1 + c_2 + \frac{3}{5}$.

Evidently, $c_1 = -c_2 = \frac{2}{5}$, and thus

$$s_n = g_n + p_n = \frac{2f_{n+1} - 2f_n}{5} + \frac{2nf_{n+1} - nf_n}{5} = \frac{2(n+1)f_{n+1} - (n+2)f_n}{5}$$

which can also be written as

$$s_n = \tfrac{2}{5}(1+n)f_{n-1} + \tfrac{1}{5}nf_n = \tfrac{2}{5}(1+n)F_n + \tfrac{1}{5}nF_{n+1} \tag{5.59}$$

The number of dominoes can be calculated similarly, starting by showing that

$$d_{n+2} = d_{n+1} + d_n + f_n \tag{5.60}$$

This is the same recurrence as (5.58) but with the initial conditions $d_1 = 0 = s_0$ and $d_2 = 1 = s_1$. Therefore, $d_{n+1} = s_n$ for all $n \geq 0$. ∎

We have proved the following interesting theorem as an outcome of Example 5.35.

Theorem 5.36 Let s_n denote the total number of squares and d_n denote the total number of dominoes in all F_{n+1} tilings of a $1 \times n$ board by dominoes and squares. Then

$$s_n = d_{n+1} = \frac{nF_{n+1} + 2(1+n)F_n}{5} \tag{5.61}$$

Problem 5.5.12 outlines how to show that $s_n = d_{n+1}$ by direct reasoning. In the next section, we will show that the sequence 0,1,2,5,10,20,... given by (5.61), is also the convolution product of the Fibonacci sequence. For example, $s_5 = 20 = \sum_{j=1}^{5} F_j F_{6-j} = 1 \cdot 5 + 1 \cdot 3 + 2 \cdot 2 + 3 \cdot 1 + 5 \cdot 1$.

PROBLEMS

5.5.1. Consider the nonhomogeneous recurrence relation $h_{n+2} = h_{n+1} + 6h_n + 6$ with the initial conditions $h_0 = 0, h_1 = 2$.

(a) Determine a homogeneous recurrence relation of order 3 satisfied by h_n.

(b) Solve the recurrence relation.

(c) Is there a homogeneous relation of order 2 satisfied by h_n?

5.5.2. Solve these nonhomogeneous recurrence relations:

(a) $h_{n+1} = 3h_n + 2n, h_0 = 0$

(b) $h_{n+1} = 3h_n + (-2)^n, h_0 = 1$

(c) $h_{n+1} = 3h_n + 3^n, h_0 = -1$

5.5.3. Solve these nonhomogeneous recurrence relations:

(a) $h_{n+1} = 2h_n + 3^n + 1, h_0 = 0$

(b) $h_{n+1} = 2h_n + n3^n, h_0 = 0$

5.5.4. At the beginning of each year, \$1000 is added to an investment that pays an annual interest of 4%. Derive and solve a recurrence relation to determine the value of the investment at the end of 40 years.

5.5.5. One line separates the plane into $h_1 = 2$ regions, and two intersecting lines separate the plane into $h_2 = 4$ regions. Three lines, none parallel and no three intersecting at a single point, separate the plane into $h_3 = 7$ regions:

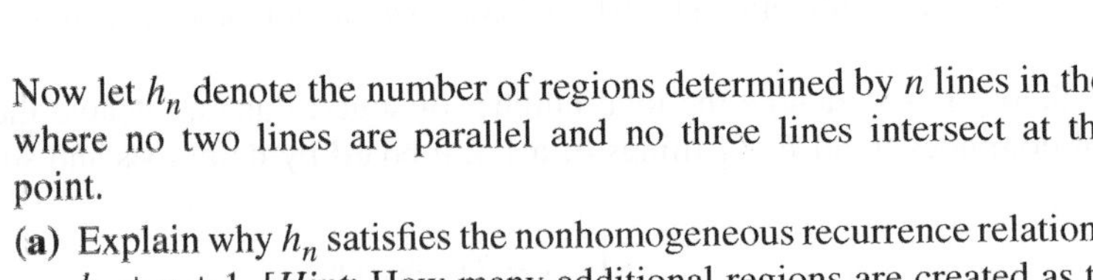

Now let h_n denote the number of regions determined by n lines in the plane, where no two lines are parallel and no three lines intersect at the same point.

(a) Explain why h_n satisfies the nonhomogeneous recurrence relation $h_{n+1} = h_n + n + 1$. [*Hint*: How many additional regions are created as the $(n + 1)$st line is drawn?]

(b) Solve the homogeneous recurrence relation $h_{n+1} = h_n$.

(c) Show that there is a particular solution of the nonhomogeneous of the form $p_n = sn^2 + tn$.

(d) Determine h_n.

5.5.6. In Example 5.17, h_n denoted the number of ways that n nonattacking kings can be placed on a $2 \times 2n$ board.

(a) Show that $h_n = 2h_{n-1} + 2^n$.

(b) Solve the nonhomogeneous recurrence relation given in part (a).

(c) Derive a formula for h_n by direct combinatorial reasoning.

5.5.7. The sequence $p_n^{(r)}$ of r-gonal numbers is obtained by starting with a single vertex point, so $p_1^{(r)} = 1$, and then successively adding additional points so that the outermost polygon contains $n + 1$ dots along each of its r sides. For example, the hexagonal numbers are shown in the diagram below. Note that $p_0^{(r)} = 0, p_1^{(r)} = 1$, and $p_2^{(r)} = r$ for all $r \geq 3$.

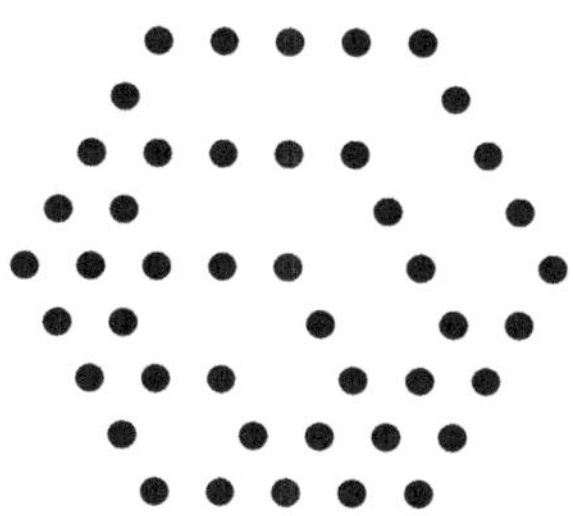

0,1,6,15,28,45, ...

(a) Show that $p_{n+1}^{(r)} = p_n^{(r)} + 1 + (r-2)\,n$.

(b) Explain why the sequence of r-gonal numbers is annihilated by the operator $C(E) = (E-1)^3$.

(c) Solve the recurrence to show that $p_n^{(r)} = n + (r-2)\binom{n}{2}$.

5.5.8. The sequence $c_n^{(r)}$ of centered r-gonal numbers is obtained by starting with a single central point, and then successively adding surrounding polygons with nr dots distributed uniformly along the r sides of the nth polygon. The centered hexagonal numbers are shown in the following diagram. Note that $c_0^{(r)} = 1, c_1^{(r)} = 1 + r$, and $c_2^{(r)} = 1 + 3r$.

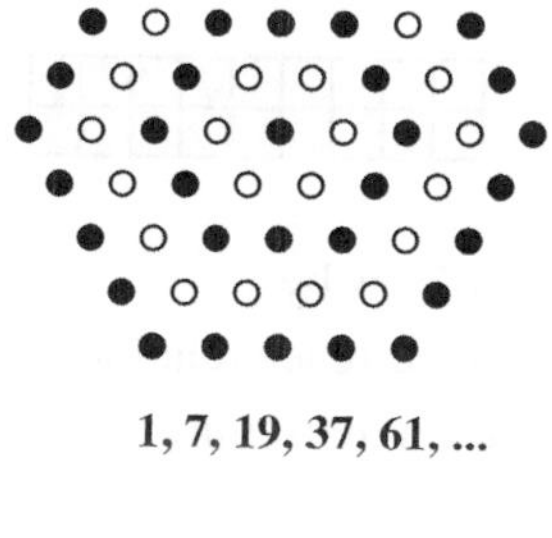

1, 7, 19, 37, 61, ...

(a) Show that $c_n^{(r)} = c_{n-1}^{(r)} + rn$, $n \geq 1$.

(b) Explain why the sequence of centered r-gonal numbers is annihilated by the operator $C(E) = (E - 1)^3$.

(c) Solve the recurrence relation to show that $c_n^{(r)} = 1 + r\binom{n+1}{2}$.

5.5.9. The number of ways to tile an n-board with red and blue squares and white dominoes is given by the Pell number P_{n+1} (see Problem 5.3.15). Let r_n and d_n denote, respectively, the number of red squares and the number of dominoes in *all* of the tilings of boards of length n. By symmetry, r_n also is the number of blue squares in all of the tilings.

(a) Create a table analogous to the table shown in Example 5.35, but for Pell tilings of length $n = 0,1,2,3$. Verify that $r_1 = 2, d_1 = 1, r_2 = 4$, $d_2 = 1$, and $r_3 = 14, d_3 = 4$.

(b) Find a recurrence relation satisfied by r_n.

(c) What is the form of the GPS for r_n, with unspecified coefficients?

(d) Verify that $r_n = \frac{1}{4}((n+1)P_n + nP_{n+1})$.

(e) Prove that $d_n = \frac{1}{4}(nP_{n-1} + (n-1)P_n)$.

(f) Prove that $r_n = d_{n+1}$.

5.5.10. A *train* of length n is a $1 \times n$ board tiled by *cars* (rectangles) of any length. For example, there are four trains of length $n = 3$:

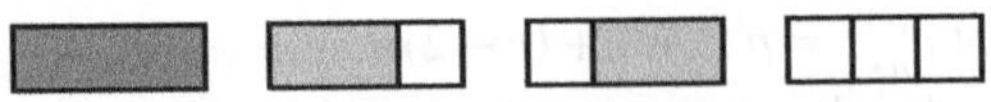

Thus, a train is an ordered sum of positive integers that sum to n, and a train therefore represents a *composition*. The four sums $3 = 2 + 1 = 1 + 2 = 1 + 1 + 1$ are the $h_3 = 4$ compositions of $n = 3$, where h_n denotes the number of compositions of n.

(a) Obtain and solve a recurrence relation for h_n.

(b) How many 1s are used to write all of the compositions of n? Equivalently, how many 1×1 unit square tiles, u_n, appear in all h_n trains of length n?

5.5.11. Suppose that a $2 \times n$ board is tiled with gray and white squares, where in each row the white squares are to the left of the gray squares. Three examples of tilings of a 2×9 board are shown here:

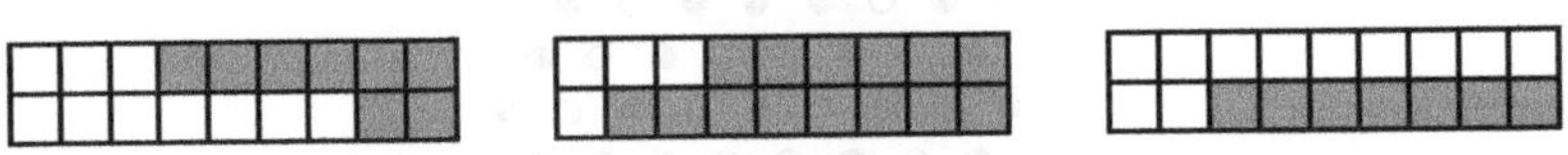

Determine the number of tilings by

(a) finding and solving a recurrence relation

(b) using direct combinatorial reasoning

5.5.12. In Example 5.35 we showed that the number of squares s_n that appear in all F_{n+1} tilings of an n-board with squares and dominoes is the same as the number of dominoes d_{n+1} that appear in all of the tilings of a board of length $n+1$. For example, there are five squares in the three tilings of 3-boards and five dominoes in the five tilings of 4-boards, as shown in the diagram below:

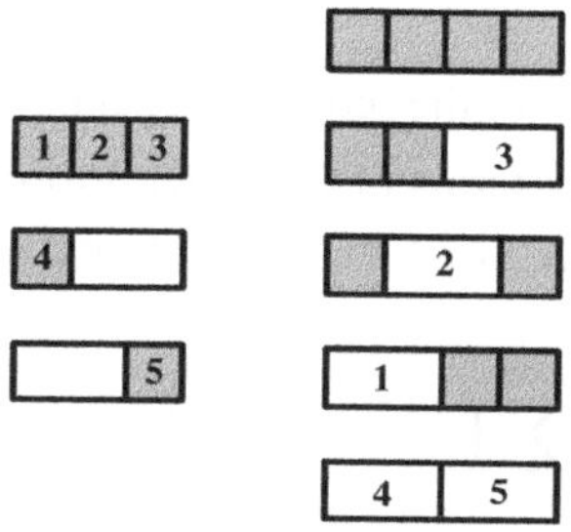

Generalize the numbering system in the diagram to create a bijective proof that $s_n = d_{n+1}$.

5.5.13. Consider all P_n of the tilings of boards of length n with red and blue squares and white dominoes, where P_n is the nth Pell number. In Problem 5.5.9, we showed that the number of red squares r_n used to tile the n-boards was the same as the number of white dominoes d_{n+1} used to create all of the tilings of boards of length $n+1$. For example, $r_2 = 4 = d_3$ as shown in the diagram below:

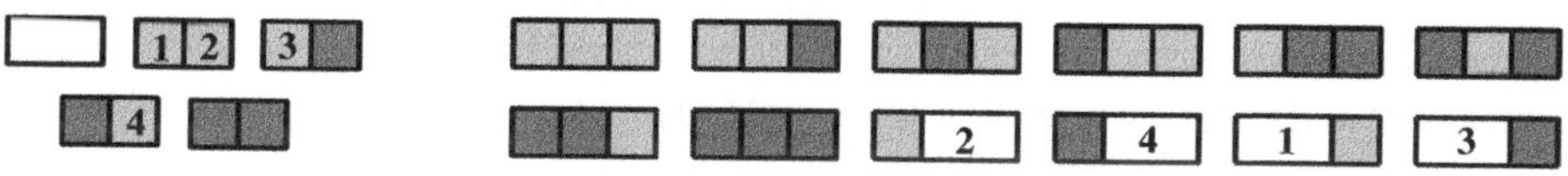

Give a bijective proof that $s_n = d_{n+1}$.

5.6 RECURRENCE RELATIONS AND GENERATING FUNCTIONS

In Chapter 3, we saw the value of associating a sequence with its corresponding generating function. In this section, we will explore how to obtain a generating function for sequences given by a recurrence relation. Once the generating function for the sequence has been determined, the recurrence relation is then solved by calculating the coefficients of the series expansion of the generating function.

We will begin our discussion by showing how the generating function of a sequence can be obtained from a linear recursion relation that is satisfied by the sequence. The section concludes by illustrating how generating functions may even be useful for more general types of recurrence relations, although there is no general theory for nonlinear recurrences.

5.6.1 Ordinary Generating Functions for Linear Recurrence Relations

Let

$$C(x) = x^k - a_1x^{k-1} - a_2x^{k-2} - \cdots - a_k = (x-\alpha_1)^{r_1}(x-\alpha_2)^{r_2}\cdots(x-\alpha_m)^{r_m} \tag{5.62}$$

be the characteristic polynomial of the sequence h_n annihilated by the operator $C(E)$. Therefore, h_n is a linear sum of terms of the general form $\binom{n}{j}\alpha_i^{n-j}$, so the ordinary generating function (OGF) of h_n is a linear sum of OGFs with the form

$$f_{ij}(x) = \sum_{n=0}^{\infty}\binom{n}{j}\alpha_i^{n-j}x^n = x^j\sum_{m=0}^{\infty}\binom{m+j}{j}(\alpha_i x)^m = \frac{x^j}{(1-\alpha_i x)^{j+1}}, \quad 0 \le j \le r_i - 1 \tag{5.63}$$

A linear sum of the terms in (5.63) can be put over the common denominator

$$Q(x) = (1-\alpha_1x)^{r_1}(1-\alpha_2x)^{r_2}\cdots(1-\alpha_mx)^{r_m} \tag{5.64}$$

with a numerator that is some polynomial $P(x)$. Since $\lim_{x\to\infty} f_{ij} = 0$, we see that $\lim_{x\to\infty}[P(x)/Q(x)] = 0$. Therefore, the degree of $P(x)$ is at most $k-1$, so it has the form

$$P(x) = b_0 + b_1x + b_2x^2 + \cdots + b_{k-1}x^{k-1} \tag{5.65}$$

From (5.62) and (5.64), we see that $Q(x) = x^kC(1/x)$; thus

$$Q(x) = 1 - a_1x - a_2x^2 - \cdots - a_kx^k \tag{5.66}$$

is a polynomial of degree k since $a_k \neq 0$.

This gives us a very useful result.

Theorem 5.37 Let the sequence h_n satisfy the homogeneous linear recurrence relation with constant coefficients given by $h_{n+k} = a_1h_{n+k-1} + a_2h_{n+k-2} + \cdots + a_kh_n$. Then the ordinary generating function $f_h(x) = \sum_{n=0}^{\infty} h_nx^n$ of the sequence is given by

$$f_h(x) = \frac{P(x)}{Q(x)} = \frac{b_0 + b_1x + b_2x^2 + \cdots + b_{k-1}x^{k-1}}{1 - a_1x - a_2x^2 - \cdots - a_kx^k} \tag{5.67}$$

where

$$b_j = h_j - h_{j-1}a_1 - h_{j-2}a_2 - \cdots - h_0a_j,\ j = 0, 1, \ldots, k-1. \tag{5.68}$$

Proof. It remains only to verify the formulas of equation (5.68). But these follow from (5.67) since

$$b_0 + b_1x + b_2x^2 + \cdots + b_{k-1}x^{k-1} = \left(1 - a_1x - a_2x^2 - \cdots - a_kx^k\right)\left(h_0 + h_1x + h_2x^2 + \cdots\right)$$
$$= h_0 + (h_1 - h_0a_1)x + (h_2 - h_1a_1 - h_0a_2)x^2 + (h_3 - h_2a_1 - h_1a_2 - h_0a_3)x^3 + \cdots.$$

Equating coefficients of powers of x shows that $b_0 = h_0, b_1 = h_1 - h_0a_1, b_2 = h_2 - h_1a_1 - h_0a_2, \ldots,$ as given by (5.68). Note that only the first k initial values $h_0 = A_0, h_1 = A_1, \ldots, h_{k-1} = A_{k-1}$ of the sequence appear in (5.68). This is expected, since these k conditions will then correspond to a unique generating function. ■

We now have a new way to obtain the OGFs of any sequence for which we know its homogeneous linear recurrence relation and initial conditions. Consider, for example, the Fibonacci recurrence relation, which has the characteristic polynomial $C(x) = x^2 - x - 1$. We see that $a_1 = a_2 = 1$, so any sequence satisfying the Fibonacci recurrence has an OGF with the form $f_h(x) = (b_0 + b_1x)/(1 - x - x^2)$.

For the Fibonacci numbers, equation (5.68) shows that $b_0 = F_0 = 0$ and $b_1 = F_1 - F_0 = 1 - 0 = 1$, so the OGF of the Fibonacci numbers is

$$f_F(x) = \frac{x}{1 - x - x^2} \tag{5.69}$$

Our next example applies Theorem 5.37 to compute the OGF of several recurrence sequences.

Example 5.38 Verify the ordinary generating functions listed in the following table:

Sequence	Recurrence Relation	Ordinary Generating Function
Lucas: 2,1,3,4,7,11, ...	$L_{n+2} = L_{n+1} + L_n$	$\frac{2 - x}{1 - x - x^2}$
Pell: 0,1,2,5,12,29, ...	$P_{n+2} = 2P_{n+1} + P_n$	$\frac{x}{1 - 2x - x^2}$
Perrin: 3,0,2,3,2,5,5,7,10, ...	$p_{n+3} = p_{n+1} + p_n$	$\frac{3 - x^2}{1 - x^2 - x^3}$
Tribonacci: 0,0,1,1,2,4,7,13,24, ...	$T_{n+3} = T_{n+2} + T_{n+1} + T_n$	$\frac{x^2}{1 - x - x^2 - x^3}$

Squares: 0,1,4,9,16,25, ...	$s_{n+3} = 3s_{n+2} - 3s_{n+1} + s_n$	$\dfrac{x+x^2}{(1-x)^3}$
Cubes: 0,1, 8,27,64,125, ...	$c_{n+4} = 4c_{n+3} - 6c_{n+2} + 4c_{n+1} - c_n$	$\dfrac{x+4x^2+x^3}{(1-x)^4}$
Triangular numbers: $t_n = \binom{n+1}{2}$ 0, 1, 3, 6, 10, ...	$t_{n+3} = 3t_{n+2} - 3t_{n+1} + t_n$, or $(E-1)^3 \binom{n+1}{2} = (E-1)^3 t_n = 0$	$\dfrac{x}{(1-x)^3}$
Squared triangular numbers: $t_n^2 = \binom{n+1}{2}^2$ $0, 1^2, 3^2, 6^2, 10^2, \ldots$	$t_{n+5}^2 = 5t_{n+4}^2 - 10t_{n+3}^2 + 10t_{n+2}^2 - 5t_{n+1}^2 + t_n^2$, or $(E-1)^5 \binom{n+1}{2}^2 = (E-1)^5 t_n^2 = 0$	$\dfrac{x+4x^2+x^3}{(1-x)^5}$

Solution. Since the denominators $Q(x) = 1 - a_1x - a_2x^2 - \cdots - a_kx^k$ are evident from the recurrence relations, we need only use equations (5.68) to confirm the coefficients $b_0, b_1, \ldots, b_{k-1}$ of the numerators $P(x) = b_0 + b_1x + b_2x^2 + \cdots + b_{k-1}x^{k-1}$ given in the table:

Lucas: $b_0 = L_0 = 2,\ b_1 = L_1 - L_0 = 2 - 1 = -1$
Pell: $b_0 = P_0 = 0,\ b_1 = P_1 - 2P_0 = 1$
Perrin: $b_0 = p_0 = 3,\ b_1 = p_1 - p_0 \cdot 0 = 0,\ b_2 = p_2 - p_1 \cdot 1 - p_0 \cdot 1 = 2 - 3 = -1$
Tribonacci: $b_0 = T_0 = 0, b_1 = T_1 - T_0 \cdot 1 = 0, b_2 = T_2 - T_1 \cdot 1 - T_0 \cdot 1 = 1 - 0 - 0 = 1$
Squares: $b_0 = s_0 = 0,\ b_1 = s_1 - s_0 \cdot 3 = 1,\ b_2 = s_2 - s_1 \cdot 3 + s_0 \cdot 3 = 4 - 3 + 0 = 1$
Cubes: $b_0 = c_0 = 0,\ b_1 = c_1 - c_0 \cdot 4 = 1,\ b_2 = c_2 - c_1 \cdot 4 + c_0 \cdot 6 = 8 - 4 = 4$
$b_3 = c_3 - c_2 \cdot 4 + c_1 \cdot 6 - c_0 \cdot 4 = 27 - 32 + 6 = 1$
Triangular numbers: $b_0 = t_0 = 0,\ b_1 = t_1 - t_0 \cdot 3 = 1,\ b_2 = t_2 - t_1 \cdot 3 + t_0 = 3 - 1 \cdot 3 + 0 = 0$
Squared triangular numbers:

$$b_0 = t_0^2 = 0, b_1 = t_1^2 - t_0^2 \cdot 5 = 1, b_2 = t_2^2 - t_1^2 \cdot 5 + t_0^2 \cdot 10 = 3^2 - 1^2 \cdot 5 = 4$$
$$b_3 = t_3^2 - t_2^2 \cdot 5 + t_1^2 \cdot 10 - t_0^2 \cdot 10 = 6^2 - 3^2 \cdot 5 + 1^2 \cdot 10 - 0 = 1$$
$$b_4 = t_4^2 - t_3^2 \cdot 5 + t_2^2 \cdot 10 - t_1^2 \cdot 10 + t_0^2 \cdot 5 = 10^2 - 6^2 \cdot 5 + 3^2 \cdot 10 - 1^2 \cdot 10 + 0 = 0$$

■

The generating functions for the cubes and squared triangular numbers in Example 5.38 show that

$$\sum_{n=0}^{\infty}\binom{n+1}{2}^2 x^n = \frac{x+4x^2+x^3}{(1-x)^5} = \frac{1}{(1-x)}\frac{x+4x^2+x^3}{(1-x)^4}$$
$$= \frac{1}{(1-x)}\sum_{n=0}^{\infty} n^3x^n = \sum_{n=0}^{\infty}(1^3+2^3+\cdots+n^3)x^n$$

since dividing an OGF by $1-x$ gives the OGF of the partial sum sequence (see Theorem 3.22). Comparing coefficients of x^n, we see that the sum of cubes is a squared triangular number:

$$1^3+2^3+\cdots+n^3 = \binom{n+1}{2}^2.$$

This identity should look familiar, since it has been derived by other means earlier in this text.

If we let $\sigma_k(n) = 1^k + 2^k + \cdots + n^k$ denote the sum of the kth powers of the first n natural numbers, then the generating functions for the sequences $\sigma_2(n)$ and $\sigma_3(n)$ were each as determined in Example 5.38. In the next chapter, we will describe the OGF of the general sequence $\sigma_k(n)$ for any k and thereby introduce a sequence of polynomials whose coefficients are known as the *Eulerian numbers.*

For our last illustration of Theorem 5.37, we will derive the OGFs for the total number of squares s_n and the total number of dominoes d_n that are required to form *all* of the f_n tilings of the $1 \times n$ boards with squares and dominoes. These sequences were introduced in Example 5.35.

Example 5.39 Let s_n and d_n denote, respectively, the number of squares and the number of dominoes that are required to form *all* of the f_n tilings of the $1 \times n$ boards. Then their ordinary generating functions are

$$f_S(x) = \sum_{n=0}^{\infty} s_nx^n = \frac{x}{(1-x-x^2)^2} \quad \text{and} \quad f_D(x) = \sum_{n=0}^{\infty} d_nx^n = \frac{x^2}{(1-x-x^2)^2} \tag{5.70}$$

Solution. From Example 5.35, we know that the fourth order operator $C(E) = (E^2 - E - 1)^2$ annihilates the nonhomogeneous recurrence relation $s_{n+2} - s_{n+1} - s_n = f_{n+1}$. Therefore, the OGF for s_n has the form

$$\sum_{n=0}^{\infty} s_nx^n = \frac{b_0 + b_1x + b_2x^2 + b_3x^3}{(1-x-x^2)^2}$$

Since $Q(x) = (1 - x - x^2)^2 = 1 - 2x - x^2 + 2x^3 + x^4$, it follows from equations (5.68) that

$$\begin{aligned} b_0 &= s_0 = 0,\ b_1 = s_1 - s_0 \cdot 2 = 1 \\ b_2 &= s_2 - s_1 \cdot 2 - s_0 \cdot 1 = 2 - 2 - 0 = 0 \\ b_3 &= s_3 - s_2 \cdot 2 - s_1 \cdot 1 - s_0 \cdot (-2) = 5 - 2 \cdot 2 - 1 + 0 = 0 \end{aligned}$$

The formula for $f_S(x)$ shown in (5.70) then follows. Since $s_n = d_{n+1}$, we also have

$$f_D(x) = \sum_{n=1}^{\infty} d_n x^n = x \sum_{n=0}^{\infty} d_{n+1} x^n = x \sum_{n=0}^{\infty} s_n x^n = \frac{x^2}{(1 - x - x^2)^2}.$$

■

An unexpected consequence of (5.70) is to see that the numbers $s_n = d_{n+1}$ are given by convolution products of the Fibonacci numbers; that is, since $x f_S(x) = f_D(x) = f_F^2(x)$, we have

$$s_n = d_{n+1} = F_n F_1 + F_{n-1} F_2 + \cdots + F_1 F_n \tag{5.71}$$

For example, $s_4 = F_4 F_1 + F_3 F_2 + F_2 F_3 + F_1 F_4 = 3 \cdot 1 + 2 \cdot 1 + 1 \cdot 2 + 1 \cdot 3 = 10$, the value shown in Example 5.35.

When (5.71) is compared with (5.61), we discover the identity $F_n F_1 + F_{n-1} F_2 + \cdots + F_1 F_n = \frac{1}{5}(n F_{n+1} + 2(1 + n) F_n)$.

5.6.2 Generating Functions for More General Recurrence Relations

Generating functions can also be used to solve recurrence relations of more general types, such as nonlinear recurrences, or linear recurrence relations with nonconstant coefficients, or recurrences with additional parameters. There is no general theory, however, and more ingenuity may be required to determine the generating function that corresponds to the recurrence relation. This is illustrated in the next example. Generating functions will also play an important role for the combinatorial number sequences investigated in Chapter 6.

Example 5.40 The Egg-Drop Numbers [3]. Let's begin with e identical eggs and assume that we are allowed d drops of the eggs. Our goal is to determine the maximum height of a building, as measured by its number of stories s, for which each floor can be checked for "egg drop safety" under the following assumptions:

- If an egg is dropped from floor n and is not broken, then all of the floors $1, 2, \ldots, n$ are safe floors. Moreover, the egg is unharmed and can be dropped again.
- If an egg dropped from floor n breaks, then all of the floors $n, n + 1, \ldots$ are unsafe and the broken egg cannot be reused.

For example, suppose that $e = 2$ eggs are available and we are allowed $d = 2$ drops. Then a three story building can be checked for egg drop safety this way:

1. Drop an egg from the second floor.
2. If it breaks, check out the first floor with our remaining egg and the second drop.
3. If the egg survives intact, check out the third floor with the second drop that is allowed.

Solution. Let $s = \left[\!\left[{d \atop e} \right]\!\right]$, $d \geq 0, e \geq 0$ denote the number of stories that can be checked out with d drops and e eggs. Clearly $\left[\!\left[{d \atop 0} \right]\!\right] = 0$ and $\left[\!\left[{0 \atop e} \right]\!\right] = 0$ since there is nothing we can do if we have no eggs and, or are not allowed any drops. Similarly, it is clear that $\left[\!\left[{1 \atop e} \right]\!\right] = \left[\!\left[{d \atop 1} \right]\!\right] = 1$, $1 \leq e, 1 \leq d$, since only a single-story building can be checked with just one drop or with one egg. Now suppose that $d \geq 2$ and $e \geq 2$. Our first drop should then be made at floor $\left[\!\left[{d-1 \atop e-1} \right]\!\right] + 1$, because

- If the egg were to break, we would be left with $d - 1$ drops and $e - 1$ eggs and we could then determine the egg drop safety of floors 1,2, ... , $\left[\!\left[{d-1 \atop e-1} \right]\!\right]$.
- If the egg survives, we are left with $d - 1$ drops and still have e eggs, and we can then determine the egg drop safety of $\left[\!\left[{d-1 \atop e} \right]\!\right]$ more stories above floor $\left[\!\left[{d-1 \atop e-1} \right]\!\right] + 1$.

We have now obtained the recurrence relation

$$\left[\!\left[{d \atop e} \right]\!\right] = \left[\!\left[{d-1 \atop e-1} \right]\!\right] + 1 + \left[\!\left[{d-1 \atop e} \right]\!\right], \quad d \geq 1, e \geq 1 \tag{5.72}$$

To solve this recurrence, introduce the generating functions given by

$$g_d(x) = \sum_{k=0}^{\infty} \left[\!\left[{d \atop k} \right]\!\right] x^k = \left[\!\left[{d \atop 0} \right]\!\right] + \left[\!\left[{d \atop 1} \right]\!\right] x + \left[\!\left[{d \atop 2} \right]\!\right] x^2 + \cdots + \left[\!\left[{d \atop k} \right]\!\right] x^k + \cdots, \quad d \geq 0 \tag{5.73}$$

Then, since

$$g_1(x) = \sum_{k=0}^{\infty} \left[\!\left[{1 \atop k} \right]\!\right] x^k = \sum_{k=1}^{\infty} \left[\!\left[{1 \atop k} \right]\!\right] x^k = \sum_{k=1}^{\infty} x^k = \frac{x}{1-x}$$

we see that

$$
\begin{aligned}
g_d(x) &= \sum_{k=1}^{\infty} \left[\!\left[{d \atop k}\right]\!\right] x^k = \sum_{k=1}^{\infty} \left(\left[\!\left[{d-1 \atop k-1}\right]\!\right] + 1 + \left[\!\left[{d-1 \atop k}\right]\!\right] \right) x^k \\
&= \sum_{k=1}^{\infty} \left[\!\left[{d-1 \atop k-1}\right]\!\right] x^k + \sum_{k=1}^{\infty} x^k + \sum_{k=1}^{\infty} \left[\!\left[{d-1 \atop k}\right]\!\right] x^k \\
&= x g_{d-1}(x) + g_1(x) + g_{d-1}(x)
\end{aligned}
$$

which can be written as

$$
g_d(x) = (1+x) g_{d-1}(x) + g_1(x), \quad d \geq 1 \tag{5.74}
$$

By iterating backward, and noting that $g_0(x) = 0$, we see that

$$
\begin{aligned}
g_d(x) &= (1+x) g_{d-1}(x) + g_1(x) \\
&= (1+x)[(1+x) g_{d-2}(x) + g_1(x)] + g_1(x) = (1+x)^2 g_{d-2}(x) \\
&\quad + [(1+x)+1]\, g_1(x) \\
&= (1+x)^2\, [(1+x) g_{d-3}(x) + g_1(x)] + [(1+x)+1] g_1(x) \\
&= (1+x)^3\, g_{d-3}(x) + [(1+x)^2 + (1+x) + 1] g_1(x) \\
&\cdots \\
&= (1+x)^d\, g_0(x) + \left[\sum_{j=0}^{d-1} (1+x)^j \right] g_1(x) = \left[\frac{(1+x)^d - 1}{(1+x)-1} \right] \frac{x}{1-x} = \frac{1}{1-x} \sum_{j=1}^{d} \binom{d}{j} x^j
\end{aligned}
$$

Recalling that division of an OGF by $1-x$ gives the OGF of the partial sum sequence, we find that

$$
\left[\!\left[{d \atop e}\right]\!\right] = [x^e]\, g_d(x) = \sum_{j\geq 1}^{e} \binom{d}{j} \tag{5.75}
$$

This formula gives the number of stories that can be checked for egg drop safety with d drops and e eggs. For example, with 7 drops and 3 eggs, a 63-story building can be checked for egg drop safety since

$$
\left[\!\left[{7 \atop 3}\right]\!\right] = \sum_{j=1}^{3} \binom{7}{j} = \binom{7}{1} + \binom{7}{2} + \binom{7}{3} = 7 + 21 + 35 = 63
$$

Similarly, with 5 drops and 7 eggs, a 31-story building can be checked since

$$
\left[\!\left[{5 \atop 7}\right]\!\right] = \sum_{j=1}^{7} \binom{5}{j} = \sum_{j=0}^{5} \binom{5}{j} - 1 = 2^5 - 1 = 31
$$

■

PROBLEMS

5.6.1. The sequence that gives the number of ways to tile a $1 \times n$ board with red squares and blue or green dominoes satisfies the recurrence relation $h_n = h_{n-1} + 2h_{n-2}$ with the initial conditions $h_0 = 1$ and $h_1 = 1$. Determine the ordinary generating function (OGF) for this sequence.

5.6.2. The sequence that gives the number of ways to tile a $3 \times 2n$ board with dominoes satisfies $u_{n+2} = 4u_{n+1} - u_n$, where $u_0 = 1$ and $u_1 = 3$. Obtain the OGF for this sequence.

5.6.3. If h_n denotes the number of ways to position n nonattacking kings on a $2 \times 2n$ board, then it has been shown that $h_n = 4h_{n-1} - 4h_{n-2}$, $n \geq 2$, where $h_0 = 1$ and $h_1 = 4$. Show that the OGF of the sequence h_n is given by $f(x) = 1/(1 - 4x + 4x^2)$.

5.6.4. Find the OGF of the sequence that satisfies the recurrence relation $h_{n+3} = 7h_{n+1} + 6h_n$ with the initial conditions $h_0 = 0, h_1 = 1, h_2 = 2$.

5.6.5. The number of ways that a $2 \times n$ board can be tiled with squares and dominoes is given by the sequence u_n that satisfies the recurrence relation $u_{n+3} = 3u_{n+2} + u_{n+1} - u_n$, with the initial conditions $u_0 = 1, u_1 = 2, u_2 = 7$. Obtain the OGF for this sequence.

5.6.6. The fewest number of moves that solves the "Tower of Brahma" puzzle with n disks is given by the solution of the recurrence relation $h_{n+1} = 2h_n + 1$, $n \geq 1$ with the initial conditions $h_0 = 0$ and $h_1 = 1$. Determine the OGF of the sequence.

5.6.7. Find the OGF of the sequence given by the nonhomogeneous recurrence relation $h_{n+2} = 4h_{n+1} - 4h_n + 5 \cdot 2^n$, $n \geq 0$ with the initial conditions $h_0 = 0, h_1 = 2$.

5.6.8. Consider the P_{n+1} tilings of an n-board with red and blue squares and white dominoes (see Problem 5.5.9), where P_n is the nth Pell number. If r_n and d_n denote, respectively, the number of red squares and the number of white dominoes used in all of the tilings, then it can be shown that $r_n = d_{n+1}$ and $r_n = 2r_{n-1} + r_{n-2} + P_n$, $n \geq 0$.

(a) Show that the OGF of the sequences r_n and d_n are, respectively

$$f_R(x) = \sum_{n=0}^{\infty} r_n x^n = \frac{x}{(1 - 2x - x^2)^2} \quad \text{and} \quad f_D(x) = \sum_{n=0}^{\infty} d_n x^n = \frac{x^2}{(1 - 2x - x^2)^2}$$

(b) Explain why the sequences r_n and d_n are convolution products of the sequence of Pell numbers, for example, $r_4 = d_5 = P_1P_4 + P_2P_3 + P_3P_2 + P_4P_1 = 1 \cdot 12 + 2 \cdot 5 + 5 \cdot 2 + 12 \cdot 1 = 44$.

5.6.9. Recall that the sequence $p_n^{(r)}$ of r-gonal numbers satisfies the recurrence relation $p_{n+1}^{(r)} = p_n^{(r)} + 1 + (r-2)n$ with the initial conditions $p_0^{(r)} = 0, p_1^{(r)} = 1$, and $p_2^{(r)} = r$ (see Problem 5.5.7). Show that the ordinary generating function $f^{(r)}$ for the r-gonal numbers is $f^{(r)}(x) = [x + (r-3)x^2]/(1-x)^3$.

5.6.10. Recall that the sequence $c_n^{(r)}$ of centered r-gonal satisfy the recurrence relation $c_n^{(r)} = c_{n-1}^{(r)} + rn, n \geq 1$, with the initial conditions $c_0^{(r)} = 1, c_1^{(r)} = 1 + r, c_2^{(r)} = 1 + 3r$ (see Problem 5.5.8).

(a) Explain why the sequence of centered r-gonal numbers is annihilated by the operator $C(E) = (E-1)^3$.

(b) Show that the ordinary generating function $g^{(r)}$ for the centered r-gonal numbers is

$$g^{(r)}(x) = \frac{1 + (r-2)x + x^2}{(1-x)^3}$$

5.6.11. Obtain the OGF f_F of the Fibonacci numbers by working with the three series

$$\begin{aligned} f_F(x) &= F_0 + F_1x + F_2x^2 + F_3x^3 + F_4x^4 + F_5x^5 + \cdots \\ xf_F(x) &= \qquad F_0x + F_1x^2 + F_2x^3 + F_3x^4 + F_4x^5 + \cdots \\ x^2f_F(x) &= \qquad\qquad F_0x^2 + F_1x^3 + F_2x^4 + F_3x^5 + \cdots \end{aligned}$$

In particular, subtract the second and third series from the first.

5.6.12. Obtain the OGF of the **(a)** Lucas and **(b)** Pell sequences by a modification of the idea shown in Problem 5.6.11.

5.6.13. Obtain the OGF of the **(a)** Perrin and **(b)** tribonacci numbers by modifying the idea of Problem 5.6.11.

5.6.14. **(a)** Let the sequence h_n be annihilated by the operator $C(E) = (E-\alpha)^r$. Show that the exponential generating function of the sequence has the form

$$f_\alpha^{(e)}(x) = \sum_{n=0}^{\infty} h_n \frac{x^n}{n!} = p_\alpha(x)e^{\alpha x}$$

where $p_\alpha(x)$ is a polynomial of degree at most $r-1$.

(b) Let the sequence h_n be annihilated by the operator $C(E) = \prod_{i=1}^{m} (E-\alpha_i)^{r_i}$. Show that the exponential generating function of the sequence has the form

$$f_h^{(e)}(x) = p_1(x)e^{\alpha_1 x} + p_2(x)e^{\alpha_2 x} + \cdots + p_m(x)e^{\alpha_m x},$$

where $p_i(x)$ is a polynomial of degree at most $r_i - 1$.

5.6.15. Obtain the EGF of the Fibonacci sequence, using the Binet formula $F_n = (\varphi^n - \hat{\varphi}^n)/\sqrt{5}$.

5.6.16. Obtain the EGF of the Lucas sequence.

5.6.17. Suppose that there are m students $s_1, s_2, \ldots, s_m$ and n tutors $t_1, t_2, \ldots, t_n$. Let $\langle m, n\rangle$ denote the number of ways that the students can choose which tutor, if any, with whom to work. This can be illustrated with a diagram known as a *bipartite graph* as shown below for the case $m = 4$ and $n = 5$. Note that some students do not choose a tutor and some tutors have no students. However, no student has more than one tutor, and no tutor has more than one student. Such a pairing is called a *matching* in graph theory.

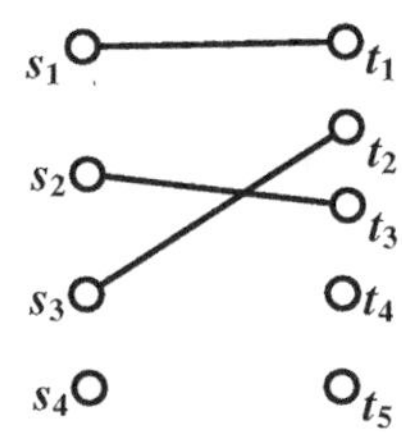

(a) Explain why $\langle m, 1\rangle = m + 1, \langle 1, n\rangle = n + 1$ and $\langle m, n\rangle = \langle n, m\rangle$.
(b) Derive the recurrence relation $\langle m, n\rangle = \langle m, n-1\rangle + m\,\langle m-1, n-1\rangle$.
(c) Explain why $\langle m, n\rangle = \sum_{k\geq 0} (m)_k \binom{n}{k}$. [Recall that $(m)_k$ is the number of k-permutations of an m set.]
(d) Show that $f_m^{(e)}(y) = \sum_{n=0}^{\infty} \langle m, n\rangle (y^n/n!) = (1+y)^m\, e^y$, where $\langle m, 0\rangle = 1$ and $\langle 0, n\rangle = 1$.
(e) Show that $f^{(e)}(x, y) = \sum_{m,n=0}^{\infty} \langle m, n\rangle [(x^m y^n)/(m!n!)] = e^{x+y+xy}$.

5.6.18. Obtain the EGF of the sequence h_n satisfying $h_{n+2} = -h_n$, $h_0 = 1$ and $h_1 = 0$.

5.6.19. Let $f^{(e)}(x) = \sum_{m=0}^{\infty} x^{3m}/(3m)!$ be the EGF of the sequence g_n satisfying $g_{n+3} = g_n$, $g_0 = 1, g_1 = 0$, $g_2 = 0$. Show that

$$f^{(e)}(x) = \frac{1}{3}(e^x + e^{\omega x} + e^{\omega^2 x}) = \frac{1}{3}\left(e^x + 2e^{-x/2}\cos\left(\frac{\sqrt{3}}{2}x\right)\right)$$

where ω is the primitive third root of unity $\omega = e^{2\pi i/3} = -\frac{1}{2} + i(\sqrt{3}/2)$; that is, $\omega^3 = 1$ and $1 + \omega + \omega^2 = 0$.

5.6.20. The *Fibonacci polynomials* $F_n(t), n \geq 0$ are defined as the solutions of the recurrence relation $F_{n+2}(t) = tF_{n+1}(t) + F_n(t)$, with $F_0 = 0$ and $F_1 = 1$.

(a) Prove that $F_{2n-1}(t)$ is an even polynomial and $F_{2n}(t)$ is an odd polynomial for all $n \geq 1$.

(b) Show that the Fibonacci polynomials are the coefficients, as functions of t, of the OGF $f_F(x,t) = \sum_{n=0}^{\infty} F_n(t)x^n = x/(1 - tx - x^2)$.

(c) Use the OGF of part (b) to show that, for all $n \geq 1$, $F_{2n}(t) = \sum_{r=0}^{n-1} \binom{n+r}{2r+1} t^{2r+1}$ and $F_{2n+1}(t) = \sum_{r=0}^{n} \binom{n+r}{2r} t^{2r}$. [*Hint*: Expand $1/(1 - x(t+x))$ as a geometric series.]

(d) Describe how the coefficients of the Fibonacci polynomials $F_n(t)$ appear in Pascal's triangle.

5.6.21. Let $f(x) = \sum_{n\geq 0} a_n x^n$ be the OGF of the sequence a_n for which $na_n = a_{n-1} + a_{n-2}, n \geq 2$ and $a_0 = 1, a_1 = 1$.

(a) Show that $f'(x) = (1 + x)f(x)$.

(b) Solve the differential equation in part (a) to show that $f(x) = e^{x+x^2/2}$.

5.7 SUMMARY AND ADDITIONAL PROBLEMS

The *homogeneous linear recurrence relation of order* k has the form $h_{n+k} = a_1 h_{n+k-1} + a_2 h_{n+k-2} + \cdots + a_k h_n$, $n \geq 0$ with constant coefficients $a_1, a_2, \ldots, a_k$. The equation can be expressed in the equivalent form $C(E)h_n = 0$, where $C(x) = x^k - a_1 x^{k-1} - a_2 x^{k-2} - \cdots - a_k$ is the *characteristic polynomial* and E is the *successor operator* defined by $Ef(n) = f(n+1)$ for any function $f(n)$. If the characteristic polynomial has the factorization $C(x) = (x - \alpha_1)^{r_1}(x - \alpha_2)^{r_2} \cdots (x - \alpha_m)^{r_m}$, then the roots of the *characteristic equation* $C(x) = 0$ are known as the *eigenvalues* $\alpha_1, \alpha_2, \ldots, \alpha_m$ of the recurrence relation, and have with the respective *multiplicities* $r_1, r_2, \ldots, r_m$. The *general solution* of the homogeneous recurrence has the form of a *generalized power sum* (GPS) of the eigenvalues, namely

$$h_n = \sum_{j=0}^{r_1-1} c_{1j} \binom{n}{j} \alpha_1^{n-j} + \sum_{j=0}^{r_2-1} c_{2j} \binom{n}{j} \alpha_2^{n-j} + \cdots + \sum_{j=0}^{r_m-1} c_{mj} \binom{n}{j} \alpha_m^{n-j}$$

The constants c_{ij} are uniquely determined by any set of initial conditions $h_0 = A_0, h_1 = A_1, \ldots, h_{k-1} = A_{k-1}$.

A nonhomogeneous linear recurrence relation has the form $C(E)h_n = q_n$, where q_n is the nonhomogeneous term of the equation. If g_n is any sequence for which $C(E)g_n = 0$, it is said that g_n is *annihilated* by $C(E)$ and g_n is called a *general solution* of the associated homogeneous recurrence relation. If $C(E)p_n = q_n$ for some sequence p_n, then p_n is a *particular solution* of the nonhomogeneous recurrence. If there is a polynomial operator $P(E)$ that annihilates p_n, so that $P(E)p_n = 0$, then $C(E)P(E)(g_n + p_n) = 0$. Therefore, $h_n = g_n + p_n$ can be expressed as a GPS of the eigenvalues of the characteristic equation $C(x)P(x) = 0$.

The ordinary generating function of a sequence h_n given by a linear homogeneous recurrence relation of order k has the form of the rational function $f_h(x) = P(x)/Q(x)$, where the numerator is a polynomial $P(x) = b_0 + b_1x + b_2x^2 + \cdots + b_{k-1}x^{k-1}$ of degree at most $k - 1$ and the denominator is the polynomial of degree k given by $Q(x) = x^k C(1/x) = 1 - a_1x - a_2x^2 - \cdots - a_kx^k$. Since many of the most common combinatorial sequences satisfy homogeneous linear recurrence relations (Fibonacci, Pell, polygonal numbers, etc.), their generating functions are easily computed. Generating functions can also be used to solve more general recurrence relations. This was illustrated with the egg drop numbers.

PROBLEMS

5.7.1. Use a piece of 8 × 8-in. squared paper to cut out the four shapes shown here:

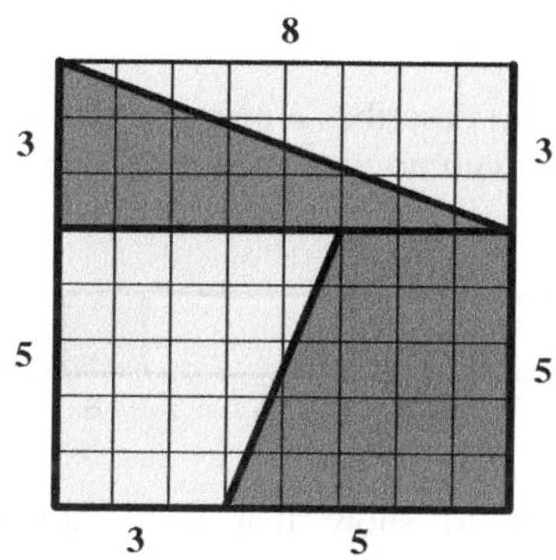

(a) Rearrange the four pieces to form this 5 × 13-in. rectangle. How does the area of the rectangle compare to the area of the square?

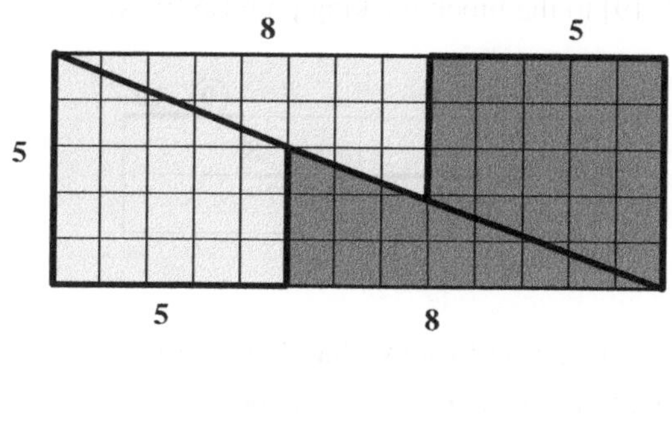

(b) Rearrange the four pieces to form this propeller shape. What is its area?

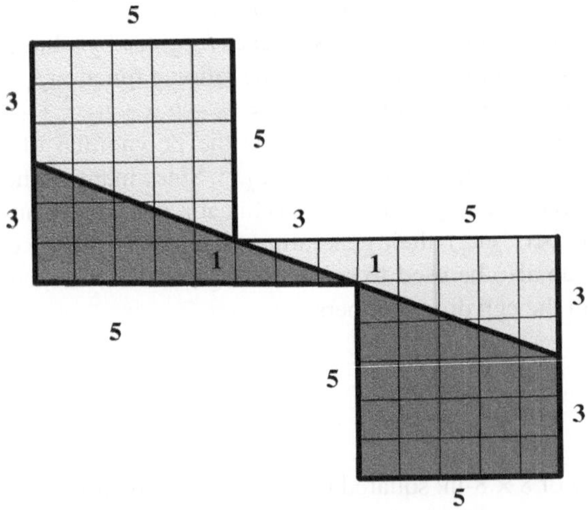

(c) Explain the excess or missing areas.

5.7.2. The following diagram describes a mapping that pairs the four-element set $S = \{1, 4, 6, 9\} \subseteq [9]$ with no two consecutive elements to the tiling *dsddsd* of a 1×10 board:

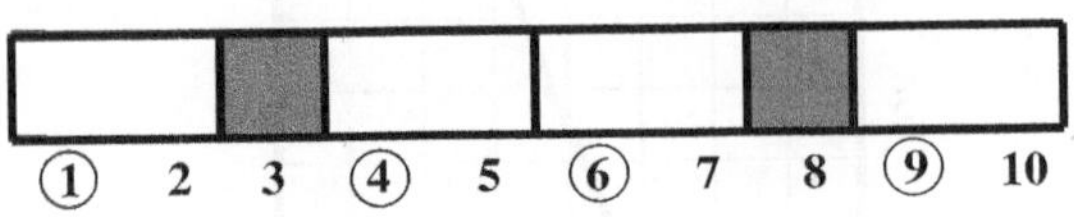

Extend this concept to show that there are $\binom{n+1-k}{k}$ subsets $S = \{s_1, s_2, \ldots, s_k\} \subseteq [n]$ with exactly k elements, with no two consecutive elements. [*Hint*: The tilings of a board of length $n + 1$ with exactly k dominoes includes $n + 1 - 2k$ squares as well.]

5.7.3. The following diagram describes a mapping that pairs the set $S = \{1, 4, 6, 9\} \subseteq [9]$ to the block walking path ENEENE:

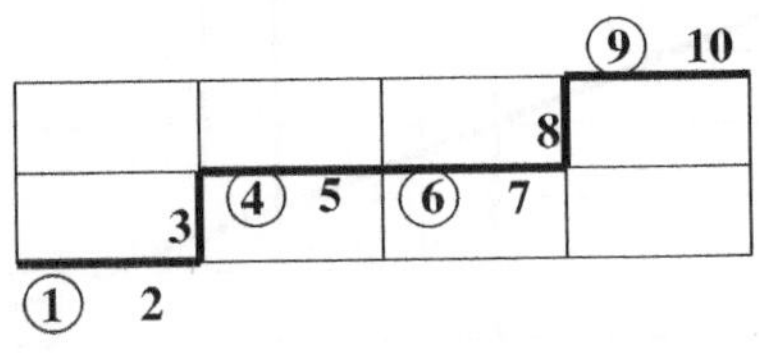

Generalize this bijection to show that there are $\binom{n+1-k}{k}$ subsets of $[n]$ with k elements, with no two consecutive elements.

5.7.4. The set $S = \{1, 4, 6, 9\} \subseteq [9]$ with no consecutive elements can be paired to the set $T = \{1, 3, 4, 6\} \subseteq [6]$, by setting $t_i = s_i - (i-1), i = 1, 2, \ldots, k = 4$. Generalize this to a bijection of the subsets $\{s_1, s_2, \ldots, s_k\} \subseteq [n]$ with no consecutive elements and the k-element subsets $\{t_1, t_2, \ldots, t_k\} \subseteq [n+1-k]$, thereby showing that there are $\binom{n+1-k}{k}$ subsets of $[n]$ with k elements, no two consecutive.

5.7.5. **(a)** Use mathematical induction to prove that any pair of successive Fibonacci numbers F_m and $F_{m+1}, m \geq 0$ have no common integer divisor larger that 1. In other words, show that $\gcd(F_m, F_{m+1}) = 1$, where gcd is the *greatest common divisor*. Note that $\gcd(a, b) = \gcd(a, ka + r) = \gcd(a, r)$ if $b = ka + r, 0 \leq r < a$.

(b) Let $0 < m < n$, so that $n = km + r$ for some $0 \leq r < m$. Prove that $\gcd(F_m, F_n) = \gcd(F_m, F_r)$ and $\gcd(m, n) = \gcd(m, r)$. [*Hint*: In Problem 5.2.3, we showned that $F_{km+r} = F_r F_{km+1} + F_{r-1} F_{km}$ and F_k divides F_{mk}.]

(c) Use part (b) to prove that $\gcd(F_m, F_n) = F_{\gcd(m,n)}$. [*Hint*: Assume that the theorem is incorrect, and let m be the smallest positive integer for which the formula doesn't hold[7] for some pair of Fibonacci numbers F_m and F_n.]

5.7.6. Prove that

$$\sum_{k \geq 0} \binom{n}{k} F_{k+r} = F_{2n+r}$$

for all integers $r \in \mathbb{Z}$ and all nonnegative integers $n \geq 0$, where F_m, $m \in \mathbb{Z}$ are the terms of the doubly infinite Fibonacci sequence $\ldots, F_{-3}, F_{-2}, F_{-1}, F_0, F_1, F_2, F_3, \ldots$ defined by the recurrence $F_{n+2} = F_{n+1} + F_n$, $n \in \mathbb{Z}$, $F_0 = 0$, $F_1 = 1$. It may be assumed that the identity has been verified for $r \geq 0$ (see Problem 5.2.10).

5.7.7. The *golden ratio* $\varphi = (1 + \sqrt{5})/2$ originated in ancient Greek mathematics with this problem. How can a line segment AB be divided into two segments AP and PB so that $(AB/AP) = (AP/PB)$?

Prove that the common ratio is given by the golden ratio.

5.7.8. An isosceles triangle ABC is tangent to the semicircle $x^2 + y^2 = 1, y \geq 0$ at point P, as shown in the diagram below. Show that the triangle has least perimeter when P separates the side AB so that $AP : PB = \varphi$, the golden ratio.

[7] This is what the British mathematician Robin Wilson would call a "minimal criminal."

[*Hint*: Show that $AP = \sqrt{\varphi}$ and $PB = (1/\sqrt{\varphi})$ for the triangle of minimum perimeter.]

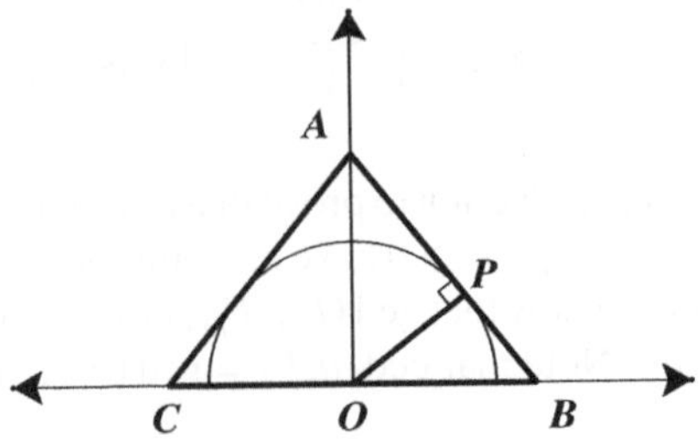

5.7.9. Prove that $F_n L_{n+1} - F_{n+1} L_n = (-1)^{n+1}\, 2$ for all $n \geq 0$.

5.7.10. How many ternary sequences of length n with the symbols 0, 1, and 2 always have a 0 immediately following every 2?

5.7.11. Let $C\,(E)$ annihilate the sequence u_n and let $D\,(E)$ annihilate the sequence v_n. Determine annihilating operators for these sequences

(a) $u_n + v_n$

(b) $w_n = \sum_{k=0}^{n} u_k$

5.7.12. It can be shown that the number of ways to tile an n-board with red and blue squares and white dominoes is given by the Pell number P_{n+1}. Use the tiling model to prove the identity $P_{m+n+1} = P_{m+1} P_{n+1} + P_m P_n$.

5.7.13. **(a)** Show that the Pell equation $x^2 - 5y^2 = -4$ is solved by the Lucas number $x = L_{2n-1}$ and the Fibonacci number $y = F_{2n-1}$ for all $n \geq 1$.

(b) Show that the Pell equation $x^2 - 5y^2 = 4$ is solved by $x = L_{2n}$ and $y = F_{2n}$ for all $n \geq 1$. [*Hint*: $x^2 - 5y^2 = (x + \sqrt{5}y)(x - \sqrt{5}y)$. *Comment*: It can be shown these are the *only* solutions of these Pell equations.]

5.7.14. Solve the recursion $(E - \alpha)^{k+1}\, h_n = 0$ given the initial conditions $h_j = 0$, $j = 0, 1, \ldots, k-1$ and $h_k = A$.

5.7.15. The "four-number game" (also known as a Ducci sequence) is played as follows. Choose a starting circular sequence of numbers (*a,b,c,d*) and at each step replace the sequence with the absolute values of the differences of adjacent entries: $(a, b, c, d) \to (|a - b|, |b - c|, |c - d|, |d - a|)$. For example, $(5, 350, 419, 37) \to (345, 69, 382, 32) \to (276, 313, 350, 313) \to (37, 37, 37, 37) \to (0, 0, 0, 0)$, which shows that (0, 0, 0, 0) was reached in just four steps.

(a) Play several games, choosing starting values (*a, b, c, d*) that you believe may take more steps before reaching (0, 0, 0, 0).

(b) If a,b,c,d are integers, prove that at least by the fourth step the 4-tuple has all even numbers. [*Hint*: Let d and e denote, respectively, odd and even numbers. For example, $(d, d, e, e) \to (e, d, e, d) \to (d, d, d, d) \to (e, e, e, e)$ shows all even numbers are reached in just three steps.]

(c) Use part (b) to explain why the game must end with (0,0,0,0) in finitely many steps. [*Hint*: The number of steps to termination at (0,0,0,0) is unchanged if (a,b,c,d) is replaced by $(a/m,\ b/m,\ c/m,\ d/m)$, where m is a common factor of a,b,c,d.]

(d) Play the four-number game starting with any sequence of four consecutive numbers from the tribonacci sequence 0, 0, 1, 1, 2, 4, 7, 13, 24, 44, 81, 149, 274, 504, 927, 1705, 3136, 5768, 10609, 19513, 35890, 66012, ... Describe the pattern you should discover, and explain why the four number game can be made arbitrarily long.

(e) Let $\alpha \in (1, 2)$ be the real eigenvalue of the tribonacci recursion discussed in Example 5.20; that is, α satisfies the equation $\alpha^3 = \alpha^2 + \alpha + 1$. Explain why the four-number game starting with $(1, \alpha, \alpha^2, \alpha^3)$ never terminates with all zeros.

5.7.16. Verify this generalized Vandermonde determinant by direct calculation:

$$V(\alpha^{(2)}, \beta, \gamma) = (\beta - \alpha)^2(\gamma - \alpha)^2(\gamma - \beta)$$

[*Hint:* Use column operations, factoring common terms from columns.]

REFERENCES

1. U. Dudley, *Numerology, or What Pythagoras Wrought*, Mathematical Association of America, 1997.
2. R. Perrin, Query 1484, *L Intermediare des Mathematiciens* **6**:76–77 (1899).
3. M. Boardman, The egg-drop numbers, *Math. Mag.* **77**(5): (Dec. 2004).

6

SPECIAL NUMBERS

6.1 INTRODUCTION

In previous chapters, we encountered the factorials, permutation numbers, the binomial and multichoose coefficients, the derangement numbers, the distribution numbers, the Fibonacci numbers, and other sequences. It is important to be familiar with these sequences since frequently a combinatorial or other mathematical question can be answered by deriving an expression that is a function of the terms of these sequences.

In this chapter, we will examine other sequences that commonly arise in mathematical problems. These include the Stirling numbers, harmonic numbers, Bernoulli numbers, Eulerian numbers, partition numbers, and Catalan numbers. Such "special" numbers allow us to answer questions by turning to this larger library of sequences whose properties are well studied and whose values are readily available.

6.2 STIRLING NUMBERS

6.2.1 Polynomials and the Difference Operator

Our work in Chapter 5 allows us to view the polynomial

$$p(n) = b_0 + b_1 n + b_2 n^2 + \cdots + b_k n^k \tag{6.1}$$

Combinatorial Reasoning: An Introduction to the Art of Counting, First Edition. Duane DeTemple and William Webb.

of degree not larger than k as a sequence that is annihilated by the operator $(E-1)^{k+1}=0$. Since $\alpha=1$ is an eigenvalue of multiplicity k, the polynomial can be expressed by a generalized power sum (GPS) of the form

$$p(n)=c_0\binom{n}{0}+c_1\binom{n}{1}+c_2\binom{n}{2}+\cdots+c_k\binom{n}{k} \tag{6.2}$$

where the constants $c_0, c_1, \ldots, c_k$ are uniquely determined by the initial values $p(0)=A_0, p(1)=A_1, \ldots, p(k)=A_k$.

To determine the coefficients $c_0, c_1, \ldots, c_k$ of the GPS given by (6.2), it is helpful to introduce the *first-order difference operator* Δ by the equation

$$\Delta=E-1. \tag{6.3}$$

The higher-order differences are then given by $\Delta^2, \Delta^3, \Delta^4, \ldots$, where $\Delta^{i+1}h_n=\Delta(\Delta^i h_n)$. It is also convenient to define $\Delta^0=1$, the identity operator.

The result of the following theorem will be quite useful.

Lemma 6.1 Let n, i, and j be nonnegative integers. Then

$$\Delta^i\binom{n}{j}=\binom{n}{j-i},\quad 0\le i\le j\le n \tag{6.4}$$

Proof. Formula (6.4) will follow by mathematical induction on i. The formula is clearly valid for $i=0$. For $i=1$, from Pascal's identity we have

$$\Delta\binom{n}{j}=\binom{n+1}{j}-\binom{n}{j}=\binom{n}{j-1}$$

Now let $i\ge 1$ and assume (6.4) holds for that value of i. Then

$$\Delta^{i+1}\binom{n}{j}=\Delta\left(\Delta^i\binom{n}{j}\right)=\Delta\binom{n}{j-i}=\binom{n}{j-i-1}$$ ■

Using formula (6.4), we see from (6.2) that

$$\Delta^i p(n)=\sum_{j=0}^{n}c_j\Delta^i\binom{n}{j}=\sum_{j=i}^{n}c_j\binom{n}{j-i}=c_i\binom{n}{0}+c_{i+1}\binom{n}{1}+\cdots+c_n\binom{n}{n-i}$$

In particular, choosing $n=0$ shows that

$$\Delta^i p(0)=\Delta^i p(n)\,|_{n=0}=c_i$$

The values of $\Delta^i p(0)$ can be calculated by a difference table that shows the successive rows of the difference sequences $\Delta^i p(n)$ for $i=0,1,2,\ldots\ n$. The example that follows illustrates the idea.

Example 6.2 Find the quadratic polynomial $p(n)$ that gives the sequence of pentagonal numbers 1,5,12,22,35,51, ...

Solution. The difference table is given by

$p(n)$	**1**		**5**		**12**		**22**		35		51
$\Delta p(n)$		**4**		**7**		**10**		13		16	
$\Delta^2 p(n)$			**3**		**3**		3		3		
$\Delta^3 p(n)$				**0**		0		0			

where it was really necessary to compute only the entries shown in bold. Since $\Delta^3 p(n) = 0$, this confirms that $p(n)$ is a quadratic polynomial $p(n) = c_0\binom{n}{0} + c_1\binom{n}{1} + c_2\binom{n}{2}$, where the coefficients are the leading entries in the rows of the difference table; that is, $c_0 = 1$, $c_1 = 4$, and $c_2 = 3$, giving us the polynomial

$$p(n) = 1\binom{n}{0} + 4\binom{n}{1} + 3\binom{n}{2} = 1 + 4n + 3\frac{n(n-1)}{2} = \frac{2 + 5n + 3n^2}{2}$$

■

The following theorem summarizes the general result.

Theorem 6.3 Let $p(n) = b_0 + b_1 n + b_2 n^2 + \cdots + b_k n^k$ be a polynomial of degree at most k. Then $p(n) = c_0\binom{n}{0} + c_1\binom{n}{1} + c_2\binom{n}{2} + \cdots + c_k\binom{n}{k}$, where $\Delta^i p(0) = c_i$ for $i = 0, 1, \ldots, k$.

An advantage of writing polynomials as sums of binomial coefficients is illustrated in the following result.

Corollary 6.4 Let $p(m) = c_0\binom{m}{0} + c_1\binom{m}{1} + c_2\binom{m}{2} + \cdots + c_k\binom{m}{k}$. Then the partial sums of the polynomial are given by

$$\sum_{m=0}^{n} p(m) = c_0\binom{n+1}{1} + c_1\binom{n+1}{2} + c_2\binom{n+1}{3} + \cdots + c_k\binom{n+1}{k+1} \quad (6.5)$$

Proof. Formula (6.5) follows directly from the hockey stick identity

$$\sum_{m=0}^{n}\binom{m}{j} = \binom{n+1}{j+1}$$

■

The following example takes advantage of Theorem 6.3 and Corollary 6.4.

Example 6.5 An $n \times n$ lattice grid, with $n + 1$ lattice points in each row and each column, is shown in the diagram below, together with some of the lattice squares that can be constructed in the grid.

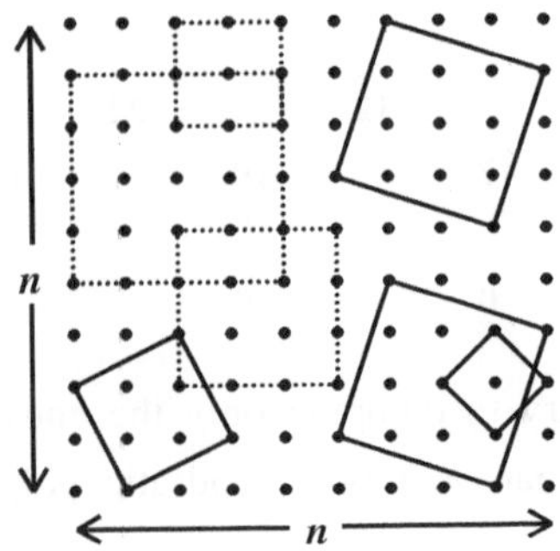

The squares with dotted sides are aligned to the grid, and the squares with solid sides are tilted to the grid. Let $p(n)$ and $q(n)$ denote, respectively, the number of aligned lattice squares and the number of both aligned and tilted squares in the grid. Show that

(a) $p(n) = \dfrac{n(n+1)(2n+1)}{6}$

(b) $q(n) = \dfrac{n(n+1)^2(n+2)}{12}$

Solution.

(a) An aligned lattice square with sides of length $n + 1 - k$, $1 \le k \le n$, must have its lower left corner at one of the k^2 lattice points in the shaded region shown in the following diagram.

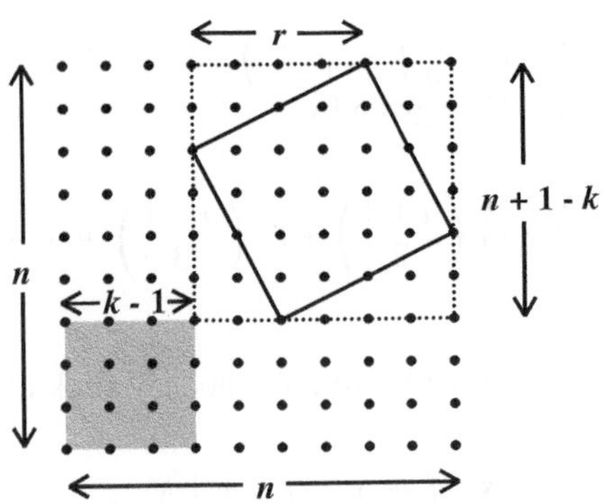

Therefore there are k^2 aligned lattice squares with sides of length $n + 1 - k$, and the total number of aligned squares is $p(n) = \sum_{k=1}^{n} k^2$. Although we derived a formula for the sum of the first n squared integers earlier, we can give a new proof by noting that

$$\Delta^4 \sum_{k=1}^{n} k^2 = \Delta^3 \left(\sum_{k=1}^{n+1} k^2 - \sum_{k=1}^{n} k^2 \right) = \Delta^3 (n+1)^2 = 0$$

Therefore, $p(n)$ is a cubic polynomial that, using Theorem 6.3, can be written in the form $p(n) = c_0 \binom{n}{0} + c_1 \binom{n}{1} + c_2 \binom{n}{2} + c_3 \binom{n}{3}$, where the constants c_0, c_1, c_2, c_3 can be determined by constructing this difference table:

$p(n)$	0		1		5		14		30
$\Delta p(n)$		1		4		11		16	
$\Delta^2 p(n)$			3		7		5		
$\Delta^3 p(n)$				2		2			
$\Delta^4 p(n)$					0				

Thus

$$p(n) = 0\binom{n}{0} + 1\binom{n}{1} + 3\binom{n}{2} + 2\binom{n}{3}$$
$$= n + 3\frac{n(n-1)}{2} + 2\frac{n(n-1)(n-2)}{6} = \frac{n(n+1)(2n+1)}{6}$$

(b) The diagram in the solution to part (a) shows that each aligned square with sides of length $n+1-k$ contains a tilted square for each value $r = 1, 2, \ldots, n-k$. Thus, including the aligned square itself, each of the k^2 aligned lattice squares circumscribes $n+1-k$ inscribed squares. This means that the total number of lattice squares, both aligned and tilted, is given by $q(n) = \sum_{k=1}^{n} (n+1-k)k^2$. Since

$$q(n) = \sum_{k=1}^{n} (n+1-k)k^2 = \sum_{k=1}^{n} \sum_{m=k}^{n} k^2 = \sum_{m=1}^{n} \sum_{k=1}^{m} k^2 = \sum_{m=1}^{n} p(m)$$

we can apply Corollary 6.4 to obtain

$$q(n) = \binom{n+1}{2} + 3\binom{n+1}{3} + 2\binom{n+1}{4}$$
$$= \frac{(n+1)n}{2} + \frac{(n+1)n(n-1)}{2} + \frac{(n+1)n(n-1)(n-2)}{12}$$

This simplifies to the expression $q(n) = n(n+1)^2(n+2)/12$. ■

6.2.2 Stirling Numbers of the Second Kind

The simplest polynomial in the variable n of degree k is surely $p(n) = n^k$. Therefore, as shown in Theorem 6.3, n^k can be expressed as a linear combination of the binomial coefficients $\binom{n}{0}, \binom{n}{1}, \ldots, \binom{n}{k}$. However, we can just as well express the polynomial as a linear combination of the Pochhammer symbols $(n)_j$, since $\binom{n}{j} = (n)_j/j!$, where $(n)_j = n(n-1)(n-2)\cdots(n-j+1)$ and $(n)_0 = 1$. The Pochhammer symbols are also known as the *falling factorials*. This prepares the way for the following definition.

Definition 6.6 The *Stirling numbers of the second kind*[1] $\left\{ {k \atop j} \right\}$, are the coefficients for which

$$n^k = \sum_{j=0}^{k} \left\{ {k \atop j} \right\} (n)_j, \quad j \geq 0, k \geq 0 \tag{6.6}$$

If we let $n = 0$, we see that $0 = \sum_{j=0}^{k} \left\{ {k \atop j} \right\} (0)_j = \left\{ {k \atop 0} \right\}$ for all $k \geq 1$. Also, comparing the coefficient n^k on both sides of (6.6) shows that $\left\{ {k \atop k} \right\} = 1$ for all $k \geq 1$. We will also let $\left\{ {0 \atop 0} \right\} = 1$. Additional values of the Stirling numbers can now be obtained by the triangle identity given in the following theorem.

Theorem 6.7 If $1 \leq j \leq k$, then

$$\left\{ {k+1 \atop j} \right\} = \left\{ {k \atop j-1} \right\} + j \left\{ {k \atop j} \right\} \tag{6.7}$$

Proof.

$$\sum_{j=1}^{k+1} \left\{ {k+1 \atop j} \right\} (n)_j = n^{k+1} = n(n^k) = n \sum_{j=1}^{k} \left\{ {k \atop j} \right\} (n)_j$$

$$= \sum_{j=1}^{k} \left\{ {k \atop j} \right\} (n - j + j)(n)_j$$

$$= \sum_{j=1}^{k} \left\{ {k \atop j} \right\} (n - j)(n)_j + \sum_{j=1}^{k} j \left\{ {k \atop j} \right\} (n)_j$$

[1] After the Scottish mathematician James Stirling (1692–1770). The Stirling numbers of the second kind are also commonly denoted with the notation $S(k, j)$. As the name suggests, there are also Stirling numbers of the first kind, and these will be introduced later in this section.

$$= \sum_{j=1}^{k} \left\{ {k \atop j} \right\} (n)_{j+1} + \sum_{j=1}^{k} j \left\{ {k \atop j} \right\} (n)_j$$

$$= \sum_{j=1}^{k} \left\{ {k \atop j-1} \right\} (n)_j + \sum_{j=1}^{k} j \left\{ {k \atop j} \right\} (n)_j$$

(replace $j + 1$ with j in the first sum)

Comparing coefficients of $(n)_j$ (see Problem 6.2.9) gives (6.7). ∎

The triangle identity (6.7), together with the initial values $\left\{ {k \atop 0} \right\} = 0$, $k \geq 1$, and $\left\{ {k \atop k} \right\} = 1$ for all $k \geq 0$, allow us to easily compute a Table 6.1 of values of the Stirling numbers where we let $\left\{ {k \atop j} \right\} = 0$ for all $j > k$. For example, the value 1701 in the bottom row of the table is given by the calculation $1701 = 301 + 4 \cdot 350$.

TABLE 6.1 Stirling Numbers of the Second Kind

	$j = 0$	1	2	3	4	5	6	7	8
$k = 0$	1	—	—	—	—	—	—	—	—
1	0	1	—	—	—	—	—	—	—
2	0	1	1	—	—	—	—	—	—
3	0	1	3	1	—	—	—	—	—
4	0	1	7	6	1	—	—	—	—
5	0	1	15	25	10	1	—	—	—
6	0	1	31	90	65	15	1	—	—
7	0	1	63	301	350	140	21	1	—
8	0	1	127	966	1701	1050	266	28	1

Although Stirling numbers have been defined algebraically, they have a useful combinatorial meaning, as shown in the following theorem. Here a *block* is a nonempty subset.

Theorem 6.8 The number of ways to partition a set of k elements into j disjoint blocks is given by the Stirling number of the second kind $\left\{ {k \atop j} \right\}$, where k and j are any positive integers.

Proof. Let $\hat{S}(k,j)$ denote the number of partitions of a k-set into j blocks. It is clear that $\hat{S}(k,j) = 0$ for $1 \leq k < j$ since there cannot be more blocks than there are elements. It is also clear that $\hat{S}(k, 1) = 1$ since there is only one way to place all of the elements into a single block—the block is the entire set. We can now proceed inductively, assuming that the number of ways to partition any k-element subset into j blocks is given by $\hat{S}(k,j)$ for each $j = 1, 2, \ldots, k$. Consider a subset with $k + 1$ elements, which we can assume is the set $[k + 1]$ of the first $k + 1$ positive integers. Two types of partitions can be considered.

Type 1: The Partition Contains the Singleton Block $\{k+1\}$. This means that the remaining elements in the set $[k]$ have been partitioned into $j-1$ blocks, and there are $\hat{S}(k, j-1)$ partitions of this type.

Type 2: The Partition Does Not Contain the Block $\{k+1\}$. All of the partitions of this type can be created in two steps. First partition $[k]$ into j blocks in $\hat{S}(k,j)$ ways, and now insert element $k+1$ into any one of these j blocks in j ways. This means that there are $j\hat{S}(k,j)$ partitions with no singleton block $\{k+1\}$.

Since all of the $\hat{S}(k+1,j)$ partitions of $[k+1]$ are of either type 1 or type 2, we find that $\hat{S}(k+1,j) = \hat{S}(k,j-1) + j\hat{S}(k,j)$. But this is exactly the triangle recurrence of the Stirling numbers shown in equation (6.7). Since the initial values $\hat{S}(k,0) = \left\{ {k \atop 0} \right\}$ and $\hat{S}(k,j) = 0, j > k$, are also in agreement, we conclude that $\hat{S}(k,j) = \left\{ {k \atop j} \right\}$ for all j and k. ■

Example 6.9 In how many ways can eight different toys be placed in five indistinguishable gift bags, with no bag left empty?

Solution. This is equivalent to a partition of the set $[8]$ into five blocks. From Table 6.1 we see that there are $\left\{ {8 \atop 5} \right\} = 1050$ ways to place the eight different toys into the five identical bags. ■

The combinatorial interpretation of the Stirling numbers allows us to discover properties that are not obvious from their implicit definition given in equation (6.6). For example, it is now easily argued that $\left\{ {k \atop 2} \right\} = 2^{k-1} - 1$ and $\left\{ {k \atop k-1} \right\} = \binom{k}{2}, k \geq 1$.

We also see that the Stirling numbers of the second kind are closely linked to the distribution numbers discussed in Section 2.6. Recall that the distribution number $T(k,j)$ is the number of ways to distribute k distinct objects to j distinct recipients in such a way that each recipient is assigned at least one object. The distribution can be carried out in two steps: first partition the k objects into j blocks in $\left\{ {k \atop j} \right\}$ ways; next, since these subsets are now distinct because they contain distinct members, assign the j blocks to the j recipients in $j!$ ways. Thus, we see that

$$T(k,j) = j! \left\{ {k \atop j} \right\} \tag{6.8}$$

Any of the results that have been obtained for distribution numbers can now be reinterpreted to yield an analogous result for Stirling numbers of the second kind. As an example, formula (4.11) shows that a Stirling number of the second kind is given by

$$\left\{ {k \atop j} \right\} = \frac{1}{j!} \sum_{i=0}^{j} (-1)^{i+j} \binom{j}{i} i^k \tag{6.9}$$

The Stirling numbers of the second kind have been defined by equation (6.6), which is a polynomial in the integer variable n. Therefore, this equation remains valid when n is replaced by the general variable x. Thus, we have the following result.

Theorem 6.10 Let $(x)_j = x(x-1)(x-2)\cdots(x-j+1)$ be the falling factorial. Then

$$x^k = \sum_{j=0}^{k} \left\{ \begin{matrix} k \\ j \end{matrix} \right\} (x)_j \tag{6.10}$$

The ordinary generating function for the Stirling numbers in column r of the Stirling number triangle of Table 6.1 will now be derived on the basis of triangle identity (6.7).

Theorem 6.11 For any $r \geq 1$,

$$f_r(x) = \sum_{k=0}^{\infty} \left\{ \begin{matrix} r+k \\ r \end{matrix} \right\} x^k = \frac{1}{(1-x)(1-2x)\cdots(1-rx)} \tag{6.11}$$

Proof. Our proof is by mathematical induction. For $r = 1$, we see that

$$f_1(x) = \sum_{k=0}^{\infty} \left\{ \begin{matrix} 1+k \\ 1 \end{matrix} \right\} x^k = \sum_{k=0}^{\infty} x^k = \frac{1}{1-x}$$

which shows that (6.11) is correct for $r = 1$. Now let $r \geq 2$, and suppose that (6.11) holds for $r - 1$. Using the triangle identity (6.7), we have

$$\begin{aligned} f_r(x) &= \sum_{k=0}^{\infty} \left\{ \begin{matrix} r+k \\ r \end{matrix} \right\} x^k = \sum_{k=0}^{\infty} \left[\left\{ \begin{matrix} r+k-1 \\ r-1 \end{matrix} \right\} + r \left\{ \begin{matrix} r+k-1 \\ r \end{matrix} \right\} \right] x^k \\ &= f_{r-1}(x) + rxf_r(x) \end{aligned}$$

which can be rearranged to show that $f_r(x) = [1/(1-rx)]f_{r-1}(x)$. Therefore, using the induction hypothesis, we obtain

$$\begin{aligned} f_r(x) &= \frac{1}{1-rx} f_{r-1}(x) = \frac{1}{1-rx} \left(\frac{1}{(1-x)(1-2x)\cdots(1-(r-1)x)} \right) \\ &= \frac{1}{(1-x)(1-2x)\cdots(1-rx)} \end{aligned}$$

which completes the induction proof. ■

To illustrate how the Stirling numbers can arise unexpectedly, let's ask this question: "What is the sum of all products of r factors, possibly repeated, taken from $[k]$, where each product is written in nondecreasing order?"

For example, take $r = 2$ and $k = 2$, so the sum is taken over all products of two numbers from the set $[r] = \{1, 2\}$, where the two factors are in nondecreasing order and the factors can be repeated. Therefore the sum is $1 \cdot 1 + 1 \cdot 2 + 2 \cdot 2 = 7$. Similarly, if $r = 3$ and $k = 2$, then the sum of the products with two factors in nondecreasing order taken from the set $\{1, 2, 3\}$ is $1 \cdot 1 + 1 \cdot 2 + 1 \cdot 3 + 2 \cdot 2 + 2 \cdot 3 + 3 \cdot 3 = 25$.

In both examples, the sum is a Stirling number:

$$\left\{ {2+2 \atop 2} \right\} = 7 \quad \text{and} \quad \left\{ {3+2 \atop 3} \right\} = 25$$

Indeed, such sums will always be Stirling numbers, as proved in the following theorem.

Theorem 6.12 The sum of all products of k factors from the set $[r]$, written in nondecreasing order and with repetition of factors allowed, is the Stirling number $\left\{ {r+k \atop r} \right\}$.

Proof. The result is a direct consequence of the generating function derived in Theorem 6.11. For example, if $r = 2$, then

$$f_2(x) = \sum_{k=0}^{\infty} \left\{ {2+k \atop 2} \right\} x^k$$

$$= \frac{1}{(1-x)(1-2x)} = \left(\sum_{i=0}^{\infty} 1^i x^i \right) \left(\sum_{j=0}^{\infty} 2^j x^j \right) = \sum_{k=0}^{\infty} \left(\sum_{\substack{i,j \geq 0 \\ i+j=k}} 1^i \cdot 2^j \right) x^k$$

so comparing coefficients of x^k, we find that

$$\left\{ {2+k \atop 2} \right\} = \sum_{\substack{i,j \geq 0 \\ i+j=k}} 1^i \cdot 2^j$$

Similarly, for $r = 3$ we have

$$f_3(x) = \sum_{k=0}^{\infty} \left\{ {3+k \atop 3} \right\} x^k = \frac{1}{(1-x)(1-2x)(1-3x)}$$

$$= \left(\sum_{i=0}^{\infty} 1^i x^i \right) \left(\sum_{j=0}^{\infty} 2^j x^j \right) \left(\sum_{m=0}^{\infty} 3^m x^m \right) = \sum_{k=0}^{\infty} \left(\sum_{\substack{i,j,m \geq 0 \\ i+j+m=k}} 1^i \cdot 2^j \cdot 3^m \right) x^k$$

so comparing coefficients gives

$$\left\{ {3+k \atop 3} \right\} = \sum_{\substack{i,j,m \geq 0 \\ i+j+m=k}} 1^i \cdot 2^j \cdot 3^m$$

The validity of the general formula is now evident. ■

6.2.3 Bell Numbers

The total number of ways to partition a k-element set into *any* number of blocks is given by the *Bell number*[2] $B(k)$. Therefore, $B(k)$ is the sum of the Stirling numbers in row k of the Stirling number triangle given in Table 6.1:

$$B(k) = \sum_{j=0}^{k} \left\{ {k \atop j} \right\} \tag{6.12}$$

The first few Bell numbers are listed in Table 6.2.

TABLE 6.2 Bell Numbers $B(k)$ for $0 \leq k \leq 10$

$B(0)$	$B(1)$	$B(2)$	$B(3)$	$B(4)$	$B(5)$	$B(6)$	$B(7)$	$B(8)$	$B(9)$	$B(10)$
1	1	2	5	15	52	203	877	4140	21,147	115,975

A recurrence relation for the Bell numbers is given in the following theorem, whose proof is obtained by combinatorial reasoning.

Theorem 6.13 For all $k \geq 1$, we obtain

$$B(k+1) = \sum_{r=0}^{k} \binom{k}{r} B(r) \tag{6.13}$$

Proof. Assume that we are counting the number of ways to partition the set $[k+1]$ into any number of blocks. For any r, with $0 \leq r \leq k$, there are $\binom{k}{r} B(r)$ ways to select an r-element subset $[k]$ and partition it into any number of blocks. The element $k+1$ and the elements that were not selected form a block from $[k+1]$ that contains the element $k+1$. Summing over r gives all of the $B(k+1)$ partitions of the set $[k+1]$ into any number of blocks. ■

[2]Eric Temple Bell (1883–1960)

Example 6.14 There are seven members of the Math Club. Assuming that a "group" is one or more members, in how many ways can the club members split into

(a) exactly four study groups?
(b) any number of study groups?

Solution.

(a) $\left\{ {7 \atop 4} \right\} = 350$
(b) $B(7) = 877$ ■

6.2.4 Stirling Numbers of the First Kind

As suggested by their names, the Stirling numbers of the first and second kinds are closely related. Indeed, the two kinds of Stirling numbers are inverses of one another in the following sense. Since, as observed in equation (6.10), the Stirling numbers of the second kind are given by the equations $x^k = \sum_{j=0}^{k} \left\{ {k \atop j} \right\} (x)_j$ that express the powers x^k as a sum of *falling factorials* $(x)_j = x\,(x-1)\,(x-2)\cdots(x-j+1)$, $(x)_0 = 1$, the Stirling numbers of the first kind are defined as the coefficients that express the falling factorial $(x)_k$ as a sum of powers of x.

Definition 6.15 The *Stirling numbers of the first kind*[3] $\left[{k \atop j} \right]$, where $j \geq 0$ and $k \geq 0$, are the coefficients for which

$$(x)_k = \sum_{j=0}^{k} (-1)^{k-j} \left[{k \atop j} \right] x^j \tag{6.14}$$

A simpler defining equation can be obtained by introducing the *rising factorials* $x^{(k)}$.

Definition 6.16 The *rising factorial* $x^{(k)}$ is the polynomial of degree k defined by

$$x^{(0)} = 1 \quad \text{and} \quad x^{(k)} = x\,(x+1)\,(x+2)\cdots(x+k-1)\,,\ k = 1,2,\ldots \tag{6.15}$$

[3]Our definition gives the *unsigned* Stirling numbers of the first kind. In some treatments, the sign variation in (6.14) given by the factor $(-1)^{k-j}$ is absorbed into the symbol to give the *signed* Stirling numbers $s\,(k,j) = (-1)^{k-j} \left[{k \atop j} \right]$. However, $s\,(k,j)$ may also denote the unsigned Stirling numbers, so the reader must be careful to check which notation has been adopted.

From equations (6.14) and (6.15), we see that

$$x^{(k)} = x(x+1)\cdots(x+k-1) = (-1)^k(-x)(-x-1)\cdots(-x-k+1)$$
$$= (-1)^k(-x)_k = (-1)^k\sum_{j=0}^{k}(-1)^{k-j}\begin{bmatrix}k\\j\end{bmatrix}(-x)^j = \sum_{j=0}^{k}\begin{bmatrix}k\\j\end{bmatrix}x^j$$

This shows that the Stirling numbers of the first kind are the coefficients of the generating function of the rising factorial, as stated in the next theorem. This result parallels Theorem 6.10, where it was shown that the Stirling numbers of the second kind are the coefficients of the generating function of the falling factorial.

Theorem 6.17 Let $x^{(k)} = x\,(x+1)\,(x+2)\cdots(x+k-1)$, $k = 1, 2, \ldots$ be the rising factorial; then

$$x^{(k)} = \sum_{j=0}^{k}\begin{bmatrix}k\\j\end{bmatrix}x^j \tag{6.16}$$

A short list of the Stirling numbers of the first kind can be obtained by expanding the rising factorials, as shown below. The rows are indexed $k = 0,1,2, \ldots$ and the value of the coefficient of x^j is the Stirling number $\begin{bmatrix}k\\j\end{bmatrix}$, $j = 0, 1, 2, \ldots$.

$$\begin{aligned}
x^{(0)} &= 1\\
x^{(1)} &= x\\
x^{(2)} &= x\,(x+1) = x + x^2\\
x^{(3)} &= x\,(x+1)\,(x+2) = 2x + 3x^2 + x^3\\
x^{(4)} &= x\,(x+1)\,(x+2)\,(x+3) = 6x + 11x^2 + 6x^3 + x^4\\
x^{(5)} &= x\,(x+1)\,(x+2)\,(x+3)\,(x+4) = 24x + 50x^2 + 35x^3 + 10x^4 + x^5
\end{aligned}$$

For example, the entry $\begin{bmatrix}5\\3\end{bmatrix} = 35$ is the coefficient of x^3 in row $k = 5$. The formula $x^{(5)} = x\,(x+1)\,(x+2)\,(x+3)\,(x+4) = (x+4)\,x^{(4)}$ shows that the entries in row 5 can be obtained from the entries in row 4. For example

$$\begin{bmatrix}4+1\\3\end{bmatrix} = \begin{bmatrix}4\\2\end{bmatrix} + 4\begin{bmatrix}4\\3\end{bmatrix}$$

More generally, we can use the identity $x^{(k+1)} = x\,(x+1)\cdots(x+k-1)\,(x+k) = (x+k)\,x^{(k)}$ to obtain a triangle identity for the Stirling numbers of the first kind.

Theorem 6.18 Let $1 \le j \le k$. Then

$$\begin{bmatrix}k+1\\j\end{bmatrix} = \begin{bmatrix}k\\j-1\end{bmatrix} + k\begin{bmatrix}k\\j\end{bmatrix} \tag{6.17}$$

Proof.

$$\begin{aligned}\begin{bmatrix}k+1\\j\end{bmatrix} &= [x^j]x^{(k+1)} = [x^j](x\cdot x^{(k)} + kx^{(k)})\\ &= [x^{j-1}]x^{(k)} + k[x^j]x^{(k)}\\ &= \begin{bmatrix}k\\j-1\end{bmatrix} + k\begin{bmatrix}k\\j\end{bmatrix}\end{aligned}$$

■

The values of the Stirling numbers of the first kind are easily computed row by row using the triangle identity (6.17), giving us Table 6.3. For example, the entry 6769 shown in the last row is obtained from the previous row $k = 7$ by the calculation $6769 = 1624 + 7 \cdot 735$.

TABLE 6.3 The Stirling Numbers of the First Kind

	$j=0$	1	2	3	4	5	6	7	8
$k=0$	1	—	—	—	—	—	—	—	—
1	0	1	—	—	—	—	—	—	—
2	0	1	1	—	—	—	—	—	—
3	0	2	3	1	—	—	—	—	—
4	0	6	11	6	1	—	—	—	—
5	0	24	50	35	10	1	—	—	—
6	0	120	274	225	85	15	1	—	—
7	0	720	1764	1624	735	175	21	1	—
8	0	5040	13068	13132	6769	1960	322	28	1

The following theorem gives a combinatorial meaning for the Stirling numbers of the first kind. Its proof is analogous to the proof of Theorem 6.8, which gave us a combinatorial interpretation of the Stirling numbers of the second kind.

Theorem 6.19 The Stirling number $\begin{bmatrix}k\\j\end{bmatrix}$ of the first kind counts the number of arrangements of k distinct objects into j nonempty circular permutations.

Proof. Suppose that $\hat{s}(k,j)$ counts the number of ways that k people numbered 1 through k can be seated around j identical circular tables, where no table is left unoccupied. No arrangements are possible if there is at least one person but no tables, so $\hat{s}(k,0) = 0$ for all $k \geq 1$. We also see that $\hat{s}(k,1) = (k-1)!$, since this is the number of ways that $k \geq 1$ people can be seated around one table. Now consider $k+1$ people, so that they can be seated at j identical tables in $\hat{s}(k+1,j)$ ways that leave no table unoccupied. These seating arrangements are of two types.

Type 1: Person k + 1 Seated Alone at a Table. There are $\hat{s}(k,j-1)$ ways to seat all except person $k+1$ at $j-1$ tables, with $k+1$ seated alone at a table.

Type 2: Person k + 1 is Not Seated Alone at a Table. If element $k + 1$ is temporarily overlooked, we are left with a seating of k people at j tables with no table unoccupied. There are $\hat{s}(k,j)$ such seatings. Person $k + 1$ can then be seated immediately to the right of any of the k people already seated, giving us $k\hat{s}(k,j)$ seatings of the $k + 1$ people at j identical tables with no table unoccupied and with person $k + 1$ not alone at a table.

Since all permissible seatings of $k + 1$ people are accounted for in the two types of seating arrangements, there are $\hat{s}(k,j-1) + k\hat{s}(k,j)$ seatings of $k + 1$ people at j identical tables that leave no table unoccupied. Thus $\hat{s}(k+1,j) = \hat{s}(k,j-1) + k\hat{s}(k,j)$, which is exactly the triangle identity (6.17) satisfied by the Stirling numbers of the first kind. Moreover, the initial values of $\hat{s}(k,1)$ also coincide with the corresponding Stirling numbers, and so we conclude that $\hat{s}(k,j) = \left[{k \atop j}\right]$. ■

Example 6.20 In how many ways can seven people be seated at

(a) three identical circular tables, so that no table is unoccupied?

(b) any number of identical circular tables, where it is permissible to leave tables unoccupied?

Solution.

(a) There are $\left[{7 \atop 3}\right] = 1624$ seatings.

(b) There are $\left[{7 \atop 1}\right] + \left[{7 \atop 2}\right] + \cdots + \left[{7 \atop 7}\right] = 720 + 1764 + 1624 + 735 + 175 + 21 + 1 = 5040 = 7!$ seatings at any number of tables. (Theorem 6.23 below explains why the sum is simply 7!) ■

Any seating of seven people at three circular tables corresponds to a permutation of [7] that contains three cycles. For example, the permutation

$$\begin{pmatrix} 1 & 2 & 3 & 4 & 5 & 6 & 7 \\ 7 & 5 & 6 & 4 & 1 & 3 & 2 \end{pmatrix}$$

corresponds to the seating (1 7 2 5)(3 6)(4), where the four persons 1, 7, 2, and 5, in that counterclockwise order, are seated at a table; persons 3 and 6 are seated at a second table; and person 4 is alone at a third table.

Since any permutation can be uniquely written as a product of cycles that corresponds to a seating at circular tables, the Stirling numbers of the first kind count the number of permutations with a specified number of cycles. The gives us the following theorem, which is sometimes taken as the definition of the Stirling numbers.

Theorem 6.21 The Stirling number of the first kind $\left[{k \atop j}\right]$ is the number of permutations of $[k]$ with exactly j cycles.

Example 6.22 Write the permutation below as a product of cycles and describe the corresponding seating at identical circular tables:

$$\begin{pmatrix} 1 & 2 & 3 & 4 & 5 & 6 & 7 & 8 & 9 & 10 \\ 3 & 8 & 1 & 4 & 10 & 2 & 9 & 7 & 6 & 5 \end{pmatrix}$$

Solution. The permutation is (1 3)(2 8 7 9 6)(4)(5 10) when written as a product of cycles. It corresponds to seating persons 1 and 3 at a table; persons 2, 8, 7, 9, and 6 at a table in that counterclockwise order; person 4 alone at a table; and persons 5 and 10 at a table. ■

6.2.5 Identities for Stirling Numbers

There are a large number of identities satisfied by both kinds of Stirling numbers. Just a few identities are shown in the next two theorems, but more identities are included in the accompanying problem set.

Example 6.20(b) suggests the identity in the following theorem. It tells us that k people can be seated at any number of identical circular tables in $k!$ ways, where it is permissible to have some of the tables remain unoccupied.

Theorem 6.23 The sum of the Stirling numbers of the first kind in row k of the Stirling number triangle shown in Table 6.3 is $k!$:

$$\sum_{j=0}^{k} \begin{bmatrix} k \\ j \end{bmatrix} = k!, \quad k \geq 0 \tag{6.18}$$

Proof. For $k \geq 1$, row k of the Stirling number triangle lists the number of permutations of $[k]$, arranged according to the number of cycles. Therefore, all $k!$ permutations of $[k]$ are counted by the row sum $\sum_{j=0}^{k} \begin{bmatrix} k \\ j \end{bmatrix}$. This is also true for $k = 0$ since $1 = \begin{bmatrix} 0 \\ 0 \end{bmatrix}$ and $1 = 0!$. ■

The identities in the next theorem emphasize the inverse nature of the two kinds of Stirling numbers. Recall that the Kronecker delta function is defined by $\delta_{k,k} =$ 1and $\delta_{j,k} = 0$ for $j \neq k$.

Theorem 6.24

(a)

$$\sum_{i=0}^{k} (-1)^i \begin{bmatrix} k \\ i \end{bmatrix} \begin{Bmatrix} i \\ j \end{Bmatrix} = (-1)^k \, \delta_{j,k} \tag{6.19}$$

(b)

$$\sum_{i=0}^{k} (-1)^i \begin{Bmatrix} k \\ i \end{Bmatrix} \begin{bmatrix} i \\ j \end{bmatrix} = (-1)^k \, \delta_{j,k} \tag{6.20}$$

(c) For any two sequences f_n and g_n,

$$g_n = \sum_{k=0}^{n} (-1)^k \left\{ n \atop k \right\} f_k \quad \text{if and only if} \quad f_n = \sum_{k=0}^{n} (-1)^k \left[n \atop k \right] g_k \qquad (6.21)$$

Proof. Only part (a) will be proved, with the proofs of (b) and (c) left as exercises. The proof of (b) is similar to the one below for (a), and the proof of (c) is a consequence of parts (a) and (b). To prove (6.19), insert equation (6.6) into (6.14) to get

$$(n)_k = \sum_{i=0}^{k} (-1)^{k-i} \left[k \atop i \right] n^i = \sum_{i=0}^{k} (-1)^{k-i} \left[k \atop i \right] \sum_{j=0}^{i} \left\{ i \atop j \right\} (n)_j$$

$$= \sum_{j=0}^{k} (-1)^k \left(\sum_{i=0}^{k} (-1)^i \left[k \atop i \right] \left\{ i \atop j \right\} \right) (n)_j$$

We can equate coefficients of $(n)_j$ for $j \geq 0$ to obtain identity (6.19). ■

PROBLEMS

6.2.1. Let $p(n) = 3n^2 - n + 2$.

(a) Calculate the difference table.

(b) Express $p(n)$ as a linear sum of binomial coefficients.

(c) Express $\sum_{m=0}^{n} p(m)$ as a linear sum of binomial coefficients.

6.2.2. Find the quadratic polynomial that gives the hexagonal numbers 0,1,6,15,28,45,... by constructing a difference table.

6.2.3. The sequence of centered triangular numbers is 1,4,10,19,31,.... Determine the quadratic polynomial that gives this sequence.

6.2.4. The diagram below shows the maximum number of pieces of pizza made by n straight cuts across a circular pizza pie:

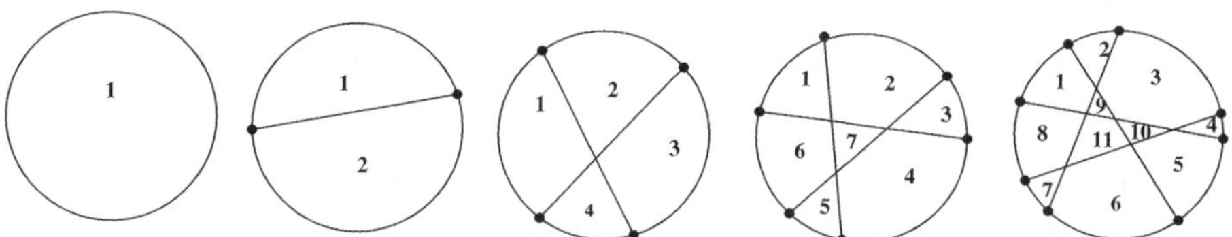

$p(0) = 1$ $p(1) = 2$ $p(2) = 4$ $p(3) = 7$ $p(4) = 11$

(a) explain why $\Delta p(n) = n + 1$.

(b) find the polynomial $p(n)$ that gives the maximum number of pieces formed by n straight cuts.

6.2.5. The diagram below shows the interior regions formed by the diagonals of convex n-gons for $n = 1,2,3,4,5,6$. Find a quartic polynomial $p(n)$ that gives the maximum number of regions formed by the diagonals of a convex n-gon.

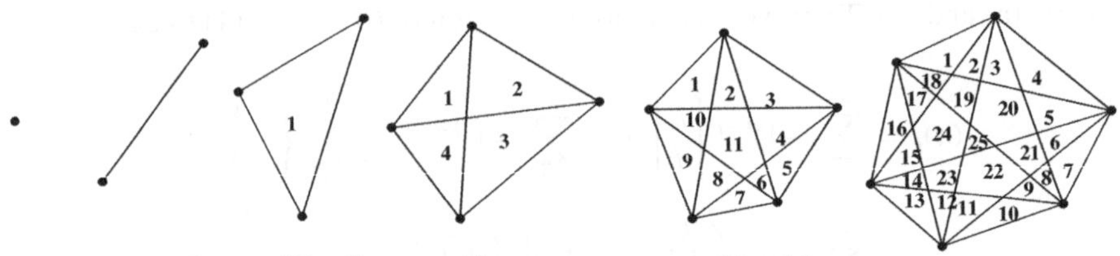

$p(1) = 0$ $p(2) = 0$ $p(3) = 1$ $p(4) = 4$ $p(5) = 11$ $p(6) = 25$

6.2.6. In 1772, Euler discovered a quadratic polynomial $P(n)$ whose beginning values for $n = 0,1,2,\ldots, 10$ were 41,41,43,47,53,61,71,83,97,113,131. Amazingly, these are all prime numbers.

(a) Construct a difference table and determine Euler's polynomial.

(b) Extend the difference table to determine $P(n)$ for all $0 \leq n \leq 15$. Are these all prime numbers?

(c) For which value of n is it obvious that $P(n)$ is not a prime number?

6.2.7. Euler's polynomial in Problem 6.2.6 often gives a prime number. However, prove that there is no nonconstant polynomial $P(n)$ whose values for $n = 0, 1, 2, \ldots$ are all prime numbers. [*Hint:* Let $P(1) = p$, where p is a prime number. Explain why $P(1 + kp)$ is divisible by p for any positive integer k.]

6.2.8. In Example 6.5, formulas were derived that gave $p(n)$, the number of aligned lattice squares, and $q(n)$, the number of aligned and tilted lattice squares in a lattice grid of size n. Show algebraically that

(a) $p(n) = \frac{1}{4}\binom{2n+2}{3}$ **(b)** $q(n) = \frac{1}{3}\binom{n+2}{2}\binom{n+1}{2}$.

(The reader is challenged to derive these formulas by direct combinatorial reasoning.)

6.2.9. If $p(n) = \sum_{j=0}^{k} c_j (n)_j$ is 0 for all integers $n \geq 0$, prove that all of the coefficients c_j are 0.

6.2.10. Recall that $p_n^{(r)} = p^{(r)}(n)$ denotes the nth r-gonal number, where $p^{(r)}(0) = 0$ and $p^{(r)}(1) = 1$. For example, the pentagonal numbers $r = 5$ are shown for $n = 1,2,3,4,5$.

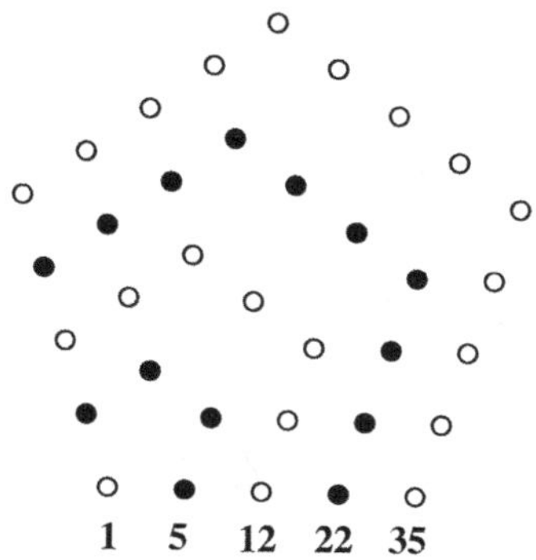

(a) Explain why $\Delta p^{(r)}(n) = 1 + (n-2)r$.

(b) Use part (a) to prove that $p_n^{(r)} = \frac{n}{2}[n(r-2) - (r-4)]$.

6.2.11. Recall that $c_n^{(r)} = c^{(r)}(n)$ denotes the nth centered r-gonal number, where $c^{(r)}(0) = 1$. For example, the centered pentagonal numbers $r = 5$ are shown for $n = 0,1,2,3,4,5$.

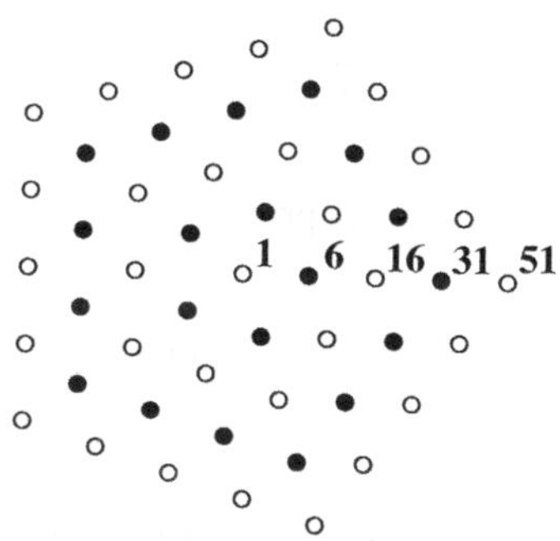

(a) Explain why $\Delta c^{(r)}(n) = (n+1)r$.

(b) Use part (a) to prove that

$$c^{(r)}(n) = 1 + r\binom{n+1}{2} = 1 + \frac{rn(n+1)}{2}$$

6.2.12. Use combinatorial reasoning to prove that

(a) $\left\{ {k \atop 2} \right\} = 2^{k-1} - 1, k \geq 1$

(b) $\left\{ {k \atop k-1} \right\} = \binom{k}{2}, k \geq 1$

(c) $\left\{ {k \atop k-2} \right\} = \binom{k}{3} + 3\binom{k}{4}, k \geq 2$

6.2.13. Apply the principle of inclusion–exclusion (PIE) to prove that

$$T(k,j) = j! \left\{ {k \atop j} \right\} = \sum_{t=0}^{j} (-1)^t \binom{j}{t} (j-t)^k$$

and show that this is equivalent to the identity

$$\left\{ {k \atop j} \right\} = \frac{1}{j!} \sum_{i=0}^{j} (-1)^{i+j} \binom{j}{i} i^k$$

given by equation (6.9).

6.2.14. **(a)** What seatings of four people at two identical circular tables are described by these two permutations, each written as a product of two cycles: (1 4)(2 3) and (1 4 2)(3)?

(b) List all permutations of [4] that contain exactly two cycles.

6.2.15. Prove that

$$\sum_{i=0}^{k} (-1)^i \left\{ {k \atop i} \right\} \left[{i \atop j} \right] = (-1)^k \, \delta_{j,k}$$

6.2.16. For any two sequences f_n and g_n, prove that

$$g_n = \sum_{k=0}^{n} (-1)^k \left\{ {n \atop k} \right\} f_k \quad \text{if and only if} \quad f_n = \sum_{k=0}^{n} (-1)^k \left[{n \atop k} \right] g_k$$

6.2.17. Prove the hockey stick–like identity for Stirling numbers of the second kind:

$$\left\{ {j+k+1 \atop j} \right\} = \sum_{i=0}^{j} i \left\{ {k+i \atop i} \right\}$$

6.2.18. Note that the identity of Theorem 6.23 can be stated as

$$\sum_{j=0}^{k} \left[{k \atop j} \right] = k! = \left[{k+1 \atop 1} \right]$$

for $k \geq 0$. Use this to give an induction proof of the identity

$$\sum_{j\geq 0}^{k} j \left[{k \atop j} \right] = \left[{k+1 \atop 2} \right], \quad k \geq 0$$

6.2.19. Prove the hockey stick–like identity for Stirling numbers of the first kind:

$$\left[{j+k+1 \atop j}\right] = \sum_{i=0}^{j} (k+i) \left[{k+i \atop i}\right]$$

6.2.20. Give a combinatorial proof that

$$\left[{k+1 \atop j+1}\right] = \sum_{i=j}^{k} \frac{k!}{i!} \left[{i \atop j}\right]$$

6.2.21. Give a combinatorial proof that

$$\left\{{k+1 \atop j+1}\right\} = \sum_{i=j}^{k} \binom{k}{i} \left\{{i \atop j}\right\}$$

6.2.22. Give an induction proof that

$$\left[{k+1 \atop n+1}\right] = \sum_{j=n}^{k} \binom{j}{n} \left[{k \atop j}\right]$$

for all integers n and k with $0 \le n \le k$.

6.2.23. Give a combinatorial proof that

$$\binom{r+b}{r} \left\{{k \atop r+b}\right\} = \sum_{j=0}^{k} \binom{k}{j} \left\{{j \atop r}\right\} \left\{{k-j \atop b}\right\}$$

[*Hint*: Ask a question about red and blue toy boxes that can be answered in two ways.]

6.2.24. Give a combinatorial proof that

$$\binom{r+b}{r} \left[{k \atop r+b}\right] = \sum_{j=0}^{k} \binom{k}{j} \left[{j \atop r}\right] \left[{k-j \atop b}\right]$$

[*Hint*: Ask a question about the number of seatings at circular tables that are identical except for the color, red or blue, of the tablecloth. Then answer your question in two ways.]

6.2.25. Give a combinatorial proof using the DIE method of Section 4.2 to prove that

$$\sum_{j=1}^{k} (-1)^j \begin{bmatrix} k \\ j \end{bmatrix} = 0, \ k \geq 2$$

6.3 HARMONIC NUMBERS

The harmonic numbers are first encountered most often in elementary calculus when the question of the convergence of the harmonic series $1 + \frac{1}{2} + \frac{1}{3} + \cdots + \frac{1}{n} + \cdots = \sum_{k=1}^{\infty} 1/k$ is considered. The partial sums of the series are the harmonic numbers.[4]

Definition 6.25 The *n*th *harmonic number* H_n is given by

$$H_0 = 0 \quad \text{and} \quad H_n = 1 + \frac{1}{2} + \frac{1}{3} + \cdots + \frac{1}{n}, \quad n \geq 1 \tag{6.22}$$

The harmonic series is divergent, as the proof below by contradiction shows.

Theorem 6.26 The harmonic numbers are an unbounded increasing sequence.

Proof. Assume that $\lim_{n\to\infty} H_n = H = 1 + \frac{1}{2} + \frac{1}{3} + \cdots + \frac{1}{n} + \cdots$ is a finite real number. Then

$$\begin{aligned} H &= 1 + \frac{1}{2} + \frac{1}{3} + \frac{1}{4} + \frac{1}{5} + \frac{1}{6} + \cdots + \frac{1}{2n-1} + \frac{1}{2n} + \cdots \\ &> \frac{1}{2} + \frac{1}{2} + \frac{1}{4} + \frac{1}{4} + \frac{1}{6} + \frac{1}{6} + \cdots + \frac{1}{2n} + \frac{1}{2n} + \cdots \\ &= 1 + \frac{1}{2} + \frac{1}{3} + + \cdots + \frac{1}{n} + \cdots = H. \end{aligned}$$

But no real number can satisfy $H > H$, so we have reached a contradiction. ∎

The beginning values of the harmonic sequence are listed in Table 6.4.

TABLE 6.4 Harmonic Numbers H_n , $0 \leq n \leq 10$

H_0	H_1	H_2	H_3	H_4	H_5	H_6	H_7	H_8	H_9	H_{10}
0	1	$\frac{3}{2}$	$\frac{11}{6}$	$\frac{25}{12}$	$\frac{137}{60}$	$\frac{49}{20}$	$\frac{363}{140}$	$\frac{761}{280}$	$\frac{7129}{2520}$	$\frac{7381}{2520}$

[4] As the name suggests, there is a connection with music: a tone of wavelength $1/n$ is the *n*th harmonic of a tone of unit wavelength. A valveless trumpet is played—with great difficulty—by "buzzing" with the lips pressed against a mouthpiece at a frequency that energizes the correct harmonic to match the desired pitch of the note in a piece of music. It is simpler to play notes on a trumpet with valves that change the unit wavelength of the instrument.

Theorem 6.27 No harmonic number H_n is an integer for $n > 1$.

Proof. The table of harmonic numbers (Table 6.4) suggests that each harmonic number H_n for $n > 1$ is, in lowest terms, a fraction with an odd number as its numerator and an even number as its denominator. Since no integer has this property, the result follows if this property can be shown. To do so, first note that any integer $k \geq 1$ can be written uniquely in the form $k = 2^{d_k} m_k$ where the exponent $d_k \geq 0$ is the *degree of evenness* of k and $m_k \geq 1$ is the *odd factor* of k. In particular, every odd integer has a 0 degree of evenness. We also see that 2^j is the unique integer with largest degree of evenness among the positive integers smaller than 2^{j+1}. Afterall, an integer with a degree of evenness larger than j is larger than 2^{j+1} and an integer of degree of eveness j that is not 2^j has the form $2^j m_j$ for an odd integer $m_j \geq 3$ so it is also larger that 2^{j+1}.

Now let $n > 1$, and suppose that $k = 2^j \leq n < 2^{j+1}$. Thus, k is the unique integer in $[n]$ with the largest degree of evenness. The degree of evenness of any other integer $i \in [n], i \neq k$, is smaller than j: $d_i < j$. Therefore, writing H_n with a common denominator of its terms, we have

$$H_n = \sum_{i=1}^{n} \frac{1}{i} = \sum_{i=1}^{n} \frac{1}{2^{d_i} m_i} = \sum_{i=1}^{n} \frac{2^{j-d_i}}{2^j m_i} = \frac{\sum_{i=1}^{n} 2^{j-d_i} m_i'}{2^j m_1 m_2 \cdots m_n}$$

where m_i' is the odd integer $m_i' = (m_1 m_2 \cdots m_n)/m_i$. The denominator of H_n is an even integer since $j \geq 1$. The numerator is an odd integer since it is a sum of the even integers $2^{j-d_i} m_i'$ for $i \neq j$ and the single odd integer m_j'. ■

6.3.1 Growth Rate of the Harmonic Numbers

The harmonic numbers increase at a slow logarithmic rate of growth. This can be seen geometrically from Figure 6.1. The n rectangles over the interval $[1, n+1]$, each have unit width and the respective heights $1, \frac{1}{2}, \frac{1}{3}, \ldots, 1/n$, so their total area is H_n.The area beneath the curve $f(x) = 1/x$ over the interval $[1, n+1]$ is $\int_1^{n+1} (1/x)\, dx = \ln(n+1)$. When this is combined with the shaded area γ_n just above the curve but still within

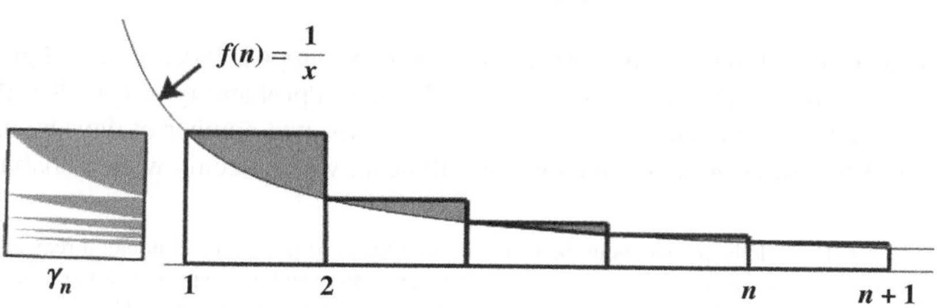

FIGURE 6.1 $H_n = \ln(n+1) + \gamma_n$

the rectangles, we see that $H_n = \gamma_n + \ln(n+1)$. At the left of the diagram, the shaded regions are shown collected inside a unit square. This shows that the sequence γ_n is increasing and bounded by 1, and therefore the sequence γ_n converges monotonically upward to a limit that we denote by γ. The number $\gamma = 0.5772156\ldots$ is called *Euler's constant*[5] and gives us the approximate formula $H_n \approx \gamma + \ln(n+1)$ for large n. For example, $H_{10} = 1 + \frac{1}{2} + \cdots + \frac{1}{10} = \frac{7381}{2520} = 2.928968\ldots$ is approximated by $\gamma + \ln\ 11 = 2.975\ldots.$

The numerical accuracy is improved by using the similar formula

$$H_n \approx \gamma + \ln\left(n + \tfrac{1}{2}\right) \tag{6.23}$$

This formula gives the better estimate $\gamma + \ln 10.5 = 2.92859\ldots$ for $H_{10} = 2.928968\ldots$.

6.3.2 Connection to Probability

The next example illustrates how the harmonic numbers can make unexpected appearances, this time in the mathematics of probability where we want to calculate an expected value.

Example 6.28 Ronnie likes green jellybeans and hates red ones. Initially, there are r red jellybeans and g green ones in his bag, and each day he randomly draws a jellybean out of the bag. If it is green he eats it, but if it is red he replaces it in the bag. What is the expected number of days until Ronnie has eaten all of the green jellybeans?

Solution. Let the expected number of days be $E(r, g)$ for a bag with r red jellybeans and g green ones. First suppose that there is only one green jellybean. On the first day, there is a probability of $r/(r+1)$ that he draws a red jellybean and therefore extends the expected time by a day. However, with a probability of $1/(r+1)$, he draws out and eats the green jellybean, which takes just one day. This gives us the recurrence relation

$$E(r, 1) = \frac{r}{r+1}(E(r, 1) + 1) + \frac{1}{r+1}$$

which can be solved to show that $E(r, 1) = 1 + r$. Next suppose there are $g \geq 2$ green jellybeans together with r red ones in the bag. There is a probability of $r/(r+g)$ that he draws a red jellybean and therefore extends the expected number of days by one. However, if he draws out and eats a green jellybean, which occurs with probability

[5] Unlike the π and e, little is known about the nature of the *third constant* γ. For example, it remains an open question if γ is a rational number. Like the first two constants π and e, γ appears in a large number of important formulas in mathematics. For information about γ, see Julian Havil's book [1].

$g/(r+g)$, this takes one day and leaves him with r red and $g-1$ green jellybeans in the bag. This reasoning gives us the recurrence relation

$$E(r,g) = \frac{r}{r+g}(E(r,g)+1) + \frac{g}{r+g}(E(r,g-1)+1)$$

When solved for $E(r,g)$, we obtain the recurrence relation

$$E(r,g) = \left(1+\frac{r}{g}\right) + E(r,g-1)$$

which can be solved by iteration to give

$$\begin{aligned} E(r,g) &= \left(1+\frac{r}{g}\right) + \left(1+\frac{r}{g-1}\right) + E(r,g-2) \\ &\cdots \\ &= \left(1+\frac{r}{g}\right) + \left(1+\frac{r}{g-1}\right) + \left(1+\frac{r}{g-2}\right) + \cdots + E(r,1) \\ &= g + r\left(1+\frac{1}{2}+\frac{1}{3}+\cdots+\frac{1}{g}\right) = g + rH_g \end{aligned}$$

so $E(r,g) = g + rH_g$.

For example, since $E(7,10) = 10 + 7H_{10} \doteq 30.50$, Ronnie can expect to take about 30 or 31 days to eat all of the green jellybeans if initially there are 10 green and 7 red ones in the bag ■

6.3.3 Connection to Stirling Numbers

The harmonic numbers have an unexpected connection with the Stirling numbers of the first kind. From the triangle identity of equation (6.17), we have

$$\begin{bmatrix} k+1 \\ 2 \end{bmatrix} = \begin{bmatrix} k \\ 1 \end{bmatrix} + k\begin{bmatrix} k \\ 2 \end{bmatrix} \tag{6.24}$$

But $\begin{bmatrix} k \\ 1 \end{bmatrix} = (k-1)!$, since this is the number of ways that k people can be seated at one circular table (or, equivalently, it is the number of permutations of $[k]$ with exactly one cycle). Therefore, if equation (6.24) is divided by $k!$, it can be rearranged as follows:

$$\frac{1}{k!}\begin{bmatrix} k+1 \\ 2 \end{bmatrix} - \frac{1}{(k-1)!}\begin{bmatrix} k \\ 2 \end{bmatrix} = \frac{1}{k}$$

By taking the sum over the range $k = 1, 2, \ldots, n$ and noting the telescoping property of the terms on the left, we see that

$$\frac{1}{n!}\begin{bmatrix} n+1 \\ 2 \end{bmatrix} = H_n \tag{6.25}$$

This formula allows us to give a combinatorial interpretation of the harmonic numbers.

Theorem 6.29 The harmonic number H_n is the average number of cycles of the $n!$ permutations of $[n]$:

$$\frac{1}{n!}\sum_{j\geq 1} j \begin{bmatrix} n \\ j \end{bmatrix} = H_n, \quad n \geq 1 \tag{6.26}$$

Proof. Since $\begin{bmatrix} n \\ j \end{bmatrix}$ permutations of $[n]$ have exactly j cycles, the sum on the left side of (6.26) is the average that we seek. The formula is certainly true when $n = 1$, so we can proceed by mathematical induction and assume that (6.26) is valid for some $n \geq 1$. Then, using equation (6.17) and (6.18), we obtain

$$\begin{aligned}
\sum_{j\geq 1} j \begin{bmatrix} n+1 \\ j \end{bmatrix} &= \sum_{j\geq 1} j \left\{ n \begin{bmatrix} n \\ j \end{bmatrix} + \begin{bmatrix} n \\ j-1 \end{bmatrix} \right\} = n \sum_{j\geq 1} j \begin{bmatrix} n \\ j \end{bmatrix} + \sum_{j\geq 1} j \begin{bmatrix} n \\ j-1 \end{bmatrix} \\
&= n \cdot n! H_n + \sum_{j\geq 1} (j-1) \begin{bmatrix} n \\ j-1 \end{bmatrix} + \sum_{j\geq 1} \begin{bmatrix} n \\ j-1 \end{bmatrix} \\
&= n \cdot n! H_n + n! H_n + n! = (n+1)! \left[H_n + \frac{1}{n+1} \right] = (n+1)! H_{n+1}
\end{aligned}$$

which shows that (6.26) also holds for $n + 1$. ■

Comparing (6.25) and (6.26), we see that we have proved this identity for Stirling numbers of the first kind.

Theorem 6.30 For all $n \geq 1$,

$$\sum_{j=1}^{n} j \begin{bmatrix} n \\ j \end{bmatrix} = \begin{bmatrix} n+1 \\ 2 \end{bmatrix} \tag{6.27}$$

6.3.4 The Domino Stacking Problem

The Greek mathematician Archimedes (ca. 287–212 B.C.) is celebrated for his work in both mechanics and mathematics. In particular, he discovered the *law of the lever*, which states that the total moment (force of rotation) generated by weights

$w_1, w_2, \ldots, w_n$ hanging at the respective signed distances $x_1, x_2, \ldots, x_n$ about a fulcrum (point of rotation) is given by the sum $x_1 w_1 + x_2 w_2 + \cdots + x_n w_n$. For example, the moment generated by the five weights shown in Figure 6.2(a) about the apex of the black triangle is $(-6) \cdot 5 + (-2) \cdot 1 + 6 \cdot 2 + 8 \cdot 6 + 10 \cdot 2 = 48$. The same moment is created when the total weight $5 + 1 + 2 + 6 + 2 = 16$ is concentrated at the distance $\bar{x} = 3$ since $48 = 3 \cdot 16$, as shown in Figure 6.2(b). This shows that moving the fulcrum to the center of gravity $\bar{x} = 3$ will balance the weights, since the torque is 0 about this point. In general, the weights are in balance if the moment about the point of rotation is 0.

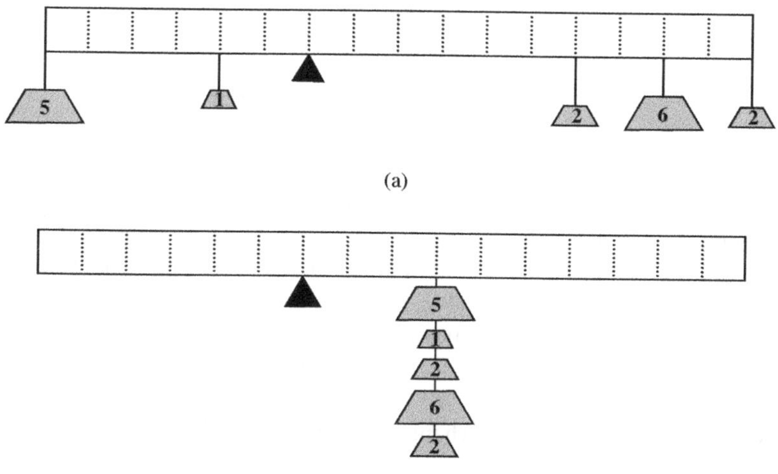

FIGURE 6.2 The moment generated by the five weights in (a) is the same as the moment generated by the five weights concentrated at their center of mass, as in (b).

R. T. Sharp [2] proposed the following problem: What is the maximum overhang possible when n dominoes of length 2 are stacked at the edge of a table? Only one domino is allowed per layer.

In Figure 6.3, assume that n dominoes are balanced over the left edge of the table. Each domino has unit weight, which we assume is concentrated at its midpoint. The

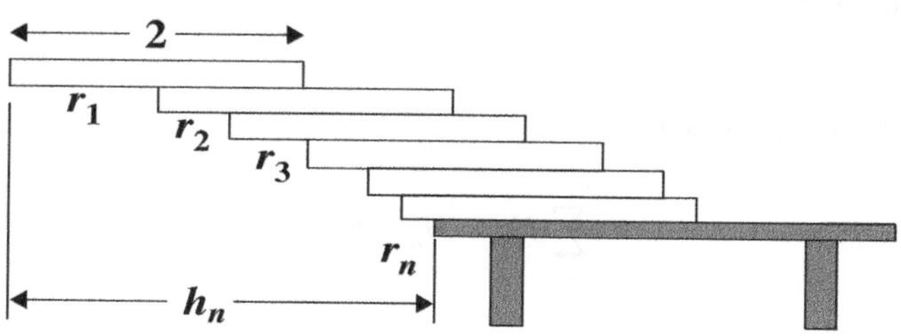

FIGURE 6.3 The domino stacking problem

overhang distances $r_1, r_2, \ldots, r_n$ are adjusted so that the dominoes above domino k are in balance over the left edge domino k. For example, the uppermost first domino overhangs the second domino at a distance of $r_1 = 1$ from the left edge of the second domino.

Since the uppermost $n - 1$ dominoes are balanced above the left edge of the nth domino, we may assume that the upper $n - 1$ dominoes are repositioned with all of their centers directly above the left edge of domino n, since the moment created by these dominoes is unchanged:

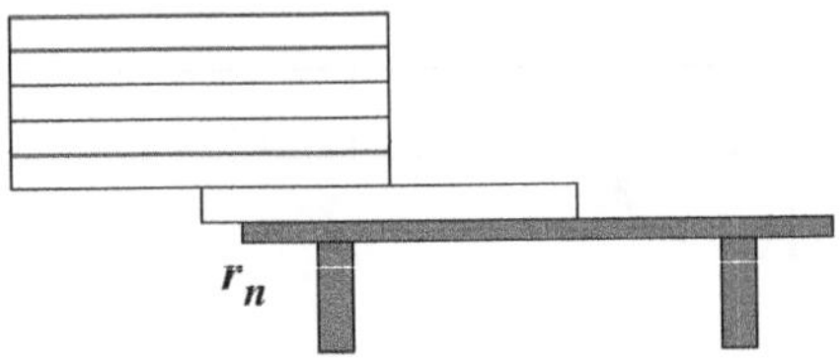

We can now determine the total moment of all n dominoes about the left edge of the table. The upper $n - 1$ dominoes have a counterclockwise moment of $(n - 1)\, r_n$ and the bottom nth domino has a clockwise moment of $1 - r_n$. For the n dominoes to be balanced about the left edge of the table, we must have $(n - 1)\, r_n = 1 - r_n$. Solving for r_n, we see that $r_n = 1/n$. The same calculation shows that $r_k = 1/k$ for all k, so the maximum overhang[6] is $1 + \frac{1}{2} + \frac{1}{3} + \cdots + \frac{1}{n}$. We have proved this theorem.

Theorem 6.31 The maximum overhang of n dominoes of length 2 is the harmonic number H_n.

An interesting consequence of this theorem is found by calculating the total moment of force about the left edge of the domino at the top of the stack. As we see from Figure 6.3, the total moment is

$$M_n = \sum_{k=0}^{n-1} (H_k + 1) = \sum_{k=0}^{n-1} H_k + n$$

But we also know that this is equal to the moment of the n dominoes when they are realigned so that their centers are all directly above the left edge of the table. This moment is $M_n = nH_n$.

Comparing the two expressions for M_n gives us the following formula for the sum of harmonic numbers.

Theorem 6.32 For all $n \geq 1$,

$$\sum_{k=0}^{n-1} H_k = H_n - n \qquad (6.28)$$

[6] If more than one domino is allowed in any layer of the stack, it is possible to achieve a larger overhang [3].

Archimedes would have easily understand this derivation of a mathematical result via mechanics, since he gave many more sophisticated proofs of this type in his book *The Method*.[7] An algebraic proof of (6.28) will be given below.

6.3.5 Summation by Parts

The integration by parts formula is well known from elementary integral calculus. The formula is obtained by integrating the derivative formula

$$f(x)g'(x) = \frac{d}{dx}(f(x)g(x)) - f'(x)g(x)$$

to obtain the following formula, using the fundamental theorem of calculus:

$$\int_a^b f(x)g'(x)\ dx = f(x)g(x)\,|_{x=a}^{x=b} - \int_a^b f'(x)g(x)\ dx$$

Less known is the analogous *summation by parts* formula, which applies to series rather than integrals. The derivative operation d/dx is replaced with the difference operator Δ, so that $\Delta w_n = (E-1)\,w_n = w_{n+1} - w_n$. Summing over the range $n = j$ to $n = k$, we see that

$$\begin{aligned}\sum_{n=j}^{k} \Delta w_n &= \sum_{n=j}^{k} (w_{n+1} - w_n) = (w_{j+1} - w_j) + (w_{j+2} - w_{j+1}) + \cdots + (w_{k+1} - w_k)\\ &= w_{k+1} - w_j = w_n\,|_{n=j}^{n=k+1} \qquad (6.29)\end{aligned}$$

This is, of course, the familiar formula for the sum of a *telescoping* sum, and is the analog of the fundamental theorem of calculus for the sum of a series. Now suppose that $w_n = u_n v_n$, so that

$$\begin{aligned}\Delta(u_n v_n) &= u_{n+1}v_{n+1} - u_n v_n = u_n(v_{n+1} - v_n) + (u_{n+1} - u_n)v_{n+1}\\ &= u_n(\Delta v_n) + (\Delta u_n)(Ev_n)\end{aligned} \qquad (6.30)$$

Summing (6.30) and using equation (6.29) we obtain

$$u_n v_n\,|_{n=j}^{n=k+1} = \sum_{n=j}^{k} u_n(\Delta v_n) + \sum_{n=j}^{k} (\Delta u_n)(Ev_n) \qquad (6.31)$$

Our calculations give us the following theorem.

[7] This book was thought to be lost until a copy made by Arabic scholars about 1000 CE was discovered in 1906 as the faint writing on pages of a medieval book that had been overwritten with religious text. The book was later lost once again, but reappeared in 1998 when it was sold at auction to an anonymous American buyer. It is still under intense study using modern imaging techniques to reveal more details of the faint mathematical writing. Reviel Netz and William Noel describe the recovery of Archimedes' works in their book [4].

Theorem 6.33

$$\sum_{n=j}^{k} \Delta w_n = w_n \Big|_{n=j}^{n=k+1} = w_{k+1} - w_j \text{ (fundamental theorem of series)} \quad (6.32)$$

$$\sum_{n=j}^{k} u_n(\Delta v_n) = u_n v_n \Big|_{n=j}^{n=k+1} - \sum_{n=j}^{k} (\Delta u_n)(E v_n) \text{ (summation by parts)} \quad (6.33)$$

This is the *summation by parts* formula, a discrete analog of the *integration by parts* formula in integral calculus.

The following example illustrates how to use summation by parts.

Example 6.34 Use summation by parts to evaluate the following sums:

(a) $\sum_{n=0}^{k} n2^{n-1} = (k-1)\,2^k + 1$

(b) $\sum_{n=0}^{k-1} H_n = kH_k - k$

Solution.

(a) Note that $\Delta 2^n = 2^{n+1} - 2^n = 2^n(2-1) = 2^n$, $\Delta n = (n+1) - n = 1$, and $E2^{n-1} = 2^n$.

Therefore

$$\sum_{n=0}^{k} n2^{n-1} = \sum_{n=0}^{k} n\Delta 2^{n-1} = n2^{n-1} \Big|_{n=0}^{n=k+1} - \sum_{n=0}^{k} (\Delta n)(E2^{n-1})$$
$$= (k+1)\,2^k - \sum_{n=0}^{k} 2^n = (k+1)\,2^k - \sum_{n=0}^{k} \Delta 2^n$$
$$= (k+1)\,2^k - (2^{k+1} - 1) = (k-1)\,2^k + 1$$

(b) Since $1 = \Delta n$ and $\Delta H_n = H_{n+1} - H_n = 1/(n+1)$, it follows that

$$\sum_{n=0}^{k-1} H_n = \sum_{n=0}^{k-1} H_n \Delta n = H_n n \Big|_{n=0}^{n=k} - \sum_{n=0}^{k-1} (\Delta H_n)(En)$$
$$= kH_k - \sum_{n=0}^{k-1} \frac{1}{n+1}(n+1) = kH_k - k$$

■

PROBLEMS

6.3.1. Suppose that the harmonic series has a finite sum H, in which case the sums of the odd terms $D = 1 + \frac{1}{3} + \frac{1}{5} + \cdots$ and even terms $E = \frac{1}{2} + \frac{1}{4} + \frac{1}{6} + \cdots$ would also be finite with $H = D + E$. Show that $2E = H$, and explain why there is a contradiction to the assumption that H is finite.

6.3.2. Prove that the harmonic series is divergent by showing that $H_{2^m} > (m+1)\frac{1}{2}$.

6.3.3. Prove that $H_N - H_n = 1/(n+1) + 1/(n+2) + \cdots + 1/N$ is never an integer for $0 < n < N$.

6.3.4. Obtain the ordinary generating functions of these sequences:

(a) $0, 1, \frac{1}{2}, \frac{1}{3}, \ldots, \frac{1}{n}, \ldots$ **(b)** $0, H_1, H_2, H_3, \ldots, H_n, \ldots$

6.3.5. Use the generating function

$$\sum_{n=0}^{\infty} H_n x^n = -\frac{\ln(1-x)}{1-x}$$

derived in Problem 6.3.4 to prove that $\sum_{n=0}^{s-1} H_n = sH_s - s$ for $s \geq 0$.

6.3.6. Let $S_n = \sum_{1 \leq r < s \leq n} (1/rs)$ and

$$H_n^{(2)} = 1 + \frac{1}{2^2} + \frac{1}{3^2} + \cdots + \frac{1}{n^2}$$

Prove that $S_n = \frac{1}{2}\left(H_n^2 - H_n^{(2)}\right)$.

6.3.7. The Euler constant was defined by $\gamma = \lim_{n\to\infty}(H_n - \ln(n+1))$. Prove that

(a) $\gamma = \lim_{n\to\infty}(H_n - \ln n)$ **(b)** $\gamma = \lim_{n\to\infty}\left(H_n - \ln\left(n + \frac{1}{2}\right)\right)$

6.3.8. The sequence $\gamma_n = H_n - \ln(n+1)$ monotonically increases to the Euler constant γ. As shown in Problem 6.3.7 the sequence $\hat{\gamma}_n = H_n - \ln\left(n + \frac{1}{2}\right)$ converges to γ. Show that $\hat{\gamma}_n$ monotonically decreases to γ. [*Hint*: Show that the function $f(x) = -[1/(x+1)] + \ln\left(x + \frac{3}{2}\right) - \ln\left(x + \frac{1}{2}\right)$ monotonically decreases to the limit 0 as $x \to \infty$.]

6.3.9. Show that

$$\sum_{n=1}^{k} \frac{2n+1}{n(n+1)} = 2H_k - \frac{k}{k+1}$$

for all $k \geq 1$.

6.3.10. Use the summation by parts formula of Theorem 6.33 to show that $\sum_{n=0}^{k-1} H_n^2 = kH_k^2 - (2k+1)H_k + 2k$ for all $k \geq 1$. [*Hint*: Use $1 = \Delta n$, $H_{n+1} - H_n = 1/(n+1)$, and the identity $\sum_{n=0}^{k-1} H_n = kH_k - k$.]

6.3.11. Use summation by parts to prove these summation formulas:

(a) $\sum\limits_{n=0}^{k-1} \frac{1}{(n+1)(n+2)} = \frac{k}{k+1}$

(b) $\sum\limits_{n=0}^{k-1} \frac{H_n}{(n+1)(n+2)} = \frac{k-H_k}{k+1}$ [*Hint*: part (a) can be helpful.]

6.3.12. Use summation by parts to prove that $\sum\limits_{n=1}^{k} (H_n/n) = \frac{1}{2}\left(H_k^2 + H_k^{(2)}\right)$, where $H_k^{(2)} = 1 + (1/2^2) + (1/3^2) + \cdots + (1/k^2)$. [*Hint*: $H_n = H_{n-1} + (1/n)$.]

6.3.13. Use summation by parts to prove that

$$\sum_{n=0}^{k-1} \binom{n}{r} H_n = \binom{k}{r+1}\left(H_k - \frac{1}{r+1}\right)$$

[*Hint*: Use $\binom{n}{r} = \Delta\binom{n}{r+1}$, the *committee selection with chair* identity $(r+1)\binom{n+1}{r+1} = (n+1)\binom{n}{r}$, and the hockey stick identity.]

6.3.14. Prove that the sum of the alternating harmonic series is ln 2; that is, show that $1 - \frac{1}{2} + \frac{1}{3} - \frac{1}{4} + \cdots = \ln 2$.

6.3.15. Jack and Jill's taste in jellybeans is exactly like Ronnie's described in Example 6.28—only green ones are eaten and red ones are returned to the bag. Jack has a bag with 10 red and 1 green jellybean, and Jill's bag has 10 green and 1 red jellybean. Who can be expected to eat all of the green jellybeans in their respective bags first?

6.3.16. Suppose that dominoes are 2 in. long and $\frac{1}{4}$ in. thick. What is the height of a stack of dominoes, one domino per layer, that extends 10 in. beyond the edge of the table?

6.4 BERNOULLI NUMBERS

The Bernoulli polynomials and Bernoulli numbers can be approached from many directions. What may be a definition and a theorem in one treatment may be reversed to become a theorem and a definition when a different starting point is taken. We will define the Bernoulli polynomials and the related number sequence with the following recursive definition.

Definition 6.35 The *Bernoulli polynomials* $B_n(t)$ are given by the recursive scheme

$$B_0(t) = 1, B_n'(t) = nB_{n-1}(t), \int_0^1 B_n(t)\, dt = 0, \ n = 1, 2, \ldots \tag{6.34}$$

The *Bernoulli numbers* are $B_n = B_n(0)$.

The Bernoulli polynomial and numbers (see Table 6.5) can therefore be calculated by successive integration, where the constant of integration is determined so that the integral over [0,1] is zero.

TABLE 6.5 Bernoulli Polynomials and Numbers

$B_n(t)$	B_n	$B_n(t)$	B_n
$B_0(t) = 1$	$B_0 = 1$	$B_5(t) = t^5 - \frac{5}{2}t^4 + \frac{5}{3}t^3 - \frac{1}{6}t$	$B_5 = 0$
$B_1(t) = t - \frac{1}{2}$	$B_1 = -\frac{1}{2}$	$B_6(t) = t^6 - 3t^5 + \frac{5}{2}t^4 - \frac{1}{2}t^2 + \frac{1}{42}$	$B_6 = \frac{1}{42}$
$B_2(t) = t^2 - t + \frac{1}{6}$	$B_2 = \frac{1}{6}$	$B_7(t) = t^7 - \frac{7}{2}t^6 + \frac{7}{2}t^5 - \frac{7}{6}t^3 + \frac{1}{6}t$	$B_7 = 0$
$B_3(t) = t^3 - \frac{3}{2}t^2 + \frac{1}{2}t$	$B_3 = 0$	$B_8(t) = t^8 - 4t^7 + \frac{14}{3}t^6 - \frac{7}{3}t^4 + \frac{2}{3}t^2 - \frac{1}{30}$	$B_8 = -\frac{1}{30}$
$B_4(t) = t^4 - 2t^3 + t^2 - \frac{1}{30}$	$B_4 = -\frac{1}{30}$	$B_9(t) = t^9 - \frac{9}{2}t^8 + 6t^7 - \frac{21}{5}t^5 + 2t^3 - \frac{3}{10}t$	$B_9 = 0$

For example, if $B_3(t) = t^3 - \frac{3}{2}t^2 + \frac{1}{2}t$ has been calculated, then $B_4(t) = 4\int \left(t^3 - \frac{3}{2}t^2 + \frac{1}{2}t\right) dt = t^4 - 2t^3 + t^2 + C$, where the constant of integration C is determined by the equation

$$0 = \int_0^1 B_4(t)\, dt = \int_0^1 \left(t^4 - 2t^3 + t^2 + C\right) dt = \frac{1}{5} - \frac{1}{2} + \frac{1}{3} + C = \frac{1}{30} + C$$

Thus, $C = -\frac{1}{30} = B_4$.

Little information about the Bernoulli polynomials and numbers is evident from the formulas, but some remarkable properties can be discovered with a closer look. Two of these are collected in the following theorem.

Theorem 6.36

$$B_n(t) = (-1)^n B_n(1-t) \tag{6.35}$$

$$B_n(t) = 2^{n-1}\left[B_n\left(\frac{t}{2}\right) + B_n\left(\frac{1+t}{2}\right)\right] \tag{6.36}$$

Proof. Only the first formula will be proved here, leaving the second derivation as an exercise. Let the polynomials $\hat{B}_n(t)$ be defined by $\hat{B}_n(t) = (-1)^n B_n(1-t)$. Then we see that

$$\hat{B}_0(t) = 1, \hat{B}'_n(t) = -(-1)^n B'_n(1-t) = (-1)^{n-1} nB_{n-1}(1-t) = n\hat{B}_{n-1}(t)$$

$$\int_0^1 \hat{B}_n(t)\,dt = (-1)^n \int_0^1 B_n(1-t)\,dt = (-1)^n \int_0^1 B_n(u)\,du = 0, \quad n = 1, 2, \ldots$$

Thus, the polynomials $\hat{B}_n(t)$ satisfy the defining recursive formulas (6.34) and we conclude that $\hat{B}_n(t) = B_n(t)$. ■

From (6.35), with $t = 1$, we see that $B_n(1) = (-1)^n B_n(0) = (-1)^n B_n$. But we also have that

$$B_n(1) - B_n(0) = \int_0^1 B'_n(t)\,dt = n \int_0^1 B_{n-1}(t)\,dt = 0, \quad n \geq 2$$

Since $B_n = B_n(0) = B_n(1) = (-1)^n B_n$, $n \geq 2$, we see that $B_{2n+1} = 0$ for all $n \geq 1$. This property was suggested in Table 6.5, since $B_3 = B_5 = B_7 = B_9 = 0$. In summary, we have

$$\begin{aligned} &B_{2n}(1) = B_{2n}, \ n \geq 0 \\ &B_1(1) = \tfrac{1}{2} = -B_1, B_{2n+1}(1) = B_{2n+1} = 0, \quad n \geq 1 \end{aligned} \tag{6.37}$$

Replacing t with $t + \frac{1}{2}$ in (6.35) shows that $B_n\left(t + \frac{1}{2}\right) = (-1)^n B_n\left(\frac{1}{2} - t\right)$; that is, $B_{2n}(t)$ has even symmetry about $t = \frac{1}{2}$ and $B_{2n+1}(t)$ has odd symmetry about $t = \frac{1}{2}$.

6.4.1 Exponential Generating Functions

Let $\boldsymbol{B}(x,t)$ be the exponential generating function for the Bernoulli polynomials:

$$\boldsymbol{B}(x,t) = \sum_{n=0}^{\infty} B_n(t)\frac{x^n}{n!} \tag{6.38}$$

Taking the derivative with respect to t and using (6.34), we see that

$$\frac{\partial}{\partial t}\boldsymbol{B}(x,t) = \sum_{n=0}^{\infty} B'_n(t)\frac{x^n}{n!} = \sum_{n=1}^{\infty} nB_{n-1}(t)\frac{x^n}{n!} = x\sum_{m=0}^{\infty} B_m(t)\frac{x^m}{m!} = x\boldsymbol{B}(x,t)$$

Thus, for some function $c(x)$ depending only on x, the EGF has the form

$$\mathcal{B}(x,t) = c(x)\, e^{xt} \tag{6.39}$$

To evaluate $c(x)$, we again turn to (6.34), but this time we integrate over the interval $0 \le t \le 1$. Integrating the EGF in (6.38) term by term, we find that

$$\int_0^1 \mathcal{B}(x,t)\, dt = \sum_{n=0}^{\infty} \left(\int_0^1 B_n(t)\, dt \right) \frac{x^n}{n!} = 1 \tag{6.40}$$

But from (6.39) we also have

$$\int_0^1 \mathcal{B}(x,t)\, dt = c(x) \int_0^1 e^{xt}\, dt = c(x)\, \frac{e^{xt}}{x} \Big|_{t=0}^{t=1} = c(x)\, \frac{e^x - 1}{x} \tag{6.41}$$

Comparing (6.41) to (6.40), we see that $c(x) = x/(e^x - 1)$. This gives us the following theorem.

Theorem 6.37 The exponential generating function of the Bernoulli polynomials is

$$\mathcal{B}(x,t) = \sum_{n=0}^{\infty} B_n(t) \frac{x^n}{n!} = \frac{xe^{xt}}{e^x - 1} \tag{6.42}$$

and the exponential generating function for the Bernoulli numbers is

$$\mathcal{B}(x) = \mathcal{B}(x,0) = \sum_{n=0}^{\infty} B_n \frac{x^n}{n!} = \frac{x}{e^x - 1} \tag{6.43}$$

As a consequence, we can give an explicit formula for the Bernoulli polynomials in which the coefficients depend on the Bernoulli numbers.

Theorem 6.38 The Bernoulli polynomials are given by

$$B_n(t) = \sum_{k=0}^{n} \binom{n}{k} B_k t^{n-k} \tag{6.44}$$

Proof. The EGF given by (6.42) is a product of the EGFs $\mathcal{B}(x)$ and e^{xt}, so

$$B_n(t) = \left[\frac{x^n}{n!}\right]\left(\frac{x}{e^x-1}e^{xt}\right) = \left[\frac{x^n}{n!}\right]\left(\sum_{r=0}^{\infty} B_r\frac{x^r}{r!}\right)\left(\sum_{s=0}^{\infty} t^s\frac{x^s}{s!}\right) = \sum_{k=0}^{n}\binom{n}{k}B_k t^{n-k}$$

■

The result of the following theorem is often used to define the Bernoulli numbers as a sequence of implicit recurrence relations.

Theorem 6.39 The Bernoulli numbers satisfy

$$B_0 = 1 \text{ and } \sum_{k=0}^{m}\binom{m+1}{k}B_k = 0, \quad m \geq 1 \tag{6.45}$$

Proof. Choose $t = 1$ and replace n with $m + 1$ in (6.44) to get

$$B_{m+1}(1) = \sum_{k=0}^{m+1}\binom{m+1}{k}B_k = \sum_{k=0}^{m}\binom{m+1}{k}B_k + B_{m+1}$$

Equation (6.45) now follows from (6.37) since $B_{m+1}(1) = B_{m+1}$ for all $m \geq 1$. ■

6.4.2 Sum of Integer *k*th Powers

The Bernoulli numbers are named for Jakob Bernoulli (1654–1705), who discovered them during his search for a formula for the sum of the kth powers of the first n nonnegative integers. If this sum is denoted by

$$\sigma_0(n) = n \text{ and } \sigma_k(n) = 0^k + 1^k + 2^k + \cdots + (n-1)^k, \; k \geq 1 \tag{6.46}$$

then, using results from our earlier work, we obtain

$$\begin{aligned}
\sigma_0(n) &= n \\
\sigma_1(n) &= 1 + 2 + \cdots + (n-1) = \binom{n}{2} = \frac{1}{2}n^2 - \frac{1}{2}n \\
\sigma_2(n) &= 1^1 + 2^2 + \cdots + (n-1)^2 = \frac{1}{6}(n-1)n(2n-1) = \frac{1}{3}n^3 - \frac{1}{2}n^2 + \frac{1}{6}n \\
\sigma_3(n) &= 1^3 + 2^3 + \cdots + (n-1)^3 = \binom{n}{2}^2 = \frac{1}{4}n^4 - \frac{1}{2}n^3 + \frac{1}{4}n^2
\end{aligned} \tag{6.47}$$

Bernoulli hoped to see a pattern and calculated additional formulas using the following idea. First, by rearranging the binomial expansion, we have that

$$(m+1)^{k+1} - m^{k+1} = \sum_{j=0}^{k} \binom{k+1}{j} m^j$$

Summing over m from $m = 0$ to $n - 1$, the terms on the left are telescoping, so we get

$$n^{k+1} = \sum_{j=0}^{k} \binom{k+1}{j} \sigma_j(n) = \sum_{j=0}^{k-1} \binom{k+1}{j} \sigma_j(n) + (k+1)\,\sigma_k(n)$$

This gives a recurrence relation for the sequence $\sigma_j(n)$:

$$\sigma_k(n) = \frac{1}{k+1}\left(n^{k+1} - \sum_{j=0}^{k-1} \binom{k+1}{j} \sigma_j(n)\right) \tag{6.48}$$

For example, when $k = 4$ in (6.48), we get

$$\begin{aligned} \sigma_4(n) &= \frac{1}{4+1}\left(n^{4+1} - \sum_{j=0}^{4-1} \binom{4+1}{j} \sigma_j(n)\right) \\ &= \frac{1}{5}\left(n^5 - \sigma_0(n) - 5\sigma_1(n) - 10\sigma_2(n) - 10\sigma_3(n)\right) \end{aligned}$$

so using the formulas shown in (6.47) and simplifying, we obtain the formula

$$\sigma_4(n) = \tfrac{1}{5}n^5 - \tfrac{1}{2}n^4 + \tfrac{1}{3}n^3 - \tfrac{1}{30}n \tag{6.49}$$

Bernoulli went on to calculate $\sigma_5(n), \sigma_6(n), \ldots,$ and $\sigma_{10}(n)$, obtaining

$$\begin{aligned} \sigma_5(n) &= \tfrac{1}{6}n^6 - \tfrac{1}{2}n^5 + \tfrac{5}{12}n^4 - \tfrac{1}{12}n^2 \\ \sigma_6(n) &= \tfrac{1}{7}n^7 - \tfrac{1}{2}n^6 + \tfrac{1}{2}n^5 - \tfrac{1}{6}n^3 + \tfrac{1}{42}n \\ \sigma_7(n) &= \tfrac{1}{8}n^8 - \tfrac{1}{2}n^7 + \tfrac{7}{12}n^6 - \tfrac{7}{24}n^4 + \tfrac{1}{12}n^2 \\ \sigma_8(n) &= \tfrac{1}{9}n^9 - \tfrac{1}{2}n^8 + \tfrac{2}{3}n^7 - \tfrac{7}{15}n^5 + \tfrac{2}{9}n^3 - \tfrac{1}{30}n \\ \sigma_9(n) &= \tfrac{1}{10}n^{10} - \tfrac{1}{2}n^9 + \tfrac{3}{4}n^8 - \tfrac{7}{10}n^6 + \tfrac{1}{2}n^4 - \tfrac{3}{20}n^2 \\ \sigma_{10}(n) &= \tfrac{1}{11}n^{11} - \tfrac{1}{2}n^{10} + \tfrac{5}{6}n^9 - n^7 + n^5 - \tfrac{1}{2}n^3 + \tfrac{5}{66}n \end{aligned} \tag{6.50}$$

Bernoulli noticed that the coefficients of the polynomials apparently could be written in terms of a sequence of numbers that begins

$$B_0 = 1, B_1 = -\tfrac{1}{2}, B_2 = \tfrac{1}{6}, B_3 = 0, B_4 = -\tfrac{1}{30}, B_5 = 0$$
$$B_6 = \tfrac{1}{42}, B_7 = 0, B_8 = -\tfrac{1}{30}, B_9 = 0, B_{10} = \tfrac{5}{66}, \ldots$$

so that the coefficient of n^{k+1-j} in $\sigma_k(n)$ is $1/(k+1)\binom{k+1}{j}B_j$. For example, the coefficient of the highest term n^{k+1} is always $1/(k+1)\binom{k+1}{0}B_0 = 1/(k+1)$, the coefficient of n^k is always $1/(k+1)\binom{k+1}{1}B_1 = B_1 = -\frac{1}{2}$, the coefficient of n^{k-1} is always $1/(k+1)\binom{k+1}{2}B_2 = (k/2)\,B_2 = (k/12)$, the coefficient of n^{k-2} is always $1/(k+1)\binom{k+1}{3}B_3 = 0$, and so on.

Athough he was unable to give a proof, Bernoulli had discovered there was a sequence of numbers (the Bernoulli numbers, of course!) that allows the sums $\sigma_k(n)$ to be expressed in the remarkable formula given in this theorem.

Theorem 6.40 If $B_0, B_1, B_2, \ldots$ are the Bernoulli numbers, then

$$\sigma_k(n) = 0^k + 1^k + 2^k + \cdots + (n-1)^k = \frac{1}{k+1}\sum_{j=0}^{k}\binom{k+1}{j}B_j n^{k+1-j} \tag{6.51}$$

Proof. The exponential generating function of the sums $\sigma_k(n)$, $k \geq 0$, is

$$S(x,n) = \sum_{k=0}^{\infty}\sigma_k(n)\frac{x^k}{k!} = \sum_{k=0}^{\infty}\sum_{t=0}^{n-1} t^k\frac{x^k}{k!} = \sum_{t=0}^{n-1}\sum_{k=0}^{\infty}\frac{(xt)^k}{k!} = \sum_{t=0}^{n-1} e^{xt} = \frac{e^{xn}-1}{e^x-1}$$

But from Theorem 6.37, we have

$$S(x,n) = \frac{e^{xn}-1}{e^x-1} = \frac{1}{x}\left(\frac{xe^{xn}}{e^x-1} - \frac{x}{e^x-1}\right) = \frac{1}{x}\sum_{k=0}^{\infty}(B_k(n) - B_k)\frac{x^k}{k!}$$

so we find that

$$\sigma_k(n) = \left[\frac{x^k}{k!}\right] S(x,n) = \frac{1}{k+1}(B_{k+1}(n) - B_{k+1})$$

Finally, we use Theorem 6.38 to obtain

$$\sigma_k(n) = \frac{1}{k+1}\left(\sum_{j=0}^{k+1}\binom{k+1}{j}B_j n^{k+1-j} - B_{k+1}\right) = \frac{1}{k+1}\sum_{j=0}^{k}\binom{k+1}{j}B_j n^{k+1-j}.$$

■

It is also possible to prove Theorem 6.40 based on the recurrence relation given in equation (6.48), but the calculation is long and difficult [5].

The evenly indexed Bernoulli numbers $B_2 = \frac{1}{6}, B_4 = -\frac{1}{30}, B_6 = \frac{1}{42}, B_8 = -\frac{1}{30}, B_{10} = \frac{5}{66}, \ldots$ suggest that they have alternating sign, and indeed it can be shown that $(-1)^{n-1} B_{2n} > 0$ for $n \geq 1$. It might also seem that they are a bounded sequence, but this is not so. More advanced techniques show that $2[(2n)!/(2\pi)^{2n}] < |B_{2n}| < 4[(2n)!/(2\pi)^{2n}]$.

6.4.3 Bernoulli Numbers and Trigonometric Taylor Series

In 1748, Euler announced his famous formula

$$e^{ix} = \cos x + i \sin x \tag{6.52}$$

that beautifully links the exponential function of a complex variable with the trigonometric functions. Replacing x with $-x$, we see that $e^{-ix} = \cos x - i \sin x$ and therefore

$$\cos x = \frac{e^{ix} + e^{-ix}}{2} \quad \text{and} \quad \sin x = \frac{e^{ix} - e^{-ix}}{2i} \tag{6.53}$$

The Taylor series for the sine and cosine functions are then found easily from the series expansion of the exponential function. We will now see how to obtain series expansions for other trigonometric functions. The coefficients of these series depend on the Bernoulli numbers, which explains why these series are not part of a beginning course in calculus in which the Bernoulli numbers have not yet been introduced.

Theorem 6.41

$$x \cot x = \sum_{n=0}^{\infty} (-4)^n B_{2n} \frac{x^{2n}}{(2n)!} \tag{6.54}$$

$$\tan x = \sum_{n=1}^{\infty} (-1)^{n-1} 4^n(4^n - 1) B_{2n} \frac{x^{2n-1}}{(2n)!} \tag{6.55}$$

$$\frac{x}{\sin x} = \sum_{n=0}^{\infty} (-1)^{n-1} (4^n - 2) B_{2n} \frac{x^{2n}}{(2n)!} \tag{6.56}$$

Proof. Only the series (6.54) will be proved here, with the proofs of the other two series expansions left as exercises. First note that

$$x \cot x = x\frac{\cos x}{\sin x} = ix\frac{e^{ix} + e^{-ix}}{e^{ix} - e^{-ix}} = ix\frac{e^{2ix} + 1}{e^{2ix} - 1} = \frac{2ix}{e^{2ix} - 1} + ix.$$

Comparing this to the EGF given by (6.43), we see that

$$x\cot x=\sum_{n=0}^{\infty}B_n\frac{(2ix)^n}{n!}+ix=B_1\frac{2ix}{1!}+ix+\sum_{n=0}^{\infty}B_{2n}(-4)^n\frac{x^{2n}}{(2n)!}=\sum_{n=0}^{\infty}B_{2n}(-4)^n\frac{x^{2n}}{(2n)!}$$

since $B_1=-\frac{1}{2}$ and $B_{2n+1}=0, n\geq 1$. ■

PROBLEMS

6.4.1. Use Definition 6.35 to derive the Bernoulli polynomial $B_5(t)$, assuming that $B_4(t)=t^4-2t^3+t^2-\frac{1}{30}$ has already been calculated.

6.4.2. Prove that

(a) $B_n(t)=2^{n-1}\left[B_n\left(\frac{t}{2}\right)+B_n\left(\frac{1+t}{2}\right)\right], \quad n\geq 0.$

(b) $B_n\left(\frac{1}{2}\right)=-\left(1-2^{1-n}\right)B_n, n\geq 0.$

6.4.3. Prove the difference formula $B_n(t+1)-B_n(t)=nt^{n-1}$. [*Hint*: Define the polynomials $p_0(t)=0$ and $p_n(t)=B_n(t+1)-B_n(t)-nt^{n-1}$ for $n\geq 1$ and take their derivatives.]

6.4.4. Use the difference formula $B_n(t+1)-B_n(t)=nt^{n-1}$ (see Problem 6.4.3) to prove that $1^k+2^k+\cdots+(n-1)^k=[1/(k+1)](B_{k+1}(n)-B_{k+1})$.

6.4.5. Prove the addition formula $B_n(t+h)=\sum_{k=0}^{n}\binom{n}{k}B_k(t)h^{n-k}$, $n\geq 0$. [*Hint*: Use the generating function for the Bernoulli polynomials given by Theorem 6.37.]

6.4.6. Prove the multiplication formula $B_n(mt)=m^{n-1}\sum_{k=0}^{m-1}B_n[t+(k/m)]$. [*Hint*: Replace mt with t and consider the polynomial $\hat{B}_n(t)=m^{n-1}\sum_{k=0}^{m-1}B_n[(t+k)/m]$.]

6.4.7. Use the multiplication formula of Problem 6.4.6 to prove that

$$B_{2n}\left(\frac{1}{3}\right)=B_{2n}\left(\frac{2}{3}\right)=\left(3^{2n-1}-1\right)B_{2n}$$

6.4.8. Prove that

$$\int_0^1 B_m(t)B_n(t)dt=(-1)^{m-1}\frac{m!n!}{(m+n)!}B_{m+n},\quad m\geq 1, n\geq 1$$

[*Hint*: Use mathematical induction and integration by parts.]

6.4.9. Use equation (6.48) to derive the formula for $\sigma_5(n)$, assuming that $\sigma_1(n), \sigma_2(n), \sigma_3(n), \sigma_4(n)$ are known.

6.4.10. Prove that

(a) $\tan x = \cot x - 2\cot 2x$. [*Hint*: Recall the double-angle formulas $\sin 2x = 2\sin x\cos x$ and $\cos 2x = \cos^2 x - \sin^2 x$.]

(b) $\tan x = \sum_{n=1}^{\infty} (-1)^{n-1}\, 4^n\, (4^n - 1)\, B_{2n} \frac{x^{2n-1}}{(2n)!}$.

6.4.11. Prove that

(a) $\cot x + \tan \frac{x}{2} = \frac{1}{\sin x}$. [*Hint*: $\cot 2y = \frac{1}{2}(\cot y - \tan y)$ by Problem 6.4.10(a).]

(b) $\frac{x}{\sin x} = \sum_{n=0}^{\infty} (-1)^{n-1}\, (4^n - 2)\, B_{2n} \frac{x^{2n}}{(2n)!}$.

6.5 EULERIAN NUMBERS

Eulerian numbers can be defined in various ways, but a very direct approach is afforded by determining the ordinary generating function $\sum_{n=0}^{\infty} n^k x^n$ for the sequence of kth powers $0^k, 1^k, 2^k, \ldots, n^k, \ldots$, where $k \geq 1$. The degree k polynomial n^k is annihilated by the operator $(E-1)^{k+1}$ and the first term of the sequence is $0^k = 0$, so we know (see Theorem 5.37) there exists a polynomial $P_k(x) = x \sum_{i=0}^{k} \left\langle {k \atop i} \right\rangle x^i$ for which

$$\sum_{n=0}^{\infty} n^k x^n = \frac{x \sum_{i=0}^{k} \left\langle {k \atop i} \right\rangle x^i}{(1-x)^{k+1}}, \quad k \geq 1 \tag{6.57}$$

The degree of $P_k(x)$ is no more than k, so we let $\left\langle {k \atop i} \right\rangle = 0$ for all $i \geq k \geq 1$.

Definition 6.42 The polynomials $P_k(x) = x \sum_{i=0}^{k-1} \left\langle {k \atop i} \right\rangle x^i$ given implicitly by (6.57) are the *Eulerian polynomials*[8] and the coefficients $\left\langle {k \atop i} \right\rangle$ are the *Eulerian numbers*.

The following theorem gives an explicit formula for the Eulerian numbers.

Theorem 6.43 The Eulerian numbers are given by

$$\left\langle {k \atop i} \right\rangle = \sum_{n=1}^{i+1} (-1)^{i+1-n} \binom{k+1}{i+1-n} n^k, \quad 0 \leq i < k \tag{6.58}$$

[8]Leonhard Euler investigated them in his 1755 book on differential calculus [6]. The Eulerian polynomials and numbers should not be confused with Euler polynomials and numbers, which are defined much like the Bernoulli polynomials and numbers and share similar properties.

Proof. Expanding $(1-x)^{k+1}$ by the binomial theorem, we see from equation (6.57) that

$$\begin{aligned}\sum_{i=0}^{k}\left\langle {k \atop i} \right\rangle x^{i+1} &= (1-x)^{k+1}\sum_{n\geq 1} n^k x^n \\ &= \left(\sum_{j\geq 0}(-1)^j \binom{k+1}{j} x^j\right)\left(\sum_{n\geq 1} n^k x^n\right) \\ &= \sum_{j\geq 0}\sum_{n\geq 1}(-1)^j\binom{k+1}{j} n^k x^{j+n} \\ &= \sum_{i\geq 0}\left(\sum_{n\geq 1}(-1)^{i+1-n}\binom{k+1}{i+1-n} n^k\right) x^{i+1} \qquad [j+n=i+1]\end{aligned}$$

so (6.58) is obtained by comparing coefficients of x^{i+1}. ■

Eulerian numbers are shown displayed in *Euler's triangle* in Table 6.6.

TABLE 6.6 Euler's Triangle[a]

k	$\left\langle {k \atop 0} \right\rangle$	$\left\langle {k \atop 1} \right\rangle$	$\left\langle {k \atop 2} \right\rangle$	$\left\langle {k \atop 3} \right\rangle$	$\left\langle {k \atop 4} \right\rangle$	$\left\langle {k \atop 5} \right\rangle$	$\left\langle {k \atop 6} \right\rangle$	$\left\langle {k \atop 7} \right\rangle$	$\left\langle {k \atop 8} \right\rangle$
1	1	0	—	—	—	—	—	—	—
2	1	1	0	—	—	—	—	—	—
3	1	4	1	0	—	—	—	—	—
4	1	11	11	1	0	—	—	—	—
5	1	26	66	26	1	0	—	—	—
6	1	57	302	302	57	1	0	—	—
7	1	120	1191	2416	1191	120	1	0	—
8	1	247	4293	15619	15619	4293	247	1	0

[a] *Caution*: Notation in this table is not standardized.

As a check, row $k=3$ gives us the OGF for the sequence of cubes

$$\sum_{n\geq 0} n^3 x^n = \frac{x+4x^2+x^3}{(1-x)^4}$$

which is the same OGF derived in the previous chapter in Example 5.38. The next two rows of Table 6.6 give us the OGFs for the 4th and 5th powers of the positive integers:

$$\sum_{n\geq 0} n^4 x^n = \frac{x+11x^2+11x^3+x^4}{(1-x)^5} \quad \text{and} \quad \sum_{n\geq 0} n^5 x^n = \frac{x+26x^2+66x^3+26x^4+x^5}{(1-x)^6}$$

6.5.1 Triangle Identity for Eulerian Numbers

From the defining equation (6.57), we see that

$$\sum_{n=1}^{\infty} n^k x^n = (1-x)^{-k-1} \sum_{i\ge 0} \left\langle {k \atop i} \right\rangle x^{i+1}$$

When this equation is differentiated and then multiplied by x, we get

$$\sum_{n\ge 1} n^{k+1} x^n = (k+1)(1-x)^{-k-2} \sum_{i\ge 0} \left\langle {k \atop i} \right\rangle x^{i+2} + (1-x)^{-k-1} \sum_{i\ge 0} \left\langle {k \atop i} \right\rangle (i+1)\, x^{i+1} \tag{6.59}$$

If k is replaced with $k+1$ in (6.57), we obtain

$$\sum_{n\ge 1} n^{k+1} x^n = (1-x)^{-k-2} \sum_{i\ge 0} \left\langle {k+1 \atop i} \right\rangle x^{i+1}$$

which shows that (6.59) can be rewritten as

$$\begin{aligned} \sum_{i\ge 0} \left\langle {k+1 \atop i} \right\rangle x^{i+1} &= (k+1) \sum_{i=0}^{k} \left\langle {k \atop i} \right\rangle x^{i+2} + \sum_{i=0}^{k} \left\langle {k \atop i} \right\rangle (i+1)\, x^{i+1} \\ &\quad - \sum_{i=0}^{k} \left\langle {k \atop i} \right\rangle (i+1)\, x^{i+2} \end{aligned} \tag{6.60}$$

Equating the coefficients of x^{i+1} in (6.60) gives

$$\left\langle {k+1 \atop i} \right\rangle = (k+1) \left\langle {k \atop i-1} \right\rangle + (i+1) \left\langle {k \atop i} \right\rangle - i \left\langle {k \atop i-1} \right\rangle$$

which simplifies to give the following result.

Theorem 6.44 The Eulerian numbers satisfy the triangle identity

$$\left\langle {k+1 \atop i} \right\rangle = (i+1) \left\langle {k \atop i} \right\rangle + (k+1-i) \left\langle {k \atop i-1} \right\rangle, \quad k \ge i \ge 1 \tag{6.61}$$

Identity (6.61) makes it easy to generate the Eulerian triangle in Table 6.6, since the entry in row $k+1$ and column i is determined by the entries in columns i and $i-1$ of row k. For example

$$\left\langle {8 \atop 3} \right\rangle = (3+1) \left\langle {7 \atop 3} \right\rangle + (8-3) \left\langle {7 \atop 2} \right\rangle = 4 \cdot 2416 + 5 \cdot 1191 = 15619$$

6.5.2 The Number of Ascents in a Permutation of $[n]$

Consider the $3! = 6$ permutations of $[3] = \{1, 2, 3\}$:

$$(1', 2', 3), (2', 3, 1), (3, 1', 2), (1', 3, 2), (3, 2, 1), (2, 1', 3) \tag{6.62}$$

Here the primes $'$ mark those terms in the permutation that are immediately followed by a larger term, what we will call an *ascent* of the permutation. Among the six permutations of [3] shown in (6.62), there is one permutation with no ascents, there are four permutations that have one ascent, and there is one permutation with two ascents. The following theorem shows that it is not an accident that the numbers 1,4,1 are precisely the first three entries of row $k = 3$ of the Eulerian triangle shown in Table 6.6.

Theorem 6.45 The number of permutations of $[k]$ with exactly i ascents is given by the Eulerian number $\left\langle {k \atop i} \right\rangle$.

Proof. Let $\left\langle\!\left\langle {k \atop i} \right\rangle\!\right\rangle$ denote the number of permutations of $[k]$ with precisely i ascents. Therefore $\left\langle\!\left\langle {k \atop i} \right\rangle\!\right\rangle = 0, i \geq k$, since no permutation of $[k]$ can have k or more ascents. Also, $\left\langle\!\left\langle {k \atop k-1} \right\rangle\!\right\rangle = 1$ since the only permutation with $k - 1$ ascents is the identity permutation $(1', 2', 3', \ldots, (k-1)', k)$. Likewise, $\left\langle\!\left\langle {k \atop 0} \right\rangle\!\right\rangle = 1$ since the only permutation with no ascents is $(k, k-1, \ldots, 3, 2, 1)$. So far, we see that $\left\langle\!\left\langle {k \atop i} \right\rangle\!\right\rangle$ matches the values of the Eulerian numbers for the $i = 0$ column and rows $k = 1$ and 2 of the Eulerian triangle. We now proceed with inductive reasoning. Consider the $\left\langle\!\left\langle {k+1 \atop i} \right\rangle\!\right\rangle$ permutations of $[k + 1]$ with i ascents. There are two types of these permutations, depending on whether the removal of element $k + 1$ leaves a permutation with the same number of ascents or one fewer.

Type 1: Removing Element $k + 1$ Retains i Ascents. This occurs when element $k + 1$ is between one of the i ascending pairs or else $k + 1$ is the first term of the permutation. Removal of $k + 1$ then gives a permutation of $[k]$ still with i ascents. Moreover, all of the $\left\langle\!\left\langle {k \atop i} \right\rangle\!\right\rangle$ permutations of $[k]$ with i ascents are obtained in this way. Since there are $i + 1$ ways to reinsert element $k + 1$ and recover a type 1 permutation of $[k + 1]$, we see that there are $(i + 1)\left\langle\!\left\langle {k \atop i} \right\rangle\!\right\rangle$ permutations of type 1.

Type 2: Removing Element $k + 1$ Decreases the Number of Ascents from i to $i - 1$. This occurs when element $k + 1$ is between a descending pair of terms or is the last term of the permutation. Removal of element $k + 1$ leaves a permutation of $[k]$ with $i - 1$ ascents, and all such $\left\langle\!\left\langle {k \atop i-1} \right\rangle\!\right\rangle$ permutations of $[k]$ are obtained this way. Since a permutation of $[k+1]$ with i ascents has $k-i$ descents, and there are $k-i+1$ places to insert element $k+1$, this means that there are $(k-i+1)\left\langle\!\left\langle {k \atop i-1} \right\rangle\!\right\rangle$ permutations of type 2.

Since all of the $\left\langle\!\left\langle {k+1 \atop i} \right\rangle\!\right\rangle$ permutations of $[k+1]$ with i ascents are either of type 1 and type 2, we have shown that

$$\left\langle\!\!\left\langle {k+1 \atop i} \right\rangle\!\!\right\rangle = (i+1)\left\langle\!\!\left\langle {k \atop i} \right\rangle\!\!\right\rangle + (k+1-i)\left\langle\!\!\left\langle {k \atop i-1} \right\rangle\!\!\right\rangle$$

This is the same triangle identity (6.61) already established for the Eulerian numbers, so we conclude that $\left\langle\!\left\langle {k \atop i} \right\rangle\!\right\rangle = \left\langle {k \atop i} \right\rangle$ for all i and k. ■

The connection just made between the Eulerian numbers and permutations with a prescribed number of ascents reveals some properties of the Eulerian numbers that are otherwise quite well hidden. For example, the permutations of $[k]$ with i ascents each have $k-1-i$ descents, and when the order of the terms is reversed, these descents become ascents. Therefore

$$\left\langle {k \atop i} \right\rangle = \left\langle {k \atop k-1-i} \right\rangle, \quad 0 \le i \le k-1 \tag{6.63}$$

which explains the symmetry evident in Table 6.6.

As a second example, notice that any row of the Eulerian triangle counts all of the permutations of $[k]$ in order according to the number of ascents from 0 to $k-1$. Therefore

$$\left\langle {k \atop 0} \right\rangle + \left\langle {k \atop 1} \right\rangle + \left\langle {k \atop 2} \right\rangle + \cdots + \left\langle {k \atop k-1} \right\rangle = k! \tag{6.64}$$

6.5.3 Sums of Integer Powers and Worpitzky's Identity

Equation (6.58) expresses Eulerian numbers as a sum of the terms n^k. The following theorem inverts this dependence, showing how to express n^k as a sum of Eulerian numbers.

Theorem 6.46 For any nonnegative integers n and k,

$$n^k = \sum_{i=0}^{k-1} \left\langle {k \atop i} \right\rangle \binom{n+i}{k} \tag{6.65}$$

Proof. Since

$$\frac{1}{(1-x)^{k+1}} = \sum_{j=0}^{\infty} \binom{k+j}{k} x^j$$

we see from (6.57) that

$$\sum_{n=1}^{\infty} n^k x^n = \left(\sum_{r\geq 0} \left\langle {k \atop r} \right\rangle x^{r+1} \right) \frac{1}{(1-x)^{k+1}} = \left(\sum_{r\geq 0}^{k} \left\langle {k \atop r} \right\rangle x^{r+1} \right) \sum_{j\geq 0} \binom{k+j}{k} x^j$$
$$= \sum_{n\geq 1} \left(\sum_{r\geq 0} \left\langle {k \atop r} \right\rangle \binom{k+n-1-r}{k} \right) x^n$$

where $n = r + 1 + j$. Equating coefficients of x^n and recalling identity (6.63) gives

$$n^k = \sum_{r=0}^{k-1} \left\langle {k \atop k-1-r} \right\rangle \binom{k+n-1-r}{k} = \sum_{i=0}^{k-1} \left\langle {k \atop i} \right\rangle \binom{n+i}{k}$$

where $i = k - 1 - r$. ■

The identities given by (6.65) give us a way to express the sequence of sums $\sigma_k(n)$ of kth powers of the first $n - 1$ positive integers in terms of Eulerian numbers. It's of interest to compare this formula with Theorem 6.40 which gave an expression for the sums $\sigma_k(n)$ in terms of the Bernoulli numbers.

Corollary 6.47

$$\sigma_k(n) = 1^k + 2^k + \cdots + (n-1)^k = \sum_{i=0}^{k-1} \left\langle {k \atop i} \right\rangle \binom{n+i}{k+1}, \quad k \geq 1 \qquad (6.66)$$

Proof. Using identity (6.65) and the hockey stick identities, we obtain

$$\sum_{j=0}^{n-1} j^k = \sum_{j=0}^{n-1} \sum_{i=0}^{k-1} \left\langle {k \atop i} \right\rangle \binom{j+i}{k} = \sum_{i=0}^{k-1} \left\langle {k \atop i} \right\rangle \sum_{j=0}^{n-1} \binom{j+i}{k} = \sum_{i=0}^{k-1} \left\langle {k \atop i} \right\rangle \binom{n+i}{k+1}$$

■

For example, if $k = 2$, then

$$\begin{aligned}
1^2 + 2^2 + \cdots + (n-1)^2 &= \sum_{i=0}^{1} \left\langle {2 \atop i} \right\rangle \binom{n+i}{2+1} \\
&= \left\langle {2 \atop 0} \right\rangle \binom{n}{3} + \left\langle {2 \atop 1} \right\rangle \binom{n+1}{3} \\
&= \frac{n(n-1)(n-2) + (n+1)n(n-1)}{6} \\
&= \frac{n(n-1)}{6}(n-2+n+1) = \frac{n(n-1)(2n-1)}{6}
\end{aligned}$$

Each side of identity (6.65) is a polynomial in the variable n, so the identity remains valid if n is replaced by a continuous variable x. This gives us the generalization known as *Worpitzky's identity* (Julius Worpitzky, 1883).

Theorem 6.48

$$x^k = \sum_{i=0}^{k-1} \left\langle {k \atop i} \right\rangle \binom{x+i}{k} \tag{6.67}$$

PROBLEMS

6.5.1. Give the formula to be entered in cell C3 of a spreadsheet that calculates the Eulerian number $\left\langle {2 \atop 1} \right\rangle$ and can then be copied into the remaining cells to generate the Eulerian numbers shown in Table 6.6:

	A	B	C	
1	$k\backslash i$	0	1	
2	1	1	–	
3	2	1	1	
4	3	1	4	
5	4	1	11	

6.5.2. Use formula (6.58) to show that $\left\langle {k \atop 1} \right\rangle = 2^k - k - 1$.

6.5.3. Give a combinatorial proof of $\left\langle {k \atop 1} \right\rangle = 2^k - k - 1$ using the fact that $\left\langle {k \atop i} \right\rangle$ is the number of permutations of $[k]$ with exactly i ascents. [*Hint*: Let A be the nonempty set of descending values preceding the one ascent, and B the set of descending values following the ascent. Now count the number of sets A and B.]

6.5.4. Show that the generating function for the Eulerian numbers $\left\langle {k \atop 1} \right\rangle$ in column $i = 1$ of Euler's triangle in Table 6.6 is

$$\sum_{k=1}^{\infty} \left\langle {k \atop 1} \right\rangle x^k = \frac{x^2}{(1-2x)(1-x)^2}$$

[*Hint*: $\left\langle {k \atop 1} \right\rangle = 2^k - k - 1$.]

6.5.5. Show that the generating function for the Eulerian numbers $\left\langle {k \atop 2} \right\rangle$ in column $i = 2$ of Euler's triangle in Table 6.6 is

$$\sum_{k=1}^{\infty} \left\langle {k \atop 2} \right\rangle x^k = \frac{x^3\left(1 + x - 4x^2\right)}{(1-3x)(1-2x)^2(1-x)^3}$$

[*Hint*: Equation (6.58) shows that $\left\langle {k \atop 2} \right\rangle = 3^k - (k+1)2^k - \binom{k+1}{2}$.]

6.5.6. Let

$$E(m,n) = \left\langle {m+n-1 \atop n-1} \right\rangle, \quad m \geq 1, n \geq 1$$

Thus

$$E(3,4) = \left\langle {3+4-1 \atop 4-1} \right\rangle = \left\langle {6 \atop 3} \right\rangle = 302$$

which is the entry at the intersection of the diagonals $m = 3$ and $n = 4$ in this form of the Eulerian triangle:

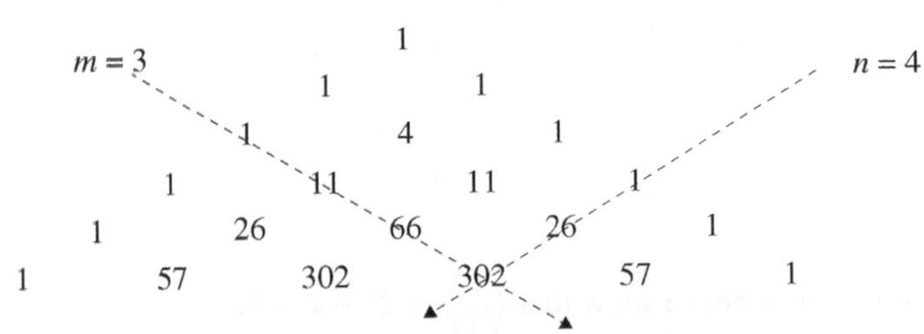

(a) prove that $E(m, n) = E(n, m)$.

(b) prove the triangle identity $E(m, n) = nE(m - 1, n) + mE(m, n - 1)$.

(c) use the table above to calculate $E(5, 3)$ using the triangle identity of part (b).

6.5.7. **(a)** Give a combinatorial proof of the binomial coefficient identity $(n+1)\binom{m+1}{n} = \binom{m}{n} + (m+2)\binom{m}{n-1}$.

(b) Define $\hat{C}(m, n) = \binom{-m-2}{n-1}$. Show that $\hat{C}(m, n) = n\hat{C}(m - 1, n) + m\hat{C}(m, n - 1)$. (This is the same triangle identity satisfied by the coefficients $E(m, n) = \left\langle {m+n-1 \atop n-1} \right\rangle$ introduced in Problem 6.5.6.)

6.5.8. Let $E(m, n) = \left\langle {m+n-1 \atop n-1} \right\rangle$, $m \geq 1, n \geq 1$, as defined in Problem 6.5.6. Prove that $E(m, n) = \sum_{j=1}^{n} jm^{n-j}E(m - 1, j)$.

6.5.9. A *run* in a permutation of $[k]$ is a maximal increasing list of one or more consecutive numbers of the permutation. For example, the runs in the six permutations of $[3]$ are shown with a vertical bar following the last number of a run:

$$(1, 2, 3|), (2, 3|, 1|), (3|, 1, 2|), (1, 3|, 2|), (2|, 1, 3|), (3|, 2|, 1|)$$

Prove that there are $\left\langle {k \atop i-1} \right\rangle$ permutations of $[k]$ with exactly i runs.

6.5.10. Use Corollary 6.47 to show that $1^3 + 2^3 + \cdots + (n-1)^3 = \binom{n}{2}^2$.

6.5.11. **(a)** Verify that the Euler polynomials $P_k(x) = (1-x)^{k+1} \sum_{n=1}^{\infty} n^k x^n$ satisfy the recursion relation $P_{k+1}(x) = x(1-x)P'_k(x) + (k+1)xP_k(x)$, $k \geq 1$.

(b) Use part (a) to obtain $P_2(x)$, $P_3(x)$, and $P_4(x)$ starting from $P_1(x) = x$.

6.5.12. Prove that the Euler polynomials $P_k(x) = (1-x)^{k+1} \sum_{n=1}^{\infty} n^k x^n$, $P_0(x) = 1$, have the exponential generating function

$$\sum_{k=0}^{\infty} P_k(x) \frac{t^k}{k!} = \frac{1-x}{1-xe^{t(1-x)}}.$$

6.6 PARTITION NUMBERS

Given a positive integer n, a *partition of* n is an unordered sum of positive integers that equals n. The *partition number*, $p(n)$, is the number of partitions of n. For example, in how many ways can a total weight of 6 lb be made from weights of 1, 2, 3, ... lb? Since the order in which the weights are selected does not matter, we will write each sum in nonincreasing order. In other words, if a 4-lb weight and two 1-lb weights are used, we would *not* express this as $1 + 4 + 1$, but instead we would write the partition as $4 + 1 + 1$. Writing the sums in this way, we find that the partitions of $n = 6$ are

$$6, 5 + 1, 4 + 2, 4 + 1 + 1, 3 + 3, 3 + 2 + 1, 3 + 1 + 1 + 1,$$
$$2 + 2 + 2, 2 + 2 + 1 + 1, 2 + 1 + 1 + 1 + 1, 1 + 1 + 1 + 1 + 1 + 1$$

This list shows all 11 partitions of 6, so the partition number of $n = 6$ is $p(6) = 11$.

This is an example of an *unrestricted* partition, since no constraints were placed either on the size of the weights, other than the requirement that they be positive integers, or on the number of weights that are allowed. However, certain constraints often apply. For example, suppose that a 3-lb weight must be used, together with any number of other weights not larger than 3, and, as before, that the weights total 6 lb. With this restriction, there are just three partitions; $3 + 3,\ 3 + 2 + 1,\ 3 + 1 + 1 + 1$.

Alternatively, suppose that we must use exactly three weights that total $n = 6$ pounds, but with no restriction on the size of the weights. This gives us the three restricted partitions $4 + 1 + 1,\ 3 + 2 + 1,\ 2 + 2 + 2$.

We will see later why it was no accident that both constraints—on the size of the largest weight or on the number of weights—resulted in the same number of partitions.

It is not difficult to imagine partition problems with various meaningful constraints. For example, in 1956 the eminent mathematician George Pólya asked this question "In how many ways can change be made for a dollar?" The terms in this partition are limited to 1s (pennies), 5s (nickels), 10s (dimes), 25s (quarters), and 50s (half-dollars). Pólya's question can be answered (see Problem 3.4.15) with ordinary generating functions, since the number of ways to make change is given by $[x^{100}](1 + x + x^2 + \cdots)(1 + x^5 + x^{10} + \cdots)(1 + x^{10} + x^{20} + \cdots)(1 + x^{25} + x^{50} + \cdots)(1 + x^{50} + x^{100} + \cdots)$, or more succinctly as

$$[x^{100}]\ \frac{1}{(1 - x)\left(1 - x^5\right)\left(1 - x^{10}\right)\left(1 - x^{25}\right)\left(1 - x^{50}\right)}$$

Using a computer algebra system (CAS; e.g., Maple, *Mathematica*, WolfamAlpha) to simplify the calculation, the answer to the making change problem is 192. Of course, this answer assumes that there is no restriction on the number of pennies, nickels, and so forth, and the answer could change if restrictions are placed on the number of coins of each denomination that are available.

Definition 6.49 A *partition* of a positive integer n is a finite sequence of positive integers $a_1, a_2, \ldots, a_k$ that is nonincreasing and sums to n:

$$a_1 \geq a_2 \geq \cdots \geq a_k \quad \text{and} \quad a_1 + a_2 + \cdots + a_k = n \tag{6.68}$$

The term a_j is the jth part of the partition, and k is the number of parts of the partition.

The following notation will be useful when restricted partitions are being counted. If C is a constraint on the type of parts $a_1, a_2, \ldots, a_k$ of the partition that are allowed, and D is a restriction on the number of parts allowed, then the number of partitions will be denoted by

$$p(n, C, D) = \text{number of partitions of } n \text{ with } \begin{cases} \text{type of parts constrained by condition } C \\ \text{number of parts constrained by condition } D \end{cases}$$

For example, using an asterisk $*$ to denote that there is no constraint, then $p(n, *, *) = p(n)$ is the number of unrestricted partitions of n. Similarly, $p(n, = j, *)$ will denote the number of partitions in which the largest part is j, and $p(n, *, = k)$ is the number of partitions with exactly k parts. We saw earlier that $p(6, = 3, *) = 3$ and $p(6, *, = 3) = 3$.

We will first investigate unrestricted partitions and then partitions that are subject to various constraints. The section ends with an application of partitions to the proof of a celebrated result of Euler, namely, the *pentagonal number theorem.*

6.6.1 Unrestricted Partitions

It is not too difficult to compute $p(n)$ directly for small values the n. For example, the unrestricted partitions of $n = 4$ are 4, 3 + 1, 2 + 2, 2 + 1 + 1, and 1 + 1 + 1 + 1, so $p(4) = 5$. With some effort, the values in Table 6.7 can be determined, although the amount of effort required increases rapidly as n grows larger.

TABLE 6.7 The First 10 Unrestricted Partition Numbers

n	1	2	3	4	5	6	7	8	9	10
$p(n)$	1	2	3	5	7	11	15	22	30	42

There is no obvious pattern to the sequence of partition numbers, and it is clear that brute-force calculation by hand is untenable for large n. Certainly no one would want to compute all $p(100) = 190{,}569{,}292$ partitions of $n = 100$ by making a list.

We need a more effective approach to compute partition numbers and understand their properties. Since generating functions solved Pólya's "making change" problem, it seems reasonable to examine the ordinary generating function of the unrestricted partition numbers. Because the number and size of the terms are not restricted, the OGF is the infinite product

$$\sum_{n\geq 0} p(n)x^n = \underbrace{(1 + x + x^2 + \cdots)}_{\text{Choose the number of 1s in the partition}} \underbrace{(1 + x^2 + x^4 + \cdots)}_{\text{Choose the number of 2s in the partition}} \underbrace{(1 + x^3 + x^6 + \cdots)}_{\text{Choose the number of 3s in the partition}} \underbrace{(1 + x^4 + x^8 + \cdots)}_{\text{Choose the number of 4s in the partition}} \cdots .$$

where we have defined $p(0) = 1$. Using product notation, the OGF is written simply as

$$\sum_{n=0}^{\infty} p(n)x^n = \prod_{j=1}^{\infty} \frac{1}{1-x^j} \tag{6.69}$$

We will see later how Euler used (6.69) to derive a remarkable recurrence relation for the partition numbers, but it is not at all clear how an explicit formula for $p(n)$ can be determined from the OGF. However, as shown in 1917 by G. H. Hardy and S. Ramanujan, with later improvements by Hans Rademacher in 1937, a formula for $p(n)$ does exist. That's the good news. The bad news is that the formula is so complicated that we won't even bother to write it down here. Suffice it to say that the "explicit" formula for $p(n)$ involves an infinite series, sums of products of complex exponentials, a hyperbolic sine function, and other strangeness.

Instead of the complicated exact formula for $p(n)$, it often suffices to use the approximate formula given in 1918 by Hardy and Ramanujan, and independently in 1920 by J. V. Uspensky, namely, the formula

$$p(n) \approx \frac{1}{4\sqrt{3}n} e^{\pi\sqrt{2n/3}} \tag{6.70}$$

This shows that $p(n)$ grows exponentially with $\sqrt{n}$.

6.6.2 Restricted Partition Numbers

The 11 partitions of $n = 6$, include 4 that have distinct (i.e., not repeated) parts, namely, 6, $5 + 1$, $4 + 2$, and $3 + 2 + 1$. There are also 4 partitions of $n = 6$ that have only odd parts: $5 + 1$, $3 + 3$, $3 + 1 + 1 + 1$, and $1 + 1 + 1 + 1 + 1 + 1$.

The following theorem, proved by Euler in 1748, shows why these two constraints have the same number of corresponding partitions.

Theorem 6.50 Let $q(n)$ denote the number of partitions of n with distinct parts. Then $q(n)$ is also equal to the number of partitions of n with odd parts:

$$q(n) = p(n, \text{distinct}, *) = p(n, \text{odd}, *) \tag{6.71}$$

Moreover, defining $q(0) = 1$, we can express the ordinary generating function by

$$\sum_{n=0}^{\infty} q(n)x^n = \prod_{j\geq 1} (1 + x^j) \tag{6.72}$$

Proof. It is clear that

$$f(x) = \prod_{j\geq 1} (1 + x^j)$$

is the OGF for the number of partitions with distinct parts, since the factor $(1 + x^j)$ determines whether the part j is included in the partition. Similarly, the OGF for the sequence of the number of partitions with odd parts is

$$g(x) = \left(1 + x + x^2 + \cdots\right)\left(1 + x^3 + x^6 + \cdots\right)\left(1 + x^5 + x^{10} + \cdots\right)\cdots$$
$$= \prod_{j\geq 1} \frac{1}{\left(1 - x^{2j-1}\right)}$$

Therefore

$$f(x) = \prod_{j\geq 1}\left(1 + x^j\right) = \frac{\prod_{j\geq 1}\left(1 + x^j\right)\prod_{k\geq 1}\left(1 - x^k\right)}{\prod_{k\geq 1}\left(1 - x^k\right)} = \frac{\prod_{j\geq 1}\left(1 - x^{2j}\right)}{\prod_{k\geq 1}\left(1 - x^k\right)}$$
$$= \frac{1}{\prod_{j\geq 1}\left(1 - x^{2j-1}\right)} = g(x)$$

Since both sequences of partition numbers have the same generating function, the sequences are identical. ■

We observed earlier that $p(6, = 3, *) = 3$ and $p(6, *, = 3) = 3$, and claimed that $p(n, = k, *) = p(n, *, = k)$ in general. However, the generating function for second of these sequences is troublesome, so we will take a different approach to finding a proof. The idea is to use a simple, yet very useful, representation of a partition called the *Ferrers diagram*.[9] For example, the partition $36 = 10 + 7 + 7 + 5 + 3 + 2 + 1 + 1$ corresponds to the Ferrers diagram shown in Figure 6.4.

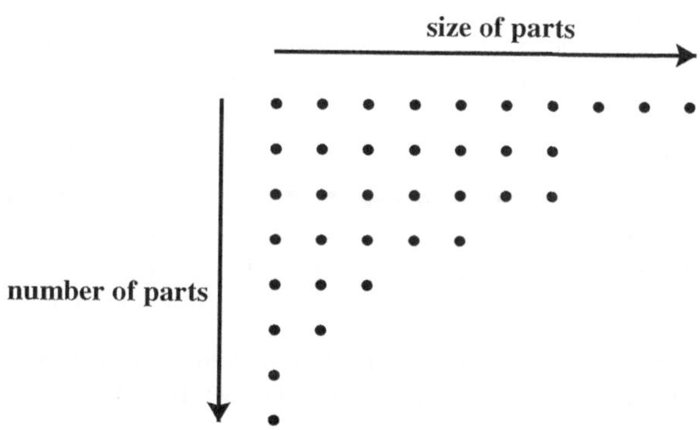

FIGURE 6.4 Ferrers diagram of the partition $10 + 7 + 7 + 5 + 3 + 2 + 1 + 1$.

[9]Devised by the British mathematician Norman Ferrers (1829–1903). The diagrams are closely related to *Young's diagrams*, in which the dots are replaced by squares.

In the general case, the partition of n given by the sum $a_1 + a_2 + \cdots + a_k = n$ is represented by the Ferrers diagram with a_j dots in row j, with the rows arranged from the top downward. Since $a_1 \geq a_2 \geq \cdots \geq a_k$, each row has either the same or fewer dots as the row above. The a_1 dots in the top row represent the largest part of the partition; and the number of rows in the diagram is k, the number of parts in the partition.

Given a partition of n, its *conjugate partition* is obtained by reflecting the dots in the Ferrers diagram over the downward 45° diagonal. This switches rows and columns, but the new pattern is still a Ferrers diagram and so it still gives a partition of n. The conjugation operation is an involution, since the conjugate of the conjugate returns the starting partition. An example is shown in Figure 6.5.

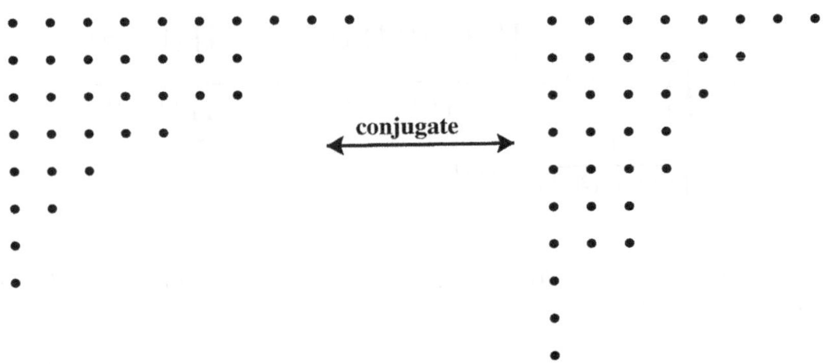

FIGURE 6.5 Conjugate of the partition $10 + 7 + 7 + 5 + 3 + 2 + 1 + 1$ is the partition $8 + 6 + 5 + 4 + 4 + 3 + 3 + 1 + 1 + 1$, as shown by these conjugate Ferrers diagrams.

Since conjugation switches rows with columns, any partition of n with k parts (the number of rows) corresponds to a conjugate partition of n whose largest part is k (the number of columns). Figure 6.6 shows this matching for $n = 6$ and $k = 3$; each conjugate pair has the partition with largest part 3 on the left and the conjugate partition with exactly 3 parts to the right. The conjugate Ferrers diagrams, and corresponding conjugate partitions, make it clear why $p(6, =3, *) = p(6, *, =3) = 3$.

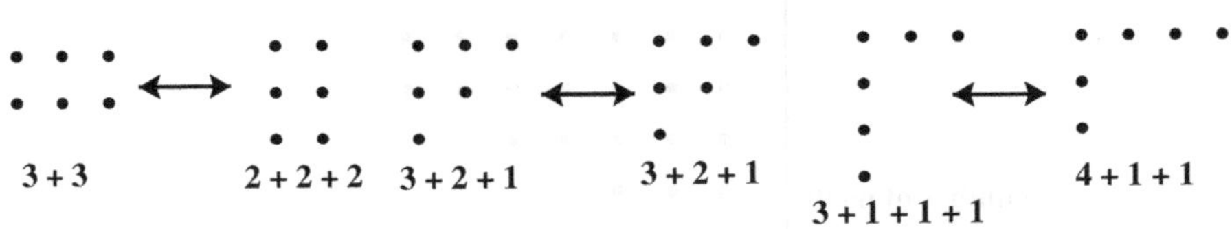

FIGURE 6.6 The three conjugate pairs of Ferrers diagrams showing that any partition of $n = 6$ with the largest part $k = 3$ is conjugate to a partition with exactly $k = 3$ parts. Therefore, $p(5, =3, *) = p(5, *, =3) = 3$.

The same reasoning applies to any choice of n and k; each partition of n with k as its largest part is the conjugate of a partition of n with exactly k parts. Therefore we have the first part of the following theorem.

Theorem 6.51 Let $p_k(n) = p(n, = k, *)$ denote[10] the number of partitions of n with largest part k. Then $p_k(n)$ is also equal to the number of partitions of n with exactly k parts:

$$p_k(n) = p(n, = k, *) = p(n, *, = k) \tag{6.73}$$

The generating function is

$$\sum_{n=0}^{\infty} p_k(n)x^n = x^k \prod_{j=1}^{k} \frac{1}{(1 - x^j)} \tag{6.74}$$

Proof. Equation (6.73) is obvious by the pairing of conjugate Ferrers diagrams. The generating function is given by

$$\begin{aligned}\sum_{n=0}^{\infty} p_k(n)x^n &= (1 + x + x^2 + \cdots)(1 + x^2 + x^4 + \cdots)(1 + x^3 + x^6 + \cdots) \cdots (x^k + x^{2k} + \cdots) \\ &= \frac{1}{(1 - x)} \frac{1}{(1 - x^2)} \frac{1}{(1 - x^3)} \cdots \frac{x^k}{(1 - x^k)}.\end{aligned}$$

■

When a Ferrers diagram is symmetric to its diagonal, the diagram and the corresponding partition are said to be *self-conjugate.* We see from Figure 6.6 that 3 + 2 + 1 is a self-conjugate partition. It can be shown that the number of self-conjugate partitions of n is the same as the number of partitions of n into distinct odd parts (see Problem 6.6.8).

6.6.3 Euler's Pentagonal Number Theorem

In Chapter 1 it was shown in Theorem 1.25 that the pentagonal numbers are given by $\frac{1}{2}k(3k - 1)$, where $k \geq 1$, giving us the sequence 1, 5, 12, 22, ... If a negative integer $-k$ is allowed, we get the similar formula $\frac{1}{2}(-k)(-3k - 1) = \frac{1}{2}k(3k + 1)$, which generates the sequence 2,7,15,27,... of *extended* pentagonal numbers. The *generalized pentagonal numbers* will now refer to the combined set {1, 2, 5, 7, 12, 15, ... } given by $\frac{1}{2}k(3k \pm 1)$ for $k \geq 1$.

Theorem 6.52 Consider all of the partitions of n into distinct parts. Let $r(n)$ be the number of partitions with an even number of parts minus the number of partitions with an odd number of parts. Then $r(n) = 0$ unless n is the generalized pentagonal number $n = \frac{1}{2}k(3k \pm 1)$, in which case $r(n) = (-1)^k$.

[10] This notation is common, but it is not entirely standardized. Sometimes $p_k(n)$ denotes the partitions of n into at most k parts.

Proof. We will use the DIE method (see Chapter 4) to evaluate $r(n) = \sum_{\pi\in Q_n} (-1)^{k_\pi}$, where the sum is taken over the set of partitions π of n into distinct parts and k_π is the number of parts of the partition π. In the Ferrers diagram of π, let the dots in the bottom row be the *base* B. Also, let the *diagonal* D be the set of rightmost dots along the 45° diagonal that begins with the last dot of the upper row of the diagram. The loops in the following diagram show how, at least for some partitions, the diagonal D of π can be moved to become the base B' of partition π', or else how the base B' of partition π' can be moved to become the diagonal D' of partition π. The two partitions of n still have distinct parts. When defined, this is a sign reversing involution since it is reversible and the number of parts (rows) changes by one:

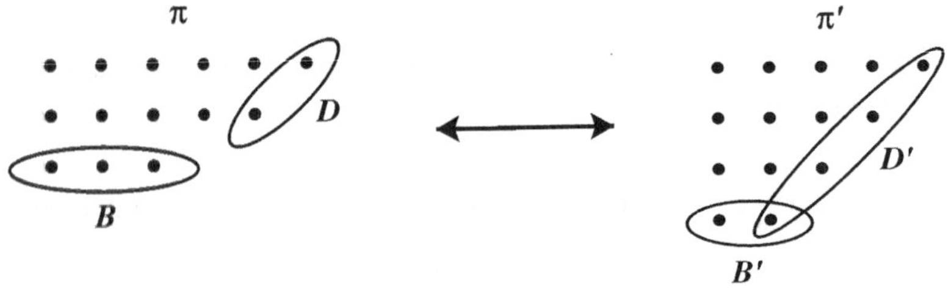

However, for some values of n there may be an exceptional partition for which neither the base B nor the diagonal D can be moved. The exceptional partition must have one of these two forms:

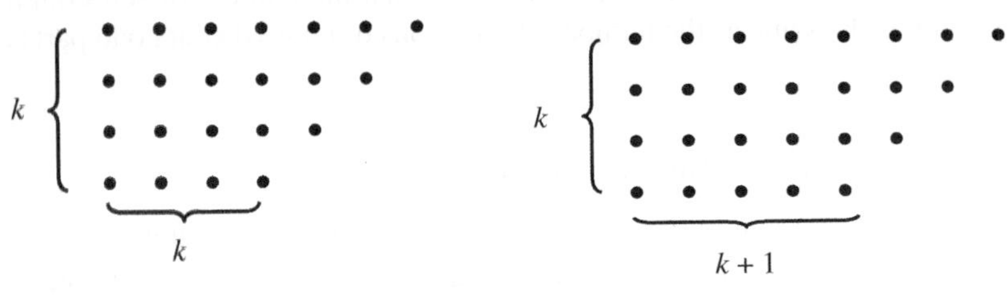

The number of dots of the exceptional partition shown to the left above is

$$n = k^2 + \frac{k(k-1)}{2} = \frac{1}{2}k(3k-1)$$

which is the kth pentagonal number. The number of dots in the exceptional partition shown above to the right is

$$n = k(k+1) + \frac{k(k-1)}{2} = \frac{1}{2}k(3k+1)$$

which is the kth extended pentagonal number. Since k is the number of parts of the partition of n, we see that $r(n) = \sum_{\pi\in Q_n} (-1)^{k_\pi} = (-1)^k$ when n is the generalized pentagonal number $n = \frac{1}{2}k(3k \pm 1)$, and is 0 for any other n. ∎

In view of Theorem 6.52, and defining $r(0) = 1$, we have the generating function

$$\sum_{n=0}^{\infty} r(n)x^n = 1 + \sum_{\substack{n=(1/2)k(3k\pm1),\\ \text{some } k\geq1}} r(n)x^n = 1 + \sum_{k=1}^{\infty} (-1)^k \left(x^{\frac{1}{2}k(3k-1)} + x^{\frac{1}{2}k(3k+1)}\right)$$
$$= \sum_{k=-\infty}^{\infty} (-1)^k x^{\frac{1}{2}k(3k-1)}$$

On the other hand, the coefficient of x^n of the product $\prod_{j\geq1}(1-x^j)$ is the algebraic sum of the number of distinct partitions of n counted positively when an even number of x^js are selected that sum to n and counted negatively when an odd number of x^js are selected. In other words, $r(n) = [x^n]\prod_{j\geq1}(1-x^j)$, giving us the identity

$$\sum_{n=0}^{\infty} r(n)x^n = \prod_{j\geq1}(1-x^j) \tag{6.75}$$

When this is compared with the generating function

$$\sum_{n=0}^{\infty} p(n)x^n = \prod_{j\geq1} \frac{1}{1-x^j}$$

given by equation (6.69), we see that $\left(\sum_{n=0}^{\infty} r(n)x^n\right)\left(\sum_{n=0}^{\infty} p(n)x^n\right) = 1$. This gives us the recurrence relation $\sum_{j=0}^{n} r(j)p(n-j) = 0, n \geq 1$, and so, solving for $p(n)$, we have proved a beautiful and surprising theorem of Leonard Euler.

Theorem 6.53 (Euler's Pentagonal Number Theorem)

$$p(n) = \sum_{\substack{k=-\infty\\ k\neq0}}^{\infty} (-1)^{k-1} p\left(n - \frac{1}{2}k(3k-1)\right)$$
$$= p(n-1) + p(n-2) - p(n-5) - p(n-7) + p(n-12) + p(n-15) - \cdots \tag{6.76}$$

Example 6.54 Verify that $p(7) = 15, p(8) = 22$ and $p(9) = 30$, using the pentagonal number theorem and the earlier entries in Table 6.6.

Solution. The sum in (6.76) extends over the generalized pentagonal numbers not larger than n, so

$$p(7) = p(6) + p(5) - p(2) - p(0) = 11 + 7 - 2 - 1 = 15,$$
$$p(8) = p(7) + p(6) - p(3) - p(1) = 15 + 11 - 3 - 1 = 22,$$
$$p(9) = p(8) + p(7) - p(4) - p(2) = 22 + 15 - 5 - 2 = 30$$

■

PROBLEMS

6.6.1. **(a)** Find all of the unrestricted partitions of $n = 5$.

(b) How many partitions of $n = 5$ have parts not larger than 3?

(c) How many partitions of $n = 5$ have no more than 3 parts?

6.6.2. **(a)** Prove that $p(n, *, \leq k) = p(n, \leq k, *)$ with Ferrers diagrams.

(b) Show that the generating function for the partitions of n into no more than k parts is given by

$$\sum_{n=0}^{\infty} p(n, *, \leq k)x^n = \prod_{j=1}^{k} \frac{1}{(1 - x^j)}$$

6.6.3. **(a)** What is the estimate of $p(100)$=190,569,292 that is given by formula (6.70)?

(b) The first five terms of the Hardy-Ramanujan series correctly give $p(200)$ = 3,972,999,029,388. What estimate is given by formula (6.70)?

6.6.4. **(a)** Draw the Ferrers diagrams of the partitions of $n = 5$, each paired to its conjugate.

(b) Which, if any, of the partitions of $n = 5$ are self-conjugate?

6.6.5. Find the generating function for the number of ways to make change for n cents using nickels, dimes, and quarters.

6.6.6. Find the generating functions for the number of partitions into

(a) positive even integers

(b) distinct squared integers

(c) integers that each appear no more than twice

6.6.7. Prove that $p(n, \text{odd parts}, = k) = p(n - k, \text{even parts}, \leq k)$.

6.6.8. **(a)** Find the two self-conjugate partitions of 10.

(b) Find the two partitions of 10 into distinct odd parts.

(c) Draw the Ferrers diagrams of the four partitions found in parts (a) and (b). What natural pairing of the partitions of part (a) with those of part (b) do you see?

(d) Explain, using the idea of part (c), why the number of self-conjugate partitions of n is the same as the number of partitions of n into distinct odd parts.

6.6.9. Show that there is always at least one self-conjugate partition of n except when $n = 2$. [*Hint*: See Problem 6.6.8(d).]

6.6.10. Recall that $p_k(n)$ is the number of partitions of n with exactly k parts (or with largest part k). Explain why

(a) $p_n(n) = 1$ **(b)** $p_1(n) = 1$
(c) $p_{n-1}(n) = 1, n > 1$ **(d)** $p_2(n) = \lfloor n/2 \rfloor$

6.6.11. Recall that $p_k(n)$ is the number of partitions of n with exactly k parts, where $p_0(0) = 0$. Prove that

(a) $p_k(n) = p_{k-1}(n-1) + p_k(n-k)$ **(b)** $p_k(n) = \sum_{j=1}^{k} p_j(n-k)$

6.6.12. The table below shows the values of $p_k(n)$ for $1 \leq n \leq 5$. Use the results of Problem 6.6.11 to extend the table three additional rows:

	$k = 1$	2	3	4	5	6	7	8
$n = 1$	1	0	0	0	0	0	0	0
2	1	1	0	0	0	0	0	0
3	1	1	1	0	0	0	0	0
4	1	2	1	1	0	0	0	0
5	1	2	2	1	1	0	0	0

6.6.13. Prove that the number of unrestricted partitions of n is equal to the number of partitions of $2n$ with exactly n parts; that is, prove that $p(n) = p_n(2n)$.

6.6.14. Prove that the number of partitions of $2m + k$ into $m + k$ parts is the same for each $k \geq 0$.

6.6.15. Prove that the number of partitions of n into 3 parts is equal to the number of partitions of $2n$ into 3 parts each of size less than n; that is, prove that $p(n, *, = 3) = p(2n, < n, = 3)$.

6.6.16. (a) Explain why $p\,(n, \text{distinct}, = k) = p\,(n - t_k)$ where $t_k = \frac{1}{2}k(k+1)$ is the kth triangular number.

(b) Determine the number of partitions of 14 into 4 distinct parts.

6.6.17. (a) Find the OGF $g(x)$ of the sequence $p\,(n,$ distinct powers of $2, *)$.

(b) What does part (a) say about the base two representation of an integer?

6.6.18. Let $S = \{1, 2, 4, 5, 7, 8, 10, 11, \ldots\}$ be the set of positive integers that are congruent to either 1 or 2 modulo 3, and let $T = \{1, 5, 7, 11, 13, 17, 19, \ldots\}$ be the set of positive integers congruent to either 1 or 5 modulo 6.

(a) Show that $p\,(8, \text{parts in } S, \text{distinct}) = p\,(8, \text{parts in } T, *)$ by making lists of the partitions of each type.

(b) Use OGFs to show that $p\,(n, \text{parts in } S, \text{distinct}) = p\,(n, \text{parts in } T, *)$.

6.6.19. Use a computer algebra system (CAS) to verify that

$$\sum_{n=0}^{\infty} p(n,*,\leq 3)x^n = \frac{1}{(1-x)}\frac{1}{(1-x^2)}\frac{1}{(1-x^3)} = \frac{1}{6}\frac{1}{(1-x)^3} + \frac{1}{4}\frac{1}{(1-x)^2} + \frac{1}{4}\frac{1}{(1-x^2)} + \frac{1}{3}\frac{1}{(1-x^3)}.$$

6.6.20. Prove that

$$p(n,*,\leq 3) = \left\{\frac{(n+3)^2}{12}\right\}$$

where $\{x\}$ denotes the *round to the nearest integer* to x function. [*Hint*: See the generating function of Problem 6.6.19.]

6.6.21. Consider the partitions of n into distinct parts. If n is not a generalized pentagonal number, then the proof of Theorem 6.52 gave a pairing of partitions with evenly many parts to partitions with an odd number of parts. However, if n is a pentagonal number, all except one exceptional partition can be paired. Using Ferrers diagrams, show

(a) the pairings for $n = 8$

(b) the pairings and the exceptional partition for $n = 5$

(c) the pairings and the exceptional partition for $n = 7$

6.6.22. Extend Example 6.54 to verify that $p(10) = 42, p(11) = 56$, and $p(12) = 77$.

6.6.23. According to Theorem 6.51, the generating function for the number of partitions of the form $a_1 + a_2 + a_3 = n$ with $a_1 \geq a_2 \geq a_3 \geq 1$ is $x^3/[(1-x)(1-x^2)(1-x^3)]$. The parts a_1, a_2, a_3 will be the sides of an integer-sided triangle of perimeter n unless $a_1 \geq a_2 + a_3$, since this would violate the triangle inequality.

(a) Show that the generating function for the number of partitions of n with three parts and for which $a_1 \geq a_2 + a_3$ is $x^4/[(1-x)(1-x^2)(1-x^4)]$.

(b) Show that the generating function for T_n, the number of noncongruent triangles of perimeter n with integer sides, is

$$\frac{x^3}{(1-x)(1-x^2)(1-x^3)} - \frac{x^4}{(1-x)(1-x^2)(1-x^4)} = \frac{x^3}{(1-x^2)(1-x^3)(1-x^4)}$$

6.7 CATALAN NUMBERS

The Catalan numbers 1,2,5,14,42,... were first encountered in Example 2.17 (of Chapter 2). With a rotation of the figure shown in that example, the Catalan number $C_n = [1/(n+1)]\binom{2n}{n}$ gives the number of block walking paths from (0,0) to (n, n) that never pass through a point above the 45° diagonal. For example, one of the $C_4 = \frac{1}{5}\binom{8}{4} = 14$ paths from (0,0) to (4,4) is shown in Figure 6.7.

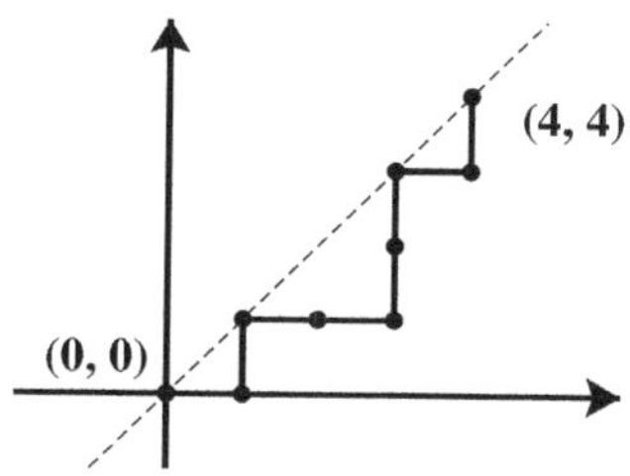

FIGURE 6.7 The path ENEENNEN is one of the $C_4 = 14$ paths from (0,0) to (4,4) that do not cross over the $y = x$ diagonal.

This is one example of a counting problem answered with Catalan numbers, but what makes the Catalan numbers of special interest is the unexpectedly high frequency with which they answer a wide variety of combinatorial questions. As an example, consider this question:

> Candidates North and East each received 25 votes, ending the election in a tie. In how many ways can the ballots, identifiable only by an N or E, be ordered so that at no time in the counting of the ballots was East ever trailing North?

The answer to the ballot problem can be obtained by simple bijective reasoning. Any sequence of counting the ballots corresponds to the unique block walking path from (0,0) to (25,25) for which a vote for candidate East is modeled by walking one block east, and a vote for North is walking one block to the north. East is never behind if the path described by the sequence in which votes are recorded corresponds to a path that never crosses the 45° line. We conclude that there are $C_{25} = 4{,}861{,}946{,}401{,}452$ ways that the 25 ballots may have been ordered.

Our next example will be solved in two ways. The first solution is again dependent on a bijection with block walking paths. The second solution is by direct combinatorial reasoning. When the two solutions are viewed together, we obtain a new derivation of the formula $C_n = [1/(n+1)]\binom{2n}{n}$ for the Catalan numbers.

Example 6.55 Determine the number of sequences $a_0, a_1, a_2 \ldots, a_{2n}$ of +1s and −1s that have these properties: their sum is 1, so that $a_0 + a_1 + a_2 + \cdots + a_{2n} = 1$, and every partial sum is positive, so that $a_0 > 0, a_0 + a_1 > 0, a_0 + a_1 + a_2 > 0, a_0 + a_1 + \cdots + a_{2n-1} > 0$.

Solution 1 (Bijection). Consider an election in which candidate E receives $n+1$ votes compared to n votes cast for candidate N. The number of sequences $a_0, a_1, a_2 \dots, a_{2n}$ of +1s and −1s with the required properties is the same as the number of paths from (0,0) to $(n+1, n)$ that stay strictly below the $y = x$ diagonal throughout the tallying of the ballots. This is equivalent to adding an additional east block E to the beginning of each of the paths from (0, 0) to (n, n), so there are C_n sequences.

Solution 2 (Direct). Suppose that $n = 3$, so we want to count the sequences $a_0, a_1, a_2 \dots, a_6$ of +1s and −1s for which $a_0 + a_1 + a_2 + \cdots + a_6 = 1$ and for which each partial sum is positive. The sequence $s = (1, -1, -1, 1, -1, 1, 1)$ sums to 1, but not all of its partial sums are positive. This is seen in the path ENNENEE that corresponds to s, which touches and even crosses above the $y = x$ line:

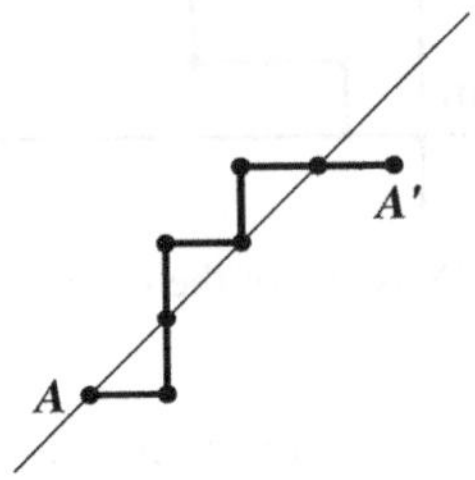

But suppose that we consider[11] extending the path by taking translates by the vector from A to A', as shown here:

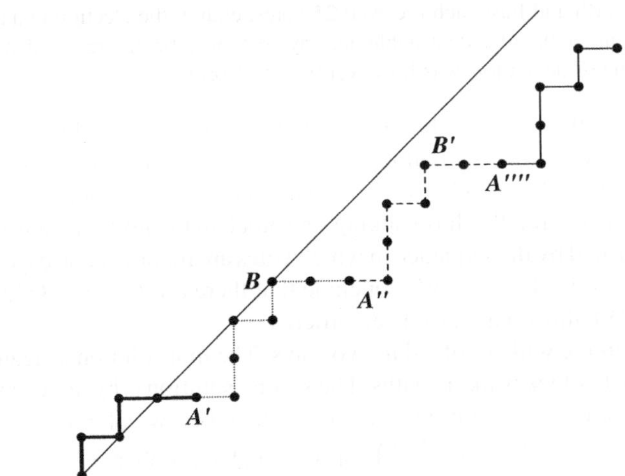

[11]This is a specialization of an idea of George Raney conceived in 1959. More generally, given any sequence of integers that sum to 1, there is a unique cyclic shift of the sequence with all of its partial sums. positive.

Each translate moves the path a unit to the right, so the extended path eventually touches the $y = x$ line for the last time, say, at point B. This means that the path from B to its translate B', namely, EEENNEN, is a path with the desired form. The corresponding sequence is $s' = (1,1,1,-1,-1,1,-1)$, a unique cyclic shift of the starting sequence s.

It is easy to see that the same reasoning applies to any of the $\binom{2n+1}{n}$ permutations of the $n + 1$ positive ones and the n negative ones. There are $2n + 1$ cyclic shifts of each of these permutations, of which exactly one has partial sums that are all positive. Thus the number of sequences $a_0, a_1, a_2 \ldots, a_{2n}$ that sum to 1 and have all positive partial sums is

$$\frac{1}{2n+1}\binom{2n+1}{n} = \frac{(2n+1)!}{(2n+1)(n+1)!n!} = \frac{(2n)!}{(n+1)n!n!} = \frac{1}{n+1}\binom{2n}{n} = C_n.$$

■

6.7.1 Triangulations of a Convex Polygonal Region by Its Diagonals

A geometric figure is *convex* if, given any two points A and B of the figure, all of the points of line segment $\overline{AB}$ are also in the figure. Consider, for example, the shaded region shown in Figure 6.8(a). This figure, a pentagon together with its interior, is called a *pentagonal region.* The two points A and B shown are in the pentagonal region, but the point $C \in \overline{AB}$ is not in the region; this means that the pentagonal region is not convex. By contrast, the hexagonal region in Figure 6.8(b) is convex; if A and B are any two points in the region, then all of the points of the line segment $\overline{AB}$ also belong to the region.

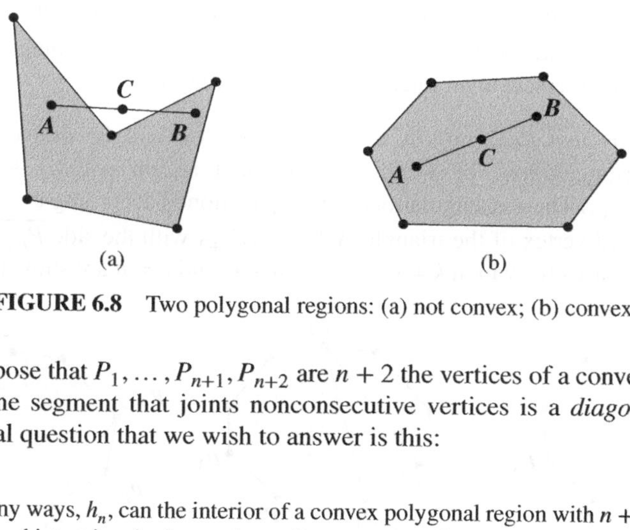

FIGURE 6.8 Two polygonal regions: (a) not convex; (b) convex.

Now suppose that $P_1, \ldots, P_{n+1}, P_{n+2}$ are $n + 2$ the vertices of a convex polygonal region. A line segment that joints nonconsecutive vertices is a *diagonal*, and the combinatorial question that we wish to answer is this:

> In how many ways, h_n, can the interior of a convex polygonal region with $n + 2$ vertices be partitioned into triangles by a subset of noncrossing diagonals?

The following diagram shows that $h_1 = 1$, $h_2 = 2$, and $h_3 = 5$:

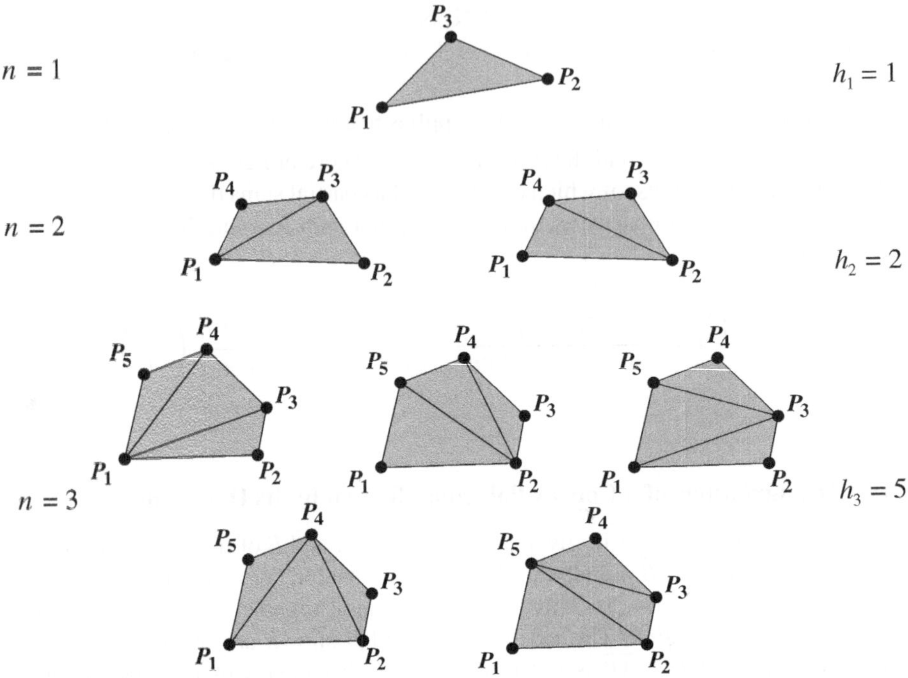

The next theorem shows why the number of triangulations of any convex polygon is a Catalan number.

Theorem 6.56 There are $C_n = [1/(n+1)]\binom{2n}{n}$ ways to partition the interior of a convex polygonal region with $n+2$ vertices into triangles using its diagonals, where no two diagonals intersect at an interior point of the region.

Solution (Recurrence Relation). A recurrence relation will be derived for h_{n+1}, the number triangulations of a convex polygonal region with $n + 3$ vertices $P_1, \ldots, P_{n+2}, P_{n+3}$. These triangulations can be partitioned according to which vertex P_{k+2} is the third vertex of the triangle $\Delta P_{n+3}\, P_1 P_{k+2}$ with the side $\overline{P_{n+3}\, P_1}$, where $0 \le k \le n$. The cases for which $k = 0$, $1 \le k \le n-1$, and $k = n$ are shown here:

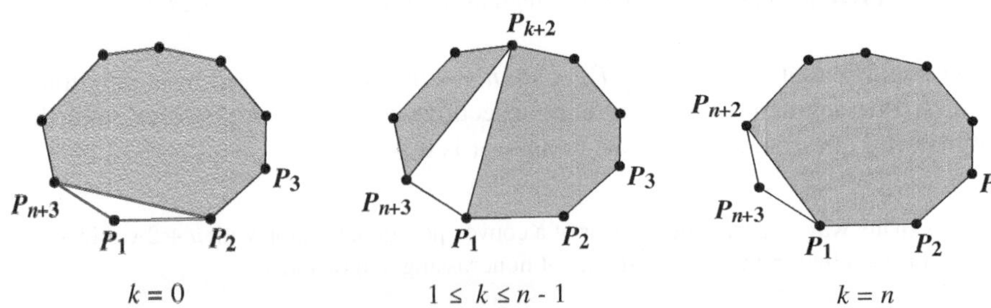

We first note that if $k = 0$, then there are h_n triangulations of the remaining shaded convex polygonal region $P_2P_3 \ldots P_{n+3}$, since it has $n + 2$ vertices. If $1 \le k \le n-1$, any triangulation that includes the triangle $\Delta P_{n+3}\, P_1 P_{k+2}$ includes the h_k triangulations of the region $P_1P_2 \cdots P_{k+2}$ and the h_{n-k} triangulations of the region $P_{k+2}P_{k+3} \ldots P_{n+3}$. Therefore, there are $h_k h_{n-k}$ triangulations that contain $\Delta P_{n+3}\, P_1 P_{k+2}$. Finally, for $k = n$, there are another h_n triangulations that contain triangle $\Delta P_{n+3}P_1P_{n+2}$. Altogether, if we define[12] $h_0 = 1$, we see that the number of triangulations of $P_1, \ldots, P_{n+2}, P_{n+3}$ is given by

$$h_{n+1} = h_0h_n + h_1h_{n-1} + h_2h_{n-2} + \cdots + h_nh_0 \tag{6.77}$$

Thus, h_{n+1} is the convolution product of the terms $h_0, h_1, \ldots, h_n$. This means that if we let $f(x) = h_0 + h_1x + h_2x^2 + \cdots + h_nx^n + \cdots$ be the generating function of the sequence $h_0, h_1, h_2, \ldots, h_n, \ldots$, then, from (6.77) and the initial values $h_0 = h_1 = 1$, we have

$$\begin{aligned} xf(x)^2 &= x\left(h_0 + h_1x + h_2x^2 + \cdots + h_nx^n + \cdots\right)^2 \\ &= x + \left(h_0h_1 + h_1h_0\right)x^2 + \left(h_0h_2 + h_1h_1 + h_2h_0\right)x^3 + \cdots \\ &= h_1x + h_2x^2 + h_3x^3 + \cdots = f(x) - 1 \end{aligned}$$

This is a quadratic equation for $f(x)$ that can be solved to show that either

$$f(x) = \frac{1 - \sqrt{1-4x}}{2x} \quad \text{or} \quad f(x) = \frac{1 + \sqrt{1-4x}}{2x} \tag{6.78}$$

However, only the first solution has the property that $f(0) = h_0 = 1$. To expand $f(x)$ in a series, recall that in Chapter 3 we had derived [see formula (3.35)] the binomial series expansion

$$\sqrt{1+x} = 1 + \sum_{k=1}^{\infty} \frac{(-1)^{k-1}}{k2^{2k-1}} \binom{2k-2}{k-1} x^k$$

When x is replaced with $-4x$, we obtain

$$\sqrt{1-4x} = 1 - 2\sum_{k=1}^{\infty} \frac{1}{k} \binom{2k-2}{k-1} x^k$$

When the terms are rearranged we find that

$$f(x) = \frac{1 - \sqrt{1-4x}}{2x} = \sum_{k=1}^{\infty} \frac{1}{k} \binom{2k-2}{k-1} x^{k-1} = \sum_{k=0}^{\infty} \frac{1}{k+1} \binom{2k}{k} x^k$$

[12]This definition means $h_0 = 1$ for the *diagon* P_1P_2, a "polygon" with two vertices. In the Harry Potter series, the train station to Hogwarts is located in Diagon Alley, which may explain why it is difficult to find.

We can now solve for h_n to see that

$$h_n = [x^n] \sum_{k=0}^{\infty} \frac{1}{k+1} \binom{2k}{k} x^k = \frac{1}{n+1} \binom{2n}{n} = C_n$$

the nth Catalan number. ■

6.7.2 Other Occurrences of Catalan Numbers

The Catalan numbers answer a surprisingly large number of combinatorial questions.[13] Here are three examples that are each counted with Catalan numbers. In each setting, the first three cases are listed explicitly to show that $C_1 = 1, C_2 = 2$, and $C_3 = 5$.

Catalan 1. C_n is the number of ways to place parentheses in an ordered list of factors $\boldsymbol{v}_0 \times \boldsymbol{v}_1 \times \boldsymbol{v}_2 \times \cdots \times \boldsymbol{v}_n$ in a nonassociative algebra (e.g., vectors in $\mathbb{R}^3$ are a nonassociative algebra under the vector cross product $\boldsymbol{v} \times \boldsymbol{w}$). Since each parenthesis encloses two terms, this is said to be a *binary bracketing* of an ordered list of symbols:

$n = 1$ $(\boldsymbol{v}_0 \times \boldsymbol{v}_1)$
$n = 2$ $((\boldsymbol{v}_0 \times \boldsymbol{v}_1) \times \boldsymbol{v}_2)$, $(\boldsymbol{v}_0 \times (\boldsymbol{v}_1 \times \boldsymbol{v}_2))$
$n = 3$ $(((\boldsymbol{v}_0 \times \boldsymbol{v}_1) \times \boldsymbol{v}_2) \times \boldsymbol{v}_3)$, $((\boldsymbol{v}_0 \times (\boldsymbol{v}_1 \times \boldsymbol{v}_2)) \times \boldsymbol{v}_3)$, $(\boldsymbol{v}_0 \times ((\boldsymbol{v}_1 \times \boldsymbol{v}_2) \times \boldsymbol{v}_3))$, $(\boldsymbol{v}_0 \times (\boldsymbol{v}_1 \times (\boldsymbol{v}_2 \times \boldsymbol{v}_3)))$, $((\boldsymbol{v}_0 \times \boldsymbol{v}_1) \times (\boldsymbol{v}_2 \times \boldsymbol{v}_3))$

Catalan 2. C_n is the number of ways of connecting n points on the x axis by noncrossing arcs in the upper half-plane such that if two arcs share an endpoint p, then p is a left endpoint of both arcs:

$n = 1$

$n = 2$

$n = 3$

Catalan 3. C_n is the number of ways $2n$ people seated at a circular table can shake hands in n pairs, with no arms crossing:

[13]R. P. Stanley, a combinatorialist at MIT, maintains an ever-growing list of problems answered by Catalan numbers; see his book, [7] and the Website http://www-math.mit.edu/~rstan/ec/catadd.pdf.

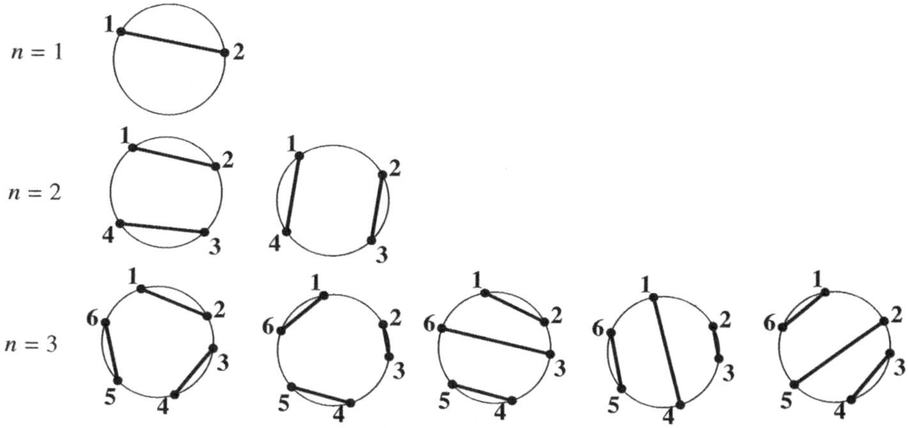

There are three common ways to show that a combinatorial question is answered by a Catalan number:

Direct—derive the binomial coefficient formula $C_n = [1/(n+1)]\binom{2n}{n}$ by a direct combinatorial calculation. For example, the "tail swap" argument given in Example 2.17 and the cyclic shift argument in the second solution of Example 6.55 are direct derivations.

Recurrence relation—derive the recurrence relation, and show that it is the one given in (6.77).

Bijection—find a bijective mapping between the arrangements of the new problem and the arrangements of a setting already known to be counted by Catalan numbers.

It is an interesting, and often difficult, challenge to find and describe a bijection between arrangements. Here is an example of a bijective proof of the Catalan 1 setting described above.

Example 6.57 Devise a bijective proof to show that C_n is the number of binary bracketings of an ordered list of $n + 1$ symbols $v_0 v_1 v_2 \cdots v_n$.

Solution. It will be shown that any triangulation of convex polygonal region with vertices $P_1, \ldots, P_{n+1}, P_{n+2}$ corresponds to a unique binary bracketing of the factors $v_0 v_1 v_2 \cdots v_n$. The edges and diagonals of the triangulation are labeled sequentially by the following algorithm:

1. Let $v_j = \overline{P_{j+1}P_{j+2}}$, $j = 0, 1, \ldots, n$ label the first $n + 1$ sides of the boundary polygon, leaving side $\overline{P_{n+2}P_1}$ unlabeled.

2. If v_j and v_{j+1} are the sides of a triangle of the triangulation, then the diagonal forming the third side of the triangle is assigned the label $(v_j v_{j+1})$.
3. If two sides of a triangle of the triangulation have been assigned labels, concatenate the labels to give a label to the third side. For example, the third side of a triangle with the labeled $((v_1v_2)v_3)$ and (v_4v_5) sides is assigned the label $(((v_1v_2)v_3)(v_4v_5))$, as seen in the example below.
4. Continue labeling until side $\overline{P_{n+2}P_1}$ is assigned a label, which will be the binary bracketing of the symbols $v_0v_1v_2 \cdots v_n$ that corresponds to the triangulation. In the example shown below, the triangulation is mapped to the bracketing $((v_0(((v_1v_2)v_3)(v_4v_5)))v_6)$:

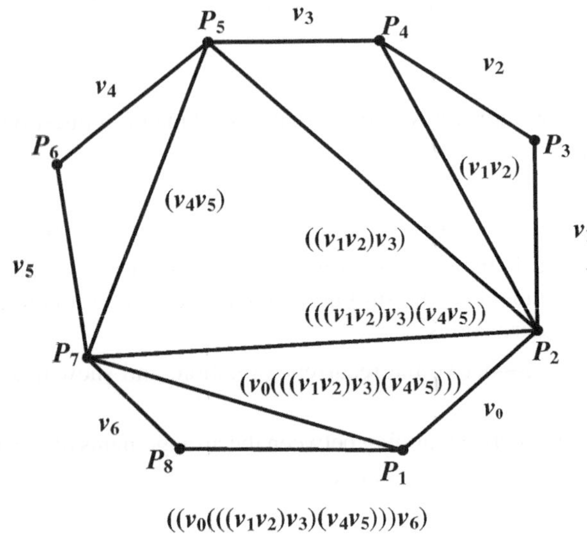

It can be checked that the mapping from a triangulation to an assignment of parentheses is reversible, so the mapping is a bijection. Since the Catalan number C_n is the number of triangulations of the polygonal region, this is also the number of binary bracketings. ■

PROBLEMS

6.7.1. What is the number of mountain ranges across the Isthmus of Panama (sea level to sea level) that can be drawn with n upstrokes and n downstrokes? The elevation gain or loss of each stroke is the same, and no pass between adjacent mountains is below sea level.

6.7.2. In how many ways can n voters favoring candidate A and n other voters favoring candidate B line up so that as the votes are recorded one by one candidate A never trails candidate B?

6.7.3. Raffle tickets are \$5 each. What is the probability that 12 people can buy a ticket, where 6 people have a \$5 bill and the other 6 people have a \$10 bill, and enough change is on hand for each successive transaction? At the beginning of the ticket sales, there is no money in the collection box.

6.7.4. **(a)** Prove that the Catalan numbers satisfy the recurrence relation $(n+2)\,C_{n+1} = (4n+2)\,C_n$, $C_0 = 1$.

(b) Use the recurrence relation proved in part (a) to calculate a table of Catalan numbers for $0 \le n \le 12$.

6.7.5. The sum of the terms of the sequence $(-1, -1, 1, 1, -1, -1, 1, 1, 1)$ is 1. Graph the path generated by the sequence, and use it to identify the unique cyclic shift of the sequence with all positive partial sums.

6.7.6. The terms of the sequence $(-2, 4, -5, 2, -2, 3, -1, 2)$ sum to $+1$.

(a) Write all of the cyclic shifts of the sequence.

(b) Graph the lattice walk generated by the sequence, and use it to identify the cyclic shift with positive partial sums.

6.7.7. Given any sequence of integers $(a_1, a_2, \ldots, a_n)$ for which $a_1 + a_2 + \cdots + a_n = 1$, prove there is a unique cyclic shift with all positive partial sums.

6.7.8. Prove that any triangulation of a convex polygonal region with $n + 2$ sides uses $n - 1$ diagonals for any $n \ge 1$.

6.7.9. Draw all of the triangulations determined by the diagonals of a convex hexagonal region that do not intersect at an interior point of the region.

6.7.10. Derive a recursion relation to verify the statement of Catalan 1; that is, prove that C_n is the number of ways to create a binary bracketing of an ordered list of symbols $x_0, x_1, \ldots, x_n$, $n \ge 1$.

6.7.11. Prove the statement of Catalan 2; that is, prove that C_n is the number of ways of connecting n points on the x-axis by noncrossing arcs in the upper half-plane, such that if two arcs share an endpoint p, then p is a left endpoint of both arcs.

6.7.12. Prove the statement of Catalan 3. Specifically, prove that C_n is the number of ways $2n$ people seated at a circular table can shake hands in n pairs, with no arms crossing.

6.7.13. Using the labeling algorithm described in Theorem 6.57, draw the triangulation of the polygonal region that corresponds to the parenthesis placement $(((v_0v_1)v_2)(((v_3v_4)v_5)v_6))$.

6.7.14. The *super Catalan* (or *little Schroeder*) number s_n is the number of ways that a convex polygonal region with $n+2$ vertices can be dissected into polygonal subregions by any subset of its diagonals that do not intersect in the interior of the starting region. The following diagrams show that $s_1 = 1$ and $s_2 = 3$:

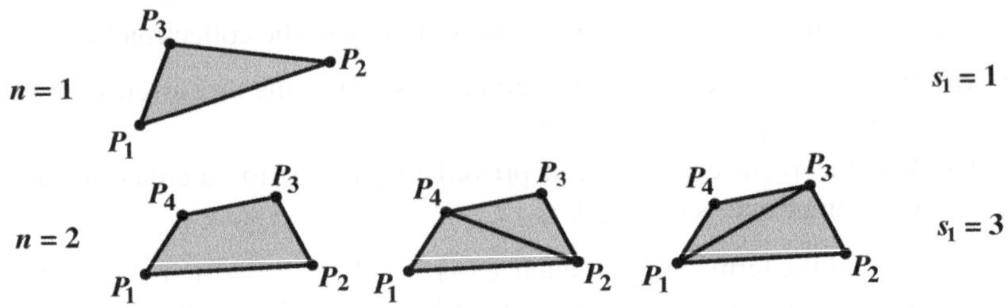

(a) draw diagrams that show $s_3 = 11$.

(b) derive the recurrence relation

$$s_{n+1} = 3s_n + 2\sum_{k=1}^{n-1} s_k s_{n-k}, \quad n \geq 2, \text{ or equivalently}$$

$$s_{n+1} + s_n = 2\sum_{k=0}^{n} s_k s_{n-k}, \quad n \geq 0$$

if we define $s_0 = 1$.

(c) show that the generating function

$$f(x) = \sum_{n=0}^{\infty} s_n x^n = 1 + x + 3x^2 + \cdots$$

is

$$f(x) = \frac{1 + x - \sqrt{1 - 6x + x^2}}{4x}$$

(d) use a CAS (computer algebra system) to verify that the sequence of super Catalan numbers s_n for $n = 0, 1, \ldots, 10$ is 1, 1, 3, 11, 45, 197, 903, 4279, 20793, 103049, 518859.

6.7.15. Example 6.57 can be modified to show that the super Catalan numbers of Problem 6.7.14 give the number of arbitrary bracketings of a list of symbols. For example, the bracketed list $(((v_0v_1v_2)(v_3v_4))v_5v_6)$ corresponds to the edge and diagonal labeling of the dissection shown below:

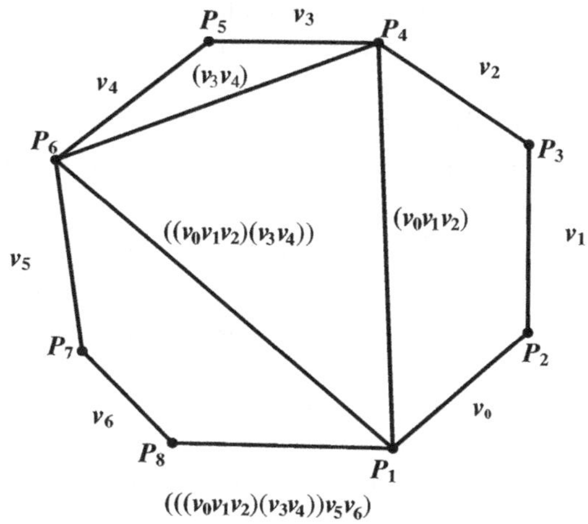

(a) give the bracketed list of symbols that corresponds to this dissection of a polygonal region:

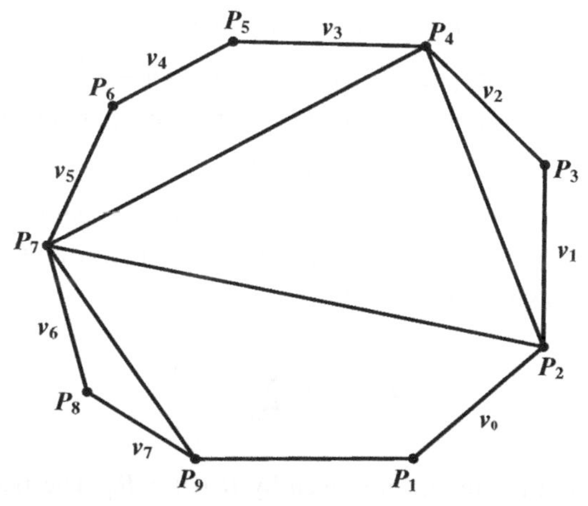

(b) draw and label the dissected polygonal region that corresponds to this bracketed list of symbols $(v_0(v_1v_2)((v_3v_4v_5)v_6)v_7v_8)$.

6.8 SUMMARY AND ADDITIONAL PROBLEMS

This chapter has defined some of the number sequences that frquently arise in various branches of mathematics, and especially often when combinatorial questions are being addressed. In particular, we have defined the following special numbers and have derived some of their most important properties:

Stirling Numbers of the First and Second Kinds. The Stirling number of the first kind, $\left[{k \atop j}\right]$, counts the number of permutations of k symbols that have j cycles; equivalently, this is the number of ways that k people can be seated at j tables with no table left unoccupied. The Stirling number of the second kind, $\left\{{k \atop j}\right\}$, counts the number of ways to partition a set of k elements into j nonempty subsets; equivalently, this is the number of ways that k people can be divided into j groups, with at least one member of each group. It was shown that if the rising and falling factorial polynomials are defined, respectively, by $x^{(k)} = x(x+1)(x+2)\cdots(x+k-1)$ and $(x)_k = x(x-1)(x-2)\cdots(x-k+1)$, then the Stirling numbers are the coefficients for which

$$x^{(k)} = \sum_{j=0}^{k} \left[{k \atop j}\right] x^j \quad \text{and} \quad x^k = \sum_{j=0}^{k} \left\{{k \atop j}\right\} (x)_j$$

Harmonic Numbers. The nth harmonic number is the sum of reciprocals $H_n = 1 + \frac{1}{2} + \frac{1}{3} + \cdots + \frac{1}{n}$, with $H_0 = 0$. The harmonic numbers increase to infinity at a logarithmic rate; more precisely, $H_n \approx \gamma + \log\left(n + \frac{1}{2}\right)$, where $\gamma = 0.5772\ldots$ is Euler's constant. It was shown that H_n is the average number of cycles in the $n!$ permutations of $[n]$, since $H_n = (1/n!)\sum_{j=1}^{n} j \left[{n \atop j}\right]$.

Bernoulli Numbers. If the *Bernoulli polynomials* are defined recursively by $B_0(t) = 1, B_n'(t) = nB_{n-1}(t), \int_0^1 B_n(t)\,dt = 0,\ n = 1, 2, \ldots,$ or equivalently by the exponential generating function

$$\boldsymbol{B}(x,t) = \sum_{n=0}^{\infty} B_n(t) \frac{x^n}{n!}$$

then the Bernoulli numbers are given by $B_n(0) = B_n$. The Bernoulli numbers appear frequently in sums and series expansions. For example

$$1^k + 2^k + \cdots + (n-1)^k = \frac{1}{k+1} \sum_{j=0}^{k} \binom{k+1}{j} B_j n^{k+1-j}$$

and

$$\tan x = \sum_{n=1}^{\infty} (-1)^{n-1}\, 4^n \left(4^n - 1\right) B_{2n} \frac{x^{2n-1}}{(2n)!}$$

Eulerian Numbers. Eularian numbers $\left\langle {k \atop i} \right\rangle$ are the coefficients of the polynomials that appear in the ordinary generating function for the sequence of kth powers, $1^k, 2^k, 3^k, \ldots, n^k, \ldots$. In particular

$$\sum_{n=0}^{\infty} n^k x^n = \frac{x \sum\limits_{i=0}^{k} \left\langle {k \atop i} \right\rangle x^i}{(1-x)^{k+1}}, \quad k \geq 1$$

Partition Numbers. A sequence of k positive integers for which $a_1 \geq a_2 \geq \cdots \geq a_k$ and $a_1 + a_2 + \cdots + a_k = n$ is a *partition* of n with k parts. A *partition number* $p(n,C,D)$ is the number of partitions of n where C is a constraint on the type of parts allowed and D is a constraint on the number of parts. If no constraints are imposed, denoted by $C = *$ and $D = *$, then $p(n)$ is the number of unrestricted partitions. Using Ferrer's diagrams, it is easily seen that $p(n,^*$, largest part $= k) = p(n$, number of parts $= k,^*)$. The unrestricted partition numbers satisfy a recursion relation known as Euler's *pentagonal number theorem*, which states that

$$\begin{aligned} p(n) &= p\,(n-1) + p\,(n-2) - p\,(n-5) \\ &\quad - p\,(n-7) + p\,(n-12) + p\,(n-15) - \cdots \end{aligned}$$

where 1, 2, 5, 7, 12, 15, ... are the generalized pentagonal numbers given by $n = \frac{1}{2}k\,(3k \pm 1),\ k \geq 1$.

Catalan Numbers. Catalan numbers are given explicitly by the formula $C_n = [1/(n+1)]\binom{2n}{n}$ and recursively by $h_{n+1} = h_0 h_n + h_1 h_{n-1} + h_2 h_{n-2} + \cdots + h_n h_0$, where $h_0 = 1$. A large number of combinatorial problems are answered in terms of the Catalan numbers. For example, C_n is the number of permutations of n ones and n negative ones with nonnegative partial sums. It is also the number of ways that $n - 3$ noncrossing diagonals of a convex polygon dissect the interior of the polygon into triangles.

The six sequences of special numbers that have been discussed are certainly not exhaustive, and other interesting but less well known sequences are often encountered. In this case, it is suggested that you search the *Online Encyclopedia of Integer Sequences* (OEIS) (http://oeis.org/). Suggestions about how to search the encyclopedia are found in Appendix B of this text.

6.8.1 Counting Functions and Distributions

The Stirling numbers and partition numbers described in this chapter can be combined with the results of Section 2.5 to count the number of mappings (functions or distributions) from X to Y, where X and Y are either sets of distinct objects or multisets of identical objects or unlabeled boxes. In Table 6.8, the unrestricted mappings are counted, as well as the number of mappings that are one-to-one (injective), onto (surjective), or both one-to-one and onto (bijective). For example, the number of ways to place m distinct toys into n identical toy boxes is the Bell number $B(m), m \leq n$, where some toy boxes can be left empty. The counts are given in terms of binomial coefficients, permutation numbers, distribution numbers, Bell numbers, Stirling numbers of the second kind, partition numbers, and the Kronecker delta function.

TABLE 6.8 The Number of Functions and Distributions from X to Y

	$Y = \{y_1, y_2, \ldots, y_n\}$		$Y = \{n \cdot y\}$	
$X = \{x_1, x_2, \ldots, x_m\}$	Total n^m	One-to-one $P(n, m)$	Total $B(m), m \leq n$	One-to-one $\begin{cases} 1, m \leq n \\ 0, m > n \end{cases}$
	Onto $T(m, n)$	Bijections $m!\delta_{m,n}$	Onto $S(m, n)$	Bijections $\delta_{m,n}$
$X = \{m \cdot x\}$	Total $\binom{m+n-1}{m}$	One-to-one $\binom{n}{m}$	Total $p(m), m \leq n$	One-to-one $\begin{cases} 1, m \leq n \\ 0, m > n \end{cases}$
	Onto $\binom{m-1}{n-1}$	Bijections $\delta_{m,n}$	Onto $p_n(m)$	Bijections $\delta_{m,n}$

PROBLEMS

6.8.1. Find the quartic polynomial $p(n)$ that gives the maximum number of interior regions within a disk formed by all of the chords drawn between pairs of the n points on the circular boundary of the disk:

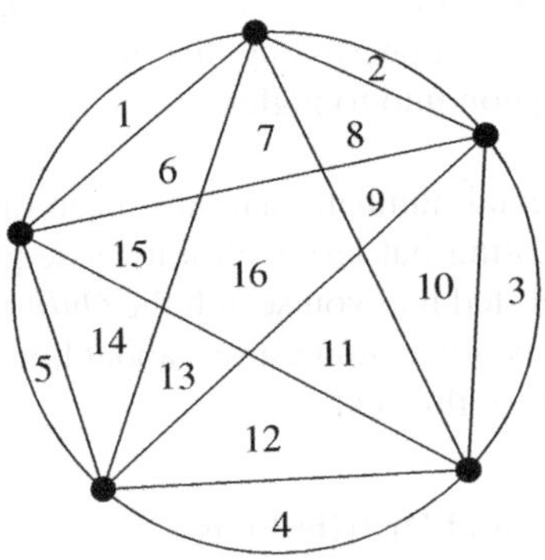

$p(5) = 16$

For example, $p(5) = 16$, as shown in the diagram. Note that $p(0) = 1, p(1) = 1$, and $p(2) = 2$.

6.8.2. Given n points in the plane, no three collinear, construct all of the lines through each pair of points. Let $p(n)$ denote the number of simply connected regions[14] determined by these lines. For example, $p(0) = 1$ since the plane is a simply connected region.

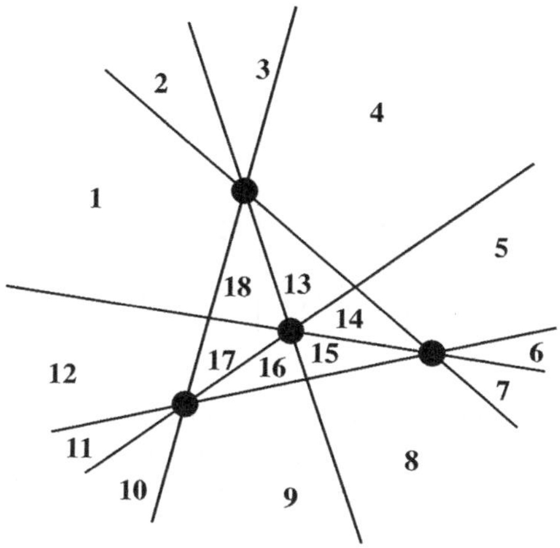

$p(4) = 18$

(a) Draw diagrams to show that $p(1) = 0$, $p(2) = 2$, and $p(3) = 7$, similar to the diagram at the right that shows that $p(4) = 18$. Be sure to explain why $p(1) = 0$.

(b) Draw a (large!) diagram to show that $p(5) = 41$.

(c) What quartic polynomial $p(n)$ gives maximum number of simply connected regions determined by the lines?

6.8.3. Let $q(0) = 0$ and $q(n) = q(n-1) + s_n$, $n \geq 1$, where s_n is the sum of the base ten digits of n. For example, $q(1) = q(0) + 1 = 1, q(2) = q(1) + 2 = 3$, and $q(3) = q(2) + 3 = 6$.

(a) Show that $q(n) = \binom{n+1}{2}$ for $0 \leq n \leq 8$.

(b) Is $q(n) = \binom{n+1}{2}$ for all $n \geq 0$?

6.8.4. The table below is constructed row by row according to the following rules:

- each new row has one more entry than the previous row.
- the first entry in a new row is the sum of the last entries of the previous two rows. For example, $20 = 5 + 15$ and $67 = 15 + 52$.

[14] A region in the plane is *simply connected* when no boundary point of the region can be enclosed by a closed curve that is inside the region.

- the remaining entries in a row are the sum of entries to the left and the upper left. For example, $151 = 37 + 114$:

1					
	2				
	3	**5**			
	7	10	**15**		
	20	27	37	**52**	
	67	87	114	151	**203**

Continue the table for three more rows, and identify the sequence of last numbers in the rows shown in bold. (A calculator—or better yet, a spreadsheet—is helpful.) Identify the entries along the main diagonal of the table shown in bold.

6.8.5. The *associated Stirling numbers of the second kind* $\left\{\left\{ {k \atop j} \right\}\right\}$ give the number of partitions of $[k]$ into j subsets that each contain two or more members.

(a) Show that

$$\left\{\left\{ {k+1 \atop j} \right\}\right\} = j \left\{\left\{ {k \atop j} \right\}\right\} + k \left\{\left\{ {k-1 \atop j-1} \right\}\right\}, \quad k \geq 3$$

(b) Extend the triangular table of the associated Stirling numbers of the second kind shown below to include the entries for k and j with $1 \leq 2j \leq k \leq 9$ (a spreadsheet is helpful):

	$j = 1$	2	3
$k = 1$	**0**	**0**	
2	**1**	**0**	
3	**1**	**0**	
4	**1**	3	**0**
5	**1**	10	**0**

6.8.6. Suppose that the lowest number in a cycle of a permutation of $[n]$ is the *rank* of that cycle. For example, the cycle (3 6 7 4) has rank 3 and the cycle (4 8 5) has rank 4. Any permutation of $[k]$ can be written as a product of cycles listed in descending order by rank. As an example, the permutation $\pi = (1\ 4\ 8)(2\ 3)(6\ 9\ 7)(5)$ would be written as $\pi = (\underline{6}\ 9\ 7)(\underline{5})(\underline{2}\ 3)(\underline{1}\ 4\ 8)$

with the descending ranks shown underlined. By introducing the 0 element, we can associate π with the one cycle permutation $\pi' = (0\,\underline{6}\,9\,7\,\underline{5}\,\underline{2}\,3\,\underline{1}\,4\,8)$. Moreover, the association is reversible; for example,

$$\pi = (\underline{5}\,8\,7)\,(\underline{2}\,4)(\underline{1}\,6\,3) \leftrightarrow \pi = (0\,5\,8\,7\,2\,4\,1\,6\,3)$$
$$\pi = (\underline{8})\,(\underline{3})\,(\underline{2}\,7)\,(\underline{1}\,5\,4\,6) \leftrightarrow \pi = (0\,8\,3\,2\,7\,1\,5\,4\,6).$$

(a) use the correspondence just described to give a combinatorial proof that $\left[{k+1 \atop 1}\right] = \sum_{j=1}^{k} \left[{k \atop j}\right]$, $k \geq 1$.

(b) extend the correspondence used in part (a) this way: distinguish one of the cycles of a permutation π of $[k]$ with an asterisk * and then apply the correspondence of part (a) to the remaining cycles of π to obtain a permutation of $\{0\} \cup [k]$ with two cycles; for example,

$$\pi = (\underline{5}\,8\,7)\,(\underline{2}\,4)^*(\underline{1}\,6\,3) \leftrightarrow \pi' = (\underline{2}\,4)^*\,(0\,5\,8\,7\,1\,6\,3)$$
$$\pi = (\underline{5}\,8\,7)\,(\underline{2}\,4)(\underline{1}\,6\,3)^* \leftrightarrow \pi' = (\underline{1}\,6\,3)^*\,(0\,5\,8\,7\,2\,4)$$
$$\pi = (\underline{6}\,8)\,(\underline{4}\,7\,5)^*\,(\underline{3})\,(\underline{1}\,2) \leftrightarrow \pi' = (4\,7\,5)^*\,(0\,6\,8\,3\,1\,2)$$

Use this correspondence to prove that

$$\left[{k+1 \atop 2}\right] = \sum_{j=1}^{k} j \left[{k \atop j}\right], \quad k \geq 1$$

6.8.7. Use the result of Problem 6.8.6(b) and Theorem 6.24(c) to prove that

$$\sum_{k=1}^{m} (-1)^{m+k} \left\{{m \atop k}\right\} \left[{k+1 \atop 2}\right] = m$$

6.8.8. Fibonacci proved that any positive fraction $(n/d) \in (0, 1)$ can be written as a sum of distinct *unit* fractions (those with 1 as the numerator), so that

$$\frac{n}{d} = \frac{1}{D_1} + \frac{1}{D_2} + \cdots + \frac{1}{D_k}, \quad D_1 < D_2 < \cdots < D_k$$

(a) prove Fibonacci's theorem by using a greedy algorithm. Subtract the unit fraction $1/D_1$ with the smallest possible denominator from n/d to leave a nonnegative fraction

$$\frac{n}{d} - \frac{1}{D_1} = \frac{nD_1 - d}{dD_1} = \frac{n_1}{d_1}$$

Show that the new fraction obtained, n_1/d_1, has a larger denominator $d_1 > d$ and a smaller numerator $n_1 < n$ than the starting fraction n/d. Repeat the process starting with n_1/d_1, and terminate when a unit numerator $n_k = 1$ is obtained.

(b) why does the algorithm necessarily terminate in no more than n_1 steps?

(c) use the greedy algorithm of part (a) to write $\frac{9}{20}$ as a sum of distinct unit fractions.

(d) are representations as sums of unit fractions unique?

6.8.9. Let $S_k = \sum_{n=0}^{k} a_n b_n$ and $B_k = \sum_{n=0}^{k} b_n$. Prove that $S_k = a_k B_k - \sum_{n=0}^{k-1} B_n(a_{n+1} - a_n)$, called the *Abel transformation.*

6.8.10. Prove that the number of partitions of $n \geq 2$ as a sum of powers of 2 (i.e., as a sum whose parts are 1, 2, 2^2, 2^3, ...) with evenly many parts is the same as the number of partitions with an odd number of parts. [*Hint*: Find the OGF of $s(n) = p(n$, powers of 2, even number of parts) $-p$ (n,powers of 2, odd number of parts).]

6.8.11. **(a)** Give the 10 partitions with no more than two parts and of size at most three, including the null partition 0.

(b) Let j and k be positive integers. Prove that there are $\binom{j+k}{k}$ partitions with no more than k parts of size at most j.

6.8.12. Let $p(n)$ denote the number of unrestricted partitions of any integer n, with $p(0) = 1$, and define $q(n) = p(n) - 4p(n-3) + 4p(n-5) - p(n-8)$. Thus, for example, $q(8) = p(8) - 4p(5) + 4p(3) - p(0) = 22 - 4 \cdot 7 + 4 \cdot 3 - 1 = 5$. Use generating functions to prove[15] that $q(n) > 0$ for all $n \geq 8$.

REFERENCES

1. J. Havil, *Gamma: Exploring Euler's Constant*, Princeton Science Library, 2009.
2. R. T. Sharp, Problem 52: Overhanging dominoes, *Pi mu Epsilon j.* **10**:9–11 (1954).
3. J. E. Drummond, On stacking tricks to achieve a large overhand, *Math. Gaz.* **65m**:40–42 (March 1981).
4. R. Netz and W. Noel, *The Archimedes Codex*, Da Capo Press, Philadelphia, 2007.
5. R. Graham, D. Knuth, and O. Patashnik, *Concrete Mathematics*, 2nd ed., Addison-Wesley, 1994.
6. L. Euler, *Institutiones Calculi Differentialis*, 1755.
7. R. P. Stanley, *Enumerative Combinatorics*, Vol. 2, Cambridge University Press, 2001.

[15] This is problem 11675 proposed and solved by Mircia Merca in the *American Mathematical Monthly* of November 2012.

PART II

TWO ADDITIONAL TOPICS IN ENUMERATION

7

LINEAR SPACES AND RECURRENCE SEQUENCES

7.1 INTRODUCTION

In this chapter we return to a discussion of *recurrence sequences*, specifically the sequences that solve a homogeneous linear recurrence relation with constant coefficients that have the form

$$h_{n+k} - a_1 h_{n+k-1} - \cdots - a_{k-1} h_{n+1} - a_k h_n = 0, a_k \neq 0 \tag{7.1}$$

Equivalently, we can recast this equation in the form

$$C(E)h_n = 0 \tag{7.2}$$

where E is the successor operator defined by $Eh_n = h_{n+1}$ and

$$C(x) = x^k - a_1 x^{k-1} - \cdots - a_{k-1}x - a_k \tag{7.3}$$

is the characteristic polynomial associated with the recurrence relation. In contrast to the development of linear recurrences given in Chapter 5, here we will more fully incorporate concepts of linear algebra. Fortunately, only the most basic ideas of a vector space are needed: subspace, spanning set, basis, dimension, and linear independence.

Combinatorial Reasoning: An Introduction to the Art of Counting, First Edition. Duane DeTemple and William Webb.

In Section 7.2, connections between vector spaces and solutions of the recurrence relation (7.1) will be established. Extensions of these ideas to nonhomogeneous recurrences and systems of recurrences are discussed in Section 7.3. Section 7.4 applies the developments of the previous sections to both obtaining and proving identities for recurrence sequences. The methods given there allow us to obtain results that would be very challenging by other means.

7.2 VECTOR SPACES OF SEQUENCES

A recurrence relation associated with a combinatorial question is usually integer-valued. However, we will be more inclusive by considering sequences all of complex numbers. If x_n and y_n are any two complex-valued sequences and c_1 and c_2 are any complex numbers, then $z_n = c_1x_n + c_2y_n$ is again a complex-valued sequence. Indeed, the set of all sequences is a vector space over the field of complex numbers, one that we will denote by $\mathcal{S}$. Of course, our primary interest is directed toward *recurrence sequences*, those sequences that solve *some* recurrence relation of the form given by (7.1), or equivalently by (7.2). If g_n and h_n are any two linear recurrence sequences, say, $C_1(E)g_n = 0$ and $C_2(E)h_n = 0$, then

$$C_1(E)C_2(E)(c_1g_n + c_2h_n) = c_1C_2(E)C_1(E)g_n + c_2C_1(E)C_2(E)h_n = 0 + 0 = 0. \tag{7.4}$$

Equation (7.4) shows that the linear combination $c_1g_n + c_2h_n$ is again a recurrence sequence, this time with $C(x) = C_1(x)C_2(x)$ as the associated characteristic polynomial. Therefore, the set of recurrence sequences that *each* solve some linear homogeneous recursion relation, not necessarily the same one, form a subspace of $\mathcal{S}_\mathcal{R}$ of $\mathcal{S}$.

7.2.1 Three Vector Spaces Associated with a Recurrence Relation

Suppose we fix our attention on the recurrence relation given by the annihilating polynomial operator $C(E)$. There are three vector subspaces of $\mathcal{S}_\mathcal{R}$ that will be associated with the corresponding recurrence relation:

V_C—*the Vector Space of Solutions.* If g_n and h_n both solve the *same* recursion relation, say, $C(E)g_n = 0$ and $C(E)h_n = 0$, then

$$C(E)(c_1g_n + c_2h_n) = c_1C(E)g_n + c_2C(E)h_n = 0 + 0 = 0 \tag{7.5}$$

In other words, the set of solutions of a given homogeneous linear recurrence relation is the vector subspace V_C of $\mathcal{S}_\mathcal{R}$ defined by

$$V_C = \{h_n | C(E)h_n = 0\} \tag{7.6}$$

In the language of linear algebra, V_C is the null space of the linear operator $C(E)$.

The set of k sequences

$$\left\{ u_n^{(j)} \in V_C | u_n^{(j)} = \delta_{n,j},\ 0 \leq j \leq k-1 \right\} \tag{7.7}$$

is a basis of V_C so V_C is a vector space of dimension $k = \dim V_C$. Of course, other bases of V_C can also be adopted, and some especially useful bases will be discussed later.

V_S—*the Vector Space of Generalized Power Sums*. Given a set of distinct complex numbers $\{\alpha_1, \alpha_2, \ldots, \alpha_m\}$ and polynomials $p_1, p_2, \ldots, p_m$, any sum of the form

$$\sum_{i=1}^{m} p_i(n)\alpha_i^n \tag{7.8}$$

is called a *generalized power sum* (GPS). The set of all generalized power sums form a vector space $\mathcal{G}$. However, our interest is in the vector subspace V_S for which each α_i is a root of the characteristic equation $C(x) = 0$. More precisely, if $C(x)$ has the factorization

$$C(x) = (x - \alpha_1)^{r_1}(x - \alpha_2)^{r_2} \cdots (x - \alpha_m)^{r_m} \tag{7.9}$$

then the multiset $\mathcal{E}_C = \{r_1 \cdot \alpha_1, r_2 \cdot \alpha_2, \ldots, r_m \cdot \alpha_m\}$ will be called the *spectrum* of the recurrence relation and $\alpha_i \in \mathcal{E}_C$ is an eigenvalue of multiplicity r_i. The vector space V_S is then defined by

$$V_S = \left\{ h_n \in S_R | h_n = \sum_{i=1}^{m} p_i(n)\alpha_i^n, \deg p_i < r_i \right\} \tag{7.10}$$

The set of $k = r_1 + r_2 + \cdots + r_m$ sequences

$$\left\{ \alpha_1^n, n\alpha_1^n, n^2\alpha_1^n, \ldots, n^{r_1-1}\alpha_1^n, \ldots, \alpha_m^n, n\alpha_m^n, n^2\alpha_m^n, \ldots, n^{r_m-1}\alpha_m^n \right\} \tag{7.11}$$

spans V_S, and therefore $\dim V_S \leq k$. Our earlier work with Vandermonde determinants proved that the set given by (7.11) is really a basis, but this will follow more easily from Theorem 7.1 (below).

V_Q—*the Vector Space of Sequences with Prescribed Generating Functions*. Let the polynomial $Q(x)$ be defined by

$$Q(x) = x^k C\left(\frac{1}{x}\right) = 1 - a_1 x - a_2 x^2 - \cdots - a_k x^k \tag{7.12}$$

where $C(x) = x^k - a_1 x^{k-1} - \cdots - a_{k-1}x - a_k$ is the characteristic polynomial of the given recursion relation. Since $a_k \neq 0$, it follows that $Q(x)$ has degree k.

If we let $f_h(x) = \sum_{n\geq 0} h_n x^n$ denote the generating function of a sequence h_n, we define the set of sequences

$$V_Q = \left\{ h_n | f_h(x) = \frac{P(x)}{Q(x)}, \deg P(x) < k \right\} \tag{7.13}$$

It is easy to verify that V_Q is a vector space of sequences. Moreover, the k sequences with the generating functions $x^j/Q(x), j = 0, 1, \ldots, k-1$, span V_Q, which shows that $\dim V_Q \leq k$.

The following theorem shows that V_C, V_S, and V_Q, although defined differently, are in fact the very same vector subspace of $S_{\mathcal{R}}$.

Theorem 7.1 The solution space V_C, the generalized power sums V_S, and the set of sequences V_Q with generating functions with the form $P(x)/Q(x)$ are identical vector spaces; that is, $V_C = V_S = V_Q$, and all three vector space have dimension k.

Proof. We first show that $V_C = V_Q$. Let $h_n \in V_C$, so that

$$C(E)h_n = h_{n+k} - a_1 h_{n+k-1} - \cdots - a_{k-1}h_{n+1} - a_k h_n = 0, n \geq 0$$

Then

$$\begin{aligned}
Q(x)f_h(x) &= (1 - a_1 x - a_2 x^2 - \cdots - a_k x^k)\left(\sum_{j\geq 0} h_j x^j\right) \\
&= \sum_{n\geq 0}\left(h_n - \sum_{j=1}^{n} a_j h_{n-j}\right)x^n \quad [\text{let } a_j = 0 \text{ for } j > k] \\
&= \sum_{n=0}^{k-1}\left(h_n - \sum_{j=1}^{n} a_j h_{n-j}\right)x^n + \sum_{n\geq k}\left(h_n - \sum_{j=1}^{k} a_j h_{n-j}\right)x^n \\
&= P(x) + 0 = P(x)
\end{aligned}$$

where $P(x)$ is a polynomial of degree less that k. Thus $f_h(x) = P(x)/Q(x)$, $\deg P(x) < k$, which shows that $h_n \in V_Q$ and therefore $V_C \subseteq V_Q$. But then $k = \dim V_C \leq \dim V_Q \leq k$, so it follows that $V_C = V_Q$.

Next we show that $V_Q = V_S$. Let $h_n \in V_Q$, so that

$$f_h(x) = \frac{P(x)}{Q(x)} = \frac{P(x)}{(1-\alpha_1 x)^{r_1}(1-\alpha_2 x)^{r_2}\cdots(1-\alpha_m x)^{r_m}}$$

The partial fraction decomposition of this rational function then has the form

$$f_h(x) = \sum_{i=1}^{m} \left(\sum_{j=0}^{r_i-1} \frac{c_{ij}}{(1-\alpha_i x)^{j+1}} \right)$$

But

$$\frac{1}{(1-\alpha_i x)^{j+1}} = \sum_{n\geq 0} \binom{n+j}{n} \alpha_i^n x^n$$

so

$$f_h(x) = \sum_{n\geq 0} \left(\sum_{i=1}^{m} \sum_{j=0}^{r_i-1} c_{ij} \binom{n+j}{j} \alpha_i^n \right) x^n = \sum_{n\geq 0} \left(\sum_{i=1}^{m} p_i(n) \alpha_i^n \right) x^n$$

where $\deg p_i < r_i$. Therefore, $h_n = \sum_{i=1}^{m} p_i(n)\alpha_i^n \in V_S$, so that $V_Q \subseteq V_S$. We also know that $\dim V_S \leq k$, so that $k = \dim V_Q \leq \dim V_S \leq k$. Thus $V_S = V_Q$. ■

Example 7.2 Let $C(x) = x^3 + x^2 - 4x + 6$. Determine the spectrum $\mathcal{E}_C$ and the vector spaces V_C, V_S, and V_Q.

Solution. Since $C(x) = (x+3)(x-1-i)(x-1+i)$, the spectrum is $\mathcal{E}_C = \{-3, 1+i, 1-i\}$. The vector spaces are

$$V_C = \{h_n | h_{n+3} + h_{n+2} - 4h_{n+1} + 6h_n = 0\}$$

$$V_S = \{c_1(-3)^n + c_2(1+i)^n + c_3(1-i)^n \,|\, c_1, c_2, c_3 \in \mathbb{C}\}$$

$$V_Q = \left\{ h_n \middle| \sum_{n\geq 0} h_n x^n = \frac{P(x)}{1+x-4x^2+6x^3}, \deg P(x) < 3 \right\}$$

Despite their very different descriptions, the three vector spaces are identical. ■

7.2.2 The Order of a Recurrence Relation

Suppose that h_n is a sequence annihilated by the operator $C(E)$. The sequence h_n can also be annihilated by other polynomial operators. Of special importance is the monic polynomial $M(x)$ of smallest degree for which $M(E)h_n = 0$. This is called the *minimal polynomial* of the sequence h_n and the degree of the minimal polynomial is called the *order* of the recurrence sequence, denoted by $\mathcal{O}(h_n)$. The *spectrum* of h_n is the multiset $\mathcal{E}_h = \mathcal{E}_M$ of the roots (eigenvalues) of $M(x) = 0$ listed with their multiplicities.

Theorem 7.3 Let $M(x)$ be the minimal polynomial of a sequence $h_n \in V_C$, where $C(E)h_n = 0$. Then $C(x)$ is a multiple of the minimal polynomial $M(x)$, and $\mathcal{E}_M \subseteq \mathcal{E}_C$.

Proof. Let $D(x) = \gcd(M(x), C(x))$ be the greatest common divisor of $M(x)$ and $C(x)$; that is, $D(x)$ is the monic polynomial of largest degree that divides both $M(x)$ and $C(x)$. A standard result of polynomial algebra guarantees there are polynomials $a(x)$ and $b(x)$ for which $D(x) = a(x)M(x) + b(x)C(x)$. But then $D(E)h_n = a(E)M(E)h_n + b(E)C(E)h_n = a(E) \cdot 0 + b(E) \cdot 0 = 0$, so $D(x)$ is also an annihilating polynomial of h_n. Since $\deg D(x) \leq \deg M(x)$ and $M(x)$ is the annihilating polynomial of smallest degree, we see that $D(x) = M(x)$. This shows that $M(x)$ is a divisor of $C(x)$, or, equivalently, $C(x)$ is a multiple of $M(x)$. ■

It is often difficult to determine the minimal polynomial $M(x)$ of a sequence h_n, and therefore the order of the sequence is not known exactly. However, if $C(E)h_n = 0$ for some annihilating operator $C(E)$, then the order of h_n is bounded by the degree of $C(x)$. Frequently, this is all the information we need. The following elementary observation shows why this is the case.

Theorem 7.4 Suppose that $\mathcal{O}(h_n) \leq k$ and $h_0 = h_1 = \cdots = h_{k-1} = 0$. Then $h_n = 0$ for all $n \geq 0$.

Proof. The sequence has an annihilating polynomial of degree no more than k, so if the initial k terms of the sequence are 0 the sequence is the identically 0 sequence. ■

Example 7.5 Let

$$s_n = \sum_j \binom{n-j}{j}, \quad n \geq 0.$$

Prove that $s_n = F_{n+1}$, the $(n+1)$st Fibonacci number.

Solution. Using Pascal's identity for the binomial coefficients, we see that

$$\begin{aligned} E^2 s_n &= \sum_j \binom{n+2-j}{j} = \sum_j \binom{n+1-j}{j} + \sum_j \binom{n+1-j}{j-1} \\ &= \sum_j \binom{n+1-j}{j} + \sum_j \binom{n-j}{j} = Es_n + s_n \end{aligned}$$

This shows that the sequences s_n and F_{n+1} are both annihilated by the Fibonacci operator $C(E) = E^2 - E - 1$, and therefore so is the sequence $h_n = s_n - F_{n+1}$. Since $\mathcal{O}(h_n) \leq \deg C(x) = 2$ and

$$h_0 = s_0 - F_1 = \binom{0}{0} - 1 = 1 - 1 = 0, \; h_1 = s_1 - F_2 = \binom{1}{0} - 1 = 1 - 1 = 0$$

this proves that $h_n = s_n - F_{n+1} = 0$ for all $n \geq 0$. ■

7.2.3 The Ring of Recurrence Sequences

We have already noted that the sum $g_n + h_n$ of two recurrence sequences is another recurrence sequence. It may be a little surprising to learn that the product $g_n h_n$ of two recurrence sequences is also a recurrence sequence. This result is most easily seen by observing that the product of two generalized power sums is again a GPS, and we know that any GPS is a recurrence sequence. Here is an example.

Example 7.6 Let P_n and F_n be, respectively, the Pell and Fibonacci sequences. Show that their product $P_n F_n$ is a recurrence sequence by finding an annihilating polynomial.

Solution. The spectrum of the Pell operator $C_P(E) = E^2 - 2E - 1$ is $\mathcal{E}_P = \{\alpha, \hat{\alpha}\} = \{1 + \sqrt{2}, 1 - \sqrt{2}\}$ and the spectrum of the Fibonacci operator $C_F(E) = E^2 - E - 1$ is $\mathcal{E}_F = \{\varphi, \hat{\varphi}\} = \{(1 + \sqrt{5})/2, (1 - \sqrt{5})/2\}$. The products of the eigenvalues that appear in the product of their GPSs are therefore $\mathcal{E}_{PF} = \{\lambda_1, \lambda_2, \lambda_3, \lambda_4\} = \{\alpha\varphi, \hat{\alpha}\varphi, \alpha\hat{\varphi}, \hat{\alpha}\hat{\varphi}\}$. Therefore, $P_n F_n = c_1\lambda_1^n + c_2\lambda_2^n + c_3\lambda_3^n + c_4\lambda_4^n$, where the constants c_1, c_2, c_3, c_4 can be determined by the initial conditions $P_0F_0 = 0, P_1F_1 = 1, P_2F_2 = 2, P_3F_3 = 5 \cdot 2 = 10$. The sequence P_nF_n is a solution of the fourth-order recurrence relation with the annihilating operator

$$
\begin{aligned}
C_{PF}(E) &= (E - \lambda_1)(E - \lambda_2)(E - \lambda_3)(E - \lambda_4) \\
&= E^4 - (\lambda_1 + \lambda_2 + \lambda_3 + \lambda_4)E^3 + (\lambda_1\lambda_2 + \lambda_1\lambda_3 + \lambda_1\lambda_4 + \lambda_2\lambda_3 + \lambda_2\lambda_4 \\
&\quad + \lambda_3\lambda_4)E^2 - (\lambda_1\lambda_2\lambda_3 + \lambda_1\lambda_2\lambda_4 + \lambda_1\lambda_3\lambda_4 + \lambda_2\lambda_3\lambda_4)E + \lambda_1\lambda_2\lambda_3\lambda_4.
\end{aligned}
$$

The coefficients of the operator can be readily calculated; for example

$$
\begin{aligned}
\lambda_1 + \lambda_2 + \lambda_3 + \lambda_4 &= \alpha\varphi + \hat{\alpha}\varphi + \alpha\hat{\varphi} + \hat{\alpha}\hat{\varphi} = (\alpha + \hat{\alpha})(\varphi + \hat{\varphi}) = 2 \cdot 1 = 2, \\
\lambda_1\lambda_2\lambda_3\lambda_4 &= (\alpha\varphi)(\hat{\alpha}\varphi)(\alpha\hat{\varphi})(\hat{\alpha}\hat{\varphi}) = (\alpha\hat{\alpha})^2(\varphi\hat{\varphi})^2 = (-1)^2(-1)^2 = 1
\end{aligned}
$$

We find that

$$
C_{PF}(E) = E^4 - 2E^3 - 7E^2 - 2E + 1
$$

so $h_n = P_nF_n$ satisfies the recurrence relation $h_{n+4} - 2h_{n+3} - 7h_{n+2} - 2h_{n+1} + h_n = 0$. Note that $C_{PF} \neq C_P C_F$. ■

Multiplication in the vector space $\mathcal{G}$ of generalized power sums is closed, commutative, and associative, and distributes over the addition operation [i.e., if f_n, g_n, and h_n are each a GPS, then so is $f_n(g_n + h_n) = f_n g_n + f_n h_n$]. Moreover, the constant sequence 1^n is a GPS and is a multiplicative identity. This gives us the following result.

Theorem 7.7 The set $\mathcal{G}$, and hence $S_{\mathcal{R}}$, is a commutative ring.

7.2.4 Upper Bounds for the Order of a Recurrence Sequence

If we are given a sequence $h_n \in V_S$ with the GPS $h_n = \sum_{i=1}^{m} p_i(n)\alpha_i^n, \deg p_i = r_i - 1$, then $M(x) = (x-\alpha_1)^{r_1}(x-\alpha_2)^{r_2} \cdots (x-\alpha_m)^{r_m}$ is its minimal polynomial. This is because any polynomial operator without the factor $(E-\alpha_i)^{r_i}$ will not annihilate h_n. If we let $\mathcal{O}(h_n)$ denote the order of the sequence h_n, we see that $\mathcal{O}(h_n) = r_1 + r_2 + \cdots + r_m = k$. The spectrum of the sequence is $\mathcal{E}_M = \{r_1 \cdot \alpha_1, r_2 \cdot \alpha_2, \ldots, r_m \cdot \alpha_m\}$, where r_i is the multiplicity of eigenvalue α_i.

Example 7.8 Determine the minimal polynomial $M(x)$, the order, and the spectrum of each of these sequences:

(a) $h_n = 2^n + 3(-1)^n + 4n^2(-1)^n$
(b) $h_n = 2 \cdot 5^n + 5n^3$

Solution.

(a) $M(x) = (x-2)(x+1)^3$, $\mathcal{O}(h_n) = 1 + 3 = 4$, $\mathcal{E}_M = \{2, -1, -1, -1\}$
(b) $M(x) = (x-5)(x-1)^4$, $\mathcal{O}(h_n) = 1 + 4 = 5$, $\mathcal{E}_M = \{5, 1, 1, 1, 1\}$

■

The theorems that follow derive upper bounds for the order of a sequence that is a sum, product, power, or subsequence of a given recurrence sequence or sequences. The value of these results will be seen in Section 7.4, where it is shown how identities for a recurrence sequence can be proved if the order of the sequence is known or estimated by an upper bound.

We begin with an example that will help us understand how to estimate the order of a sequence that is either a sum or product of two sequences.

Example 7.9 Consider the following recurrence sequences, given together with their order and spectra:

$$f_n = n^2 2^n + 3^n,\ \mathcal{O}(f_n) = 4,\ \mathcal{E}_f = \{2, 2, 2, 3\}$$
$$g_n = 6^n - 5 \cdot 4^n,\ \mathcal{O}(g_n) = 2,\ \mathcal{E}_g = \{6, 4\} \quad .$$
$$h_n = 5^n - 3^n + 7,\ \mathcal{O}(h_n) = 3,\ \mathcal{E}_h = \{5, 3, 1\}$$

Determine the order of the sequences (a) $f_n + g_n$ (b) $f_n + h_n$, and (c) $f_n g_n$.

Solution.

$$f_n + g_n = n^2 2^n + 3^n + 6^n - 5 \cdot 4^n,\ \mathcal{O}(f_n + g_n) = 6,\ \mathcal{E}_{f+g} = \{2, 2, 2, 3, 6, 4\}$$
$$f_n + h_n = n^2 2^n + 5^n + 7,\ \mathcal{O}(f_n + h_n) = 5,\ \mathcal{E}_{f+h} = \{2, 2, 2, 5, 1\}$$
$$\begin{aligned} f_n g_n &= n^2 2^n \cdot 6^n - 5n^2 \cdot 2^n \cdot 4^n + 3^n \cdot 6^n - 5 \cdot 3^n \cdot 4^n \\ &= (n^2 - 5) \cdot 12^n - 5n^2 \cdot 8^n + 18^n,\ \mathcal{O}(f_n g_n) = 7,\ \mathcal{E}_{fh} = \{12, 12, 12, 8, 8, 8, 18\} \end{aligned}$$

We note that

$$6 = \mathcal{O}(f_n + h_n) < \mathcal{O}(f_n) + \mathcal{O}(h_n) = 4 + 3 \text{ and } 7 = \mathcal{O}(f_n g_n) < \mathcal{O}(f_n) \cdot \mathcal{O}(g_n) = 4 \cdot 2$$

■

Example 7.9 suggests that the order of a sum or product of recurrence sequences is bounded, respectively, by the sum or product of their orders. The following theorem affirms this assertion.

Theorem 7.10 Let g_n and h_n be recurrence sequences and c_1, and c_2 be constants. Then

$$\mathcal{O}(c_1 g_n + c_2 h_n) \leq \mathcal{O}(g_n) + \mathcal{O}(h_n) \tag{7.14}$$

$$\mathcal{O}(g_n h_n) \leq \mathcal{O}(g_n)\mathcal{O}(h_n). \tag{7.15}$$

Proof. Just note that $\mathcal{E}_{g+h} \subseteq \mathcal{E}_g \cup \mathcal{E}_h$ and $\mathcal{E}_{gh} \subseteq \mathcal{E}_g \otimes \mathcal{E}_h$, where $\otimes$ denotes the multiset of pairwise products of the elements. ■

If $\mathcal{O}(h_n) = k$, then it follows by induction from Theorem 7.10 that $\mathcal{O}(h_n^t) \leq k^t$. However, a much better estimate can be obtained by noting that if the GPS representing h_n is the k-fold sum $h_n = x_1 + \cdots + x_k$, where each x_i represents a term in the GPS, then

$$h_n^t = (x_1 + \cdots + x_k)^t = \sum_{\substack{j_1+\cdots+j_k=t \\ j_i \geq 0}} x_1^{j_1} \cdots x_k^{j_k}.$$

Since there are $\left(\binom{k}{t}\right) = \binom{t+k-1}{t}$ terms in this sum, we get the following improved estimate.

Theorem 7.11 If $\mathcal{O}(h_n) = k$, then $\mathcal{O}(h_n^t) \leq \binom{k+t-1}{t}$.

Example 7.12 Determine the spectrum and order of the sequence F_n^3, where $F_n = (\varphi^n - \hat{\varphi}^n)/\sqrt{5}$ is the nth Fibonacci number.

Solution. First note that $F_n^3 = [(\varphi^n - \hat{\varphi}^n)/\sqrt{5}]^3 = (1/5\sqrt{5})(\varphi^{3n} - 3\varphi^{2n}\hat{\varphi}^n + 3\varphi^n\hat{\varphi}^{2n} + \hat{\varphi}^{3n})$. Therefore $\mathcal{E}_{F_n^3} = \{\varphi^3, \hat{\varphi}^3, \varphi^2\hat{\varphi}, \varphi\hat{\varphi}^2\} = \{\varphi^3, \hat{\varphi}^3, -\varphi, -\hat{\varphi}\}$, so $\mathcal{O}(F_n^3) = 4$. We see that $\mathcal{O}(F_n^3) = 4 = \binom{2+3-1}{3} < 8 = 2^3 = (\mathcal{O}(F_n))^3$, illustrating the result of Theorem 7.11. ■

If $h_n = \sum_{i=1}^{m} p_i(n)\alpha_i^n$, $\deg p_i = r_i - 1$, is a GPS with the spectrum $\{r_1 \cdot \alpha_1, r_2 \cdot \alpha_2, \ldots, r_m \cdot \alpha_m\}$, then its subsequence $h_{ns+t} = \sum_{i=1}^{m} p_i(ns+t)\alpha_i^{ns+t} = \sum_{i=1}^{m} \hat{p}_i(n)(\alpha_i^s)^n$, where

$\hat{p}_i(n) = p_i(ns+t)\alpha_i^t$, is a GPS with eigenvalues that are a subset of the multiset $\{r_1 \cdot \alpha_1^s, r_2 \cdot \alpha_2^s, \dots, r_m \cdot \alpha_m^s\}$. This observation gives us this result.

Theorem 7.13 Let h_n be a recurrence sequence of order $\mathcal{O}(h_n) = k$, with the spectrum $\mathcal{E}_h = \{r_1 \cdot \alpha_1, r_2 \cdot \alpha_2, \dots, r_m \cdot \alpha_m\}$. Then the spectrum $\mathcal{E}_g$ of the subsequence $g_n = h_{ns+t}$ is a subset of the multiset $\{r_1 \cdot \alpha_1^s, r_2 \cdot \alpha_2^s, \dots, r_m \cdot \alpha_m^s\}$ and $\mathcal{O}(h_{ns+t}) \le r_1 + r_2 + \cdots + r_m = k$.

Example 7.14 For each sequence h_n and each subsequence g_n below, find the spectra $\mathcal{E}_h$ and $\mathcal{E}_g$, determine the orders $\mathcal{O}(h_n)$ and $\mathcal{O}(g_n)$, and express g_n as a GPS.

(a) $h_n = 2^n + (-2)^n + 3^n$, $g_n = h_{2n}$

(b) $h_n = F_n, g_n = F_{3n}$, where $F_n = \frac{\varphi^n - \hat{\varphi}^n}{\sqrt{5}}$ is the nth Fibonacci number

(c) $h_n = F_n, g_n = F_{ns}$

Solution.

(a) $\mathcal{E}_h = \{2, -2, 3\}, \mathcal{O}(h_n) = 3, \mathcal{E}_g = \{2^2, (-2)^2, 3^2\} = \{4, 9\}, \mathcal{O}(g_n) = 2,$
$g_n = h_{2n} = 2^{2n} + (-2)^{2n} + 3^{2n} = 2 \cdot 4^n + 9^n$.

(b) $\mathcal{E}_{F_n} = \{\varphi, \hat{\varphi}\}, \mathcal{O}(h_n) = 2, \mathcal{E}_{F_{3n}} = \{\varphi^3, \hat{\varphi}^3\}, \mathcal{O}(F_{3n}) = 2,\ g_n = F_{3n} = \dfrac{\varphi^{3n} - \hat{\varphi}^{3n}}{\sqrt{5}}$

(c) $\mathcal{E}_{F_n} = \{\varphi, \hat{\varphi}\}, \mathcal{O}(h_n) = 2, \mathcal{E}_{F_{ns}} = \{\varphi^s, \hat{\varphi}^s\}, \mathcal{O}(F_{ns}) = 2, g_n = F_{ns} = \dfrac{\varphi^{sn} - \hat{\varphi}^{sn}}{\sqrt{5}}$

■

7.2.5 Bases and Representations

The solutions of a homogeneous linear recurrence relation $C(E)h_n = 0$ form a vector space V_C of dimension k. A unique recurrence sequence $h_n^{(A)} \in V_C$ is determined by imposing a set of k initial conditions, giving us the correspondence

$$h_n^{(A)} \leftrightarrow [A_0, A_1, \dots, A_{k-1}] \quad \text{where} \quad h_0 = A_0, h_1 = A_1, \dots, h_{k-1} = A_{k-1} \tag{7.16}$$

which matches the sequence $h_n^{(A)}$ to the initial values $A = [A_0, A_1, \dots, A_{k-1}]$. A basis $\{u_n^{(1)}, u_n^{(2)}, \dots, u_n^{(k)}\}$ of V_C therefore corresponds to a $k \times k$ matrix of initial conditions

$$A = \begin{bmatrix} A_0^1 & A_1^1 & \cdots & A_{k-1}^1 \\ A_0^2 & A_1^2 & \cdots & A_{k-1}^2 \\ \vdots & \vdots & \ddots & \vdots \\ A_0^k & A_1^k & \cdots & A_{k-1}^k \end{bmatrix} \tag{7.17}$$

where $u_n^{(i)} \leftrightarrow u_n^{(A^i)}$, for which $A^i = \left[A_0^i, A_1^i, \ldots, A_{k-1}^i\right]$ denotes the ith row of A, $i = 1, 2, \ldots, k$. The matrix A is nonsingular since any linear dependence of rows would correspond to a linear dependence of the sequences in the set $\left\{u_n^{(1)}, u_n^{(2)}, \ldots, u_n^{(k)}\right\}$.

Example 7.15 Let $C(E) = E^2 - E - 1$ be the Fibonacci recurrence. Determine the basis $\{u_n, v_n\}$ that corresponds to the given 2×2 matrix of initial conditions:

(a) $\begin{bmatrix} 0 & 1 \\ 1 & 1 \end{bmatrix}$ (b) $\begin{bmatrix} 2 & 1 \\ 1 & 3 \end{bmatrix}$ (c) $\begin{bmatrix} 1 & \varphi \\ 1 & \hat{\varphi} \end{bmatrix}$ (d) $\begin{bmatrix} 1 & 0 \\ 0 & 1 \end{bmatrix}$ (e) $\begin{bmatrix} 3 & 5 \\ 8 & 13 \end{bmatrix}$ (f) $\begin{bmatrix} 0 & 2 \\ \varphi^2 & \varphi^3 \end{bmatrix}$

Solution.

(a) $\{F_n, F_{n+1}\}$ (b) $\{L_n, L_{n+1}\}$ (c) $\{\varphi^n, \hat{\varphi}^n\}$ (d) $\{F_{n-1}, F_n\}$, where $F_{-1} = 1$

(e) $\{F_{n+4}, F_{n+6}\}$ (f) $\{2F_n, \varphi^{n+2}\}$

■

Example 7.15 illustrates why there is considerable freedom in how a solution of the Fibonacci recurrence—indeed, any recurrence—can be represented. The sequences F_nL_n, F_{2n}, $[(L_{n-1}L_n + L_nL_{n+1})/5]$, and $F_n^2 + 2F_{n-1}F_n$ look different from one another, but each is a representation of the same recurrence sequence 0,1,3,8, ...

A convenient basis for any second-order recurrence relation $h_{n+2} = ah_{n+1} + bh_n$ is the pair of *Lucas sequences*. First assume that the eigenvalues $\alpha = (a + \sqrt{a^2 - 4b})/2$ and $\beta = (a - \sqrt{a^2 - 4b})/2$ are distinct. When this is the case, the discriminant $D = a^2 - 4b \neq 0$, $\alpha + \beta = a$, and $\alpha - \beta = \sqrt{D}$. The initial conditions matrix $\begin{bmatrix} 0 & 1 \\ 2 & a \end{bmatrix}$ then generates the basis $\{u_n, v_n\}$ where $u_n = (\alpha^n - \beta^n)/\sqrt{D}$ is known as the *Lucas sequence of the first kind* and $v_n = \alpha^n + \beta^n$ is the *Lucas sequence of the second kind.* In the case of a repeated eigenvalue, so that $h_{n+2} = ah_{n+1} + (a^2/4)h_n$, we see that $D = 0$ and $\alpha = a/2$. The Lucas sequences are $u_n = n\alpha^{n-1}$ and $v_n = 2\alpha^n$, again with the initial conditions matrix $\begin{bmatrix} 0 & 1 \\ 2 & a \end{bmatrix}$.

Any sequence h_n that solves $h_{n+2} = ah_{n+1} + bh_n$ can be written as

$$h_n = \frac{h_0}{2} v_n + \left(h_1 - \frac{h_0 a}{2}\right) u_n \tag{7.18}$$

Example 7.16 Calculate the Lucas sequences for the recursion relation $h_{n+2} = 3h_{n+1} - h_n$, and use this basis to solve the recurrence relation with the initial conditions $h_0 = 2, h_1 = 1$.

Solution. The eigenvalues are $\alpha = (3 + \sqrt{9-4})/2 = 1 + (1+\sqrt{5})/2 = 1 + \varphi = \varphi^2$ and $\beta = 1/\alpha = 1/\varphi^2 = \hat{\varphi}^2$. Therefore

$$u_n = \frac{\left(\varphi^2\right)^n - \left(\hat{\varphi}^2\right)^n}{\sqrt{5}} = \frac{\varphi^{2n} - \hat{\varphi}^{2n}}{\sqrt{5}} = F_{2n}$$

$$v_n = \left(\varphi^2\right)^n + \left(\hat{\varphi}^2\right)^n = \varphi^{2n} + \hat{\varphi}^{2n} = L_{2n}$$

Using equation (7.18), we see that $h_n = \frac{2}{2}L_{2n} + [1 - (2 \cdot 3)/2]F_{2n} = L_{2n} - 2F_{2n}$. ■

A recursion relation of order k likewise has considerable freedom in the choice of a basis. If the recursion is solved by the sequence h_n of order k, then the k sequences $u_n^{(j)} = h_{n+j}$, $j = 0, 1, \ldots, k-1$ are a basis; they are linearly independent since otherwise the sequence h_n would satisfy a linear recurrence relation of order less than k, a contradiction. Therefore, any solution of the recursion has the form $\sum_{j=0}^{k-1} c_j h_{n+j}$.

PROBLEMS

In problems 1 through 6, let

$$t_{n+2} = 5t_{n+1} - 6t_n;\ t_0 = 0, t_1 = 1$$
$$u_{n+2} = 3u_{n+1} - 2u_n;\ u_0 = 0, u_1 = 1$$
$$v_{n+3} = 3v_{n+2} + 4v_{n+1} - 12v_n;\ v_0 = 1, v_1 = -1, v_2 = 9$$

7.2.1. Find the spectrum (multiset of eigenvalues) of each of these sequences:

(a) t_n **(b)** u_n **(c)** v_n

7.2.2. Find the GPS for these sequences: **(a)** t_n **(b)** u_n **(c)** v_n

7.2.3. Find the spectrum, GPS, and order of these sequences:

(a) t_{2n} **(b)** u_{3n} **(c)** t_n^2 **(d)** u_{2n}^2

7.2.4. Calculate the spectrum, GPS, and order of v_{2n}. What bound for the order is given by Theorem 7.13?

7.2.5. **(a)** What upper bounds for the orders of v_n^2 and $t_{2n} + v_n^2$ are given by Theorems 7.10, 7.11, and 7.13?

(b) What are the spectra of v_n^2 and $t_{2n} + v_n^2$? What bounds do these multisets give for the order of the sequences?

7.2.6. Find the spectrum and GPS of the sequence $t_n u_n$.

7.2.7. Find the spectrum and order of the sequence $h_n = 2^n + n^2$.

7.2.8. Explain why each of these sequences is or is not a recurrence sequence:

(a) $(-3)^n$ **(b)** $\frac{1}{n+1}$ **(c)** C_n, the sequence of Catalan numbers **(d)** $n^4 F_{3n}^5$

7.2.9. Let $C(E) = \sum_{j=0}^{k} a_j E^{k-j}$ annihilate the sequence h_n and have the spectrum $E_C = \{r_1 \cdot \alpha_1, r_2 \cdot \alpha_2, \ldots, r_m \cdot \alpha_m\}$.

(a) For any $\beta \neq 0$, show that the operator $C_\beta(E) = \beta^k \sum_{j=0}^{k} a_j((E/\beta))^{k-j}$ annihilates the sequence $\beta^n h_n$.

(b) What is the spectrum of the sequence $\beta^n h_n$?

7.2.10. The set $B_1 = \{\alpha^n, n\alpha^n, n^2\alpha^n, \ldots, n^{r-1}\alpha^n\}$ is a basis for the null space V_C of the operator $C(E) = (E - \alpha)^r$.

(a) Explain why

$$B_2 = \left\{\alpha^n, \binom{n}{1}\alpha^{n-1}, \binom{n}{2}\alpha^{n-2}, \ldots, \binom{n}{r-1}\alpha^{n-(r-1)}\right\}$$

is another basis of V_C.

(b) Determine the change of basis matrix in terms of Stirling numbers $\left\{ {i \atop j} \right\}$ of the second kind.

7.2.11. Show that the generating function of the sequence $P_n F_n$ discussed in Example 7.6 is

$$f_{PF}(x) = \frac{x(1-x^2)}{1 - 2x - 7x^2 - 2x^3 + x^4}$$

7.2.12. If u_n is a second-order recurrence sequence with eigenvalues α and β, $\alpha \neq \beta$, prove that u_n^t satisfies a recurrence relation of order $t + 1$.

7.2.13. Given $u_n = (c_0 + c_1 n)\alpha^n$, $c_1 \neq 0$, determine a homogeneous recurrence relation of order $t + 1$ that is satisfied by $u_n^t = (c_0 + c_1 n)^t \alpha^{nt}$.

7.3 NONHOMOGENEOUS RECURRENCES AND SYSTEMS OF RECURRENCES

Combinatorial problems are sometimes best approached by determining either a nonhomogeneous linear recurrence or a coupled system of linear recurrences. In this section we will see that both types of recurrence relations can be replaced with a single homogeneous recurrence relation. Thus, the combinatorial problem can be solved by applying the methods described in the preceding section.

7.3.1 Nonhomogeneous Recurrences

A sequence h_n is a *nonhomogeneous recurrence* if there is a polynomial $C(x)$ for which

$$C(E)h_n = q_n \tag{7.19}$$

for some recurrence sequence q_n. Since we are assuming that $q_n \in \mathcal{S}_{\mathcal{R}}$, there is an annihilating polynomial $P(x)$ for which

$$P(E)q_n = 0 \tag{7.20}$$

Therefore

$$P(E)C(E)h_n = P(E)q_n = 0 \tag{7.21}$$

so that $P(E)C(E)$ is an annihilating polynomial operator for h_n.

Considerable information follows from equation (7.21). First, we see that h_n is a recurrence sequence with the characteristic equation $P(x)C(x) = 0$. Therefore $h_n \in \mathcal{S}_{\mathcal{R}}$ and $\mathcal{O}(h_n) \leq \deg P + \deg C$. Moreover, h_n is a generalized power sum that can be decomposed into a sum of two sequences

$$h_n = g_n + p_n \tag{7.22}$$

In this decomposition, the first sequence $g_n \in V_C$, known as the *general solution,* is a GPS of the terms in the spectrum $\mathcal{E}_C$ of the eigenvalues given by $C(x) = 0$. We see that the general solution g_n satisfies the homogeneous *auxiliary* recurrence relation $C(E)g_n = 0$ associated with (7.19).

The second sequence $p_n = h_n - g_n$ is known as a *particular solution.* We see that p_n satisfies the nonhomogeneous recurrence since

$$C(E)p_n = C(E)(h_n - g_n) = C(E)h_n - C(E)g_n = q_n - 0 = q_n$$

Evidently, the terms of the GPS forming the particular solution p_n must include terms in h_n that are not present in g_n. These must be terms that involve eigenvalues in $\mathcal{E}_{PC}$ that are not in the spectrum $\mathcal{E}_C$. There are two ways for these new terms to occur. First, if $P(\alpha) = 0$ but $C(\alpha) \neq 0$, so that $\alpha \in \mathcal{E}_{PC}$ but $\alpha \notin \mathcal{E}_C$ then the GPS of p_n will include terms involving α even though terms in the GPS of g_n do not involve α. The second possibility is if $P(\beta) = 0$ and $C(\beta) = 0$, so that β is an eigenvalue of $\mathcal{E}_{PC}$ with a greater multiplicity than in has in $\mathcal{E}_C$, In this case, the GPS of p_n will involve higher powers of n than can occur in the GPS of g_n.

It will now be shown that the particular solution p_n is unique when it is written as a GPS of terms only from the spectrum of $\mathcal{E}_{PC}$ that to not belong to V_C. Suppose that $p_n$and $\hat{p}_n$ are both particular solutions of the nonhomogeneous recurrence relation, neither of whose GPS representations involve a term in V_C. Then

$$C(E)(\hat{p}_n - p_n) = C(E)\hat{p}_n - C(E)p_n = q_n - q_n = 0$$

which shows that $\hat{p}_n - p_n \in V_C$. But the only sequence in V_C that is not a GPS of $\mathcal{E}_C$ is the all-zero sequence, and thus $\hat{p}_n = p_n$.

We have now shown that the set of solutions of the nonhomogenous recurrence $C(E)h_n = q_n$ form an *affine space*, namely, the set

$$V_C + p_n = \{h_n = g_n + p_n | C(E)g_n = 0 \quad \text{and} \quad C(E)p_n = q_n\} \tag{7.23}$$

If the nonhomogeneous recurrence $C(E)h_n = q_n$ of order k is accompanied by k initial conditions $h_0 = A_0, h_1 = A_1, \ldots, h_{k-1} = A_{k-1}$, then the final solution is given by $h_n = g_n + p_n$, where p_n is a particular solution and $g_n \in V_C$ is the unique solution of

$$C(E)g_n = 0, g_0 = A_0 - p_0, g_1 = A_1 - p_1, \ldots, g_{k-1} = A_{k-1} - p_{k-1} \tag{7.24}$$

The following theorem summarizes the observations made above.

Theorem 7.17 Let q_n be a recurrence sequence annihilated by the polynomial operator $P(E)$. Then any solution of the nonhomogeneous recurrence relation $C(E)h_n = q_n$ has the form $h_n = g_n + p_n$, where p_n is a particular solution that can be represented as generalized power sum of the spectrum $\mathcal{E}_{PC}$ and g_n is a solution of the auxiliary homogeneous recurrence $C(E)g_n = 0$. If the recurrence relation has order k, then the nonhomogeneous recurrence is uniquely solved for any set of initial conditions $h_0 = A_0, h_1 = A_1, \ldots, h_{k-1} = A_{k-1}$.

Example 7.18 Determine the affine space of solutions of the nonhomogeneous recursion relation

$$h_n = h_{n-1} + 2h_{n-2} + (n+2)(-1)^n, \quad n \geq 2 \tag{7.25}$$

and then solve the recurrence with the initial conditions $h_0 = 0$ and $h_1 = 1$.

Solution. In operator form, the equation is $C(E)h_n = (n+2)(-1)^n$, where $C(E) = E^2 - E - 2 = (E-2)(E+1)$. The nonhomogeneous term $q_n = (n+2)(-1)^n$ is annihilated by $P(E) = (E+1)^2$. Since $C(E)P(E) = (E-2)(E+1)^3$ annihilates h_n, we know that h_n is a GPS of the spectrum $\{2, -1, -1, -1\}$, which means that h_n has the form $h_n = c_1 2^n + c_2(-1)^n + c_3 n(-1)^n + c_4 n^2(-1)^n$. We see that $h_n = g_n + p_n$, where $g_n = c_1 2^n + c_2(-1)^n$ is the general solution of the associated homogeneous recurrence relation $C(E)g_n = 0$ and $p_n = c_3 n(-1)^n + c_4 n^2(-1)^n$ is the form of a particular solution of the nonhomogeneous recurrence. A short calculation shows that $(n+2)(-1)^n = q_n = C(E)p_n = [(3c_3 + 5c_4) + 6nc_4](-1)^n$, and solving for c_3 and c_4 we find that $c_4 = \frac{1}{6}$ and $c_3 = (2 - 5c_4)/3 = \frac{7}{18}$. Therefore, a particular solution is $p_n = [(7+3n)/18]n(-1)^n$. The solutions of (7.25) form the affine space

$$\left\{h_n = g_n + p_n | g_n = c_1 2^n + c_2(-1)^n, p_n = \frac{(7+3n)}{18} n(-1)^n\right\}$$

Since

$$0 = h_0 = c_1 2^0 + c_2(-1)^0 + p_0 = c_1 + c_2 + 0$$
$$1 = h_1 = c_1 2^1 + c_2(-1)^1 + p_1 = 2c_1 - c_2 - \frac{5}{9}$$

we find that $c_1 = -c_2 = \frac{14}{27}$. Therefore the recursion relation with the initial conditions $h_0 = 0$ and $h_1 = 1$ is solved by

$$h_n = \frac{14}{27}(2^n - (-1)^n) + \frac{(7+3n)}{18}n(-1)^n$$

■

Example 7.19 Let h_n denote the number of regions determined by the diagonals of a convex polygon with $n + 2$ sides, where at most two diagonals intersect at a common interior point of the polygon. For example, the diagram below shows that the diagonals of a convex pentagon form $h_3 = 11$ regions:

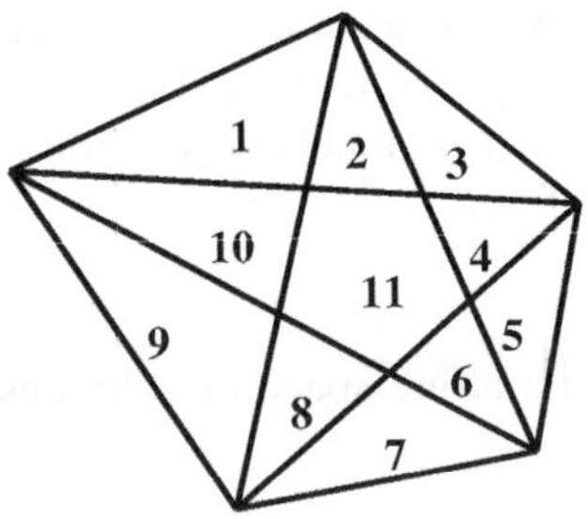

(a) Define $h_0 = 0$ and then show that

$$h_n = h_{n-1} + \binom{n+1}{3} + n, \quad n \geq 1 \tag{7.26}$$

(b) Solve the recurrence (7.26) using Theorem 7.17, showing that

$$h_n = \binom{n+1}{2} + \binom{n+1}{3} + \binom{n+1}{4} \tag{7.27}$$

which is the sum of three successive terms in row $n + 1$ of Pascal's triangle.

Solution.

(a) The following diagram shows a convex polygon with the $n + 2$ vertices $v_1, v_2, \ldots, v_{n+1}, v_{n+2}$, so its diagonals form h_n regions within the polygon:

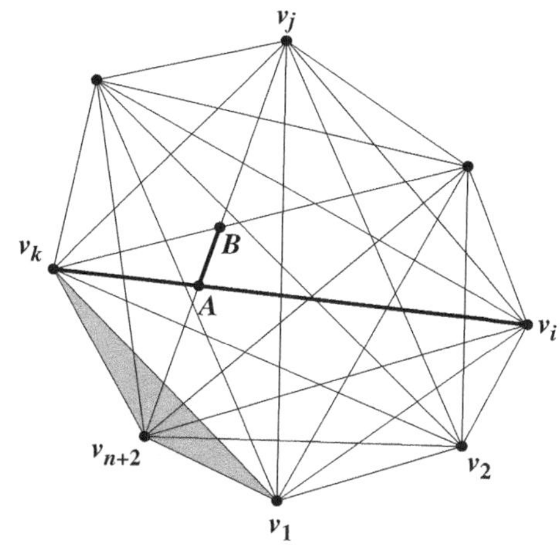

Now delete vertex v_{n+2}, which leaves a convex polygon $v_1, v_2, \ldots, v_{n+1}$ partitioned into h_{n-1} regions by its diagonals. We now count the number of regions that have been removed. The diagram shows that n shaded regions with a vertex at v_{n+2} are deleted. Also, each point of intersection A at which the diagonal $\overline{v_{n+2}v_j}$ is intersected by the diagonal $\overline{v_iv_k}$ decreases the number the regions by one since this corresponds to the deletion of some segment $\overline{AB}$ along the diagonal $\overline{v_{n+2}v_j}$. Since there are $\binom{n+1}{3}$ choices of i, j, and k, altogether the deletion of vertex v_{n+2} decreases the number of regions by $\binom{n+1}{3} + n$, which proves recursion relation (7.26)

(b) The recursion relation can then be written as $C(E)h_n = q_n$, where $C(E) = (E - 1)$ and $q_n = \binom{n+2}{3} + n + 1$. Since q_n is a polynomial in n of degree 3, it is annihilated by the operator $P(E) = (E - 1)^4$. Therefore, $(E - 1)^5 h_n = P(E)C(E)h_n = P(E)q_n = 0$, so h_n is a polynomial of degree 4, say, $h_n = c_0 + c_1\binom{n}{1} + c_2\binom{n}{2} + c_3\binom{n}{3} + c_4\binom{n}{4}$. Using the recursion relation (7.26), we have

$$h_0 = 0, h_1 = 1, h_2 = 4, h_3 = 11, h_4 = 25 \tag{7.28}$$

The initial conditions can then be used to determine the constants c_0, c_1, c_2, c_3, c_4. For example, $0 = h_0 = c_0$ and $1 = h_1 = c_1\binom{1}{1} = c_1$. Continuing in this way, we find that $h_n = \binom{n}{1} + 2\binom{n}{2} + 2\binom{n}{3} + \binom{n}{4}$, which is equivalent to (7.27) by repeated use of Pascal's identity. ■

It should be noted that the special form of the recursion relation allows it to be solved by a simple calculation that depends on the hockey stick identities. Indeed, using telescoping series, we have

$$h_n = \sum_{k=1}^{n} (h_k - h_{k-1}) = \sum_{k=1}^{n} \binom{n}{1} + \sum_{k=1}^{n} \binom{n+1}{3} = \binom{n+1}{2} + \binom{n+2}{4}$$

which is equivalent to (7.27). A straightforward calculation shows that another expression for h_n is

$$h_n = \frac{n(n+1)(1+n+n^2)}{24}$$

7.3.2 Systems of Linear Recursion Relations

The next theorem shows that a system of homogeneous linear recursion relations with constant coefficients can be solved by replacing the system with a single recurrence relation that is satisfied by each sequence of the system.

Theorem 7.20 Let the k sequences $u_n^{(1)}, u_n^{(2)}, \dots, u_n^{(k)}$ satisfy the system of k linear recurrence relations given by

$$\begin{aligned} P_{11}(E)u_n^{(1)} + P_{12}(E)u_n^{(2)} + \cdots + P_{1k}(E)u_n^{(k)} &= 0 \\ P_{21}(E)u_n^{(1)} + P_{22}(E)u_n^{(2)} + \cdots + P_{2k}(E)u_n^{(k)} &= 0 \\ \cdots \\ P_{k1}(E)u_n^{(1)} + P_{k2}(E)u_n^{(2)} + \cdots + P_{kk}(E)u_n^{(k)} &= 0 \end{aligned} \tag{7.29}$$

where $P_{ij}(E)$, $1 \le i, j \le k$ are polynomials. Then the operator $C(E) = \det[P_{ij}(E)]$ annihilates each of the sequences; that is, each sequence satisfies the same linear recurrence

$$C(E)u_n^{(j)} = 0, \; j = 1, 2, \dots, k \tag{7.30}$$

so that the determinant $C(x) = \det[P_{ij}(x)]$ is the characteristic polynomial of each sequence. The order of the sequences is no more than the degree of $C(x)$.

Proof. Recall that a determinant of a matrix is given by the sum of the products of the elements and corresponding cofactors of any row of a matrix. The sum of products of the elements in a row with the cofactors of a different row is zero, since this is the same as the determinant of a matrix with two identical rows. Thus, we have this general result about determinants: If c_{ij} is the cofactor of the element P_{ij} of the matrix $M = [P_{ij}]$, then

$$\sum_{r=1}^{k} c_{ri} P_{rj}(E) = \delta_{i,j} \det M(E) \tag{7.31}$$

We can apply this result to the matrix corresponding to the system shown in (7.29). By multiplying the rth equation $\sum_{j=1}^{k} P_{rj}(E)u_n^{(j)} = 0$ of the system by c_{rj} and summing over r, $1 \le r \le k$, we see that

$$\begin{aligned} 0 &= \sum_{r=1}^{k} c_{ri} \sum_{j=1}^{k} P_{rj}(E)u_n^{(j)} = \sum_{j=1}^{k}\left(\sum_{r=1}^{k} c_{ri}P_{rj}(E)\right)u_n^{(j)} = \sum_{j=1}^{k} \delta_{i,j}(\det M(E))u_n^{(j)} \\ &= (\det M(E))u_n^{(i)} \end{aligned}$$

for every i, $1 \le i \le k$. ■

The examples below illustrate how the theorem is applied.

Example 7.21 Tiling a Rectangular Board In how many ways can a $3 \times 3n$ rectangular board be tiled with horizontal 1×3 trominoes and vertical 2×1 dominos? For example, one of the tilings of a 3×9 board is shown by this diagram:

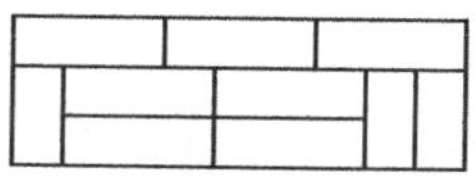

Solution. Although our main interest is to determine the sequence u_n that counts the number of tilings of the complete rectangular board, it is helpful to introduce the auxiliary sequences v_n and w_n that count the tilings of the pruned boards shown below. Therefore, let the sequences u_n, v_n, and w_n count the tilings of the boards shown in the following diagram:

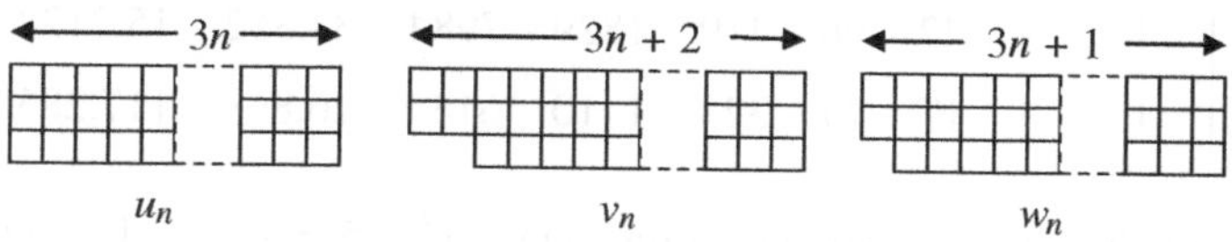

A system of recurrence relations for the sequences can be obtained by taking cases that depend on how the cells in the leftmost column are covered. For example, if a domino covers two cells of the left column of the rectangular board, the remaining cell is necessarily covered by a tromino and we are left with a trimmed board of the type counted by the sequence v_n. Proceeding in this way, we obtain the system of linear recurrence relations

$$u_{n+1} = u_n + 2v_n, v_{n+1} = w_{n+1} + v_n, w_{n+1} = u_{n+1} + w_n \tag{7.32}$$

One approach to solving this system is to substitute one equation into the other repeatedly, hoping to uncouple the system and obtain a higher-order recurrence for the sequence u_n. However, Theorem 7.20 above offers a more convenient approach. First, note that the system (7.32) can be written in the matrix operator form

$$\begin{bmatrix} E-1 & -2 & 0 \\ 0 & E-1 & -E \\ -E & 0 & E-1 \end{bmatrix} \begin{bmatrix} u_n \\ v_n \\ w_n \end{bmatrix} = \begin{bmatrix} 0 \\ 0 \\ 0 \end{bmatrix} \tag{7.33}$$

The determinant of this matrix is the operator $C(E) = E^3 - 5E^2 + 3E - 1$, so all three sequences u_n, v_n, and w_n are annihilated $C(E)$. That is, each sequence satisfies the *same* third-order linear recurrence relation, so that

$$\begin{aligned} u_{n+3} &= 5u_{n+2} - 3u_{n+1} + u_n,\ v_{n+3} = 5v_{n+2} - 3v_{n+1} + v_n, \\ w_{n+3} &= 5w_{n+2} - 3w_{n+1} + w_n. \end{aligned} \tag{7.34}$$

By a direct count, it is seen that $u_1 = 3, v_0 = 1$, and $w_0 = 1$. Using equations (7.32), we then find (in the following order) that

$$w_1 = u_1 + w_0 = 3 + 1 = 4, v_1 = w_1 + v_0 = 4 + 1 = 5, u_2 = u_1 + 2v_1 = 3 + 2 \cdot 5 = 13,$$
$$w_2 = u_2 + w_1 = 13 + 4 = 17, v_2 = w_2 + v_1 = 17 + 5 = 22$$

The equation $3 = u_1 = u_0 + 2v_0 = u_0 + 2$ shows that we should define $u_0 = 1$. This gives us the first three values of all three of the sequences. The following table is then generated using the recursion shown in equation (7.34) for each sequence:

	$n = 0$	1	2	3	4	5	6	7	8
u_n	1	3	13	57	249	1087	4745	20,713	90,417
v_n	1	5	22	96	419	1829	7984	34,852	15,2137
w_n	1	4	17	74	323	1410	6155	26,868	117,285

The roots of the characteristic equation $C(x) = x^3 - 5x^2 + 3x - 1 = 0$ are a real eigenvalue $\alpha = 4.365\ldots$ and the complex conjugate pair $\beta, \bar{\beta} = 0.317\ldots \pm i\,0\,358\ldots$. Each of our three sequences can then be written as a GPS of the form $c_1\alpha^n + c_2\beta^n + c_3\bar{\beta}^n$ where the constants c_1, c_2, and c_3 can be calculated from the initial conditions that correspond to the given sequence. ■

Example 7.22 (Pell–Tribonacci–Fibonacci Recurrence) Determine the sequence given by

$$g_{n+3} = \begin{cases} 2g_{n+2} + g_{n+1}, & n \equiv 0 \pmod 3 \\ g_{n+2} + g_{n+1} + g_n, & n \equiv 1 \pmod 3 \\ g_{n+2} + g_{n+1}, & n \equiv 2 \pmod 3 \end{cases}, \quad n \geq 0 \tag{7.35}$$

with the initial conditions

$$g_0 = 0, g_1 = 1 \tag{7.36}$$

Solution. Specifically, g_n is the sequence that cyclically interlaces the Pell–tribonacci–Fibonacci recurrence relations. If we let $u_n = g_{3n}, v_n = g_{3n+1}$, and $w_n = g_{3n+2}$, then

$$u_{n+1} = v_n + 2w_n, v_{n+1} = u_{n+1} + v_n + w_n, w_n = u_n + v_n, n \geq 0, \tag{7.37}$$

with the initial conditions $u_0 = 0$ and $v_0 = 1$. We also see from (7.37) that $w_0 = u_0 + v_0 = 0 + 1 = 1$.

By Theorem 7.20, all three sequences u_n, v_n, and w_n are annihilated by

$$C(E) = \det \begin{bmatrix} E & -1 & -2 \\ -E & E-1 & -1 \\ -1 & -1 & 1 \end{bmatrix} = E^2 - 7E + 1$$

This gives us the characteristic equation $C(x) = x^2 - 7x + 1$, which can be factored easily by noting that $7 = L_4 = \varphi^4 + \hat{\varphi}^4$ and $1 = (-1)^4 = (\varphi\hat{\varphi})^4$. Therefore, $C(x) = x^2 - 7x + 1 = (x - \varphi^4)(x - \hat{\varphi}^4)$, which means that each of the sequences u_n, v_n, and w_n is a GPS of the eigenvalues φ^4 and $\hat{\varphi}^4$. Equivalently, using the Binet formula for the Fibonacci numbers, each sequence is a sum of the form $c_1F_{4n} + c_2F_{4n+1}$. Since $u_0 = 0 = F_0$ and $u_1 = 2w_0 + v_0 = 2 \cdot 1 + 1 = 3 = F_4$, we see that $u_n = F_{4n}$. Similarly, $v_n = F_{4n+1}$ and $w_n = F_{4n+2}$, and we get the sequence

n	0	1	2	3	4	5	6	7	8	9	10	11	$\cdots$
$\cdots g_n$	0	1	1	3	5	8	21	34	55	144	233	377	$\cdots$

Surprisingly, the sequence g_n defined by (7.35) and (7.36) is the subsequence of the Fibonacci numbers obtained by removing every fourth term; that is, $g_{3n+r} = F_{4n+r}, r = 0, 1, 2.$ ■

Example 7.23 (Tribonacci–Pell–Fibonacci Recurrence) Determine the sequence given by

$$h_{n+3} = \begin{cases} h_{n+2} + h_{n+1} + h_n, & n \equiv 0 \pmod 3 \\ 2h_{n+2} + h_{n+1}, & n \equiv 1 \pmod 3 \\ h_{n+2} + h_{n+1}, & n \equiv 2 \pmod 3 \end{cases}, \qquad n \geq 0 \tag{7.38}$$

where $h_0 = 0, h_1 = 1, h_2 = 1$.

Solution. As in Example 7.22, this sequence cyclically interlaces the tribonacci–Pell–Fibonacci recurrences, albeit in a different cyclical order. We now have $u_{n+1} = u_n + v_n + w_n, v_{n+1} = 2u_{n+1} + w_n, w_n = u_n + v_n, n \geq 0$, which gives us the common annihilating operator

$$C(E) = \det \begin{bmatrix} E-1 & -1 & -1 \\ -2E & E & -1 \\ -1 & -1 & 1 \end{bmatrix} = E(E-7)$$

for all three sequences u_n, v_n, and w_n In particular, we see that each sequence u_{n+1}, v_{n+1}, and w_{n+1} is annihilated by $E - 7$, so it is a geometric sequence with a common ratio 7. Therefore, we get $h_{3n+3+r} = c_r 7^n$, $r = 0, 1, 2$, and $n \geq 0$ for the constants $c_0 = h_3 = 0 + 1 + 1 = 2, c_1 = h_4 = 2 \cdot 2 + 1 = 5, c_2 = h_5 = 5 + 2 = 7$ determined by the initial conditions for h_0, h_1, and h_2. See the following table:

n	0	1	2	3	4	5	6	7	8	9	10	11	$\cdots$
h_n	0	1	1	2	5	7	$2 \cdot 7$	$5 \cdot 7$	$7 \cdot 7$	$2 \cdot 7^2$	$5 \cdot 7^2$	$7 \cdot 7^2$	$\cdots$

■

PROBLEMS

7.3.1. Find a polynomial $f(E)$ that annihilates h_n given

(a) $h_{n+2} - 3h_{n+1} + 7h_n = (-5)^n$

(b) $h_{n+2} - 3h_n = n^2$

(c) $h_{n+3} - h_{n+2} + 2h_n = 3 \cdot 2^n + 7$

(d) $h_{n+2} + h_{n+1} - 4h_n = 5n + 2(-1^n) + 3^n$

(e) $h_{n+2} - h_{n+1} - h_n = 3F_n$

7.3.2. Determine the spectrum associated with each of these nonhomogeneous recurrence relations:

(a) $x_{n+2} - 5x_{n+1} + 6x_n = 7$

(b) $y_{n+2} - 5y_{n+1} + 6y_n = 5 \cdot 2^n$

(c) $z_{n+2} - 3z_{n+1} + 2z_n = 3n^2$

7.3.3. Let g_n denote the number of regions formed inside a circle when n points are placed on the circumference and all chords are drawn, where it is assumed that no three chords intersect at a common point interior to the circle. Thus, $g_0 = 1, g_1 = 1$, and $g_2 = 2$.

(a) Obtain a nonhomogeneous recursion relation for g_n.

(b) Use the recurrence to compute g_n for $n = 2,3,4,5,6$.

(c) Determine the formula for g_n.

(d) Determine the formula for g_n by summing the telescoping sum $\sum_{k=0}^{n-1} (g_{k+1} - g_k)$.

7.3.4. Find an annihilating polynomial for h_n when

(a) $C(E)h_n = 2^n$

(b) $C(E)h_n = n2^n + n^2$

7.3.5. Suppose that $C(E)u_n = 0$ and $D(E)v_n = 0$. Find an annihilating polynomial for h_n when

(a) $C(E)h_n = u_n$

(b) $C(E)h_n = u_n + nv_n$.

7.3.6. Let h_n count the number of ways to tile a $1 \times n$ board using red, blue, and green squares, and red dominoes, where a red tile can be followed by a tile of any color but a blue or green square can be followed only by other blue and green tiles.

(a) Obtain a recurrence for h_n.

(b) Solve the recurrence for h_n.

7.3.7. Let h_n count the number strings of the digits $0,1,2,\ldots,9$ of length n, where a digit following a prime digit must also be a prime digit; a 4, 6, or 8 can be followed only by a 0; and a 9 can be followed only by a 9. For example, 11409 and 60172 are permissible strings of length 5, but 01754 and 80599 are not allowed.

(a) Obtain a recurrence for h_n.

(b) Solve the recurrence for h_n.

7.3.8. Find a recurrence relation that is satisfied by both u_n and v_n if the sequences satisfy the system $u_{n+1} = 2u_n + v_n, v_{n+1} = u_n - v_n$.

7.3.9. What recurrence is satisfied by both of the sequences x_n and y_n if $x_n = x_{n-1} + x_{n-2} - y_{n-1}, y_n = 2x_{n-2} + 3y_{n-2}$?

7.3.10. Let u_n denote the number of ways to tile a $3 \times 2n$ board with dominoes, and let v_n denote the number of ways to tile a $3 \times (2n + 1)$ board with a corner 1×1 cell removed.

(a) Derive and solve a coupled system of recurrences to find a GPS for each sequence u_n and v_n.

(b) Give the generating function of each sequence.

7.3.11. How many block walking paths of length n move along blocks to the east, west, or north, never returning to the same intersection a second time? For example, NEENWW and WNNEEN are permissible walks of length 6, but NWEEWN is not allowed since the WE pair returns the path to a previously occupied position, as does the EW pair. [*Hint*: Let u_n denote the number of paths of length n that begin with a block to the north and let v_n denote the number of paths of length n that begin with a block to the east or, equivalently, with a block to the west.]

7.4 IDENTITIES FOR RECURRENCE SEQUENCES

In previous chapters of this book, we have discovered and proved identities for recurrence sequences by various techniques, including combinatorial models, mathematical induction, the principle of inclusion–exclusion, DIE method, manipulations of a GPS, generating functions, and iteration. Different methods worked better on some identities than others, and there was no one approach superior to another. However, identities that involve complicated expressions present a great challenge if we are limited to these techniques.

In this section, we will develop a more universally applicable method that can be used to prove identities for recurrence sequences in the vector space $\mathcal{S}_{\mathcal{R}}$. For convenience, we will refer to this approach as the *operator method*, since its core idea is to determine the nature of an annihilating polynomial operator $f(E)$.

7.4.1 The Operator Method

Suppose that we wish to prove an identity of the form $A = B$, where A and B are each algebraic expressions that involve recurrence sequences. By Theorem 7.4, if there is an annihilating polynomial $f(E)$ of $A - B$ and it is shown that $A - B = 0$ for $n = 0,1,2,\ldots,\deg f - 1$ (or any consecutive list of terms of length $\deg f$), then $A - B = 0$ for all $n \geq 0$. The remarkable thing about the operator method is that it is not necessary to know $f(E)$ explicitly; it suffices only to know that it exists and to have an upper bound for its degree. Theorems 7.10, 7.11, and 7.13 are useful for obtaining such a bound.

7.4.2 Free and Bound Variables

To understand how to use the operator method, it is useful to distinguish the role of the variables that appear in A and B as free or bound. A variable n is *free* if its domain is unrestricted, most often the set of positive integers or the set of nonnegative integers. By way of contrast, a variable j is *bound* if it is restricted to a domain determined by other free variables. For example, suppose that n is a free variable in the sum $\sum_{j=0}^{n} h_j$. The index of summation j is restricted to the set $\{0,1,2,\ldots,n\}$ that is dependent on the free variable n, so it is a bound variable. If k is a fixed value, say, the order of a recurrence sequence, then the index of summation j in the sum $\sum_{j=0}^{k} h_j$ is neither free nor bound. Indeed, the sum is simply a constant with no dependence on any variable.

To see how both free and bound variables may occur, consider these five identities. Here F_n and L_n are the Fibonacci and Lucas sequences, and P_n is the Pell sequence defined by $P_0 = 0, P_1 = 1, P_{n+2} = 2P_{n+1} + P_n$:

(I) $$F_{2n} = F_n L_n, \quad n \geq 0$$

(II) $$L_n^2 = 5F_n^2 + 4(-1)^n, \quad n \geq 0$$

(III) $$\sum_{j=1}^{n} F_{2j} = F_{2n+1} - 1, \quad n \geq 1$$

(IV) $$\sum_{j=0}^{n} F_{n+j} = F_{2n+2} - F_{n+1}, \quad n \geq 0$$

(V) $$\sum_{j=0}^{n} 2^{n-j} P_{2j} = \frac{1}{7}\left(P_{2n} + 4P_{2n+1} - 4 \cdot 2^n\right), \quad n \geq 0$$

The variable n is a free variable in all five identities, and the variable j is a bound variable in identities (III), (IV), and (V). The expressions in the first two identities are said to be *closed forms,* since no bound variables appear. However, the sums on the left sides of identities (III), (IV), and (V) include bound variables, so they will be called *open forms.*

Identities (I) through (V) can be proved by induction, generating functions, and other approaches that we have examined earlier, but we will now prove them with the operator method. The advantage of the operator method is that it continues to be effective for even quite complicated identities that would be a severe challenge if we were limited to our earlier methods of proof.

7.4.3 Elementary Applications of the Operator Method

Example 7.24 Prove identity (I):

$$F_{2n} = F_n L_n, \quad n \geq 0$$

Proof. By Theorems 7.10 and 7.13, $\mathcal{O}(F_{2n}) \leq 2$ and $\mathcal{O}(F_n L_n) \leq 2^2 = 4$. Therefore, $\mathcal{O}(F_{2n} - F_n L_n) \leq 2 + 4 = 6$. But $F_{2n} - F_n L_n = 0$ for $n = 0, 1, \ldots, 5$. For example, $F_6 - F_3 L_3 = 8 - 2 \cdot 4 = 0$, $F_8 - F_4 L_4 = 21 - 3 \cdot 7 = 0$, and $F_{10} - F_5 L_5 = 55 - 5 \cdot 11 = 0$. By Theorem 7.4, $F_{2n} - F_n L_n = 0$ for all n. ■

The proof just given is very short, since there was no need to know that the spectrum of F_{2n} is $\{\varphi^2, \hat{\varphi}^2\}$ and the spectrum of $F_n L_n$ is $\{\varphi^2, \hat{\varphi}^2, -1\}$, so that $\mathcal{O}(F_{2n} - F_n L_n) = 3$. If we had taken these spectra into account, this would show that the identity is proved by the three initial cases $F_0 - F_0 L_0 = 0 - 0 = 0, F_2 - F_1 L_1 = 1 - 1 \cdot 1 = 0$, and $F_4 - F_2 L_2 = 3 - 1 \cdot 3 = 0$.

Example 7.25 Prove identity (II):

$$L_n^2 = 5F_n^2 + 4(-1)^n, \quad n \geq 0$$

Proof. The spectrum is a subset of $\{\varphi^2, \hat{\varphi}^2, -1\}$, so $\mathcal{O}\left(L_n^2 - 5F_n^2 - 4(-1)^n\right) \leq 3$. Therefore, it suffices to verify the identity for $n = 0,1,2$. We see that $L_0^2 - 5F_0^2 - 4(-1)^0 = 2^2 - 0 - 4 = 0$, $L_1^2 - 5F_1^2 - 4(-1)^1 = 1 - 5 + 4 = 0$, and $L_2^2 - 5F_2^2 - 4(-1)^2 = 3^2 - 5 \cdot 1^2 - 4 = 0$, so the identity is proved. ■

The next example illustrates the proof of an identity with a summation that involves a bound variable.

Example 7.26 Prove identity (III):

$$\sum_{j=1}^{n} F_{2j} = F_{2n+1} - 1, \quad n \geq 1$$

Proof. The free variable n appears only as the upper limit of the summation, and the terms F_{2j} depend only on the bound variable j. Therefore

$$(E - 1) \sum_{j=1}^{n} F_{2j} = \sum_{j=1}^{n+1} F_{2j} - \sum_{j=1}^{n} F_{2j} = F_{2n+2}$$

The sequence F_{2n} is annihilated by $(E - \varphi^2)(E - \hat{\varphi}^2) = E^2 - 3E + 1$, so $(E^2 - 3E + 1)(E - 1) \sum_{j=1}^{n} F_{2j} = 0$. The sequence F_{2n+1} is also annihilated by $E^2 - 3E + 1$, and the constant 1 is annihilated by $E - 1$. Altogether we see that $(E^2 - 3E + 1)(E - 1)\left(\sum_{j=1}^{n} F_{2j} - F_{2n+1} + 1\right) = 0$. Since the annihilating operator has degree 3, it suffices to verify that the proposed identity holds for the three initial cases $n = 1,2,3$. We see that

$$\sum_{j=1}^{1} F_{2j} - F_3 + 1 = 1 - 2 + 1 = 0$$

$$\sum_{j=1}^{2} F_{2j} - F_5 - 1 = 4 - 5 + 1 = 0$$

$$\sum_{j=1}^{3} F_{2j} - F_7 + 1 = 12 - 13 + 1 = 0$$

so the identity is proved. ■

Example 7.27 Prove identity (IV):

$$\sum_{j=0}^{n} F_{n+j} = F_{2n+2} - F_{n+1}, \quad n \geq 0$$

Proof. Identity (IV) is similar to identity (III), but the terms F_{n+j} of the sum depend on both the free variable n and the bound variable j. Since the operator $E^2 - E - 1$ annihilates the Fibonacci sequence F_n, it also annihilates any shifted sequence F_{n+j}. Thus

$$\begin{aligned}(E^2 - E - 1)\sum_{j=0}^{n} F_{n+j} &= \sum_{j=0}^{n+2} F_{n+2+j} - \sum_{j=0}^{n+1} F_{n+1+j} - \sum_{j=0}^{n} F_{n+j}\\ &= \sum_{j=0}^{n} F_{n+2+j} + F_{2n+3} + F_{2n+4} - \sum_{j=0}^{n} F_{n+1+j} - F_{2n+2} - \sum_{j=0}^{n} F_{n+j}\\ &= \sum_{j=0}^{n} (F_{n+2+j} - F_{n+1+j} - F_{n+j}) + F_{2n+3} + F_{2n+4} - F_{2n+2}\end{aligned}$$

Each term in the last sum above is 0 and therefore $(E^2 - E - 1)\sum_{j=0}^{n} F_{n+j} = F_{2n+3} + F_{2n+4} - F_{2n+2}$. We also know that F_{2n} and any of its shifts are annihilated by $E^2 - 3E + 1$, so we have now shown that $(E^2 - E - 1)(E^2 - 3E + 1) \sum_{j=0}^{n} F_{n+j} = 0$.

This means that $\sum_{j=0}^{n} F_{n+j}$ is annihilated by an operator of degree 4, an operator that also annihilates the right side of the prosed identity. Therefore the sequence $\sum_{j=0}^{n} F_{n+j} - F_{2n+2} + F_{n+1}$ is a recurrence sequence of order no more than 4. Since an easy calculation shows that the first four terms $n = 0,1,2,3$ of the sequence are 0, the sequence is the 0 sequence, proving the identity. ■

Suppose that the preceding example is modified to finding a closed-form expression for the sum $\sum_{j=0}^{n} F_{n+j}$.

By the preceding calculations, we know the sum is a GPS of the spectrum $\{\varphi, \hat{\varphi}, \varphi^2, \hat{\varphi}^2\}$. Since $\{F_n, F_{n+1}, F_{2n}, F_{2n+1}\}$ spans the same vector space, we know that there are constants for which $\sum_{j=0}^{n} F_{n+j} = c_1 F_n + c_2 F_{n+1} + c_3 F_{2n} + c_4 F_{2n+1}$. The constants can be determined from the first four values of the sum, showing that $c_1 = 0, c_2 = -1, c_3 = 0$, and $c_4 = 1$. Thus, the operator method has simultaneously derived and proved identity (IV).

In the next example, we will use the operator method both to derive and simultaneously to prove identity (V).

Example 7.28 Let $s_n = \sum_{j=0}^{n} 2^{n-j} P_{2j}$. Find a closed form for s_n.

Solution. The operator $f(E) = E - 2$ annihilates the sequence 2^n, so it is then easy to check that $(E - 2)s_n = P_{2n+2}$. This sequence is annihilated by some second-order operator $g(E)$ whose null space is spanned by the sequences P_{2n} and P_{2n+1}. Since $f(E)g(E)s_n = 0$, we know that the set of sequences $\{2^n, P_{2n}, P_{2n+1}\}$ spans the null space of $f(E)g(E)$. Thus, there are constants for which $s_n = c_1P_{2n} + c_2P_{2n+1} + c_3 2^n$. These constants are determined by the initial values $s_0 = 0, s_1 = 2, s_2 = 16$, and we find that $s_n = \frac{1}{7}(P_{2n} + 4P_{2n+1} - 4 \cdot 2^n)$. ■

In the solution just given, it was not necessary to know the explicit formula $g(E) = E^2 - 6E + 1$, nor was it necessary to determine the spectrum $\mathcal{E}_g = \{(1 + \sqrt{2})^2, (1 - \sqrt{2})^2\} = \{3 + 2\sqrt{2}, 3 - 2\sqrt{2}\}$.

7.4.4 Expressing Sums in Closed Form

We begin with the simplest case where the free variable n appears only as the upper limit of the sum s_n, so the sum has the form $s_n = \sum_{j=0}^{n} h_j$. Since $(E - 1)s_n = \sum_{j=0}^{n+1} h_j - \sum_{j=0}^{n} h_j = h_{n+1}$, we see that s_n satisfies the first-order nonhomogeneous recurrence $C(E)s_n = (E - 1)s_n = h_{n+1}$. Next suppose that the nonhomogeneous term of the equation $q_n = h_{n+1}$ is annihilated by an operator $P(E)$ of order k. Then $P(E)C(E)s_n = P(E)(E - 1)s_n = P(E)h_{n+1} = 0$. This tells us that s_n is a GPS of the spectrum $\mathcal{E}_{PC}$. If $P(1) \neq 0$, then 1 is a simple eigenvalue of $\mathcal{E}_{PC}$ and therefore s_n is in the vector space spanned by the sequences $\{1, h_n, h_{n+1}, \ldots, h_{n+k-1}\}$. In other words, there are constants for which $s_n = c_0 + c_1h_n + \cdots + c_{k-1}h_{n+k-1}$.

However, suppose that $P(1) = 0$. Then $1 \in \mathcal{E}_P$ is an eigenvalue of multiplicity $r \geq 1$. This means 1 is an eigenvalue of multiplicity $r + 1$ in $\mathrm{E}_{CP} = \mathrm{E}_{(E-1)P}$, and therefore, there are constants for which $s_n = c_0n^r + c_1h_{n+1} + \cdots + c_{k-1}h_{n+k-1}$, where the last k terms span the subspace generated by the sequences $1, n, n^2, \ldots, n^{r-1}$.

We have proved the following theorem.

Theorem 7.29 Let h_n be a recurrence sequence for which $P(E)h_n = 0$, where the annihilating operator $P(E)$ has order k and either 1 is an eigenvalue of multiplicity $r \geq 1$ or 1 is not an eigenvalue, so that $r = 0$. Then there are coefficients $c_0, c_1, \ldots c_k$ for which the sum $\sum_{j=0}^{n} h_j$ has the closed form

$$\sum_{j=0}^{n} h_j = c_0n^r + c_1h_n + c_2h_{n+1} + \cdots + c_kh_{n+k-1} \tag{7.39}$$

Example 7.30 Prove that $\sum_{j=0}^{n} F_{3j+2} = \frac{1}{2}(F_{3n+4} - 1)$.

Solution. The second-order operator $P(E) = (E - \varphi^3)(E - \hat{\varphi}^3)$ annihilates the sequence F_{3n+2}. By Theorem 7.29, since $P(1) \neq 0$, there are constants c_0, c_1, c_2 for which

$$\sum_{j=0}^{n} F_{3j+2} = c_0 + \sum_{j=0}^{1} c_jF_{3(n+j)+2} = c_0 + c_1F_{3n+2} + c_2F_{3n+5}$$

The equations for $n = 0,1,2$ show that $c_0 = -\frac{1}{2}, c_1 = c_2 = \frac{1}{4}$, so

$$\sum_{j=0}^{n} F_{3j+2} = \tfrac{1}{4}(F_{3n+2} + F_{3n+5}) - \tfrac{1}{2} = \tfrac{1}{2}(F_{3n+4} - 1)$$

■

Example 7.31 Express the sum $s_n = F_1^2 + F_3^2 + \cdots + F_{2n-1}^2$ in closed form.

Solution. The second-order operator $(E - \varphi^2)(E - \hat{\varphi}^2)$ annihilates the sequence F_{2n}, so there is a third-order operator $P(E)$ with spectrum $\{(\varphi^2)^2, (\hat{\varphi}^2)^2, \varphi^2\hat{\varphi}^2\} = \{\varphi^4, \hat{\varphi}^4, 1\}$ that annihilates F_{2n+i}^2. Since 1 is an eigenvalue of $P(E)$ of multiplicity $r = 1$, it is then an eigenvalue of multiplicity 2 of the sum s_n. Therefore, there are constants c_0, c_1, c_2, c_3 for which

$$s_n = F_1^2 + F_3^2 + \cdots + F_{2n-1}^2 = c_0 n + c_1 F_{2n}^2 + c_2 F_{2n+1}^2 + c_3 F_{2n+2}^2.$$

The constants are determined by the values $s_1 = 1, s_2 = 5, s_3 = 30, s_4 = 199$, showing that

$$F_1^2 + F_3^2 + \cdots + F_{2n-1}^2 = \frac{2n - 2F_{2n}^2 - F_{2n+1}^2 + F_{2n+2}^2}{5}$$

A simpler but equivalent form for the sum is

$$F_1^2 + F_3^2 + \cdots + F_{2n-1}^2 = \frac{2n + F_{4n}}{5}$$

■

A sum of the form $s_n = \sum_{j=a}^{bn+c} h_{dn+ej}$ where the free variable n appears in both the limit of the sum and in the terms of the sum can still be simplified to a closed form with the operator method. To see why, suppose that $f(E)$ is an annihilating operator of h_{dn}. Then $f(E) \sum_{j=a}^{bn+c} h_{dn+ej}$ is a sum of terms of the form $h_{(d+eb)n+i}$. Therefore, if $g(E)$ annihilates $h_{(d+eb)n}$, the operator $f(E)g(E)$ annihilates s_n. If it can be shown that $\mathcal{O}(f(E)g(E)) \leq k$, then s_n is a linear combination of at most k sequences that span the null space of $f(E)g(E)$.

The procedure just outlined was employed in Examples 7.27 and 7.28 to give proofs of identities (IV) and (V). The next three examples give more advanced applications of the operator method. Note that little explicit information about the annihilating operators is required.

Example 7.32 Determine a closed form for the sum $s_n = \sum_{j=n}^{2n} F_{n+j}$.

Solution. First note that F_n, and therefore F_{n+j}, is annihilated by a second-order operator $f(E)$ [namely, the Fibonacci operator $f(E) = E^2 - E - 1$, although we don't need to have the specific formula]. Since

$$s_n = \sum_{j=n}^{2n} F_{n+j} = \sum_{j=0}^{2n} F_{n+j} - \sum_{j=0}^{n-1} F_{n+j}$$

it follows that $f(E)s_n$ is a sum of terms of the form F_{3n+a} and F_{2n+b}. The terms of the form F_{3n+a} are annihilated by some second-order operator $g(E)$, and similarly the terms of the form F_{2n+b} are annihilated by another second-order operator $h(E)$. Altogether the sequence s_n is annihilated by some operator $f(E)g(E)h(E)$ of order at most 6. Moreover, for any constants $c_1, c_2, c_3, c_4, c_5, c_6$, this operator annihilates both sides of the equation

$$s_n = c_1F_{3n} + c_2F_{3n+1} + c_3F_{2n} + c_4F_{2n+1} + c_5F_n + c_6F_{n+1}$$

Since $s_0 = 0, s_1 = 3, s_2 = 16, s_3 = 76, s_4 = 343, s_5 = 1508$, we can solve for the constants to satisfy the equation for $n = 0,1,\ldots,5$. We find that $c_1 = 1, c_2 = 1, c_3 = 0$, $c_4 = -1, c_5 = 0, c_6 = 0$, giving us the identity $s_n = F_{3n} + F_{3n+1} - F_{2n+1}$, or equivalently, $s_n = F_{3n+2} - F_{2n+1}$. ■

Example 7.33 Prove the identity $\sum_{j=0}^{n} 2^{n-j}P_{n+j} = P_{2n+2} - 2^{n+1}P_{n+1}$.

Solution. The sequence 2^nP_n is the product of the first-order sequence 2^n and the second-order sequence P_n, so its order is no more than $1 \cdot 2 = 2$ by Theorem 7.10. This means that there is a second-order operator $f(E)$ that annihilates the sequence 2^nP_n. Moreover, $f(E)\sum_{j=0}^{n} 2^{n-j}P_{n+j}$ is a sum of terms of the form P_{2n+a}, each of which is annihilated by an operator $g(E)$ of second order. Altogether, we see that the operator $f(E)g(E)$ of degree at most 4 annihilates the sequence $u_n = \sum_{j=0}^{n} 2^{n-j}P_{n+j} - P_{2n+2} + 2^{n+1}P_{n+1}$. Since $u_n = 0$ for $n = 0,1,2,3$, we see that $u_n = 0$ for all $n \geq 0$, which verifies the identity. ■

The next example illustrates how the operator method can be applied to multiple sums in which more than one bound variable appears.

Example 7.34 Calculate a closed form expression for the sum $s_n = \sum_{i=0}^{n}\sum_{j=0}^{i} F_{i-j}$.

Solution. Since

$$(E-1)s_n = \sum_{i=0}^{n+1}\sum_{j=0}^{i} F_{i-j} - \sum_{i=0}^{n}\sum_{j=0}^{i} F_{i-j} = \sum_{j=0}^{n+1} F_{n+1-j} = \sum_{r=0}^{n+1} F_r$$

we now have a single sum for which the free variable appears only in the upper limit and there is just one bound variable, r. By Theorem 7.29, there are constants for which $(E-1)s_n = \hat{c}_0 + \hat{c}_1F_n + \hat{c}_2F_{n+1}$. Thus, $(E-1)^2(E^2-E-1)s_n = 0$, so we know there are constants for which $s_n = c_0 + c_1n + c_2F_n + c_3F_{n+1}$. The constants can be evaluated by using the initial conditions $s_0 = 0, s_1 = 1, s_2 = 3, s_3 = 7$, showing that

$$s_n = \sum_{i=0}^{n}\sum_{j=0}^{i} F_{i-j} = -3 - n + 3F_{n+1} + 2F_n = F_{n+4} - n - 3$$

■

7.4.5 Sums with Binomial Coefficients

Let Σ denote a sum over all integers j. We know of several sums involving binomial coefficients that, when expressed in closed form, are a recurrence sequence. Two examples are $\sum \binom{n}{j} = 2^n$ and $\sum \binom{n}{j} F_j = F_{2n}$, where we note that all but finitely many terms in the sum are zero. It can be shown that if h_n is a recurrence sequence, then so is the sequence of sums of the form

$$s_n = \sum \binom{dn + ej + f}{an + bj + c} h_j$$

in the case that only finitely many binomial coefficients in the sum are not zero. We will not prove this in general, but a simple case of the result is easy to obtain as a consequence of the following lemma.

Lemma 7.35

$$\sum \binom{n}{j} \binom{j}{r} \alpha^{j-r} = \binom{n}{r} (1 + \alpha)^{n-r} \tag{7.40}$$

for all integers n and r, and any real or complex variable α.

Proof. We will give two derivations, one algebraic and another combinatorial:

Algebraic Derivation. The formula is obvious $(0 = 0)$ except when $0 \le r \le n$. In these cases, we obtain

$$\sum \binom{n}{j} \binom{j}{r} \alpha^{j-r} = \frac{n!}{r!} \sum \frac{\alpha^{j-r}}{(n-j)!(j-r)!} = \frac{n!}{r!} \sum \frac{\alpha^j}{(n-j-r)!j!}$$
$$= \binom{n}{r} \sum \binom{n-r}{j} \alpha^j = \binom{n}{r} (1 + \alpha)^{n-r}$$

Combinatorial Derivation. Consider counting the number of tilings of a $1 \times n$ board with exactly r red square tiles and with the other cells covered by square tiles that are either white or else one of α colors other than red or white. The number of tilings is evidently $\binom{n}{r}(1 + \alpha)^{n-r}$, since we can first choose, in $\binom{n}{r}$ ways, r of the n cells to be covered with a red tile and then tile the remaining $n - r$ cells in $(1 + \alpha)^{n-r}$ ways. Alternatively, in $\binom{n}{j}$ ways, choose j cells of the board to be covered with a nonwhite tile, and from among these choose in $\binom{j}{r}$ ways the cells to be tiled red. This leaves the other $j - r$ cells to be tiled with one of the α colors in α^{j-r} ways. Thus, there are $\binom{n}{j}\binom{j}{r}\alpha^{j-r}$ tilings with j colored (i.e., not white) tiles. Summing over j gives $\sum \binom{n}{j}\binom{j}{r}\alpha^{j-r}$ as the total number of tilings.

■

Theorem 7.36 Let h_n be a recurrence sequence annihilated by the operator $C(E)$ with eigenvalues α_i of multiplicity r_i. Then the sums $s_n = \sum \binom{n}{j} h_j$ are a recurrence sequence annihilated by the operator $C(E-1)$ with the eigenvalues $1+\alpha_i$ of multiplicity r_i.

Proof. Let $C(E) = (E-\alpha_1)^{r_1}(E-\alpha_2)^{r_2}\cdots(E-\alpha_m)^{r_m}$ be the annihilating operator of the sequence h_n. Therefore, the GPS representing h_j is a sum of terms of the form $\binom{j}{r}\alpha_i^{n-r}$, $0 \le r < r_i$. It then follows from Lemma 7.35 that s_n is a sum of terms of the form $\binom{n}{r}(1+\alpha_i)^{n-r}$. Thus, s_n is annihilated by the operator $(E-\alpha_1-1)^{r_1}(E-\alpha_2-1)^{r_2}\cdots(E-\alpha_m-1)^{r_m} = C(E-1)$. ■

Example 7.37 Prove that $\sum \binom{n}{j} F_j = F_{2n}$.

Solution. The eigenvalues of F_n are $\{\varphi, \hat{\varphi}\}$, so by Theorem 7.36, the sum $s_n = \sum \binom{n}{j} F_j$ is a recurrence sequence with the eigenvalues $\{1+\varphi, 1+\hat{\varphi}\} = \{\varphi^2, \hat{\varphi}^2\}$. Therefore, there are constants such that $s_n = c_1 F_{2n} + c_2 F_{2n+1}$. Since $s_0 = 0$ and $s_1 = 1$, we see that $s_n = F_{2n}$. ■

In many cases, we must search for an annihilating operator of the sum, as in the next example.

Example 7.38 Find a closed-form expression for the sum $s_n = \sum \binom{n-j}{j}$.

Solution. Since $s_0 = 1, s_1 = 1, s_2 = 2$, we see that s_n is not a geometric sequence. so we will search for an annihilating operator $f(E) = E^2 - aE - b$ of order 2. Using Pascal's identity, we see that

$$\begin{aligned}
(E^2 - aE - b)\sum\binom{n-j}{j} &= \sum\binom{n+2-j}{j} - a\sum\binom{n+1-j}{j} - b\sum\binom{n-j}{j} \\
&= \sum\left[\binom{n+1-j}{j} + \binom{n+1-j}{j-1}\right] - a\sum\left[\binom{n-j}{j}\right. \\
&\quad \left. + \binom{n-j}{j-1}\right] - b\sum\binom{n-j}{j} \\
&= \sum\left[\binom{n-j}{j} + \binom{n-j}{j-1} + \binom{n-(j-1)}{j-1}\right] \\
&\quad -a\sum\left[\binom{n-j}{j} + \binom{n-j}{j-1}\right] - b\sum\binom{n-j}{j} \\
&= (2-a-b)\sum\binom{n-j}{j} + (1-a)\sum\binom{n-j}{j-1}
\end{aligned}$$

where it has been noted that

$$\sum \binom{n-(j-1)}{j-1} = \sum \binom{n-j}{j}.$$

This shows that $(E^2 - aE - b)s_n = 0$ if $a = 1$ and $b = 1$. In other words, the Fibonacci operator $f(E) = E^2 - E - 1$ annihilates s_n, so there are constants for which $s_n = c_1 F_n + c_2 F_{n+1}$. Using the initial conditions $s_0 = 1$ and $s_1 = 1$, we get $s_n = F_{n+1}$. The operator method therefore gives a new proof of the identity $\sum \binom{n-j}{j} = F_{n+1}$ that had been proved earlier by other means. ■

A similar approach can be taken to solve higher order recurrences, as shown in our last example.

Example 7.39 Find a closed-form expression for the sum $s_n = \sum \binom{n}{j} F_{n+j} F_{2j}$, assuming that it is a recurrence of order 4.

Solution. Searching for an annihilating operator of the form $f(E) = E^4 + aE^3 + bE^2 + cE + d$, we find, with considerable algebraic manipulation, that the sequence s_n is annihilated by the operator $f(E) = E^4 - 11E^3 + 21E^2 + 4E - 4 = (E^2 - 3E + 1)(E^2 - 8E - 4)$.

The eigenvalues can be calculated from the quadratic formula, showing that the spectrum is $\mathcal{E}_f = \{\varphi^2, \hat{\varphi}^2, 2\varphi^3, 2\hat{\varphi}^3\}$. The null space of $f(E)$ is therefore spanned by the basis of sequences $\{F_{2n}, F_{2n+1}, 2^n F_{3n}, 2^n F_{3n+1}\}$ so there are constants for which $s_n = c_1 F_{2n} + c_2 F_{2n+1} + c_3 2^n F_{3n} + c_4 2^n F_{3n+1}$.

Using the initial conditions $s_0 = 0, s_1 = 1, s_2 = 13$, and $s_3 = 118$ to solve for the constants, we find that $c_1 = \frac{1}{5}, c_2 = -\frac{2}{5}, c_3 = -\frac{1}{5}$, and $c_4 = \frac{2}{5}$. Therefore, $\sum \binom{n}{j} F_{n+j} F_{2j} = \frac{1}{5}\left(F_{2n} - 2F_{2n+1} - 2^n F_{3n} + 2 \cdot 2^n F_{3n+1}\right)$. ■

In the preceding example, we assumed that the order of the recurrence sequence $s_n = \sum \binom{n}{j} F_{n+j} F_{2j}$ is 4. This is a consequence of a theorem that we neither state nor prove here, but would tell us that the order is at most $1 \cdot 2 \cdot 2 = 4$ since the coefficient of j in the binomial is 1 and both Fibonacci sequences F_n and F_{2n} are of order 2. In general, the sums of the form $s_n = \sum \binom{n}{bj} h_j$ have order at most bk if h_n is a recurrence sequence of order k. Knowing the order can simplify finding the annihilating polynomial of the sum. For instance, if we know that there is a fourth-order annihilating polynomial $f(E) = E^4 - aE^3 - bE^2 - cE - d$ for the sequence $s_n = \sum \binom{n}{j} F_{n+j} F_{2j}$, then the eight initial values

n	0	1	2	3	4	5	6	7
s_n	0	1	13	118	1021	8705	73894	626417

give us the equations

$$1021 = 118a + 13b + c + 0$$

$$8705 = 1021a + 118b + 13c + d$$

$$73894 = 8705a + 1021b + 118c + 13d$$

$$626417 = 73894a + 8705b + 1021c + 118d$$

This linear system is readily solved (a CAS is helpful) to give $f(E) = E^4 - 11E^3 + 21E^2 + 4E - 4$.

In practice, it is usually effective to guess an upper bound k for the order of a sequence of sums $s_n = \sum h_j$, where h_j is a recurrence sequence or is a product of a binomial coefficient and a recurrence sequence. Computing $2k$ values $s_0, s_1, s_2, \dots, s_{2k-1}$ gives a system of k linear equations for the coefficients $a_1, a_2, \dots a_k$ of a recurrence relation of the form $s_{n+k} = a_1 s_{n+k-1} + a_2 s_{n+k-2} + \cdots + a_k s_n$ by setting $n = 0,1,2, \dots, k-1$. The solution can be checked by evaluating more terms in the sequence of sums. If the check fails, it is necessary to assume a larger value of the order k.

PROBLEMS

Unless stated otherwise, use the operator method to prove identities and express sums in closed form. Recall that Σ and Σ_j both denote a sum over all integers j.

7.4.1. Prove that

(a) $F_n^2 + F_{n+1}^2 = F_{2n+1}$

(b) $F_{n+1}^2 - F_{n-1}^2 = F_{2n}$

(c) $F_{n+1}F_{n-1} - F_n^2 = (-1)^n$ (Simson's identity)

(d) $F_n F_{n+1} = \frac{1}{5}\left(2F_{2n} + F_{2n+1} - (-1)^n\right)$

7.4.2. Prove the Gehn–Cesàro identity $F_n^4 - F_{n-2}F_{n-1}F_{n+1}F_{n+2} = 1$.

7.4.3. Prove that $F_{m+n} = \frac{1}{2}(F_m L_n + F_n L_m)$.

7.4.4. Prove that $F_{m+n} = F_{m+1}F_n + F_m F_{n-1}$.

7.4.5. Let $t_{n+2} = 5t_{n+1} - 6t_n, t_0 = 0, t_1 = 1$.

(a) What is the spectrum associated with the sum $s_n = \sum_{i=0}^n t_i$?

(b) Express s_n as a GPS.

(c) Give another form of s_n as an expression involving the terms of the sequence t_n.

7.4.6. Let $u_{n+2} = 3u_{n+1} - 2u_n, u_0 = 0, u_1 = 1$.

(a) Why is $c_0 + c_1 u_n + c_2 u_{n+1}$ not a correct form for the sum $s_n = \sum_{i=0}^n u_i$?

(b) Express s_n as an expression involving the terms of the sequence u_n.

(c) Express s_n as a GPS.

7.4.7. Let $h_{n+2} - ah_{n+1} - bh_n = 0$, where $h_0 = 0, h_1 = 1$, and $a + b \neq 1$.
(a) Show that 1 is not an eigenvalue of h_n.
(b) Find a closed form of the sum $s_n = \sum_{i=0}^{n} h_i$.

7.4.8. Let $h_{n+2} - ah_{n+1} - bh_n = 0$, where $h_0 = 0, h_1 = 1, b \neq -1$, and $a + b = 1$.
(a) Show that 1 is a simple (i.e., nonrepeated) eigenvalue of h_n.
(b) Find a closed form of the sum $s_n = \sum_{i=0}^{n} h_i$.

7.4.9. Prove that:
(a) $\sum_j \binom{n}{j} F_{1+j} = F_{2n+1}, \quad n \geq 0$
(b) $\sum_j \binom{n}{j} F_{2+j} = F_{2n+2}, \quad n \geq 0$
(c) $\sum_j \binom{n}{j} F_{m+j} = F_{2n+m}, \quad m, n \geq 0$

7.4.10. Let L_n be the extended sequence of Lucas numbers $\ldots, 7, -4, 3, -1, 2, 1, 3, 4, 7, \ldots$. Prove that $\sum_j \binom{m}{j} L_{n+j} = L_{2m+n}, \quad n \in \mathbb{Z}$, by these two methods:
(a) by the result of Problem 7.4.9, recalling that $L_n = \frac{1}{5}(F_{n-1} + F_{n+1})$.
(b) by the Binet formula $L_n = \varphi^n + \hat{\varphi}^n$.

7.4.11. **(a)** Verify that the operator $g(E) = E^2 + E - 1$ annihilates the sequence $(-1)^n F_n$.
(b) Prove that $\sum_{j=0}^{n} (-1)^j F_{j+1} = 1 + (-1)^n F_n$.

7.4.12. **(a)** Verify that the operator $f(E) = E^2 - 2E - 4$ annihilates the sequence $2^n F_n$.
(b) Express the sum $s_n = \sum_{j=0}^{n} 2^{n-j} F_{n+j}$ in closed form.

7.4.13. Express the sum $t_n = F_0^2 + F_2^2 + F_4^2 + \cdots + F_{2n}^2, n \geq 0$ in closed form.

7.4.14. Prove the identity $\sum_{j=0}^{n} 2^{n-j} P_j = P_{n+2} - 2^{n+1}$ for the Pell numbers using
(a) the operator method
(b) ordinary generating functions

7.4.15. Express the sum $\sum_{j=0}^{n} j^3$ in closed form.

7.4.16. Find a closed-form expression for the sum $s_n = \sum_{j=0}^{n} T_j$, where T_n is the tribonacci sequence defined by $T_0 = 0, T_1 = 1, T_2 = 1, T_{n+3} = T_{n+2} + T_{n+1} + T_n$.

7.4.17. Let s_n be the number of ways that a $1 \times n$ board can be tiled with squares and dominoes, where exactly one red domino is used and any number of dominoes and squares can be used, all of which are white.
(a) Show that $s_n = \sum_{j=1}^{n-1} F_j F_{n-j}$.
(b) Show that $s_n = \frac{1}{5}((n-1)F_n + 2nF_{n-1})$ in closed form.

7.4.18. Give a proof of the identity

$$\sum \binom{n}{j}\binom{j}{r}\alpha^j = \binom{n}{r}\alpha^r(1+\alpha)^{n-r}$$

(see Lemma 7.35) by mathematical induction.

7.4.19. Let $s_n = \sum \binom{n}{j} F_{2j}$.

(a) Find a recursion relation satisfied by s_n, and use it to compute the first six values of the sequence.

(b) Show that $\sum \binom{n}{j} F_{2j} = (\alpha^n - \beta^n)/\sqrt{5}$, where $\alpha, \beta = (5 \pm \sqrt{5})/2$.

7.4.20. Let $s_n = \sum \binom{n}{j} P_j$, where P_n denotes the Pell sequence $P_0 = 0, P_1 = 1$, $P_{n+2} = 2P_{n+1} + P_n$.

(a) Find a recursion relation satisfied by s_n, and use it to compute the first six values of the sequence.

(b) Prove that $\sum \binom{n}{j} P_j = (\alpha^n - \beta^n)/2\sqrt{2}$, where $\alpha, \beta = 2 \pm \sqrt{2}$.

7.4.21. It can be proved that $s_n = \sum_j \binom{n}{j-n} P_j$ is a second-order recurrence sequence. Find this recurrence relation using the values $s_0 = 0, s_1 = 3, s_2 = 24, s_3 = 198$.

7.5 SUMMARY AND ADDITIONAL PROBLEMS

This chapter has applied concepts from linear spaces to investigate solutions of the homogeneous linear recurrence $C(E)h_n = 0$, where $C(x) = x^k - a_1x^{k-1} - \cdots - a_{k-1}x - a_k$. Three vector spaces of sequences—the null space of $C(E)$, the set of GSPs of eigenvalues of $C(x) = 0$, and the set of sequences with OGFs of the form of a rational function $P(x)/Q(x)$, where $Q(x) = x^kC(1/x)$ and degree $P(x) < k$, were shown to be identical. This structure was utilized to solve nonhomogeneous recurrences and systems of linear recurrences, and to both derive and prove identities for recurrence sequences.

PROBLEMS

7.5.1. Let $s_n = \sum_j \binom{n}{j-n} h_j$, where h_n is a recurrence sequence of order 2 that satisfies the recurrence relation $h_{n+2} = ah_{n+1} + bh_n$.

(a) Given $f(E) = E^2 - cE - d$, show that

$$f(E)s_n = \sum_j \binom{n}{j-n}((a^2 + 2a + b - c + 1)h_{j+2} + (ab + 2b - c)h_{j+1} - dh_j)$$

(b) Show that $f(E)s_n = 0$ if $c = a^2 + a + 2b$ and $d = b(a - b + 1)$.

(c) Prove that $s_{n+2} = (a^2 + a + 2b)s_{n+1} + (b + ab - b^2)s_n$.

7.5.2. Let

$$s_n = \sum_j \binom{n}{j-n} L_j$$

where L_n is the Lucas sequence.

(a) Use the result of Problem 7.5.1 to show that $s_{n+2} - 4s_{n+1} - s_n = 0$.

(b) Show that $s_n = L_{3n}$.

7.5.3. Let

$$s_n = \sum_j \binom{n}{j-n} P_j$$

where P_n is the Pell sequence. Use the result of Problem 7.5.1 to show that $s_{n+2} - 8s_{n+1} - 2s_n = 0$.

7.5.4. Let $s_n = \sum_{j=0}^{n} 2^{n+j} P_{3n-j}$, where P_n is the Pell sequence.

(a) If $f(E)2^n P_{3n} = 0$, and $g(E)2^{2n} P_{2n} = 0$, show that $g(E)f(E)S_n = 0$.

(b) Show that S_n is a linear combination of the sequences $2^n P_{3n}$, $2^n P_{3n+1}$, $2^{2n} P_{2n}$, $2^{2n} P_{2n+1}$.

7.5.5. Prove that the type of partial fraction decomposition used in the proof of Theorem 7.1 exists for all rational functions for which the degree of the denominator $Q(x)$ exceeds the degree of the numerator $P(x)$.

8

COUNTING WITH SYMMETRIES

8.1 INTRODUCTION

Imagine that a 2×2-*ft* square tabletop is to be covered with four 1×1-*ft* ceramic tiles that are either white or blue. Since each of the four corners of the table can be covered in two ways, there are $2^4 = 16$ ways to tile the table, all of which are shown in Figure 8.1. We see that there is one way if all four tiles are white, four tilings with one blue tile and three white tiles, six tilings with two tiles of each color, and so on. But are there really 16 different tilings? Each row of tilings in Figure 8.1 can be viewed to be the same pattern, because a rotation of the table by 90°, or 180°, or 270° transforms one pattern into the other. When rotational symmetries are considered, it seems that a better answer is to say that there are six ways to tile the table, where no two coloring patterns are equivalent to one another under any rotation.

This chapter develops methods of counting that take symmetries into account. Since symmetries are described by an algebraic group of actions that permute the parts of the arrangement from one position to another, a remarkable interplay between group theory and combinatorics will emerge. Little background will be required from group theory. It suffices to know that a group G is a set that is closed under some associative binary operation $\circ$ [for all $\gamma, \eta, \tau \in G$, $\gamma \circ \eta \in G$, and $(\gamma \circ \eta) \circ \tau = \gamma \circ (\eta \circ \tau)$], there is an identity e [$e \circ \gamma = \gamma \circ e = \gamma$ for all $\gamma \in G$], and each element has an inverse [for all $\gamma \in G$, there is an element γ^{-1} for which $\gamma \circ \gamma^{-1} = \gamma^{-1} \circ \gamma = e$].

To ensure that our understanding is not obscured by abstraction, our discussion is kept concrete and visual. In particular, in Section 8.2 we will show how the number of ways to color some familiar objects—a tabletop, a square stained-glass window, and

Combinatorial Reasoning: An Introduction to the Art of Counting, First Edition. Duane DeTemple and William Webb.

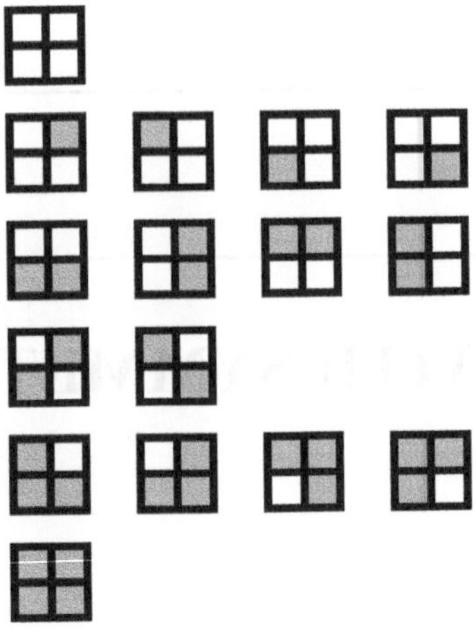

FIGURE 8.1 There are 16 2-colorings of a square, grouped into 6 rows of equivalent color patterns.

a beaded necklace—is related to the group of symmetries of the object. Section 8.3 formalizes the discoveries made in Section 8.2, and gives a proof of Burnside's lemma. In Section 8.4, a polynomial called the *cycle index* is described, which, in turn, leads to Pólya's enumeration formula. For example, Pólya's formula inventories all of the nonequivalent ways to tile the tabletop by the generating function

$$Z = w^4 + w^3b + 2w^2b^2 + wb^3 + b^4 \tag{8.1}$$

The term $2w^2b^2$ tells us that there are two ways to title the tabletop with two white and two blue tiles, and neither pattern can be rotated to coincide with the other. Figure 8.1 shows the two patterns in rows three and four; either the two white tiles are adjacent or opposite one another.

8.2 ALGEBRAIC DISCOVERIES

8.2.1 Groups of Actions

Let $X = \{x_1, x_2 \ldots, x_m\}$ be any set, which will often be a set of labels of some geometric figure. Any permutation $\gamma : X \to X$ is called an *action* on the set X. If $\gamma(x_i) = x_j$, then there is a corresponding permutation $\hat{\gamma} : [m] \to [m]$ given by $\hat{\gamma}(i) = j$. As an example, suppose that $X = \{x_1, x_2, x_3, x_4\}$ is the set of corners of the square tabletop shown below:

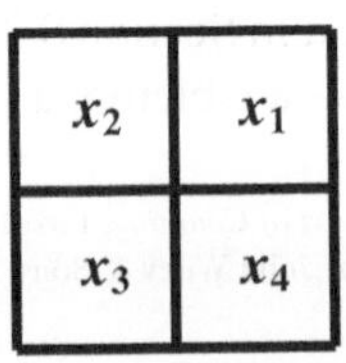

A 90° counterclockwise rotation of the square is given by the action $\rho : x_1 \to x_2, x_2 \to x_3, x_3 \to x_4, x_4 \to x_1$ on X. This corresponds to the permutation $\hat{\rho}$ on $[4] = \{1, 2, 3, 4\}$ given by

$$\hat{\rho} = \begin{pmatrix} 1 & 2 & 3 & 4 \\ 2 & 3 & 4 & 1 \end{pmatrix} \tag{8.2}$$

Frequently the set of actions is determined by the group of symmetries of the underlying geometric figure. In this case, the group of symmetries of the figure induces a corresponding group of actions on X, which, in turn, can be viewed as a permutation group of $[m]$. All three groups—the group of symmetries, the group of actions on X, and the permutation group of $[m]$—all have the same group structure, so there is no need to distinguish between these three isomorphic groups. In particular, we can identify ρ and $\hat{\rho}$.

8.2.2 Cycle Decomposition

There are several different ways to write a permutation γ of $[m]$. For example, the permutation

$$\gamma = \begin{pmatrix} 1 & 2 & 3 & 4 & 5 & 6 & 7 & 8 \\ 3 & 6 & 8 & 4 & 7 & 2 & 5 & 1 \end{pmatrix}$$

for which $\gamma(1) = 3, \gamma(2) = 6, \ldots, \gamma(8) = 1$ can also be written

$$\gamma : 1 \to 3, 3 \to 8, 8 \to 1, 2 \to 6, 6 \to 2, 4 \to 4, 5 \to 7, 7 \to 5$$

This form reveals that the permutation consists of disjoint cycles: a 3-cycle (1 3 8), the 2-cycle (2 6), a 1-cycle (4), and another 2-cycle (5 7). We will see throughout this chapter that the most useful expression for a permutation is its *cycle decomposition*, in which the permutation is written as a product of disjoint cycles: $\gamma = (138)(26)(4)(57)$.

We could have written the 3-cycle (1 3 8) as either (3 8 1) or (8 1 3), but it is better to write the smallest integer of the cycle first. The cycle decomposition can permute the order of the cycles, but it is helpful to write them in ascending lexicographic order with the leading terms in the cycles in increasing order.

If γ and η are two permutations, their composition will be written as a product; that is, the group operation $\gamma \circ \eta$ will be written as $(\gamma \circ \eta)(i) = \gamma\eta(i) = \gamma(\eta(i))$. For example, if $\gamma = (1\ 3)(2\ 5)(4)$ and $\eta = (1\ 4\ 2)(3\ 5)$, then

$$\gamma\eta = (1\ 3)(2\ 5)(4)(1\ 4\ 2)(3\ 5) = (1\ 4\ 5)(2\ 3)$$
$$\eta\gamma = (1\ 4\ 2)(3\ 5)(1\ 3)(2\ 5)(4) = (1\ 5)(2\ 3\ 4)$$

where the cycle decompositions of the two products are given.

The cycle decomposition of the 90° rotation ρ of the tabletop is $\rho = (1\ 2\ 3\ 4)$, a single cycle of length 4. Similarly, the rotation of the table by 180° that rotates each square of the tabletop to the diagonally opposite square is given by the permutation $\rho^2 = (1\ 3)(2\ 4)$, a product of two disjoint 2-cycles. A rotation by 270° is given by $\rho^3 = (1\ 4\ 3\ 2)$, and $\rho^4 = e = (1)(2)(3)(4)$ is the identity transformation that fixes each element of X and consists of four 1-cycles.

The rotational symmetries of the tabletop are therefore given by the group $C_4 = \{e, \rho, \rho^2, \rho^3\}$, which is called the *cyclic group of order 4*. We write $|C_4| = 4$ to indicate that there are four members of the group (see Figure 8.2):

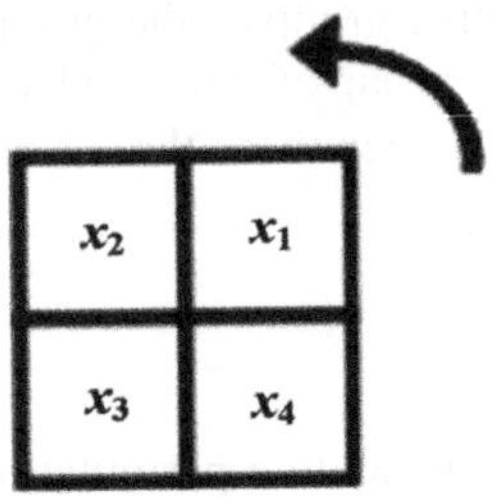

FIGURE 8.2 Rotational symmetries of a square are given by the cyclic group C_4.

$$\begin{aligned}
C_4 &= \{e, \rho, \rho^2, \rho^3\}, \text{ rotations of the square} \\
e &= (1)(2)(3)(4), \text{ the identity (rotation by } 0°) \\
\rho &= (1\ 2\ 3\ 4), \text{ rotation by } 90° \\
\rho^2 &= (1\ 3)(2\ 4), \text{ rotation by } 180° \\
\rho^3 &= (1\ 4\ 3\ 2), \text{ rotation by } 270°
\end{aligned}$$

The cyclic group of rotational symmetries of the square tabletop induces the group of actions given by the permutation group

$$\{(1)(2)(3)(4), (1\ 2\ 3\ 4), (1\ 3)(2\ 4), (1\ 4\ 3\ 2)\}$$

8.2.3 A Preview of the Cycle Index and Pattern Inventory

The importance of the cyclic structure of the group of actions on X will become clear later, but as a preview, let's *code* the identity $e = (1)(2)(3)(4)$ action on the rotating tabletop by the monomial t_1^4 in the variable t_1. This monomial informs us that there are four 1-cycles in that group action. Similarly, the cycle decomposition of the 90° rotation $\rho = (1\ 2\ 3\ 4)$ is one 4-cycle, so it is coded by the monomial t_4^1. The 180° rotation $\rho^2 = (1\ 3)(2\ 4)$ has two 2-cycles, so its corresponding monomial is t_2^2, telling us that this action has two 2-cycles. Finally, the 270° rotation $\rho^3 = (1\ 4\ 3\ 2)$ is a single 4-cycle, so its corresponding polynomial is t_4^1, the same as for the 90° rotation. By adding the polynomials for each action in the group C_4 and dividing by the group order, we obtain

$$Z = \frac{1}{|C_4|}\left(t_1^4 + t_4^1 + t_2^2 + t_4^1\right) = \frac{1}{4}\left(t_1^4 + t_2^2 + 2t_4^1\right) \tag{8.3}$$

a polynomial called the *cycle index*.

Here is why the cycle index is amazing. If we set $t_1 = w + b, t_2 = w^2 + b^2, t_4 = w^4 + b^4$, then

$$\begin{aligned} Z &= \tfrac{1}{4}\left(t_1^4 + t_2^2 + 2t_4^1\right) = \tfrac{1}{4}((w + b)^4 + (w^2 + b^2)^2 + 2(w^4 + b^4)) \\ &= w^4 + w^w b + 2w^2b^2 + wb^3 + b^4 \end{aligned} \tag{8.4}$$

This is Pólya's formula, a generating function whose coefficients enumerate the nonequivalent colorings of the tabletop (8.1) according to the number of white and the number of blue tiles used. Formula (8.4) is therefore a *pattern inventory* for the 2-colorings of the tabletop. If we set $w = b = 1$, then we see there are $1 + 1 + 2 + 1 + 1 = 6$ nonequivalent tilings altogether. An inventory of the 3-colorings of the tabletop is just as easy: if there are now red, white, and blue tiles available, we set $t_1 = r + w + b, t_2 = r^2 + w^2 + b^2, t_4 = r^4 + w^4 + b^4$ in the cycle index to get the pattern inventory

$$\begin{aligned} Z &= \tfrac{1}{4}\left(t_1^4 + t_2^2 + 2t_4^1\right) = \tfrac{1}{4}((r + w + b)^4 + (r^2 + w^2 + b^2)^2 + 2(r^4 + w^4 + b^4)) \\ &= r^4 + w^4 + b^4 + rw^3 + rb^3 + wr^3 + wb^3 + br^3 + bw^3 \\ &\quad + 2r^2w^2 + 2r^2b^2 + 2w^2b^2 + 3r^2wb + 3w^2rb + 3b^2rw \end{aligned} \tag{8.5}$$

For example, the last term $3b^2rw$ informs us that there are three nonequivalent tilings using two blue tiles, one red tile, and one white tile. If we let $r = w = b = 1$, then $Z = 1 + 1 + 1 + 1 + \cdots + 2 + 3 + 3 + 3 = 24$, and so there are 24 rotationally nonequivalent tabletops with red, white, and blue tiles.

8.2.4 More Groups of Symmetries

Only rotational symmetries are appropriate for the tabletop, but suppose that the square is allowed to float in space. For example, imagine that the square is a frame that holds four square stained-glass panes, and the frame can be flipped over as well as rotated. The group of symmetries of the floating square is known as the *dihedral group* D_4.

Figure 8.3 shows the four flips (reflections) $\varphi_1, \varphi_2, \varphi_3, \varphi_4$ of the dihedral group D_4.

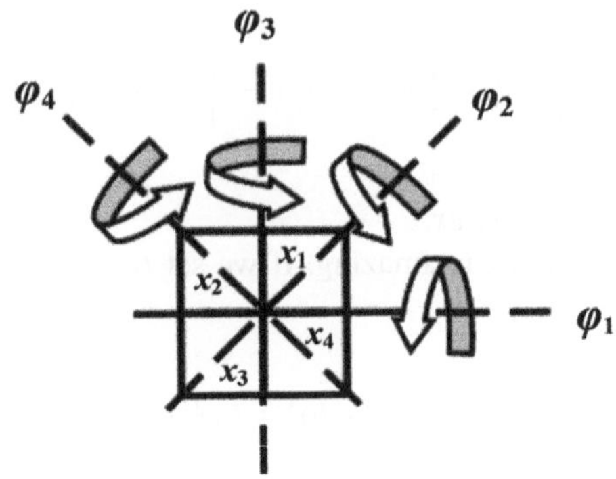

FIGURE 8.3 The symmetries of a floating square are given by the dihedral group D_4.

The flips, like the rotations, can be written as a product of cycles:

$$\varphi_1 = (1\ 4)(2\ 3)$$
$$\varphi_2 = (1)(2\ 4)(3)$$
$$\varphi_3 = (1\ 2)(3\ 4)$$
$$\varphi_4 = (1\ 3)(2)(4)$$

When the flips are combined with the four rotations, we obtain the dihedral group $D_4 = \{e, \rho, \rho^2, \rho^3, \varphi_1, \varphi_2, \varphi_3, \varphi_4\}$. Since the group has eight elements, it has order 8, and we write $|D_4| = 8$.

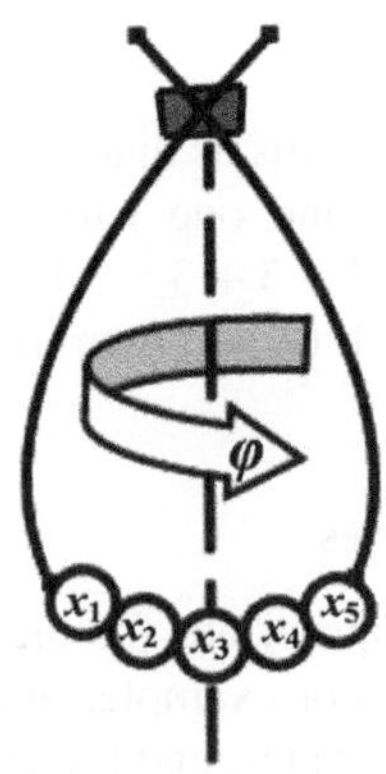

FIGURE 8.4 The group of symmetries on a beaded necklace with a clasp is $C_2 = \{e, \varphi\}$, where φ is a flip.

Another useful example of a set and its symmetries is shown in Figure 8.4. The necklace has five spherical beads and a fixed clasp. The group of actions on the necklace describes the two ways to wear the necklace, each way the "flip" of the other. Therefore, the group of symmetries on $X = \{x_1, x_2, x_3, x_4, x_5\}$ is the cyclic group $C_2 = \{e, \varphi\}$. This in turn induces the permutation group $G = \{(1)(2)(3)(4)(5), (1\ 5)(2\ 4)(3)\}$ on $[5] = \{1, 2, 3, 4, 5\}$.

In summary, the symmetries of the beaded necklace are described by

$$G = \{e, \varphi\}, \text{ the permutation group on [5] given by}$$

$$e = (1)(2)(3)(4)(5), \text{ the identity}$$

$$\varphi = (15)(24)(3), \text{ the flip}$$

The reader should note that the cycle structure of the group actions is dependent on the set X. For example, suppose that the necklace has six beads. Now the group of symmetries $C_2 = \{e, \varphi\}$ induces a permutation group on [6], where the identity $e = (1)(2)(3)(4)(5)(6)$ has six 1-cycles and the flip $\varphi = (1\ 6)(2\ 5)(3\ 4)$ has three 2-cycles.

Yet another group of symmetries is shown in Figure 8.5, in which the members of the set $X = \{x_1, x_2, x_3, x_4, x_5\}$ are the five vertices of the complete graph K_5. If the vertices are allowed to be permuted with no restrictions, then the group of actions is the set of all permutations of the five vertices in X. This is known as the *symmetric group* S_5 of degree 5. The group of actions induces the set of all of the of all permutations of $\{1, 2, 3, 4, 5\}$, and therefore the order of the group is $|S_5| = 5!$.

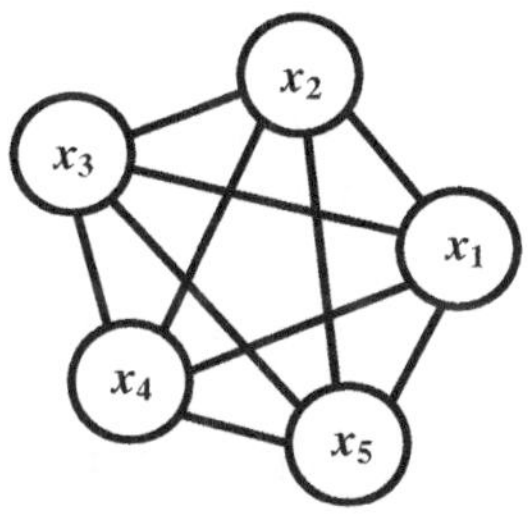

FIGURE 8.5 The group of symmetries on the labels of a complete graph on five vertices is the symmetric group S_5.

8.2.5 Fixed Points and Stabilizer Subgroups

Let $\gamma \in G$ be an action taken from a group of actions G on a set X. We say that $x \in X$ is a *fixed point* under γ if $\gamma x = x$. The subset of all points of X that are fixed by the action γ is denoted by

$$X_\gamma = \{x \in X \mid \gamma x = x\} \tag{8.6}$$

The identity action e fixes every element of X, so $X_e = X$ Other actions may have no fixed points at all. For the square tabletop, $X_\rho = X_{\rho^2} = X_{\rho^3} = \varnothing$, since rotations of 90°, 180°, and 270° move each corner to a new corner. For the floating square stained-glass frame, $X_{\varphi_2} = \{x_1, x_3\}$ and $X_{\varphi_4} = \{x_2, x_4\}$. The flip φ in Figure 8.4 of the beaded necklace fixes the middle bead, so $X_\varphi = \{x_3\}$. If $G = S_5$ is the symmetric group of 120 permutations, there are 44 actions with no fixed points since this is the number of derangements in the group.

If $x \in X$, we can also consider the set of actions $\gamma \in G$ that fix x. Define

$$G_x = \{\gamma \in G \mid \gamma x = x\} \tag{8.7}$$

Certainly $e \in G_x$ since $ex = x$ for any $x \in X$. Moreover, if $\gamma_1 \in G_x$ and $\gamma_2 \in G_x$, then $(\gamma_1\gamma_2)x = \gamma_1(\gamma_2 x) = \gamma_1 x = x$, which shows that $\gamma_1\gamma_2 \in G_x$. This is enough to guarantee that G_x is a subgroup of G. It is called the *stabilizer subgroup of* x, or simply the *stabilizer of* x.

For the tabletop, the stabilizer of each square corner x_i is the trivial subgroup $G_{x_i} = \{e\}$ since none of the rotations by 90°, 180°, or 270° fix any corner. The stabilizers of the floating square shown in Figure 8.3 are $G_{x_1} = G_{x_3} = \{e, \varphi_2\}$ and $G_{x_2} = G_{x_4} = \{e, \varphi_4\}$. The stabilizers of the beaded necklace in Figure 8.4 are the trivial groups $G_{x_i} = \{e\}$ when $i = 1, 2, 4$, or 5, but $G_{x_3} = \{e, \varphi\}$ since the flip φ fixes the middle bead. The stabilizers of the complete graph in Figure 8.5 are the groups each isomorphic to the symmetric group S_4 that fix one x_i and are a permutation of the remaining vertices of the graph.

8.2.6 Orbits

Let G be a group of actions on set X. If $x \in X$, then the set of images of x under all of the actions from G on x is called the *orbit of* x. That is, the orbit of $x \in X$ under the actions of the group G is the subset of X defined by

$$\text{orb}(x) = \{\gamma x \in X \mid \gamma \in G\} \tag{8.8}$$

For example, each corner of the tabletop can be rotated to any other corner, so the orbit of any corner is all of X. Similarly, the orbit of any pane of the floating square is all of X, and the orbit of any label of the complete graph is X. In each of these three examples, the orbit of each $x \in X$ is all of X, and the group of actions is said to be *transitive*.

Things are different for the beaded necklace. Here the group of actions $C_2 = \{e, \varphi\}$ on X is not transitive. The necklace has three different orbits:

$$\text{orb}(x_1) = \text{orb}(x_5) = \{x_1, x_5\},\ \text{orb}(x_2) = \text{orb}(x_4) = \{x_2, x_4\},\ \text{orb}(x_3) = \{x_3\} \tag{8.9}$$

We see that the orbits of X partition $X = \{x_1, x_2, x_3, x_4, x_5\}$ into disjoint subsets.

More generally, any set X is partitioned into disjoint orbits determined by the group of actions on X. To see why, first note that each $x \in X$ belongs to at least one orbit, namely, orb(x), since $x = ex$. If it is assumed that x belongs to another orbit, say,

$x \in \text{orb}(y)$, then $x = \gamma y$ for some $\gamma \in G$. But then $y = \gamma^{-1}x$, which shows $y \in \text{orb}(x)$ and so $\text{orb}(x) = \text{orb}(y)$.

8.2.7 Comparing the Number of Fixed Points, Stabilizers, and Orbits

It is logical to ask this question: "What connections exist between the sets of fixed points, the stabilizers, and the number of orbits?"

As an example, consider the floating square stained-glass frame, with the panes $X = \{x_1, x_2, x_3, x_4\}$ under the group of actions induced by the dihedral group $D_4 = \{e, \rho, \rho^2, \rho^3, \varphi_1, \varphi_2, \varphi_3, \varphi_4\}$. Its sets of fixed points are

$$X_e = \{x_1, x_2, x_3, x_4\}$$
$$X_\rho = X_{\rho^2} = X_{\rho^3} = X_{\varphi_1} = X_{\varphi_3} = \varnothing$$
$$X_{\varphi_2} = \{x_1, x_3\}, X_{\varphi_4} = \{x_2, x_4\}$$

which shows us that the number of times some action fixes some $x \in X$ is

$$\sum_{\gamma \in D_4} |X_\gamma| = 4 + 0 + 0 + 0 + 0 + 0 + 2 + 2 = 8$$

The stabilizers are $G_{x_1} = G_{x_3} = \{e, \varphi_2\}$, $G_{x_2} = G_{x_4} = \{e, \varphi_4\}$, so $\sum_{x \in X} |G_x| = 2 + 2 + 2 + 2 = 8$. Finally, there is just one orbit, X itself.

If we let k denote the number of orbits, then the floating square example satisfies the relation

$$\sum_{\gamma \in G} |X_\gamma| = \sum_{x \in X} |G_x| = k|G| \tag{8.10}$$

This result will be proved to hold in general in the next section. It is known as *Burnside's lemma.*

More evidence supporting formula (8.10) is given in the following example.

Example 8.1 (Verifying Burnside's Lemma for a Beaded Necklace) Verify these two formulas for the beaded necklace shown in Figure 8.4.

$$\sum_{\gamma \in \{e, \varphi\}} |X_\gamma| = 5 + 1 = 6$$

$$\sum_{x \in X} |G_x| = 1 + 1 + 2 + 1 + 1 = 6$$

Solution. Since $|G| = |\{e, \varphi\}| = 2$ and there are $k = 3$ orbits, we see that Burnside's lemma (8.10) holds for the beaded necklace. ■

8.2.8 Colorings and Labelings

Let $Y = \{y_1, y_2, \dots, y_n\}$ be a set a colors or labels that can be assigned to the elements of $X = \{x_1, x_2, \dots, x_m\}$. Any function $f : X \to Y$ then defines a *coloring* or *labeling* of X, where $f(x_i) = y_j$ is the color or label assigned to element x_i. For example, the 2-colorings, white or blue, of the tiles of the tabletop are given by the set of functions from $X = \{x_1, x_2, x_3, x_4\}$ into $Y = \{w, b\}$. The function that maps (x_1, x_2, x_3, x_4) to (b, w, b, w) is the tiling by blue-white-blue-white tiles in that order, which appears as the first coloring in the fourth row of Figure 8.1.

It will be convenient to let T denote the set of all colorings of elements X with colors chosen from a set Y: $T = \{f \mid f : X \to Y\}$. Since there are n choices of color in Y for each of the m objects in X, we see that $|T| = n^m$.

The coloring $g \in T$ is said to be *equivalent* to the coloring $f \in T$ if there is some group action γ for which $g(x) = f(\gamma x)$ for all $x \in X$. We will write $g = \gamma f$ to show that the coloring g is the result of the action by the group element $\gamma \in G$ on coloring f. For example, suppose that g and f are these two of the colorings of the tabletop with one blue tile and three white tiles:

The formulas for these colorings are $g : (x_1, x_2, x_3, x_4) = (w, b, w, w)$ and $f : (x_1, x_2, x_3, x_4) = (b, w, w, w)$. These are equivalent colorings since the coloring given by g is the same as the coloring ρf given by the rotation of the coloring f by the group action $\rho \in C_4$.

In the following section, we will prove that the relation $g \sim f$ on T defined by $g \sim f$ if and only if $g = \gamma f$ for some $\gamma \in G$ is an equivalence relation on the set of colorings. Therefore, the set of actions in the group G partitions the set of colorings T into disjoint equivalence classes. The number of coloring patterns is then the number of equivalence classes.

Burnside's lemma, given by equations (8.10), gives us a way to count the number of nonequivalent colorings. That is, each formula gives us a way to count the number of different equivalence classes. To see how, view G as a group of actions, not on X, but now on the set of colorings T. The set $T_\gamma = \{f \in T \mid \gamma f = f\}$ is the set of colorings fixed by the action $\gamma \in G$. For example, each of the four-colorings (b,b,b,b), (b,w,b,w), (w,b,w,b), and (w,w,w,w) of the tabletop is fixed by the identity e and by the 180° rotation $\rho^2 \in C_4$. The complete list of the size of the sets of fixed 2-colorings is $|T_e| = 2^4 = 16$, $|T_\rho| = 2$, $|T_{\rho^2}| = 4$, $|T_{\rho^3}| = 2$. Then, by Burnside's lemma, the number of nonequivalent colorings is

$$k = \frac{1}{|G|} \sum_{\gamma \in C_4} |T_\gamma| = \frac{1}{|C_4|}(16 + 2 + 4 + 2) = \frac{24}{4} = 6$$

This confirms why we discovered that there are six nonequivalent tilings (or colorings) of the tabletop, as shown in the six rows in Figure 8.1.

The number of orbits (i.e., the number of equivalence classes of colorings) could also be computed by listing the sizes of the stabilizer subgroups: $G_f = \{\gamma \in G | \gamma f = f\}$. For example, the coloring $f = (b,w,b,w)$ of the tabletop is fixed by the members of the stabilizer $G_f = \{e, \rho^2\}$, so $|G_f| = |G_{(b,w,b,w)}| = |\{e, \rho^2\}| = 2$. The complete list of the sizes of the stabilizers is

$$
\begin{aligned}
&|G_{(b,b,b,b)}| = |G_{(w,w,w,w)}| = |C_4| = 4, |G_{(b,w,b,w)}| = |G_{(w,b,w,b)}| = 2 \\
&|G_{(b,w,w,w)}| = \cdots = |G_{(w,w,w,b)}| = |G_{(b,b,w,w)}| = \cdots = |G_{(w,w,b,b)}| \\
&\quad = |G_{(b,b,b,w)}| = \cdots = |G_{(w,b,b,b)}| = |\{e\}| = 1
\end{aligned}
$$

so by Burnside's lemma the number of nonequivalent 2-colorings of the tabletop is

$$
\begin{aligned}
k &= \frac{1}{|G|} \sum_{f \in T} |G_f| \\
&= \frac{1}{|C_4|}(4+4+2+2+1+1+1+1+1+1+1+1+1+1+1+1) = \frac{24}{4} = 6
\end{aligned}
$$

Example 8.2 (Number of n-Colorings of a Necklace with m Beads) What is the number of beaded necklaces, each with m spherical beads of one of n colors?

Solution. The set of beads is $X = \{x_1, x_2, \ldots, x_m\}$, the set of colors is $Y = \{y_1, y_2, \ldots, y_n\}$, and the group of actions is induced by the cyclic group $C_2 = \{e, \varphi\}$. For the identity action, each of the m beads can be any of n colors, so $|T_e| = n^m$. If m is even, there are $n^{m/2}$ ways to color the first $m/2$ beads, and these account for all of the colorings that are fixed by the flip φ. Similarly, if m is odd, there are $n^{(m+1)/2}$ colorings fixed by φ. Applying Burnside's lemma, we then see that the number of nonequivalent colorings of a necklace of m beads chosen from n colors is

$$
k = \frac{1}{|\{e, \varphi\}|}(|T_e| + |T_\varphi|) = \begin{cases} \dfrac{n^m + n^{m/2}}{2}, & m \text{ even} \\[2ex] \dfrac{n^m + n^{(m+1)/2}}{2}, & m \text{ odd} \end{cases}
$$

For example, there are $(2^5 + 2^3)/2 = \frac{40}{2} = 20$ nonequivalent necklaces with five beads of two colors, and there are $(3^6 + 3^3)/2 = (729 + 27)/2 = \frac{756}{2} = 378$ different necklaces with six beads of three colors. ■

PROBLEMS

8.2.1. Let γ and η be the permutations

$$\gamma = \begin{pmatrix} 1 & 2 & 3 & 4 & 5 \\ 4 & 1 & 5 & 2 & 3 \end{pmatrix} \quad \text{and} \quad \eta = \begin{pmatrix} 1 & 2 & 3 & 4 & 5 \\ 5 & 2 & 4 & 3 & 1 \end{pmatrix}$$

(a) write γ and η as products of cycles.

(b) compute the products $\gamma\eta, \eta\gamma, \gamma^2, \eta^2$, expressing your answers as a product of cycles. [*Note*: $(\gamma\eta)(1) = \gamma(\eta(1)) = \gamma(5) = 3$.]

8.2.2. Let $\gamma = (1\ 3\ 4)(2\ 5), \eta = (1\ 5)(2\ 3)(4)$. Find the following permutations, written as a product of disjoint cycles:

(a) $\gamma\eta$ **(b)** $\eta\gamma$ **(c)** γ^{-1}

8.2.3. **(a)** What is the order of a k-cycle?

(b) What is the order of a product of a 3-cycle and a 4-cycle?

(c) What is the order of a product of a 4-cycle, a 5-cycle, and a 6-cycle?

8.2.4. Tabulate all of the products $\gamma\eta$, where $\gamma, \eta \in C_4 = \{e, \rho, \rho^2, \rho^3\}$.

8.2.5. Tabulate all of the products $\gamma\eta$, where $\gamma, \eta \in D_3 = \{e, \rho, \rho^2, \varphi_1, \varphi_2, \varphi_3\}$. For example, since $\rho = (1\ 2\ 3)$ and $\varphi_1 = (1)(2\ 3)$, it follows that $\rho\varphi_1 = (1\ 2\ 3)(1)(2\ 3) = (1\ 2)(3) = \varphi_3$.

8.2.6. In Problems 8.2.4 and 8.2.5, the set of products in each row and in each column formed the entire set of elements of the group. Does this property hold for the table of products of any group G?

8.2.7. The top of a square table will be covered with nine congruent ceramic tiles, $X = \{x_1, x_2, \ldots, x_9\}$, so its group of symmetries is the cyclic group $G = C_4 = \{e, \rho, \rho^2, \rho^3\}$:

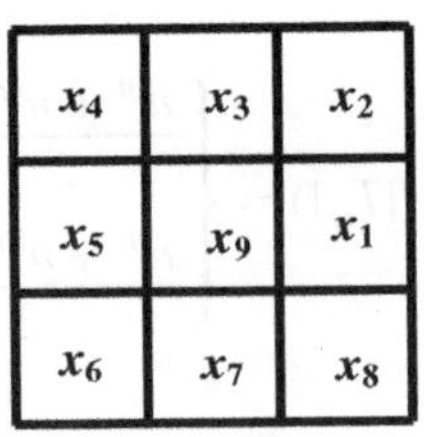

(a) determine the set of fixed points X_γ for each $\gamma \in G$.

(b) determine the stabilizer subgroup G_x for each $x \in X$.

(c) determine the orbits orb(x) for each $x \in X$.

(d) verify that the length of each orbit is given by $|\text{orb}(x)| = |G|/|G_x|$.

8.2.8. A floating square stained-glass window frame with nine square panes $X = \{x_1, x_2, \ldots, x_9\}$, has the symmetries of the dihedral group $D_4 = \{e, \rho, \rho^2, \rho^3, \varphi_1, \varphi_2, \varphi_3, \varphi_4\}$ as shown in Figure 8.3.

(a) Determine the set of fixed points X_γ for each $\gamma \in G$.

(b) Determine the stabilizer subgroup G_x for each $x \in X$.

(c) Determine the orbits, orb (x) for each $x \in X$.

(d) Verify that the length of each orbit is given by $|\text{orb}(x)| = |G|/|G_x|$.

8.2.9. The triangular tabletop shown will be tiled with six triangles that are white or blue:

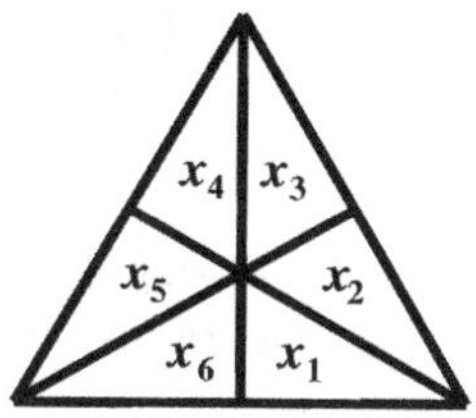

The group of symmetries is $C_3 = \{e, \rho, \rho^2\}$. This induces the permutation group G on [6] for which $e = (1)(2)(3)(4)(5)(6)$, $\rho = (1\ 3\ 5)(2\ 4\ 6)$.

(a) Write ρ^2 as a product of disjoint cycles.

(b) Determine the cycle index of G.

(c) Determine the number of nonequivalent 2-colorings of the triangular tabletop.

(d) Determine the number of nonequivalent 2-colorings with three tiles of each color.

8.2.10. The square tabletop is divided into eight regions that will be tiled with a pattern of blue and white triangular tiles:

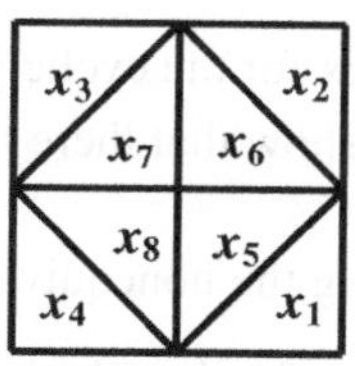

Let G be the group of actions on [8] induced by the cyclic group C_4.

(a) Write each group action as a product of disjoint cycles.

(b) Determine the cycle index.

(c) Determine the number of nonequivalent 2-colorings of the tabletop.

(d) Determine the number of nonequivalent 2-colorings with four tiles of each color.

In the remaining problems, $X = \{x_1, x_2, x_3, x_4, x_5\}$ denotes the set of five disks shown here:

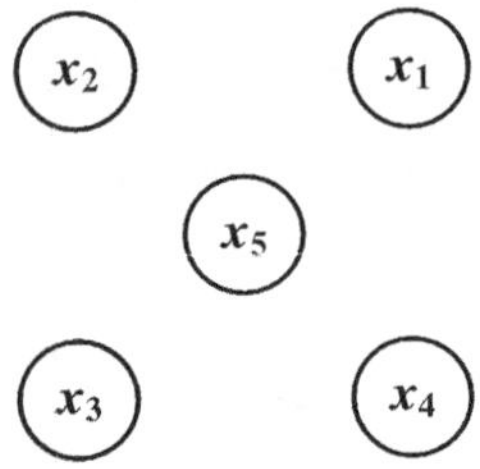

8.2.11. Write each action $\gamma \in C_4 = \{e, \rho, \rho^2, \rho^3\}$ on the rotationally symmetric arrangement of disks X as a product of disjoint cycles.

8.2.12. Compute the set of fixed points X_γ for each $\gamma \in C_4 = \{e, \rho, \rho^2, \rho^3\}$.

8.2.13. Compute the stabilizer subgroups G_x of $G = C_4$ for each $x_i \in X$.

8.2.14. Let f be the coloring of X given by $f = (w, b, w, b, w)$. Calculate the stabilizer G_f and the orbit $\langle f \rangle$ under the group of actions C_4.

8.2.15. Let $G = D_4 = \{e, \rho, \rho^2, \rho^3, \varphi_1, \varphi_2, \varphi_3, \varphi_4\}$ be the group of actions on the floating pattern of five disks X. Write each action $\gamma \in D_4$ as a product of disjoint cycles. For example, $\varphi_1 = (1)(2\ 4)(3)(5)$.

8.2.16. Compute the set of fixed points X_γ for each $\gamma \in G = D_4 = \{e, \rho, \rho^2, \rho^3, \varphi_1, \varphi_2, \varphi_3, \varphi_4\}$.

8.2.17. Compute the stabilizer subgroups G_x of $G = D_4$ for each $x_i \in X$.

8.2.18. Let f be the coloring of X given by $f = (w, b, b, b, b)$. Calculate the stabilizer G_f and the orbit $\langle f \rangle$ under the group of actions D_4.

8.2.19. (a) Compute the cycle index for the cyclic group C_4 acting on X.

(b) Use the cycle index to show that there are 12 nonequivalent 2-colorings of the five disks.

(c) Draw a diagram showing the nonequivalent 2-colorings of X.

8.2.20. (a) Compute the cycle index for the dihedral group D_4 acting on X.

(b) Use the cycle index to show that there are 63 nonequivalent 3-colorings of the five disks.

8.3 BURNSIDE'S LEMMA

In the previous section, we introduced the following sets of interest:

$X = \{x_1, x_2, \dots, x_m\}$, a set of objects to be colored or labeled

$Y = \{y_1, y_2, \dots, y_n\}$, a set of colors or labels

$T = \{f : X \to Y\}$, a set of colorings or labelings of X

$G =$ a group of actions on either

X, if $\gamma \in G$ is a permutation of X

T, where $\gamma f \in T$ is the coloring $(\gamma f)(x) = f(\gamma x)$

$X_\gamma = \{x \in X | \gamma x = x\}$, the subset of points of X fixed under $\gamma \in G$

$G_x = \{\gamma \in G | \gamma x = x\}$, the stabilizer subgroup of G determined by $x \in X$

$T_\gamma = \{f \in T | \gamma f = f\}$, the subset of colorings of T fixed $\gamma \in G$

$G_f = \{\gamma \in G | \gamma f = f\}$, the stabilizer subgroup of G determined by $f \in T$

Any group G of actions is isomorphic to a permutation group of $[m]$, so each action $\gamma \in G$ can be expressed as a cycle decomposition, that is, as a product of disjoint cycles. For example, if X is a set of m objects that can be permuted with no restrictions imposed, then the group of actions is the symmetric group S_m of all permutations of $[m]$, a group of order $m!$.

More often, X represents the parts of some geometric figure and the symmetries of the figure are described by a group. As an example, suppose that X labels the vertices of a regular polygon of m vertices in the sequential counterclockwise order $x_1, x_2, \dots, x_m$. The rotational symmetries of the polygon are described by the *cyclic group of order m*, $C_m = \{e, \rho, \rho^2, \dots, \rho^{m-1}\}$, $\rho^m = e$, where ρ is the counterclockwise rotation by $360°/m$. The corresponding action on X is $\rho = (1\ 2\ \dots\ m)$, a single cycle of length m. If m is even, then $\rho^2 = (1\ 3\ \dots\ )(2\ 4\ \dots\ )$ is a product of two cycles of length $m/2$, but when m is odd, $\rho^2 = (1\ 3\ \dots\ m\ 2\ 4\ \dots\ m-1)$, a single cycle of length m.

If the regular m-gon is allowed to float, then there are additional symmetries since reflections (flips) are allowed. As shown in Figure 8.6, the symmetries of a floating regular polygon with m vertices are described by the *dihedral group* $D_m = \{e, \rho, \rho^2, \dots, \rho^{m-1}, \varphi_1, \varphi_2, \dots, \varphi_m\}$, which has order $2m$.

8.3.1 Equivalence Classes of Colorings

The result of the following theorem was stated but not proved in the previous section.

Theorem 8.3 The relation relation $g \sim f$ on T defined by $g \sim f$ if and only if $g = \gamma f$ for some $\gamma \in$ G is an equivalence relation.

Proof. We must show that $\sim$ is reflexive, symmetric, and transitive.

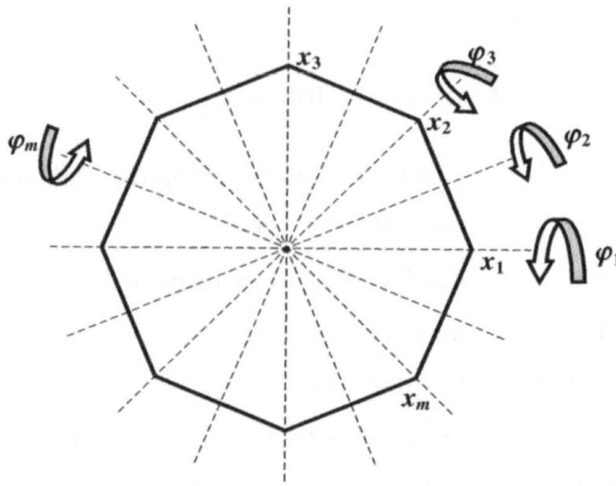

FIGURE 8.6 The regular polygon with m vertices has the rotational symmetries of the cyclic group C_m and the floating symmetries of the dihedral group D_m. For odd m, each flip fixes one vertex. For even m, flips alternate between fixing two vertices and fixing no vertex.

Reflexive Property. If f is any coloring, then $f = ef$ for the identity e, and so $f \sim f$.

Symmetric Property. Suppose that $f \sim g$. Then $f = \gamma g$ for some $\gamma \in G$. If γ^{-1} is the inverse of γ, we see that $g = eg = (\gamma^{-1}\gamma)g = \gamma^{-1}(\gamma g) = \gamma^{-1}f$, which shows that $g \sim f$.

Transitive Property. Suppose that $f \sim g$ and $g \sim h$. Then $f = \gamma g$ and $g = \eta h$ for some $\gamma, \eta \in G$. Then $f = \gamma g = \gamma(\eta h) = (\gamma\eta)h$, which shows that $f \sim h$ since $\gamma\eta \in G$. ■

The relation $\sim$ partitions T into disjoint equivalence classes. Any coloring $f \in T$ belongs to a unique equivalence class, namely, its orbit $\langle f \rangle = \{f_1, f_2, \ldots, f_i\}$, where $f_1 = ef, f_2 = \gamma_2 f, \ldots, f_i = \gamma_i f$ for some set of distinct actions $\{e, \gamma_2, \gamma_3, \ldots \gamma_i\}$. If $\gamma f = \eta f$, only one representative group action is chosen for this set.

Example 8.4 Determine the orbit $\langle f \rangle$ and stabilizer G_f of the coloring (b,b,b,w,w,w) of the triangular arrangement of the six disks shown below, whose symmetries are those of the dihedral group $D_3 = \{e, \rho, \rho^2, \varphi_1, \varphi_2, \varphi_3\}$, $\rho^3 = e$; then show that $|\langle f \rangle||G_f| = |G|$:

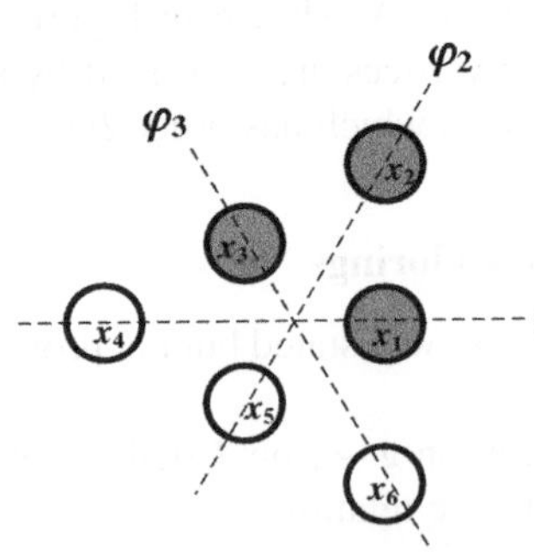

$f = (b, b, b, w, w, w)$

Solution. The coloring is fixed under the reflection φ_2, so the stabilizer is $G_f = \{e, \varphi_2\}$. The orbit $\langle f \rangle = \{f_1, f_2, f_3\}$ is the set of three 2-colorings shown here:

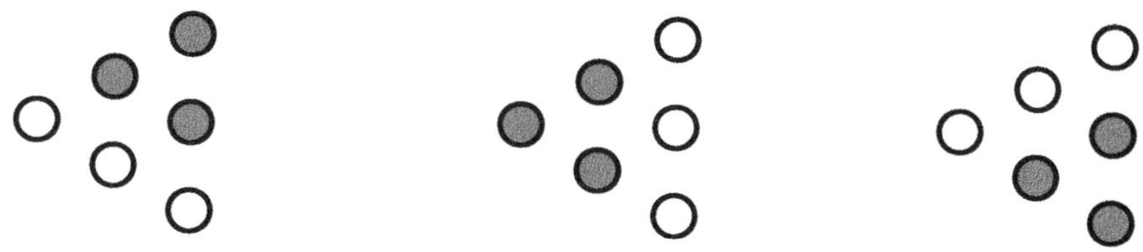

Therefore, $|\langle f \rangle||G_f| = 2 \cdot 3 = 6 = |D_3|$. This can also be shown by constructing a table of group products in this way. Along the left edge of the table are the distinct group actions $\{e, \rho, \rho^2\}$ that determine the orbit $\langle f \rangle$; that is, $f_1 = ef, f_2 = \rho f$, and $f_3 = \rho^2 f$. The members of the stabilizer $G_f = \{e, \varphi_2\}$ of f are placed at the top of the table. We then get this table of products:

	e	φ_2
e	$ee = e$	$e\varphi_2 = \varphi_2$
ρ	$e\rho = \rho$	$\rho\varphi_2 = \varphi_3$
ρ^2	$e\rho^2 = \rho^2$	$\rho^2\varphi_2 = \varphi_1$

Each member of $G = D_3$ appears exactly one time as a product, so $|\langle f \rangle||G_f| = |G|$. ■

The following theorem states that the formula $|\langle f \rangle||G_f| = |G|$ holds in general, and gives us a way to calculate the length of the orbit of a coloring f.

Theorem 8.5 Let G be a group of actions on a set of colorings T, and let $f \in T$ be a coloring. If $\langle f \rangle$ is the orbit of colorings equivalent to f, then the length of the orbit is

$$|\langle f \rangle| = \frac{|G|}{|G_f|} \tag{8.11}$$

Proof. Example 8.4 suggests that we consider a table of products constructed in the following way. If $\langle f \rangle = \{\gamma_1 f, \gamma_2 f, \dots, \gamma_i f\}, \gamma_1 = e$, is the orbit of f, place the group elements $\{\gamma_1, \gamma_2, \dots, \gamma_i\}$ along the left edge of the table. We assume no coloring in the orbit is repeated, so these group elements are all different. Next, place the elements of the stabilizer of f, $G_f = \{\hat{\gamma}_1, \hat{\gamma}_2, \dots, \hat{\gamma}_j\}, \hat{\gamma}_1 = e$, along the top edge of the table. The entries within the table are the $|\langle f\rangle||G_f| = ij$ products $\gamma_r\hat{\gamma}_s$, $1 \le r \le i, 1 \le s \le j$:

	$\hat{\gamma}_1 = e$	$\hat{\gamma}_2$	$\cdots$	$\hat{\gamma}_j$
$\gamma_1 = e$	ee	$e\hat{\gamma}_2$	$\cdots$	$e\hat{\gamma}_j$
γ_2	$\gamma_2 e$	$\gamma_2\hat{\gamma}_2$	$\cdots$	$\gamma_2\hat{\gamma}_j$
$\vdots$	$\vdots$	$\vdots$	$\ddots$	$\vdots$
γ_i	$\gamma_i e$	$\gamma_i\hat{\gamma}_2$	$\cdots$	$\gamma_i\hat{\gamma}_j$

Formula (8.11) follows once it is shown that every $\gamma \in G$ appears exactly once as one of these products, since this shows that $|\langle f\rangle||G_f| = |G|$. There are two claims here:

Claim 1: The entries in the table are distinct. Suppose that $\gamma_r\hat{\gamma}_s = \gamma_t\hat{\gamma}_u$. Then $\gamma_r\hat{\gamma}_s f = \gamma_t\hat{\gamma}_u f$. Since $\hat{\gamma}_s f = \hat{\gamma}_u f = f$, this gives $\gamma_r f = \gamma_t f$, so $\gamma_r = \gamma_t$ since the entries of the left column of the table appear just one time each. We now have that $r = t$, so our original assumption becomes $\gamma_r\hat{\gamma}_s = \gamma_r\hat{\gamma}_u$. Multiplying by γ_r^{-1} on the left then shows us that $\hat{\gamma}_s = \hat{\gamma}_u$, and so $s = u$.

Claim 2: Every action $\gamma \in G$ appears in the table. Let $\gamma \in G$. Then $\gamma f \in \langle f\rangle$, so there is some $\gamma_r \in \{\gamma_1, \gamma_2, \dots, \gamma_i\}$ for which $\gamma f = \gamma_r f$. Therefore, $\gamma_r^{-1}\gamma f = f$, so $\gamma_r^{-1}\gamma \in G_f$ is a member of the stabilizer of f. Then, for some $\hat{\gamma}_s \in G_f$, we have $\gamma_r^{-1}\gamma = \hat{\gamma}_s$. Left multiplication by γ_r then shows that $\gamma = \gamma_r\hat{\gamma}_s$.

■

In Theorem 8.5 we assumed that the group G acted on a set of colorings, but we could just as well have assumed that G acts on a set X. In that case, the set $\langle x\rangle$ would be the orbit of an element $x \in X$, and we would obtain the following corollary of Theorem 8.5.

Corollary 8.6 Let G be a group of actions on a set X, and let $x \in X$. Then the length of the orbit of x is

$$|\text{orb}(x)| = \frac{|G|}{|G_x|} \tag{8.12}$$

In particular, the length of any orbit is a divisor of $|G|$, the order of the group.

8.3.2 The Proof of Burnside's Lemma

Many well-known results in mathematics are misnamed, since they are named for someone other than its originator. For example, it was pointed out earlier in this text that the Binet formulas for the Fibonacci numbers were first derived by Abraham

DeMoivre. The English mathematician William Burnside (1852–1927) quoted what is now known as "his lemma" in his 1897 book on group theory. There he attributes the result to Frobenius even though the result originated in 1845 with Augustin Cauchy. In view of this history, the theorem proved next is sometimes called "the lemma that is not Burnside's."

Theorem 8.7 (Burnside's Lemma) Let G be a group of actions on a set of colorings T, partitioned into disjoint equivalence classes by the equivalence relation $g \sim f$ if and only if $g = \gamma f$ for some $\gamma \in G$. Then each equivalence class is an orbit of equivalent colorings, and the number of orbits (i.e., the number of nonequivalent colorings) is given by these two formulas:

$$k = \frac{1}{|G|}\sum_{\gamma \in G} |T_\gamma| \tag{8.13}$$

$$k = \frac{1}{|G|}\sum_{f \in T} |G_f| \tag{8.14}$$

Proof. Let $\xi : G \times T \to \{0, 1\}$ be the indicator function defined by

$$\xi(\gamma, f) = \begin{cases} 1, & \text{if } \gamma f = f \\ 0, & \text{otherwise} \end{cases}$$

Then the total number of instances that some coloring f is fixed by some action γ is the sum

$$|F_{G,T}| = \sum_{\gamma \in G, f \in T} \xi(\gamma, f) = \sum_{\gamma \in G}\left(\sum_{f \in T} \xi(\gamma, f)\right) = \sum_{\gamma \in G} |T_\gamma|$$

The same sum, iterated in the other order, is

$$|F_{G,T}| = \sum_{\gamma \in G, f \in T} \xi(\gamma, f) = \sum_{f \in T}\left(\sum_{\gamma \in G} \xi(\gamma, f)\right) = \sum_{f \in T} |G_f| = \sum_{f \in T} \frac{|G|}{|\langle f \rangle|}$$

where the last step is a consequence of (8.11). Now suppose that T has k equivalence classes, say, $T = \langle f_1 \rangle \cup \langle f_2 \rangle \cup \cdots \cup \langle f_k \rangle$. Then

$$|F_{G,T}| = |G| \sum_{f \in T} \frac{1}{|\langle f \rangle|} = |G| \sum_{j=1}^{k} \sum_{f \in \langle f_j \rangle} \frac{1}{|\langle f \rangle|} = |G| \sum_{j=1}^{k} |\langle f_j \rangle| \frac{1}{|\langle f_j \rangle|} = k|G|$$

which completes the derivation of (8.13) and (8.14). ■

An interesting example of Burnside's lemma is the case where $X = [m]$ and $G = S_m$ is the symmetric group. The total number of times that any integer $i \in [m]$ is fixed by some permutation $\pi \in S_m$ is given by $\sum_{j=0}^{k} j \binom{m}{j} D_{m-j}$ since there are $\binom{m}{j}$ ways to choose j integers that are fixed and D_{m-j} permutations (the derangement number) that do not fix any of the $m - j$ integers not chosen. The symmetric group S_m has order $| S_m | = m!$ and the orbit of every integer is all of $[m]$. Therefore, there is just one orbit, so $k = 1$ and formula (8.13) gives us the identity

$$\frac{1}{m!} \sum_{j=0}^{k} j \binom{m}{j} D_{m-j} = 1 \tag{8.15}$$

Formula (8.15) tells us that the average number of fixed points over all $m!$ permutations of the symmetric group is one.

Some examples of counting nonequivalent colorings using Burnside's lemma were given in the preceding section, and here are two additional examples.

Example 8.8 Given a prime number p, determine the number of nonequivalent n-colorings of the p vertices of a regular polygon under the cyclic group C_p.

Solution. There are $|T_e| = n^p$ colorings fixed by the identity e. The $p - 1$ rotations in C_p are all are cycles of length p, so each of these fixes only the constant coloring in which all vertices have the same color, one of the n available colors; that is, $|T_\gamma| = n$ for each of the $p - 1$ rotations $\rho, \rho^2, \ldots, \rho^{p-1}$ in C_p. By Burnside's lemma, the number of different colorings is

$$k = \frac{1}{|C_p|} \sum_{\gamma \in C_p} |T_\gamma| = \frac{1}{p}(n^p + (p-1)n) = \frac{n(n^{p-1} - 1)}{p} + n$$

What is interesting is that k must be an integer. Therefore, if n is not divisible by the prime number p, the formula shows us that $n^{p-1} - 1$ is divisible by p. This celebrated result from number theory is known as *Fermat's little theorem*. For example, if $n = 10$, the theorem tells us that $10^{100} - 1$ is divisible by the prime number $p = 101$. ∎

Example 8.9 Determine the number of nonequivalent 2-colorings of the six disks in the triangular arrangement shown on the next page and in Example 8.4, with the symmetries induced by the dihedral group D_3:

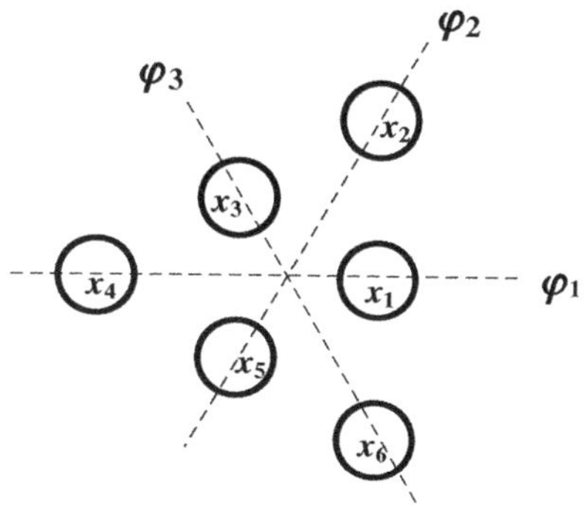

Solution. To apply formula (8.13), we must count how many 2-colorings are fixed by the action of each $\gamma \in G = D_3 = \{e, \rho, \rho^2, \varphi_1, \varphi_2, \varphi_3\}$. Since the identity $e = (1)(2)(3)(4)(5)(6)$ fixes each of the disks $x_1, x_2, \ldots, x_6$, and each disk can be either of two colors, it follows that $|T_e| = 2^6$. The 120° rotation $\rho = (1\ 3\ 5)(2\ 4\ 6)$, however, has two cycles, each of which can be of either color, so $|T_\rho| = 2^2$. Likewise, $|T_{\rho^2}| = 2^2$ for the 270° rotation. The reflection $\varphi_1 = (1)(2\ 6)(3\ 5)(4)$ has four cycles, and the discs corresponding to each cycle can be one the two colors, so $|T_{\varphi_1}| = 2^4$, and similarly $|T_{\varphi_2}| = |T_{\varphi_3}| = 2^4$. Therefore, by Burnside's lemma, there are

$$k = \frac{1}{|D_3|}(|T_e| + |T_\rho| + |T_{\rho^2}| + |T_{\varphi_1}| + |T_{\varphi_2}| + |T_{\varphi_3}|)$$
$$= \frac{1}{6}(2^6 + 2^2 + 2^2 + 2^4 + 2^4 + 2^4) = \frac{120}{6} = 20$$

nonequivalent 2-colorings of the pattern of six disks. We also see that the number of 3-colorings is $k = \frac{1}{6}(3^6 + 3^2 + 3^2 + 3^4 + 3^4 + 3^4) = \frac{990}{6} = 165$, with similar formulas for the number of n-colorings obtained by replacing the 3 with n. ■

Burnside's lemma gives us a way, although not an easy way, to count the total number of nonequivalent colorings. In the next section, we will introduce an easier approach by means of the *cycle index*. This sets us up to understand Pólya's method of counting in which a generating function can be computed that will answer questions such as these:

1. How many nonequivalent 2-colorings have two blue disks and four white disks?
2. How many 3-colorings have one red disk, two blue disks, and three white disks?

PROBLEMS

8.3.1. Let $X = \{x_1, x_2, x_3, x_4, x_5, x_6\}$ be the pattern of six disks shown below, with the symmetries induced by the dihedral group $D_3 = \{e, \rho, \rho^2, \rho^3, \varphi_1, \varphi_2, \varphi_3\}$, where ρ is a rotation by 120° and the flips φ_i are shown in Example 8.4. The disks have been 2-colored by the function $f = (b,w,b,w,b,w)$.

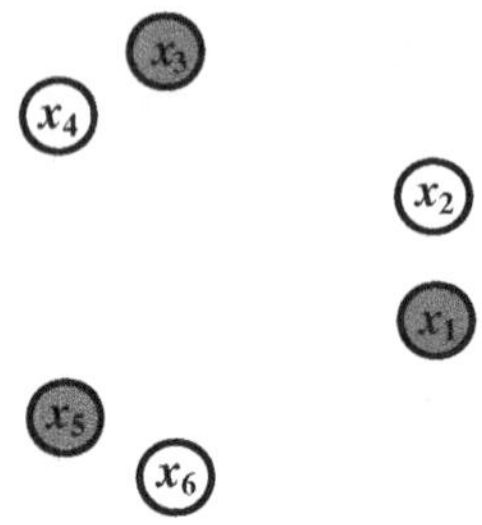

(a) Show that there are 3 ways to express the orbit $\langle f \rangle$ using different sets of group actions that give the equivalent colorings of f.

(b) Determine the stabilizer subgroup of f.

(c) Use each of the forms of the group actions determining the orbit of f to make a multiplication table with the stabilizer of f.

8.3.2. Let $X = \{x_1, x_2, x_3, x_4, x_5, x_6, x_7, x_8\}$ be the pattern of eight discs shown below symmetric with the symmetries of the dihedral group $D_4 = \{e, \rho, \rho^2, \rho^3, \rho^4, \varphi_1, \varphi_2, \varphi_3, \varphi_4\}$. Let ρ be a rotation by 90° and let flip φ_i fix disk x_i. The disks have been 2-colored by the function $f = (w, b, b, b, w, b, b, b)$.

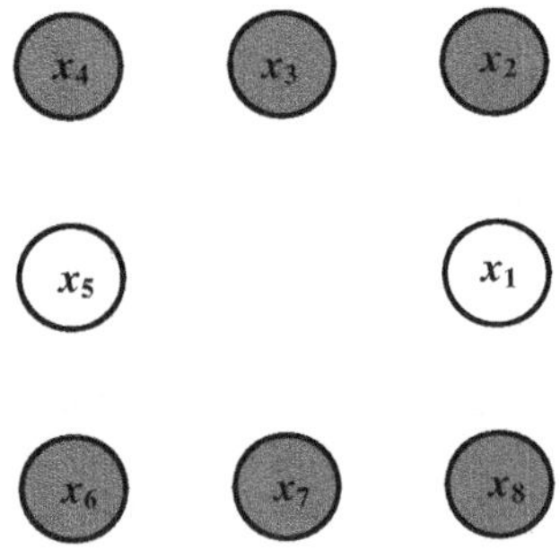

(a) Determine four different ways to express the orbit $\langle f \rangle$, using different sets of actions to describe the equivalent colorings of the orbit.

(b) Determine the stabilizer of f.

(c) Make four multiplication tables similar to the one in Example 8.4, using each of the four ways to write the orbit $\langle f \rangle$.

8.3.3. Let $X = \{x_1, x_2, \ldots, x_5\}$ be the vertices of a regular pentagon allowed by float so that its group of actions is induced by the symmetries of the dihedral group D_5.

(a) Determine the set $X_\gamma, \gamma \in D_5$.

(b) Determine the set $G_x, x \in X = \{x_1, x_2, x_3, x_4, x_5\}$.

(c) Verify Burnside's lemma using part (a).

(d) Verify Burnside's lemma using part (b).

8.3.4. Repeat Problem 8.3.3 but for the vertices $X = \{x_1, x_2, \ldots, x_6\}$ of a regular hexagon under the actions induced by the dihedral group D_6.

8.3.5. Let $X = [24]$ be the set of vertical and horizontal edges that join the vertices of the square grid in the following diagram Assume that X has the symmetries induced by the dihedral group $D_4 = \{e, \rho, \rho^2, \rho^3, \rho^4, \varphi_1, \varphi_2, \varphi_3, \varphi_4\}$.

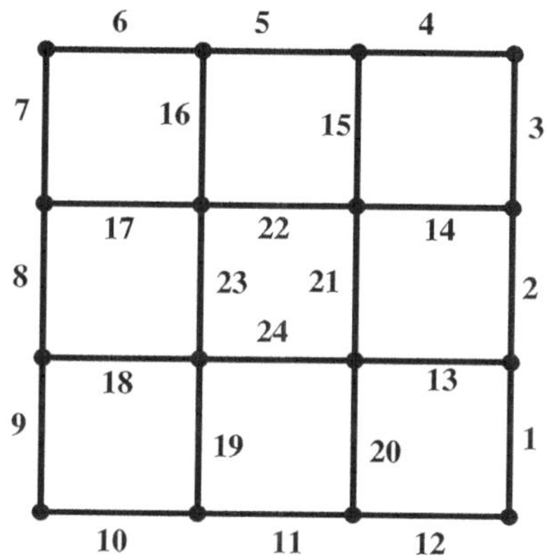

(a) Partition X into equivalence classes.

(b) Verify that the number of equivalence classes obtained in part (a) is given by the formula $k = (1/|G|) \sum_{\gamma \in G} |X_\gamma|$.

8.3.6. Repeat the Problem 8.3.5, but for the pattern of 40 vertical and horizontal edges joining adjacent vertices in a 5×5 square grid.

(a) Draw a diagram with all of the edges in an orbit labeled a, all of the edges in a different orbit with b, and so on.

(b) What is the number of orbits?

(c) Verify that $k = (1/|G|) \sum_{\gamma \in G} |X_\gamma|$ is the number of orbits found in part (b).

8.3.7. A 2×3-*ft* rectangular table is tiled with 1×1-*ft* square tiles. Use Burnside's lemma to determine the number of nonequivalent tables given that the tiles are white or blue.

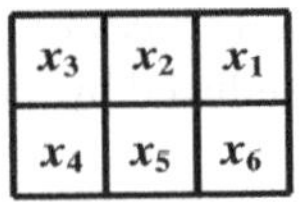

That is, show that the number of nonequivalent 2-colorings is given by each of these formulas.

(a) $k = \frac{1}{|G|} \sum_{\gamma \in G} |T_\gamma|$ **(b)** $k = \frac{1}{|G|} \sum_{f \in T} |G_f|$.

8.3.8. Let $X = \{x_1, x_2, x_3, x_4, x_5, x_6\}$ be the floating pattern of six disks shown below:

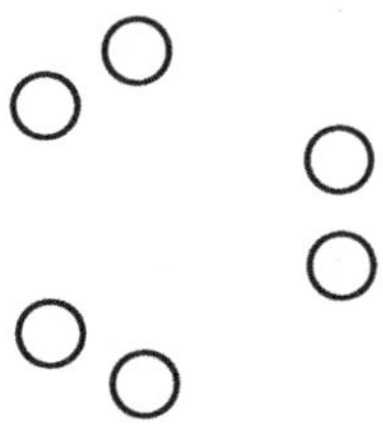

Determine the number of nonequivalent 2-colorings as given by $k = (1/|G|) \sum_{\gamma \in G} |T_\gamma|$.

8.3.9. Complete the following table, listing all of the possible monomials for the permutations $\gamma \in S_4$ and the number of permutations with that monomial. For example, the table shows that 6 permutations include two 1-cycles and one 2-cycle.

Monomial	Number of Permutations
t_1^4	1
$t_1^2 t_2^1$	6

8.3.10. A café has two identical small tables that each seat one person, three identical circular tables that each seat two people, and two identical circular tables that each seat three people.

(a) Explain why the number of ways 14 people can be seated at the tables is given by the expression $14!/(2!3!2!1^2 2^3 3^2)$.

(b) At the General Café, there are $c(1), c(2), \ldots, c(m)$ identical circular tables that seat $1, 2, \ldots, m$ people, respectively, where $c(i) \geq 0$ and $c(1) + 2c(2) + \cdots + mc(m) = m$ is the seating capacity. Explain why the number of ways to seat m people at the tables is given by the expression

$$\sigma(c) = \frac{m!}{\prod_{k=1}^{m} c(k)!k^{c(k)}}$$

(c) Explain why the equation in part (b), known as *Cauchy's formula*, is the number of permutations in the symmetric group S_m whose cycle decomposition has $c(1)$ cycles of length 1, $c(2)$ cycles of length 2, ..., and $c(m)$ cycles of length m. That is, there are $\sigma(c)$ permutations of $[m]$ with the cycle index $t_1^{c(1)} t_2^{c(2)} \cdots t_m^{c(m)}$.

8.4 THE CYCLE INDEX AND PÓLYA'S METHOD OF ENUMERATION

As in the preceding sections, we wish to count the number of colorings of a set of objects $X = \{x_1, x_2, \ldots, x_m\}$ that are nonequivalent to one another under any action γ from a group G. There are n colors available taken from the set $Y = \{y_1, y_2, \ldots, y_n\}$.

We have seen that the total number of colorings can be determined with Burnside's lemma, but in this section we will see that there is a more direct way to count the total number of colorings by means of a polynomial in several variables called the *cycle index*. The cycle index prepares us to derive a powerful method of enumeration known as *Pólya counting*. Pólya's formula is a generating function that gives us an inventory of colorings organized by the number of times each color appears.

8.4.1 Monomials and the Cycle Index

As its name suggests, the cycle index is determined by the cycle structure of the actions of a group G on a set X. The cycle index is defined this way.

Definition 8.10 (Monomials and the Cycle Index) Let G be a group of permutations on a set $X = \{x_1, x_2, \ldots, x_m\}$. Given $\gamma \in G$, let $c_\gamma(1)$ denote the number of cycles in the cycle decomposition of γ of length 1, $c_\gamma(2)$ denote the number of cycles of length 2, ..., and $c_\gamma(m)$ denote the number of cycles of length m. Then the *monomial* of γ is the polynomial

$$p_\gamma(t_1, t_2, \ldots, t_m) = t_1^{c_\gamma(1)} t_2^{c_\gamma(2)} \cdots t_m^{c_\gamma(m)} \tag{8.16}$$

in the m variables $t_1, t_2, \ldots, t_m$. The *cycle index* of G is the polynomial

$$Z(G) = \frac{1}{|G|} \sum_{\gamma \in G} p_\gamma(t_1, t_2, \ldots, t_m) = \frac{1}{|G|} \sum_{\gamma \in G} t_1^{c_\gamma(1)} t_2^{c_\gamma(2)} \cdots t_m^{c_\gamma(m)} \tag{8.17}$$

The letter Z is very standard, since it is a reference to *Zycklenzeiger*, meaning *cycle indicator* in German.

Example 8.11 The cycle index for the tabletop was derived earlier, as given by the polynomial (8.3). Now determine the cycle index for the floating square stained-glass frame, the beaded necklace with m beads, and the complete graph on m vertices.

Solution.

Floating Square. The monomials that correspond to each of the actions from the dihedral group D_4 are evident from Figure 8.3: $e \to t_1^4$; $\rho, \rho^3 \to t_4^1$; $\rho^2 \to t_2^2$, $\varphi_1, \varphi_3 \to t_2^2$, $\varphi_2, \varphi_4 \to t_1^2 t_2^1$. Therefore the cycle index is

$$Z(D_4) = \frac{1}{8}\left(t_1^4 + 2t_4^1 + 3t_2^2 + 2t_1^2 t_2^1\right) \tag{8.18}$$

Even though the tabletop and window frame both have the same underlying set $X = \{x_1, x_2, x_3, x_4\}$, the different symmetry groups C_4 and D_4 give rise to different cycle indices.

Beaded Necklace. A necklace with a fixed clasp and m beads always includes the monomial t_1^m and either the monomial $t_2^{m/2}$ if m is even or otherwise the monomial $t_1^1 t_2^{(m-1)/2}$ when m is odd. Therefore the cycle index is

$$Z(C_2) = \begin{cases} \frac{1}{2}\left(t_1^m + t_2^{m/2}\right), & m \text{ even} \\ \frac{1}{2}\left(t_1^m + t_1^1 t_2^{(m-1)/2}\right), & m \text{ odd} \end{cases} \tag{8.19}$$

Symmetric Group S_m. Every permutation γ from the symmetric group S_m is allowed. The monomials therefore have all of the possible forms

$$p_\gamma = t_1^{c(1)} t_2^{c(2)} \cdots t_m^{c(m)} \tag{8.20}$$

Here the exponents must satisfy the constraint

$$c(1) + 2c(2) + \cdots + mc(m) = m \tag{8.21}$$

since each of the integers $1, 2, \ldots m$ occurs in exactly one cycle of some length between 1 and m. Next, we need to determine the number monomials that have this form. Imagine filling a $1 \times m$ rectangle with $c(1)$ boxes of size 1×1, $c(2)$ boxes of size 1×2, and so on. Now insert the terms of the $m!$ permutations of $[m]$

into the row of boxes. This means that there are $\prod_{1 \le k \le m} k^{c(k)} c(k)!$ monomials of the form (8.20) since the $c(k)$ boxes of length k can be permutated in $c(k)!$ ways and the entries in each box can be rotated in k ways. Therefore, the number of monomials of the form (8.20) that satisfy (8.21) is

$$\sigma(c) = \frac{m!}{\prod\limits_{1 \le k \le m} k^{c(k)} c(k)!}, c(1) + 2c(2) + \cdots + mc(m) = m \qquad (8.22)$$

and the cycle index is

$$Z(S_m) = \frac{1}{m!} \sum_c \sigma(c) t_1^{c(1)} t_2^{c(2)} \cdots t_m^{c(m)}, \qquad (8.23)$$

where the sum is over all $c = (c(1), c(2), \ldots, c(m))$ satisfying (8.21). ■

Example 8.12 Determine the cycle index for the group of actions induced by the dihedral group D_3 on the triangular arrangement of six disks shown below.

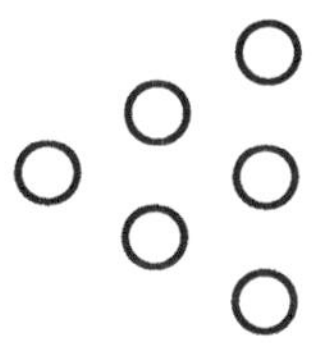

Solution. The diagram makes it apparent that the monomials are $e \to t_1^6$; $\rho, \rho^2 \to t_3^2$; $\varphi_1, \varphi_2, \varphi_3 \to t_1^2 t_2^2$. Therefore the cycle index is

$$Z(D_3) = \tfrac{1}{6}\left(t_1^6 + t_3^2 + t_3^2 + t_1^2 t_2^2 + t_1^2 t_2^2 + t_1^2 t_2^2\right) = \tfrac{1}{6}\left(t_1^6 + 2t_3^2 + 3t_1^2 t_2^2\right) \qquad (8.24)$$

■

8.4.2 Total Number of Colorings Determined from the Cycle Index

The cycle index neatly summarizes essential information about the symmetries of a set X under a group of actions G. The total number of nonequivalent n-colorings can be determined easily from the cycle as we now show.

Suppose, for example, that we wish to count the total number of nonequivalent 2-colorings of the triangular pattern of six disks in Example 8.12, which has the cycle index $Z(D_3) = \frac{1}{6}\left(t_1^6 + 2t_3^2 + 3t_1^2 t_2^2\right)$. We know from Burnside's lemma, formula

(8.13), that the total number of nonequivalent colorings can be calculated once we know the number of colorings fixed by the actions in D_3. The monomial t_1^6 shows that one action has six 1-cycles. Each 1-cycle can be either of two colors, so these cycles can be colored in any of 2^6 ways. This is the number obtained by setting $t_1 = 2$ in the monomial t_1^6. Similarly, there are two monomials of the form t_3^2, corresponding to two permutations having two 3-cycles. Each cycle has a choice of two colors, so there are 2^2 colorings fixed by these permutations. Once again, this is the number of fixed colorings obtained by setting $t_3 = 2$ in the term $2t_3^2$ appearing in the cycle index. By the same reasoning, the last term $3t_1^2t_2^2$ of the cycle index informs us that three permutations each have two 1-cycles that can be colored in 2^2 ways and two 2-cycles that can also be colored in 2^2 ways. Therefore, each of the three permutations with the monomial $t_1^2t_2^2$ has $2^2 \cdot 2^2$ colorings fixed by the permutation. Finally, setting $t_1 = t_2 = 2$ in the term $3t_1^2t_2^2$ of the cycle index shows that there are $3 \cdot 2^2 \cdot 2^2$ colorings fixed by these three permutations.

In view of the discussion above, we see that by setting $t_1 = t_2 = t_3 = 2$ in the cycle index there are

$$Z(D_3)|_{t_i=2} = \frac{1}{6}(2^6 + 2 \cdot 2^2 + 3 \cdot 2^2 \cdot 2^2) = \frac{64 + 8 + 48}{6} = \frac{120}{6} = 20 \tag{8.25}$$

nonequivalent 2-colorings of the six-disk triangular pattern.

Similarly, the number of nonequivalent 3-colorings is obtained by letting $t_1 = t_2 = t_3 = 3$ in the cycle index, and we see that there are $Z(D_3)|_{t_i=3} = \frac{1}{6}(3^6 + 2 \cdot 3^2 + 3 \cdot 3^2 \cdot 3^2) = \frac{990}{6} = 165$ nonequivalent 3-colorings.

More generally, we have the following.

Theorem 8.13 Let $Z(G) = (1/|G|)\sum_{\gamma\in G} p_\gamma(t_1, t_2, \ldots, t_m) = (1/|G|)\sum_{\gamma\in G} t_1^{c_\gamma(1)} t_2^{c_\gamma(2)} \cdots t_m^{c_\gamma(m)}$ be the cycle index for a set $X = \{x_1, x_2, \ldots, x_m\}$ under the group of actions G. Then the number of nonequivalent n-colorings of X is

$$Z(G)|_{t_1=t_2=\cdots t_m=n} = \frac{1}{|G|}\sum_{\gamma\in G} p_\gamma(n, n, \ldots, n) = \frac{1}{|G|}\sum_{\gamma\in G} n^{c_\gamma(1)+c_\gamma(2)+\cdots+c_\gamma(m)} \tag{8.26}$$

Proof. Any power t_i^j in the cycle index corresponds to j cycles for which there are n^j choices of colors. Therefore, the total number of n-colorings fixed by the group of actions is obtained by setting $t_i = n$ for every i, $i = 1, 2, \ldots, m$. Formula (8.13) given by Burnside's lemma then gives us (8.26). ■

Since the cycle index of the rotating tabletop is $Z(C_4) = \frac{1}{4}\left(t_1^4 + t_2^2 + 2t_4^1\right)$, the number of nonequivalent 2-colorings is $\frac{1}{4}(2^4 + 2^2 + 2 \cdot 2) = \frac{24}{4} = 6$, agreeing with our earlier count. The number of nonequivalent 3-colorings is $\frac{1}{4}(3^4 + 3^2 + 2 \cdot 3) = \frac{96}{4} = 24$, and the number of nonequivalent n-colorings is $\frac{1}{4}(n^4 + n^2 + 2n)$.

Example 8.14 Determine the cycle index for the group of actions D_8 on a bracelet with 8 spherical beads (see diagram), and determine the number of different bracelets if the beads can be any of n different colors.

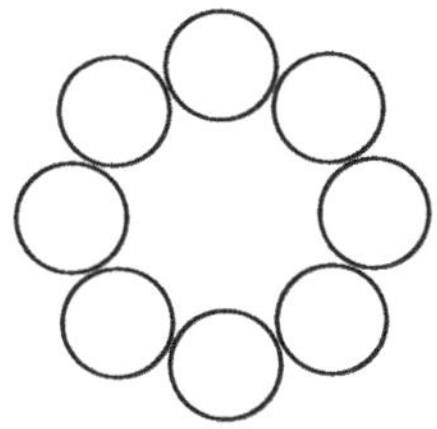

Solution. It should be easy to verify the cycle structure and corresponding monomials of the actions in D_8 as given in this table:

e	t_1^8
$\rho, \rho^3, \rho^5, \rho^7$	t_8^1
ρ^2, ρ^6	t_4^2
ρ^4	t_2^4
$\varphi_1, \varphi_3, \varphi_5, \varphi_7$	$t_1^2 t_2^3$
$\varphi_2, \varphi_4, \varphi_6, \varphi_8$	t_2^4

Adding the eight monomials gives us the cycle index $Z(D_8) = \frac{1}{16}\left(t_1^8 + 4t_1^2 t_2^3 + 5t_2^4 + 2t_4^2 + 4t_8^1\right)$, and the number of nonequivalent n colorings is therefore $\frac{1}{16}(n^8 + 4n^2n^3 + 5n^4 + 2n^2 + 4n) = \frac{1}{16}(n^8 + 4n^5 + 5n^4 + 2n^2 + 4n)$.

This formula shows there are 30, 498, and 4435 different bracelets, respectively, when up to 2, 3, or 4 choices of colors for the beads are available. ■

8.4.3 Pólya's Inventory Function

Let's return once again to the rotating square tabletop, which has the cycle index $Z(C_4) = \frac{1}{4}\left(t_1^4 + t_2^2 + 2t_4^1\right)$ given in equation (8.3). It was shown in Theorem 8.13 that the total number of 2-colorings is determined by setting $t_1 = t_2 = t_4 = 2$ in the cycle index. However, suppose that we want a more detailed description of the 2-colorings, say, one that gives us the number of nonequivalent colorings that use two white and two blue tiles. We are therefore no longer interested in the full set T of all of the two colorings, but instead we are interested only in the subset $T^{(2,2)} \subset T$ of colorings that use two tiles of each color. The number of nonequivalent colorings will again be determined from Burnside's lemma, but now the group of actions will be applied to $T^{(2,2)}$. Just as before, two colorings $g, f \in T^{(2,2)}$ are equivalent to one another if and only there is some action $\gamma \in C_4$ for which $g = \gamma f$. By Burnside's lemma, the number of nonequivalent colorings of the tabletop that each use two tiles of each color is given by

$$\frac{1}{|C_4|} \sum_{\gamma \in C_4} \left|T_\gamma^{(2,2)}\right| = \frac{1}{4}\left(\left|T_e^{(2,2)}\right| + \left|T_\rho^{(2,2)}\right| + \left|T_{\rho^2}^{(2,2)}\right| + \left|T_{\rho^3}^{(2,2)}\right|\right) \tag{8.27}$$

We see that we need to count the number of colorings in $T^{(2,2)}$ that are fixed by each action $\gamma \in C_4$.

Consider first the identity e, which fixes every coloring of the four 1-cycles. However, since only the colorings in $T^{(2,2)}$ are considered, there must be two 1-cycles that are white and two 1-cycles that are blue. These can be counted with an ordinary generating function this way. Each 1-cycle is white or blue, so it contributes a factor $w + b$ to the generating function. Since there are four 1-cycles, the generating function for all of the 2-colorings of the four 1-cycles is $(w + b)^4$. The generating function corresponding to all of the 2-colorings in T that are fixed by the identity e is therefore

$$(w + b)(w + b)(w + b)(w + b) = (w + b)^4 = w^4 + 4w^3b + 6w^2b^2 + 4wb^3 + b^4$$

We see that the generating function is obtained from the monomial t_1^4 associated with the identity e if we set $t_1 = w + b$. The number of colorings in $T^{(2,2)}$ fixed by e is the coefficient of w^2b^2 of $t_1^4 = (w + b)^4$, namely, 6.

Next consider the 90° rotation ρ whose monomial is t_4^1. The 4-cycle fixed by ρ must be all white or all blue, so its generating function is $w^4 + b^4$. This is obtained from the monomial t_4^1 by setting $t_4 = w^4 + b^4$. We see that no coloring in $T^{(2,2)}$ is fixed by this rotation since the coefficient of w^2b^2 in $t_4^1 = (w^4 + b^4)^1$ is 0. Similarly, the rotation ρ^3 by 270° has the same monomial t_4^1 so it does not fix any coloring from $T^{(2,2)}$.

Finally, consider the 180° rotation ρ^2. Its monomial t_2^2 informs us that ρ^2 has two 2-cycles. A 2-cycle must use two colors simultaneously, so each 2-cycle introduces the factor $w^2 + b^2$ into the generating function. Therefore, all of the two colorings of the two 2-cycles have the generating function $t_2^2 = (w^2 + b^2)^2 = w^4 + 2w^2b^2 + b^4$. The coefficient of w^2b^2 is 2, so there are two colorings from $T^{(2,2)}$ that are fixed by ρ^2.

From (8.27) we now see that the number of nonequivalent colorings of the tabletop that each use two white and two blue tiles is the coefficient of w^2b^2 in the generating function

$$\begin{aligned} Z(C_4)|_{t_i=w^i+b^i} &= \tfrac{1}{4}\left(t_1^4 + t_2^2 + 2t_4^1\right)\big|_{t_i=w^i+b^i} \\ &= \tfrac{1}{4}((w+b)^4 + (w^2+b^2)^2 + 2(w^4+b^4)) \\ &= w^4 + w^3b + 2w^2b^2 + wb^3 + b^4 \end{aligned}$$

A generating function of this type is known as a *pattern inventory*.

We have concentrated on seeing why the coefficient of w^2b^2 gives the number of colorings with two tiles of each color, but the same reasoning applies to counting the number of other combinations of colors. Indeed, the pattern inventory is a complete listing of the nonequivalent 2-colorings organized by the number of colors of each type that are used. For example, the coefficient of w^3b is one, so there is one nonequivalent tiling of the tabletop that uses three white tiles and one blue tile.

If a general set of colors $Y = \{y_1, y_2, \dots, y_n\}$ is available, the pattern inventory can be derived from the cycle index by setting each variable t_i equal to the corresponding factor $t_i = y_1^i + y_2^i + \cdots + y_n^i$ of the associated generating function. For example, let's derive the pattern inventory of the 3-colorings of the triangular pattern of six disks considered earlier in Example 8.12. The three colors red, white, and blue form the set of colors $Y = \{r, w, b\}$, and the cycle index of the arrangement of disks is

$$Z(D_3) = \frac{1}{6}\left(t_1^6 + 2t_3^2 + 3t_1^2t_2^2\right) \tag{8.28}$$

as shown in Example 8.12. It is then evident how each monomial corresponds to a term added to the generating function. For instance, the monomial $t_1^2t_2^2$ shows that there are two 1-cycles, each with the generating function factor $t_1 = r + w + b$, and that there are two 2-cycles that each contribute the factor $t_2 = r^2 + w^2 + b^2$ to the pattern inventory. Similar considerations apply to each monomial. Altogether, by setting $t_1 = r + w + b, t_2 = r^2 + w^2 + b^2$, and $t_3 = r^3 + w^3 + b^3$ in the cycle index, we obtain the pattern inventory

$$Z(D_3) = \frac{1}{6}((r+w+b)^6 + 2(r^3+w^3+b^3)^2 + 3(r+w+b)^2(r^2+w^2+b^2))$$

When this expression is expanded, either by hand or better yet with some help from a CAS such as Maple or *Mathematica* we obtain

$$\begin{aligned} &r^6 + w^6 + b^6 + 2r^5b + 2r^5w + 2w^5r + 2w^5b + 2b^5r + 2b^5w + 4r^4w^2 + 4r^4b^2 + 4w^4r^2 \\ &+4w^4b^2 + 4b^4r^2 + 4b^4w^2 + 6r^3w^3 + 6r^3b^3 + 6w^3b^3 + 6r^4wb + 6w^4rb + 6b^4rw \\ &+12r^3w^2b + 12r^3b^2w + 12w^3r^2b + 12w^3b^2r + 12b^3r^2w + 12b^3w^2r + 18r^2w^2b^2 \end{aligned}$$

The term $12rw^3b^2$ in the pattern inventory informs us that there are 12 nonequivalent colorings of the six disks that each include one red, two blue, and three white disks. These are shown in Figure 8.7.

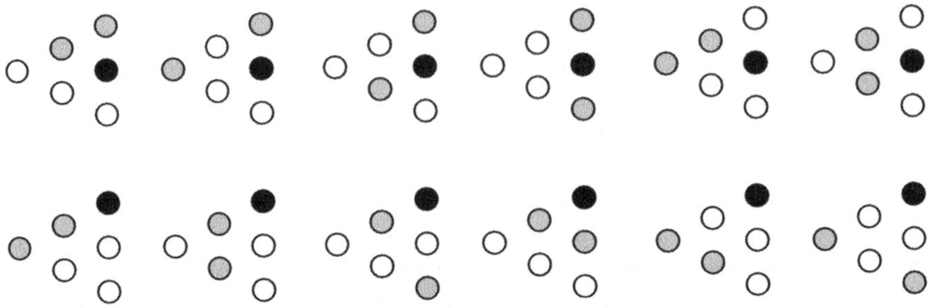

FIGURE 8.7 The 12 nonequivalent colorings of the triangular pattern of six disks that include one red, two blue, and three white disks.

The following theorem describes in general how the cycle index is used to derive the generating function that gives a complete inventory of colorings. The result is widely known as *Pólya's enumeration formula*, since it was used by Pólya [1] to count chemical compounds in his paper of 1937. It was later discovered that his formula had been derived about 10 years earlier by J. H. Redfield [2]. Owing to this history, the theory is sometimes called *Pólya-Redfield counting*.

Theorem 8.15 (Pólya's Enumeration Formula) Let G be a group of actions on a set $X = \{x_1, x_2, \dots, x_m\}$, let $Y = \{y_1, y_2, \dots, y_n\}$be a set of colors or labels, let $T = \{f \mid f : X \to Y\}$ be the set of all colorings of X, and let $Z(G) = (1/|G|)\sum_{\gamma\in G} p_\gamma(t_1, t_2, \dots, t_m)$ be the cycle index of G acting on X. Then the pattern inventory enumerating the nonequivalent colorings of X is the generating function obtained by setting

$$t_1 = y_1 + y_2 + \cdots + y_n, t_2 = y_1^2 + y_2^2 + \cdots + y_n^2, \dots, t_m = y_1^m + y_2^m + \cdots + y_n^m$$

in the cycle index. The coefficient of $y_1^{Y_1} y_2^{Y_2} \cdots y_n^{Y_n}$ in $Z(G)|_{t_i = y_1^i + y_2^i + \cdots + y_n^i}$ is the number of nonequivalent colorings of X that include Y_1 colors y_1, Y_2 colors $y_2, \dots,$ and Y_n colors y_n.

Proof. Let $T^{(Y_1, Y_2, \dots, Y_n)}$ be the subset of colorings of X that include Y_i instances of color y_i, where $Y_1 + Y_2 + \cdots + Y_n = m$. Any i-cycle uses one of the n colors i times, so it contributes a factor $t_i = y_1^i + y_2^i + \cdots + y_n^i$ to the generating function. Therefore, the number of colors in $T^{(Y_1, Y_2, \dots, Y_n)}$ fixed by the action $\gamma \in G$ is the coefficient of $y_1^{Y_1} y_2^{Y_2} \cdots y_n^{Y_n}$ in $p_\gamma(y_1 + y_2 + \cdots, y_1^2 + y_2^2 + \cdots, \dots)$, where p_γ is the monomial of γ. Each coefficient of the pattern inventory now follows as a consequence of Burnside's Lemma. ■

Example 8.16 (Floating Square Window with Eight Panes) Determine the pattern inventory for the 2-colorings of the floating square window pane shown below:

Solution. The monomials of the actions induced by the dihedral group D_4 are shown in this table:

e	ρ	ρ^2	ρ^3	φ_1	φ_2	φ_3	φ_4
t_1^8	t_4^2	t_2^4	t_4^2	t_2^4	t_2^4	t_2^4	t_2^4

Therefore the cycle index is $Z(D_4) = \frac{1}{8}\left(t_1^8 + 2t_4^2 + 5t_2^4\right)$, and the pattern inventory is $\frac{1}{8}((w+b)^8 + 2(w^4+b^4)^2 + 5(w^2+b^2)^4) = w^8 + w^7b + 6w^6b^2 + 7w^5b^3 + 13w^4b^4 + 7w^3b^4 + 6w^2b^6 + wb^7 + b^7$. Setting $w = b = 1$ shows that there are $1 + 1 + 6 + 7 + 13 + 7 + 6 + 1 + 1 = 43$ nonequivalent 2-colorings, including 13 colorings with four white and four blue panes. ■

Example 8.17 (Face Colorings of a Floating Cube) In how many different ways can the six faces of a cube be painted white or blue? How many of these ways have three faces of each color?

Solution. The diagram below shows that there are three types of rotations of a cube:

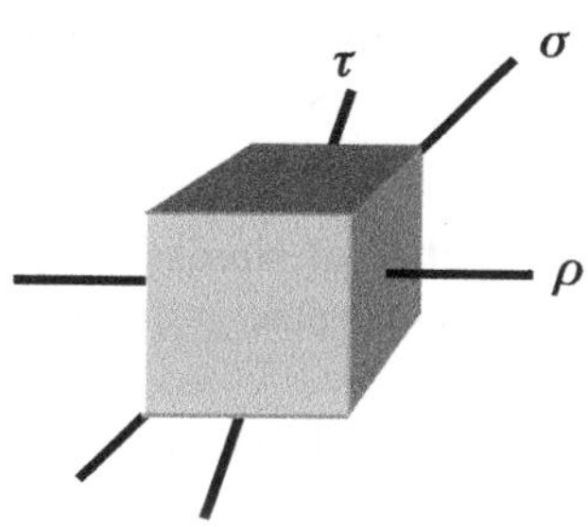

First, there are the rotations ρ, ρ^2, and ρ^3 about the axis through any pair of centers of opposite faces, where ρ is a 90° rotation. There are three such rotation axes since there are three pairs of opposite faces. Next, there are the rotations σ and σ^2 about each pair of opposite corners, where σ is a rotation by 120°. There are four rotations of this type since there are four pairs of opposite corners. Finally, there is the 180° rotation τ about the axis through midpoints of opposite edges of the cube. There are six rotations of this type, since the 12 edges of the cube form six opposite pairs. The monomials for the types of actions are shown in this table:

e	ρ	ρ^2	ρ^3	σ	σ^2	τ
t_1^6	$t_1^2t_4^1$	$t_1^2t_2^2$	$t_1^2t_4^1$	t_3^2	t_3^2	t_2^3

Therefore, since the group has order $1 + 3 \cdot 3 + 4 \cdot 2 + 6 \cdot 1 = 24$, we get the cycle index $Z = \frac{1}{24}\left(t_1^6 + 6t_1^2t_4^1 + 3t_1^2t_2^2 + 8t_3^2 + 6t_2^3\right)$.

Setting $t_1 = t_2 = t_3 = t_4 = 2$ shows that there are $\frac{1}{24}(2^6 + 6 \cdot 2^2 \cdot 2 + 3 \cdot 2^2 \cdot 2^2 + 8 \cdot 2^2 + 6 \cdot 2^3) = \frac{240}{24} = 10$ nonequivalent 2-colorings of the faces of the cube. Setting $t_i = w^i + b^i, i = 1, 2, 3, 4$ gives us, after some algebraic reduction, the pattern inventory $w^6 + w^5b + 2w^4b^2 + 2w^3b^3 + 2w^2b^4 + wb^5 + b^5$. Therefore, there are 2 nonequivalent colorings of the cube with three white and three blue faces. ■

For our last example, we need a little background from graph theory. A *graph* is a set of vertices together with a set of edges that connect some of the pairs of the vertices. For example, the diagram below shows four graphs on three vertices:

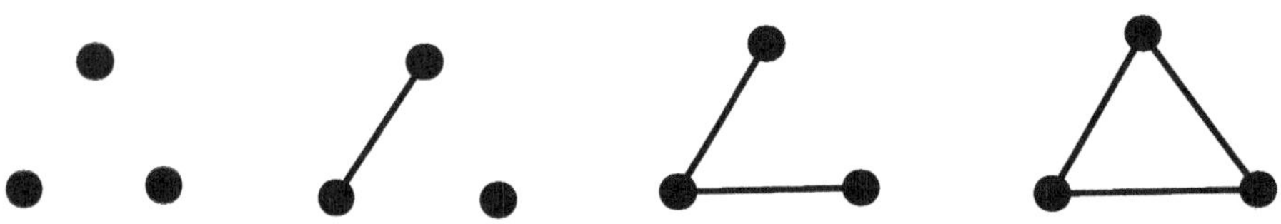

No two of these graphs are isomorphic since there is no permutation of vertices that gives a corresponding permutation of edges. Moreover, this is a complete listing of the nonisomorphic graphs on three vertices.

Example 8.18 (Counting the Number of Nonisomorphic Graphs on Four Vertices) What is the number of nonisomorphic graphs on four vertices?

Solution. The following diagram depicts the graph on four vertices with all six edges $X = \{x_1, x_2, x_3, x_4, x_5, x_6\}$:

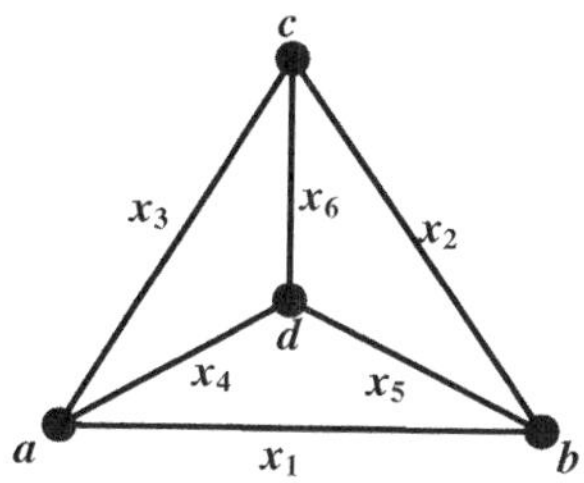

The group of actions on X will be the permutation group induced by the symmetric group of all 24 permutations of the set of vertices $\{a,b,c,d\}$. For example, the permutation $(abc)(d)$ rotates the edges by 120° about d, so the induced permutation of edges is $(x_1x_2x_3)(x_4x_5x_6)$, which we write simply as (1 2 3)(4 5 6). A 2-coloring of the edges will correspond to removing an edge that is colored white and retaining an edge colored black, so the number of nonequivalent colorings is the same as the number of nonisomorphic graphs.

We must first determine the monomials corresponding to each of the 24 permutations. These are listed in the following table:

Type of Permutation	Number of Permutations of that Type	Induced Permutation on Edges	Monomial
e	1	(1) (2) (3) (4) (5) (6))	t_1^6
$(a)(bcd)$	8	(2 6 5)(1 3 4)	t_3^2
$(ab)(cd)$	3	(1)(2 4)(3 5)(6)	$t_1^2t_2^2$
$(a)(b)(cd)$	6	(1)(2 5)(3 4)(6)	$t_1^2t_2^2$
$(abcd)$	6	(1 2 6 4)(3 5)	$t_2^1t_4^1$

This gives us the cycle index $Z = \frac{1}{24}\left(t_1^6 + 8t_3^2 + 9t_1^2t_2^2 + 6t_2^1t_4^1\right)$, and then, by setting $t_i = w^i + b^i, i = 1, 2, 3, 4$, we obtain the pattern inventory $w^6 + w^5b + 2w^4b^2 + 3w^3b^3 + 2w^2b^4 + wb^5 + b^6$.

The $1 + 1 + 2 + 3 + 2 + 1 + 1 = 11$ nonisomorphic graphs on four vertices are shown below, grouped by the number of edges:

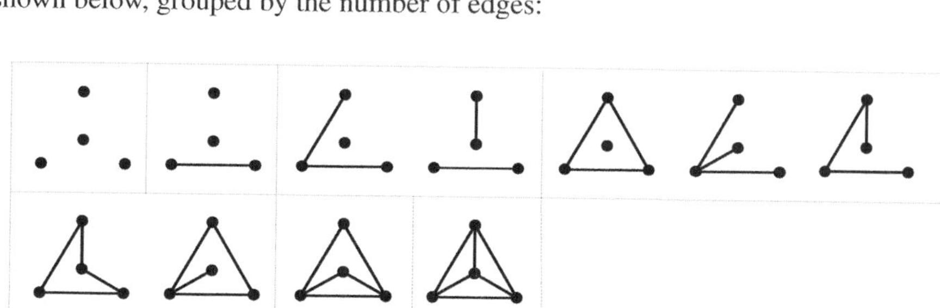

■

PROBLEMS

8.4.1. Determine the cycle index of the vertices of each of these polygons under rotation:

(a) a regular pentagon

(b) a regular heptagon (7-gon)

(c) a regular hexagon

(d) a regular dodecagon (12-gon)

8.4.2. Repeat Problem 8.4.1, but view the polygons as floating by allowing flips.

8.4.3. Let $X = \{x_1, x_2, x_3\}$ be the vertices of an equilateral triangle, which is therefore symmetric under the set of actions $\{e, \varphi_1, \varphi_2, \varphi_3\}$, where φ_i is the flip that fixes vertex x_i and exchanges the other two vertices. Is $Z = \frac{1}{4}\left(t_1^3 + 3t_1^1t_2^1\right)$ the cycle index? If so, the number of 3-colorings is $\frac{1}{4}(3^3 + 3 \cdot 3 \cdot 3) = \frac{54}{4} = 13\frac{1}{2}$. Explain what is wrong and determine the correct number of 3-colorings of the vertices of the triangle.

8.4.4. **(a)** What is the number of nonequivalent 3-colorings of the vertices of a regular heptagon (7-gon) symmetric under rotations?

(b) How many 3-colorings have 2 red, 2 white, and 3 blue vertices?

8.4.5. **(a)** What is the number of nonequivalent 2-colorings of the vertices of a regular hexagon symmetric under rotations?

(b) How many 2-colorings have 3 white and 3 blue vertices?

8.4.6. The regular tetrahedron shown has four faces $X = \{x_1, x_2, x_3, x_4\}$. There are two types of symmetry:

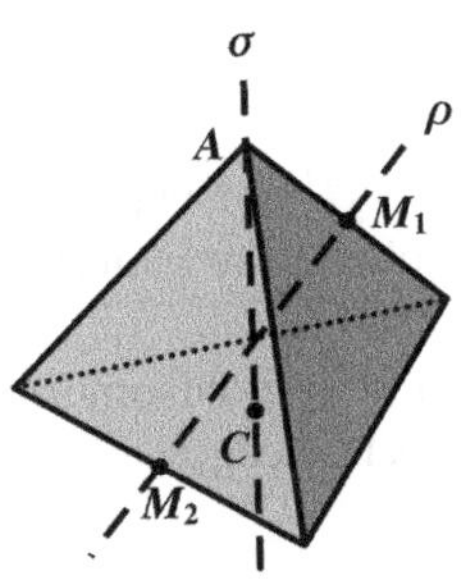

Type ρ, a 180° rotation about the axis through the midpoints M_1 and M_2 of a pair of opposite edges. The two faces meeting at each edge are exchanged.

Types σ and σ^2, the 120° and 240° rotations about an axis through a vertex A and the center C of the opposite face. Each rotation fixes one face and cycles the three faces that share the vertex A.

(a) Verify that the cycle index for the faces of the tetrahedron is $Z = \frac{1}{12}\left(t_1^4 + 3t_2^2 + 8t_1^1t_3^1\right)$.

(b) Calculate the pattern inventory for the 3-colorings of the faces of the tetrahedron.

(c) How many 3-colorings of the faces are nonequivalent?

(d) How many nonequivalent colorings of the faces have one red, one white, and two blue faces?

8.4.7. A carbon atom is at the center of a regular tetrahedron, and any combination of hydrogen (H), chlorine (Cl), ethyl (C_2H_5), and methyl (CH_3) can occur at the four vertices $X = \{x_1, x_2, x_3, x_4\}$.

(a) Why is the cycle index for the set of vertices also $Z = \frac{1}{12}(t_1^4 + 3t_2^2 + 8t_1^1t_3^1)$, the same as for the faces of the tetrahedron as noted Problem 8.4.6?

(b) What is the number of different nonequivalent molecules?

8.4.8. **(a)** Determine the cycle index for the six edges of a regular tetrahedron under rotation (see Problem 8.4.6).

(b) Find the number of 3-colorings of the edges of the tetrahedron.

(c) How many nonequivalent 3-colorings of the edges of the tetrahedron have two edges of each color?

8.4.9. The group of rotational symmetries of the cube was described in Example 8.17, where the cycle index for the set of six faces was derived.

(a) Derive the cycle index for the set of the eight vertices of a cube.

(b) What is the number of nonequivalent 2-colorings of the eight vertices?

(c) How many nonequivalent ways can the eight vertices of a cube be colored with four white and four black vertices?

8.4.10. The group of rotational symmetries of a cube was described in Example 8.17.

(a) Derive the cycle index for the set of 12 edges of a cube.

(b) What is the number of nonequivalent 3-colorings of the 12 edges?

(c) In how many nonequivalent ways can the 12 edges of a cube be colored with six white and six black edges?

8.4.11. The following diagram shows two 2-colorings of the vertices of a binary tree on seven vertices:

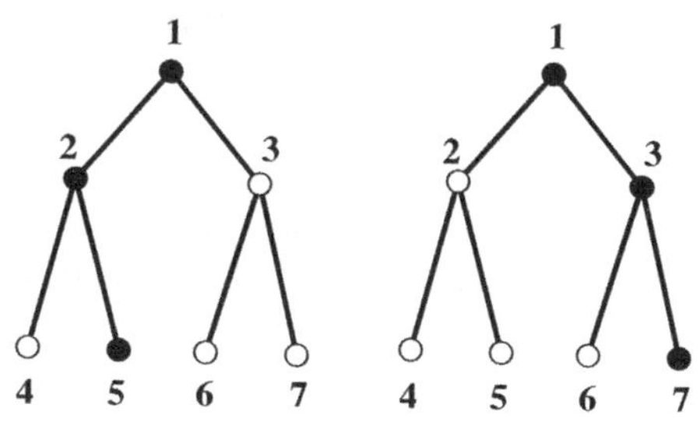

The colorings are equivalent since the direction of branching can be switched to obtain an isomorphic tree. This switching shown in the diagram above is given by the action

$$\gamma = \begin{pmatrix} 1 & 2 & 3 & 4 & 5 & 6 & 7 \\ 1 & 3 & 2 & 6 & 7 & 4 & 5 \end{pmatrix} = (1)(2\ 3)(4\ 6)(5\ 7)$$

(a) derive the group of symmetries of the tree, showing each action as a product of disjoint cycles.
(b) derive the cycle index for the group.
(c) derive the 2-color pattern inventory.
(d) in how many nonequivalent ways can the vertices of the tree by colored white or black?
(e) how many 2-colorings of the vertices have four white and three black vertices?

8.4.12. (a) Show that the cycle index of the cyclic group is $Z(C_m) = (1/m) \sum_{k|m} \varphi(k) t_k^{m/k}$, where $k|m$ means that k is a divisor of m and $\varphi(k)$ is Euler's φ-function that counts the number of positive integers in $[k]$ that are relatively prime to k; that is, $\varphi(k)$ is the number of integers i, $1 \leq i \leq k$, for which i and k have only 1 as a common divisor. For example, $\varphi(6) = |\{1, 5\}| = 2$, $\varphi(7) = |\{1, 2, \dots, 6\}| = 6$, and $\varphi(12) = |\{1, 5, 7, 11\}| = 4$.
(b) Prove the identity $m = \sum_{k|m} \varphi(k)$.

8.4.13. Show that the cycle index of the dihedral group acting on the vertices of a regular m-gon is

$$Z(D_m) = \begin{cases} \frac{1}{2} Z(C_m) + \frac{1}{2} t_1^1 t_2^{(m-1)/2}, & m \text{ odd} \\ \frac{1}{2} Z(C_m) + \frac{1}{4}\left(t_2^{m/2} + t_1^2 t_2^{(m-2)/2}\right), & m \text{ even} \end{cases}$$

where $Z(C_m)$ is the cycle index of the cyclic group C_m.

8.5 SUMMARY AND ADDITIONAL PROBLEMS

This chapter has developed formulas to count colorings or labelings of a set X of m objects that take symmetries into account. If one coloring arrangement is the same as another under a given symmetry, the two coloring patterns are equivalent. The goal is to count the number of nonequivalent colorings. The group of symmetries induces a group of actions on X that can be identified with a group of permutations of the set $[m]$.

Each action γ corresponds to a monomial of the form $t_1^{c_\gamma(1)} t_2^{c_\gamma(2)} \cdots t_m^{c_\gamma(m)}$, which indicates that the cycle decomposition of the action has $c(i)$ cycles of length i. Summing over all of the monomials and dividing by the order of the group of actions gives a polynomial known as the *cycle index*, namely, $Z = (1/|G|) \sum_{\gamma \in G} t_1^{c_\gamma(1)} t_2^{c_\gamma(2)} \cdots t_m^{c_\gamma(m)}$. If $t_i = w^i + b^i$, a generating function known as *Pólya's pattern inventory* is obtained, whose coefficient of $w^j b^k$ is the number of nonequivalent colorings with color w used j times and color b used k times. Similarly, setting $t_i = r^i + w^i + b^i$ gives the pattern inventory for three colors r, w, and b, and so on for any number of colors.

PROBLEMS

The following problems show how to obtain the cycle index for the alternating group A_m. The definition of the alternating group, given below, depends on the concept of a transposition; the 2-cycle $(i\ j)$, $i \neq j$, which swaps i and j, is called a *transposition*.

8.5.1. The two permutations of S_2 can be written as a product of transpositions, namely, $(1)(2) = (1\ 2)(1\ 2)$ and $(1\ 2)$.

(a) Prove that the six permutations in S_3 can each be written as a product of transpositions. For example, $(1\ 2\ 3) = (1\ 2)(2\ 3)$ and $(1)\ (2\ 3) = (1\ 2)$ $(1\ 2)(2\ 3)$.

(b) Prove that each of the $m!$ permutations of S_m for $m \geq 2$ can be written as a product of transpositions. [*Hint*: Show that any cycle can be written as a product of transpositions.]

8.5.2. It can be shown that if a permutation can be written as an even (odd) number of transpositions, then every representation of the permutation necessarily has an even (odd) number of transpositions. Thus, the permutations of $[m]$ are of two types: *even* if it is a product of an even number of transpositions, and *odd* otherwise.

(a) Prove that a cycle of even length is odd, necessarily the product of an odd number of transpositions, and a cycle of odd length is even and must be a product of an even number of transpositions.

(b) Prove that if an even permuation π is written as a product of disjoint cycles, then an even number of these cycles must be even.

(c) Prove that the even permutations are a subgroup A_m of the symmetric group S_m. It is called the *alternating group* A_m on $[m]$.

(d) Prove that $|A_m| = m!/2$, $m \geq 2$, so half of the permutations of S_m are even and half are odd. For example, $|A_3| = |\{(1)(2)(3), (1\ 2\ 3), (1\ 3\ 2\}| = 3 = 3!/2$. This result is usually proved as a consequence of Lagrange's theorem from group theory, but show that it can also be derived with mathematical induction.

8.5.3. Verify the following cycle indices:

(a) $Z(A_2) = t_1^2$

(b) $Z(A_3) = \frac{1}{3}\left(t_1^3 + 2t_3^1\right)$

(c) $Z(A_4) = \frac{1}{12}\left(t_1^4 + 8t_1^1t_3^1 + 3t_2^2\right)$

8.5.4. Show that $Z(A_m) = (1/m!) \sum_c \sigma(c)[1 + (-1)^{c(2)+c(4)+\cdots}]t_1^{c(1)}t_2^{c(2)} \cdots t_m^{c(m)}$, where $\sigma(c) = m!/[\prod_{1\leq k\leq m} k^{c(k)}c(k)!]$ is the number of m-tuples $c = (c(1), c(2), \ldots, c(m))$ of nonnegative integers for which $c(1) + 2c(2) + \cdots + mc(m) = m$.

REFERENCES

1. G. Pólya (1937). "Kombinatorische Anzahlbestimmungen für Gruppen, Graphen und chemische Verbindungen". *Acta Mathematica* 68 (1): 145–254.
2. Redfield, J. Howard (1927). "The Theory of Group-Reduced Distributions". American Journal of Mathematics 49 (3): 433–455.

PART III

NOTATIONS INDEX, APPENDICES, AND SOLUTIONS TO SELECTED ODD PROBLEMS

INDEX OF NOTATIONS

Notation	Name	Description
$\mathbb{N}$	Natural numbers	$\{1, 2, \ldots\}$
$\mathbb{Z}$	Integers	$\{\ldots, -2, -1, 0, 1, 2, 3, \ldots\}$
$\mathbb{R}$	Real numbers	
$\mathbb{C}$	Complex numbers	$\{x + iy \| x, y \in \mathbb{R}\}$
$[n]$	First n natural numbers	$[n] = \{1, 2, \ldots, n\}$
$\lfloor x \rfloor$	Floor function	Greatest integer $\leq x$
$\lceil x \rceil$	Ceiling function	Least integer $\geq x$
$\|S\|$	Cardinality of a finite set S	Number of elements in set S
$n!$	Factorial	Number of ordered arrangements of elements of an n-element set: $n! = n \cdot (n-1) \cdot \cdots \cdot 3 \cdot 2 \cdot 1,\ \ 0! = 1$
$P(n, r)$	Permutation	Number of ordered arrangements of r objects from an n-element set: $P(n, r) = n \cdot (n-1) \cdots (n - r + 1s)$
F_n	nth Fibonacci number	$F_0 = 0, F_1 = 1, F_2 = 1, F_3 = 2, \ldots$ $F_{n+2} = F_{n+1} + F_n, n \geq 0$
L_n	nth Lucas number	$L_0 = 2, L_1 = 1, L_2 = 3, L_3 = 4, \ldots$ $L_{n+2} = L_{n+1} + L_n, n \geq 0$
f_n	nth combinatorial Fibonacci number	Number of ways to tile a $1 \times n$ chessboard with 1×1 squares and 1×2 dominoes: $f_n = F_{n+1}$

(*continued*)

Combinatorial Reasoning: An Introduction to the Art of Counting, First Edition. Duane DeTemple and William Webb.

(*Continued*)

Notation	Name	Description
$(x)_r$	Falling factorial (Pochhammer symbol)	$(x)_r = x \cdot (x-1) \cdot \cdots \cdot (x-r+1)$
$x^{(r)}$	Rising factorial	$x^{(r)} = x \cdot (x+1) \cdots\cdot (x+r+1)$
$C(n, r)$ or $\binom{n}{r}$	Binomial coefficient	Number of r-element subsets of an n-element set: $\binom{n}{r} = \frac{n!}{r!\,(n-r)!} = \frac{n_{(r)}}{r!}$
$\binom{x}{r}$	Generalized binomial coefficient	$\binom{x}{r} == \frac{x_{(r)}}{r!}$
$\{r_1 \cdot a_1, r_2 \cdot a_2, \ldots, r_k \cdot a_k\}$	Multiset	Generalized set that contains r_j indistinguishable elements of type a_j, where $j = 1, 2, \ldots, k$
$\left(\!\binom{k}{r}\!\right)$	Multichoose coefficient	Number of ways to choose an r-element multiset from a multiset of k types: $\left(\!\binom{k}{r}\!\right) = \binom{r+k-1}{r}$
$\binom{n}{r_1, r_2, \ldots, r_k}$, $r_1 + r_2 + \ldots + r_k = n$	Multinomial coefficient	Number of permutations of an n-element multiset $\{r_1 \cdot a_1, r_2 \cdot a_2, \ldots r_k \cdot a_k\}$: $\binom{n}{r_1, r_2, \ldots, r_k} = \frac{n!}{r_1! \cdot r_2! \cdots \cdot r_k!}$, $r_1 + r_2 + \cdots + r_k = n$
$T(m, n)$	Distribution number	Number of ways to distribute n distinct objects into r distinct boxes, with no box left empty
D_n	nth derangement number	Number of permutations of $(1, 2, \ldots, n)$ that leave no element in its natural position: $D_0 = 1, D_1 = 0, D_2 = 1, D_3 = 2, D_4 = 9, \ldots$ $D_n = n!\left(1 - \frac{1}{1!} + \frac{1}{2!} - \frac{1}{3!} + \cdots + (-1)^n \frac{1}{n!}\right)$ $= nD_{n-1} + (-1)^n, \quad n \geq 2$ $= (n-1)(D_{n-1} + D_{n-2}), \quad n \geq 2$
E	Successor operator	$Eh_n = h_{n+1}$
$\delta_{j,k}$	Kronecker delta	$\delta_{j,k} = \begin{cases} 1, & \text{if } j = k \\ 0, & \text{if } j \neq k \end{cases}$
$s(k,j)$ or $\begin{bmatrix} k \\ j \end{bmatrix}$	Stirling number of first kind (unsigned)	Number of permutations of $1, 2, \ldots, k$ with j cycles; number of ways to arrange k people around j indistinguishable circular tables, with no table unoccupied

(*Continued*)

Notation	Name	Description
$S(k,j)$ or $\left\{ {k \atop j} \right\}$	Stirling number of second kind	Number of ways to partition a k-element set into j nonempty subsets; number of ways to distribute k distinct objects into j identical boxes, with no box left empty
B_n	Bernoulli numbers	$B_n = B_n(0)$
$B_n(x)$	Bernoulli polynomials	$B_0 = 1, B'_n(x) = nB_{n-1}(x)$ $\int_0^1 B_n(x)\, dx = 0,\ n = 1, 2, \ldots$
B_n	Bernoulli numbers	$B_n = B_n(0)$
$\left\langle {n \atop k} \right\rangle$	Eulerian number	Number of permutations π of $[n]$ with exactly k ascents: $\pi_j < \pi_{j+1}$
$B(n)$	Bell number	Number of partitions of $[n]$ into any number of nonempty subsets
H_n	nth harmonic number	$H_0 = 1, H_n = 1 + \frac{1}{2} + \frac{1}{3} + \cdots + \frac{1}{n}$
$p(n)$	Unrestricted partitions of n	Number of ways to express n as a nonincreasing sum of positive integers
$q(n)$	Partitions of n into distinct summands	Number of ways to express n as a decreasing sum of positive integers
$p_k(n)$	Restricted partition number	Number of ways to express n as a sum of k positive integers, or as a sum of integers with largest part k
C_n	nth Catalan number	Number of ways to dissect a convex polygon with $n + 2$ sides into triangles using the diagonals of the polygon; number of lattice paths from (0, 0) to $(2n, 2n)$ that never go above the main $x = y$ diagonal: $C_n = \frac{1}{n+1}\binom{2n}{n}$
t_j^i	monomial	factor indicating there are i cycles of length j
$Z(G)$	cycle index of group G	$Z[G] = \frac{1}{\lvert G\rvert} \sum_{\gamma \in G} t_1^{c_\gamma(1)} t_2^{c_\gamma(2)} L t_m^{c_\gamma(m)}$, where $c_\gamma(j) = i$ if γ is a permutation with i cycles of length j

APPENDIX A

MATHEMATICAL INDUCTION

A.1 PRINCIPLE OF MATHEMATICAL INDUCTION

Mathematical induction is one of the most common and useful methods of proof, not only in combinatorics but also in all branches of mathematics. In its most basic form, the goal is to prove the truth of a sequence of statements S_n for all $n \geq 0$. The principle of mathematical induction shows us how to do this in two steps.

Let $S_0, S_1, \ldots, S_n, \ldots$ be a sequence of statements for which the following two steps can be completed:

Basis Step—prove that S_0 is true.

Induction Step—for any $n \geq 0$, prove that if S_n is true, then so is S_{n+1}.

Then the statement S_n is true for every $n \geq 0$.

Statement S_0 must be proved in the basis step, but often this is an easy case that depends on a simple algebraic calculation or other straightforward reasoning. The important thing to notice about the induction step is that neither S_n nor S_{n+1} is proved. What is proved is the implication that *if* we assume that S_n is true, *then* S_{n+1} must also be true. The assumption that S_n is true is called the *induction hypothesis.*

It seems obvious that once the basis and induction steps of mathematical induction are shown, then every statement S_n is true. After all, we know that S_0 is true, so the

---

Combinatorial Reasoning: An Introduction to the Art of Counting, First Edition. Duane DeTemple and William Webb.

case $n = 1$ in the induction step tells us that S_1 is true. Taking $n = 1$ in the induction step next tells us that S_2 is true, and so on. However, we can't really prove the "and so on", but instead we can ground our reasoning on the defining axioms of the nonnegative integers. In particular, if a subset of integers contains 0 and contains $n+1$ whenever it contains n, then the set is the entire set of nonnegative integers.

The examples that follow illustrate how to apply the principle of mathematical induction and some of its logically equivalent variations.

Example A.1 Prove that $n! > 2^n$ for all $n \geq 4$.

Proof by Mathematical Induction. The inequality is false for $n = 0,1,2,3$, so our basis case is $n = 4$ since then $4! = 24 > 16 = 2^4$. Now let $n \geq 4$ and assume that $n! > 2^n$. Then $(n+1)! = (n+1)n! > (n+1)2^n > (1+1)2^n = 2^{n+1}$ since $n \geq 4 > 1$. Since both the basis and induction steps have been completed, we have proved that $n! > 2^n$ for all $n \geq 4$. ■

The key insight in the induction step is to make a connection between S_n and S_{n+1}. A common instance of this is when the statements concern a sequence of sums $s_n = \sum_{k=0}^{n} a_k$, since then

$$s_{n+1} = \sum_{k=0}^{n+1} a_k = \sum_{k=0}^{n} a_k + a_{n+1} = s_n + a_{n+1}$$

Example A.2 If F_k is the kth Fibonacci number, then $\sum_{k=0}^{n} F_k = F_{n+2} - 1$ for all $n \geq 0$.

Proof by Mathematical Induction. Let $s_n = \sum_{k=0}^{n} F_k$. Then $s_0 = F_0 = 0 = 1 - 1 = F_2 - 1$, which proves the basis step. Now let $n \geq 0$ and assume as the induction hypothesis that $s_n = F_{n+2} - 1$. Then

$$s_{n+1} = \sum_{k=0}^{n+1} F_k = s_n + F_{n+1} = (F_{n+2} - 1) + F_{n+1} = (F_{n+2} + F_{n+1}) - 1 = F_{n+3} - 1$$

which completes the induction step. ■

Note that in Example A.2 we had to first guess that the sum is $F_{n+2} - 1$ before we used mathematical induction to prove that this is the sum for all n. It was also important to use the recurrence formula for the Fibonacci numbers to provide a connection between S_n and S_{n+1}. The following example also utilizes the connection provided by a recurrence relation.

Example A.3 Prove that the Fibonacci number F_{3k} is an even number for all $k \geq 0$.

Proof by Mathematical Induction. The recursion $F_{3k} = F_{3k-1} + F_{3k-2}$ does not provide a connection between $F_{3(k+1)}$ and F_{3k}. Hence, it is more natural to prove a stronger statement: F_{3k} is even and F_{3k+1} and F_{3k+2} are both odd for all $k \geq 0$.

Basis step—$F_0 = 0$ is even, and $F_1 = F_2 = 1$ are odd.

Induction step—let $k \geq 0$ and assume that F_{3k} is even and F_{3k+1} and F_{3k+2} are odd. Then $F_{3(k+1)} = F_{3k+2} + F_{3k+1}$ is even since it is a sum of two odd numbers, $F_{3(k+1)+1} = F_{3k+3} + F_{3k+2}$ is odd since it is a sum of an even and an odd number, and $F_{3(k+1)+2} = F_{3k+4} + F_{3k+3}$ is odd since it is a sum of an odd and an even number. ■

In Example A.3, we needed to assume that three successive cases were true to prove the next case. Sometimes we may need to assume that *all* of the previous statements $S_0 S_1, \ldots, S_n$ are true, and use this hypothesis to show that the next statement S_{n+1} is true. This is called the *principle of strong induction*, discussed below.

A.2 PRINCIPLE OF STRONG INDUCTION

Let $S_0, S_1, \ldots, S_n, \ldots$ be a sequence of statements for which the following two steps can be completed:

Basis step—prove that S_0 is true.

Induction step—for any $n \geq 0$, prove that if the statements $S_0 S_1, \ldots, S_n$ are assumed to be true, then so is S_{n+1}.

Then the statement S_n is true for every $n \geq 0$.

The value of strong induction is evident in the following example.

Example A.4 Every natural number $n > 1$ is either a prime number or a product of prime numbers.

Proof by Strong Induction.

Basis step—the number $n = 2$ is a prime number since it is divisible only by itself and by 1.

Induction step—let $n \geq 2$. Assume that every integer $2, 3, \ldots, n$ is either a prime number or a product of prime numbers. Now consider $n + 1$. If it is prime, we are done, so suppose not. If $n + 1$ is not a prime number, it is divisible by some smaller integer and so $n + 1 = ab$, where a and b are integers $2 \leq a, b \leq n$. By the strong induction hypothesis, both a and b are each either a prime number or a product of prime numbers. Since $n + 1 = ab$, we see that $n + 1$is a product of prime numbers.

■

One final variation on mathematical induction is the well ordering principle, discussed next.

A.3 WELL ORDERING PRINCIPLE

It can be shown that the following simple assertion is logically equivalent to the principle of mathematical induction.

Every nonempty subset C of natural numbers has a smallest member.

The next example shows how a sequence of statements can be proved by the well ordering principle.

Example A.5 Any amount of postage 6 cents or more can be made with 3¢ and 4¢ stamps.

Proof by Well Ordering Principle. Let C be the set of postage amounts 6¢ or more that cannot be made with some combination of 3¢ and 4¢ stamps. If C is a nonempty set, then it has a smallest amount n that cannot be made with 3¢ and 4¢ stamps. Clearly n is not 6, 7, or 8, since these amounts of postage can be made with 3¢ and 4¢ stamps: $6 = 3 + 3, 7 = 3 + 4, 8 = 4 + 4$.

Borrowing language from British mathematician Robin Wilson, our "minimal criminal" n in the set of criminals C is an amount at least 9. But then $6 \leq n - 3 < n$ so $n - 3$ is at least 6 and is not a criminal. Therefore, some combination of 3¢ and 4¢ stamps make $n - 3$¢ of postage. Adding a 3¢ stamp then shows that n cents postage can be made, contradicting our assumption n is a criminal. We conclude that C must be the empty set and there are no criminals. That is, every amount of postage 6 cents or more can be made with 3¢ and 4¢ stamps. ■

APPENDIX B

SEARCHING THE *Online Encyclopedia of Integer Sequences* (OEIS)

The *Online Encyclopedia of Integer* Sequences (OEIS®, http://oeis.org/) is a remarkable resource for anyone working with sequences and arrays of integers. Once some of the early terms of the sequence or array have been calculated, they can be entered in the search box and the OEIS will return a list of sequences or arrays that include the terms that were entered. Each matching sequence or array is shown with additional terms as well as combinatorial interpretations, formulas, generating functions, properties, and references. This information helps to identify which sequence or array is the exact match for the sequence under investigation.

B.1 SEARCHING A SEQUENCE

Suppose that you wish to investigate the number of ways, say, h_n, that 1×2 dominoes and 1×3 trominoes can tile a $1 \times n$ board. It is easy to determine the values $h_1 = 0, h_2 = 1, h_3 = 1, h_4 = 1, h_5 = 2, h_6 = 2, h_7 = 3, h_8 = 4$, so the list 0,1,1,1,2,2, 3,4 can be inserted in the search box of the OEIS. There will yield 42 sequences which each contain the entered terms. Thee first of these is the Padovan sequence 1, 0, 0, **0, 1, 1, 1, 2, 2, 3, 4**, which is assigned the reference number A000931. This sequence starts differently but then agrees at the fourth term and beyond. Since this happens often, it is usually best to omit the first term or two of the sequence being searched. It is also best to avoid having too many sequences returned, so it is a good idea to search on about six terms. The sequence of tilings by dominoes and trominoes

Combinatorial Reasoning: An Introduction to the Art of Counting, First Edition. Duane DeTemple and William Webb.

clearly satisfies the recursion relation $h_n = h_{n-2} + h_{n-3}$. The OEIS informs us that this is also a property of the Padovan numbers, so we can be sure that the tiling problem is solved by that sequence.

B.2 SEARCHING AN ARRAY

An array is searched by entering the sequence obtained by reading rows of a triangular array or the upward diagonals of a square array. For example, entering 1,2,1,1,3,3,1,1,4,6,4,1 will return "Pascal's triangle read by rows," A007318. Note that it wasn't necessary to enter the first two rows. For a square array, do a search on the sequence of entries along the upward diagonals (the "antidiagonals"). For example, searching on 1,2,0,0,1,3,1,0,0,1,4,3,0,0,0 returns "the square array of Pascal's triangle, A052553.

B.3 OTHER SEARCHES

Even a search for fractions is possible on OEIS; just search on the integer sequence of numerators. For example, a search on 3, 11, 25, 137, 49 returns the numerators of the harmonic numbers, sequence A001008. It is also possible to search on keywords such as *Stirling*, which yields 1150 results. The first of these is the Stirling numbers of the second kind, and the second is the Stirling numbers of the first kind.

B.4 BEGINNINGS OF OEIS

The OEIS was begun as a database by Neil J. A. Sloane in 1965. In 1973, he published a book, *The Handbook of Integer Sequences,* which contained 2372 entries. By 1995, it had become the *Encyclopedia of Integer Sequences* with 5487 entries. The following year the encyclopedia was published online, where it continues to thrive. By the beginning of 2013, it contained over 220,000 sequences. There is now a nonprofit OEIS Foundation that manages the website.

APPENDIX C

GENERALIZED VANDERMONDE DETERMINANTS

In Chapter 5, it was shown that any linear sum h_n (GPS) of the terms of the set $S = \left\{ \binom{n}{j} \alpha_i^{n-j} : i = 1, 2, \ldots, m,\ j = 1, 2, \ldots, r_i - 1 \right\}$ is a sequence that is annihilated by the operator $C(E) = (E - \alpha_1)^{r_1}(E - \alpha_2)^{r_2} \cdots (E - \alpha_m)^{r_m}$. Here $\alpha_1, \alpha_2, \ldots, \alpha_m$ are the distinct eigenvalues of the operator, and $r_1, r_2, \ldots, r_m$ are their respective multiplicities. To show that the sequence h_n is determined uniquely by any set of k initial conditions $h_i = A_i, i = 0, 1, \ldots, k-1$, it must be shown that the Vandermonde determinant $V(\alpha_1^{(r_1)}, \alpha_2^{(r_2)}, \ldots, \alpha_m^{(r_m)})$ is nonzero, where

$$V\left(\alpha_1^{(r_1)}, \alpha_2^{(r_2)}, \ldots, \alpha_m^{(r_m)}\right) = \det \begin{bmatrix} 1 & 0 & \cdots & 1 & 0 & \cdots \\ \alpha_1 & 1 & \cdots & \alpha_2 & 1 & \cdots \\ \alpha_1^2 & 2\alpha_1 & \cdots & \alpha_2^2 & 2\alpha_2 & \cdots \\ \alpha_1^3 & 3\alpha_1^2 & \cdots & \alpha_2^3 & 3\alpha_2^2 & \cdots \\ \vdots & \vdots & \vdots & \vdots & \vdots & \vdots \\ \alpha_1^{k-1} & \binom{k-1}{1}\alpha_1^{k-2} & \cdots & \alpha_2^{k-1} & \binom{k-1}{1}\alpha_2^{k-2} & \cdots \end{bmatrix} \tag{C.1}$$

Combinatorial Reasoning: An Introduction to the Art of Counting, First Edition. Duane DeTemple and William Webb.

The following theorem evaluates the Vandermonde determinant, and shows it is not zero for distinct eigenvalues.

Theorem C.1 For $m \geq 2$, and any positive integers $r_1, r_2, \ldots, r_m$, $r_1 + r_2 + \cdots + r_m = k$, we get

$$V\left(\alpha_1^{(r_1)}, \alpha_2^{(r_2)}, \ldots, \alpha_m^{(r_m)}\right) = \prod_{1 \leq i < j \leq m} (\alpha_j - \alpha_i)^{r_i r_j}$$

Proof (Outline). Rather than give a general proof, it will suffice to describe how to obtain the formula for the $k = 6$ case $V(\alpha^{(2)}, \beta, \gamma^{(3)})$ once it is assumed we have *all* the Vandermonde expansions of order 5. It shouldn't be difficult to understand how the induction step is carried out in general.

To derive the formula for $V(\alpha^{(2)}, \beta, \gamma^{(3)})$, assume that $V(\alpha^{(2)}, \beta, \gamma^{(2)}) = (\beta - \alpha)^2 (\gamma - \alpha)^4 (\gamma - \beta)^2$. Now consider the function

$$p(x) = V(\alpha^{(2)}, \beta, \gamma^{(2)}, x) = \det \begin{bmatrix} 1 & 0 & 1 & 1 & 0 & 1 \\ \alpha & 1 & \beta & \gamma & 1 & x \\ \alpha^2 & 2a & \beta^2 & \gamma^2 & 2\gamma & x^2 \\ \alpha^3 & 3\alpha^2 & \beta^3 & \gamma^3 & 3\gamma^2 & x^3 \\ \alpha^4 & 4\alpha^3 & \beta^4 & \gamma^4 & 4\gamma^3 & x^4 \\ \alpha^5 & 5\alpha^4 & \beta^5 & \gamma^5 & 5\gamma^4 & x^5 \end{bmatrix}$$

which is a polynomial of degree 5 whose leading term is the cofactor $c = V\left(\alpha^{(2)}, \beta, \gamma^{(2)}\right)$ of x^5; that is, $p(x) = cx^5 +$ (terms of lower degree in x). If $x \in \{\alpha, \beta, \gamma\}$, two columns of the determinant are identical, so we know that α, β, and γ are roots of $p(x)$. Next, apply the differentiation operator d/dx to get

$$p'(x) = \det \begin{bmatrix} 1 & 0 & 1 & 1 & 0 & 0 \\ \alpha & 1 & \beta & \gamma & 1 & 1 \\ \alpha^2 & 2\alpha & \beta^2 & \gamma^2 & 2\gamma & 2x \\ \alpha^3 & 3\alpha^2 & \beta^3 & \gamma^3 & 3\gamma^2 & 3x^2 \\ \alpha^4 & 4\alpha^3 & \beta^4 & \gamma^4 & 4\gamma^3 & 4x^3 \\ \alpha^5 & 5\alpha^4 & \beta^5 & \gamma^5 & 5\gamma^4 & 5x^4 \end{bmatrix}$$

When x is either α or γ, the last column is identical to column 2 or 5, so we see that $p'(\alpha) = 0$ and $p'(\gamma) = 0$. This means that α and γ are at least double roots of $p(x)$, so we now know that the degree 5 polynomial has the form $p(x) = c\,(x-\alpha)^2\,(x-\beta)\,(x-\gamma)^2$. Finally, consider $\frac{1}{2}(d/dx)^2 p(x)$, which is

$$\frac{1}{2}p''(x) = \det \begin{bmatrix} 1 & 0 & 1 & 1 & 0 & 0 \\ \alpha & 1 & \beta & \gamma & 1 & 0 \\ \alpha^2 & 2a & \beta^2 & \gamma^2 & 2\gamma & 1 \\ \alpha^3 & 3\alpha^2 & \beta^3 & \gamma^3 & 3\gamma^2 & 3x^2 \\ \alpha^4 & 4\alpha^3 & \beta^4 & \gamma^4 & 4\gamma^3 & 6x^2 \\ \alpha^5 & 5\alpha^4 & \beta^5 & \gamma^5 & 5\gamma^4 & 10x^3 \end{bmatrix}$$

and so $\frac{1}{2}p''(\gamma) = V(\alpha^{(2)}, \beta, \gamma^{(3)})$. But we also have

$$\begin{aligned} &V(\alpha^{(2)}, \beta, \gamma^{(3)}) \\ &= \frac{1}{2}p''(x)\Big|_{x=\gamma} = \frac{1}{2}\left(\frac{d}{dx}\right)^2 c\,(x-\alpha)^2\,(x-\beta)\,(x-\gamma)^2\Big|_{x=\gamma} \\ &= \frac{1}{2}c2(x-\alpha)^2(x-\beta)\Big|_{x=\gamma} = (\beta-\alpha)^2(\gamma-\alpha)^4(\gamma-\beta)^2(\gamma-\alpha)^2(\gamma-\beta) \\ &= (\beta-\alpha)^2(\gamma-\alpha)^6(\gamma-\beta)^3 \end{aligned}$$

which is formula (C.1) when $r_1 = 2, r_2 = 1$, and $r_3 = 3$. ■

For a generalized solution formed as a linear sum of the sequences in the set $\{n^j\alpha_i^n : i = 1, \ldots, m, j = 0, 1, \ldots, r_{i-1}\}$, the corresponding Vandermonde determinant takes the form

$$\begin{aligned} &\hat{V}\left(\alpha_1^{(r_1)}, \alpha_2^{(r_2)}, \ldots, \alpha_m^{(r_m)}\right) \\ &= \det \begin{bmatrix} 1 & 0 & \cdots & 1 & 0 & \cdots \\ \alpha_1 & \alpha_1 & \cdots & \alpha_2 & \alpha_2 & \cdots \\ \alpha_1^2 & 2\alpha_1^2 & \cdots & \alpha_2^2 & 2\alpha_2^2 & \cdots \\ \alpha_1^3 & 3\alpha_1^3 & \cdots & \alpha_2^3 & 3\alpha_2^3 & \cdots \\ \vdots & \vdots & \vdots & \vdots & \vdots & \vdots \\ \alpha_1^{k-1} & (k-1)\,\alpha_1^{k-1} & \cdots & \alpha_2^{k-1} & (k-1)\,\alpha_2^{k-1} & \cdots \end{bmatrix} \end{aligned}$$

The following theorem evaluates this determinant, though we will not give a proof.

Theorem C.2 Let $\rho_n = (n-1)!\,(n-2)! \cdots 1!0!$. Then, for all $k \geq 2$, and positive integers $r_1, r_2, \ldots, r_m$ for which $r_1 + r_2 + \cdots + r_m = k$,

$$\hat{V}\left(\alpha_1^{(r_1)}, \alpha_2^{(r_2)}, \ldots, \alpha_m^{(r_m)}\right) = \prod_{i=1}^{m} \rho_{r_i} \alpha_i^{\binom{r_i}{2}} \prod_{1 \leq i < j \leq m} (\alpha_j - \alpha_i)^{r_i r_j} \tag{C.2}$$

An interesting special case is the determinant

$$\hat{V}\left(1^{(n)}\right) = \hat{V}_n = \det \begin{bmatrix} 1 & 0 & \cdots & 0 & 0 \\ 1 & 1 & \cdots & 1 & 1 \\ 1 & 2 & \cdots & 2^{n-1} & 2^n \\ \vdots & \vdots & \vdots & \ddots & \vdots \\ 1 & n & \cdots & n^{n-1} & n^n \end{bmatrix} = n!(n-1)! \cdots 2!1! \tag{C.3}$$

This result can be given an independent proof as follows.

Proof of Formula (C.3). Recall from definition 6.15 that

$$(m)_r = \sum_{j=0}^{r} (-1)^{r-j} \begin{bmatrix} r \\ j \end{bmatrix} m^j,$$

where $\begin{bmatrix} r \\ j \end{bmatrix}$ denote the Stirling numbers of the first kind. Therefore

$$(m)_n = \sum_{j=0}^{n-1} (-1)^{n-j} \begin{bmatrix} n \\ j \end{bmatrix} m^j + m^n = \begin{cases} n! \text{ for } m = n \\ 0 \text{ for } m < n \end{cases}$$

This shows that there is a linear combination of the first n columns of the determinant that, when added to the last column, gives a column of all 0s except for the term $n!$ in the bottom row. Then $\hat{V}_n = n!\hat{V}_{n-1}$, giving formula C.3 by induction. ■

HINTS, SHORT ANSWERS, AND COMPLETE SOLUTIONS TO SELECTED ODD PROBLEMS

Answers or solutions to selected problems in various sections of the book are provided here.

PROBLEM ANSWERS AND SOLUTIONS

Chapter 1

1.2.1. **(a)** 18 **(b)** 4 **(c)** 10 **(d)** 7 **(e)** 17

1.2.3. Each person can know anywhere from zero (no one) to nine (everyone) people. But if someone knows no one, there cannot be someone who knows everyone, and vice versa. Thus, place the 10 people into the 9 boxes that are labeled $1, 2, \ldots, 8$, and $0|9$. By the pigeonhole principle, some box has at least two members; that is, there are at least two people at the party with the same number of acquaintances.

1.2.7. Pick any two of the five points and draw a great circle through them. At least two of the remaining three points belong to the same closed hemisphere determined by the great circle. These two points, and the two starting points, are four points in the same closed hemisphere.

1.2.9. Each of the 51 numbers belongs to one of the 50 sets $\{1,101\}$, $\{2,99\}, \ldots,$ $\{50,51\}$. Some set contains two of the chosen numbers, and these sum to 101.

1.2.11. Place each of the 51 numbers into one of the 50 sets $\{1,2\}$, $\{3,4\}, \ldots,$ $\{99,100\}$. One set contains a pair of consecutive integers that are relatively prime.

1.2.15. Suppose that the $3n$ consecutive numbers are a, $a + 1, \ldots,$ b. Each of the $n + 1$ numbers in the given subset belongs to one of the sets $\{a, a + 1, a + 2\}$, $\{a + 3, a + 4, a + 5\}, \ldots,$ $\{b - 2, b - 1\ b\}$. According to the pigeonhole principle, one of these sets has two members of the subset, and these differ by at most 2.

1.2.19. **(a)** Any path returns to its starting position in m steps, so there are m spirals each covering m squares.

(b) Since m/d and n/d are relatively prime, there is a unique spiral with mn/d^2 steps that covers a $d \times d$ square at each step. By part (a) we now see there are d nonintersecting spirals on the torus.

1.3.1. The left and right columns of height $n - 1$ of the trimmed board can each be tiled with vertical dominoes. The remaining board is has all of its rows of even length $m - 2$, so it can be tiled with horizontal dominoes.

1.3.5. **(a)** Since the colors alternate, all eight corners of the solid are black.

(b) The cube is black if and only if the sum $j + k + h$ is odd. For example, the cube in column 1, row 1, and layer 1 is black since $1 + 1 + 1 = 3$, an odd number. Thus, j, k, and h must all be odd, or one must be odd and the other two even.

(c) If the cube that is removed is black, and is in column j, row k, and layer h. Then j, k, and h are all odd or two are even and one is odd. With no loss in generality, assume that $j + k$ is even and h is odd. Theorem 1.21 tells us that layer h can be tiled with dominoes confined to that layer. When layer h is removed it leaves two (possibly one if $h = 1$ or n) rectangular solids with an even dimension and so it can be tiled with solid dominoes.

1.4.3. The answer is $s_{2n+1} = 8t_n + 1$, since $(2n + 1)^2 = 4n^2 + 4n + 1 = 8\{[n(n + 1)]/2\} + 1 = 8t_n + 1$. See the following diagram:

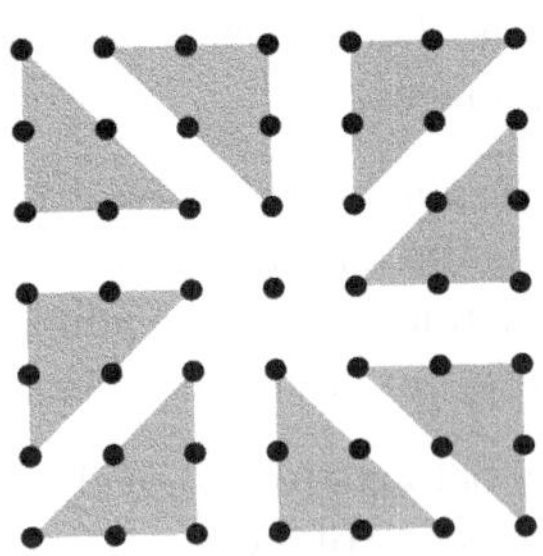

1.4.9. **(a)** Consider the array of dots in the x,y-coordinate plane with a dot at (m,n) that represents the p–q domino, with $0 \leq q \leq p \leq n$. This array is a triangle with $n+1$ dots per side, so there are $t_{n+1} = \frac{1}{2}(n+1)(n+2)$ dominoes in a double-n set. For $n = 3,6,9,12,15,18$, the number of dominoes are the triangular numbers 55, 91, 136, and 190.

(b) Imagine that you have two double-n sets, so that each p–q domino from one set can be paired with the $(n-p)$–$(n-q)$ complementary domino from the second set. For example, in a double-15 set, pair the 11–6 domino from one set with the complementary 4–9 domino from the second set. Each pair of complementary dominoes has a total of $2n$ pips, so according to part (a) there are $(2n)t_{n+1} = (2n)\frac{1}{2}(n+1)(n+2) = n(n+1)(n+2)$ pips in the two double-n sets. Therefore, a single double-n set has a total of $\frac{1}{2}n(n+1)(n+2)$ pips. *Alternate solution.* Each half-domino with k pips, $0 \le k \le n$, occurs $n+2$ times in a double-n set, so the total number of pips is given by

$$\begin{aligned}(n+2)(0+1+2+\cdots+n) &= (n+2)t_n = (n+2)\frac{n(n+1)}{2}\\ &= \frac{1}{2}n(n+1)(n+2)\end{aligned}$$

1.5.3. $$\begin{aligned}\binom{n-1}{k-1} + \binom{n-1}{k} &= \frac{(n-1)!}{(k-1)!(n-k)!} + \frac{(n-1)!}{k!(n-k-1)!}\\ &= \frac{(n-1)!}{k!(n-k)!}[k+n-k] = \frac{n!}{k!(n-k)!} = \binom{n}{k}.\end{aligned}$$

1.5.5. (a) Five ways:

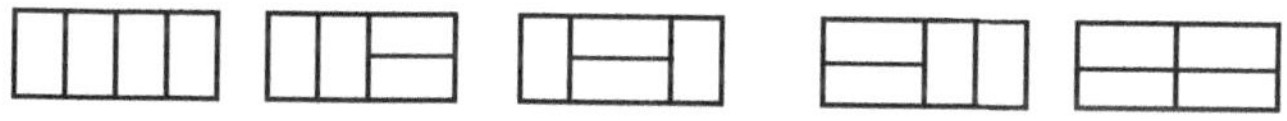

(b) Draw a horizontal midline through each tiling by dominoes. The lower $1 \times n$ row board is equivalent to a tiling by squares and dominoes, showing that there are f_n tilings of a $2 \times n$ board with dominoes.

1.5.7. (a) Any one of the 12 vertical dashed lines can be chosen to end a car and start a new one, so there are $\binom{12}{1} = 12$ ways to form a train of length 13 with 2 cars.

(b) Any choice of 4 of the 12 vertical dashed lines form a train with 5 cars. Thus, there are

$$\binom{12}{4} = \frac{12 \cdot 11 \cdot 10 \cdot 9}{4 \cdot 3 \cdot 2 \cdot 1} = 495$$

trains of length 13 and 5 cars.

1.5.9. **(a)** Six ways.

(b) $\binom{n-1}{k-1}$

since the summands can be viewed as the length of k cars that form a train of length n.

1.6.1. When truncated, the six corners of the octahedron are replaced with squares, and the eight triangles become hexagons. The sets of squares and hexagons partition the set of all of the ball's panels, so there are $8 + 6 = 14$ panels in all by the addition principle.

1.6.3. [*Hint*: Each pentagon is bordered by 5 seams that each separate panels of different types.]

1.6.5. Let $A = \{C, S\}$ and $B = \{C, V, M\}$ be the sets of flavor choices. Then Maria will have $|A \times B| = |A||B| = 2 \cdot 3 = 6$ types of cones to her liking.

1.6.11. $3 \cdot 4 \cdot 2 = 14$

Chapter 2

2.2.1. **(a)** 26! **(b)** 22! **(c)** $5! \cdot 22!$ **(d)** $\binom{26}{5} 21!$

2.2.5. There will be one pole with 3 flags, and 4 flags on the other two poles. There are 3 ways to choose the 3-flag flagpole, and 11! orders to arrange the flags, giving $3 \cdot 11!$ ways to fly the flags.

2.2.9. There are $6! = 720$ permutations, so some arrangement has been repeated at least after 721 weeks by the pigeonhole principle. This means Jane might have a new arrangement for nearly 14 years with careful planning!

2.2.15. **(a)** $\binom{5}{2} = (5 \cdot 4)/2 = 10$, so there are 10 different blocks, just right to code the 10 digits $1, 2, 3, \ldots, 9, 0$.

(b) The missing permutations are shown below. Note that the sequence of digits $1, 2, \ldots, 9, 0$ is obtained by moving a long bar one step at a time:

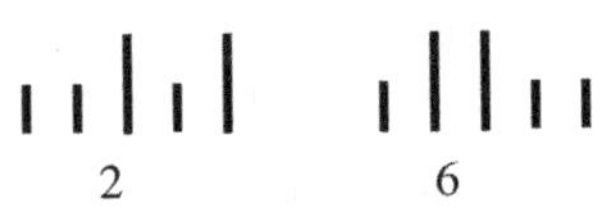

2.2.17. **(a)** Every pair of points determines a chord, so there are $\binom{n}{2}$ chords.

(b) Every crossing point corresponds to 4 of the n points, and conversely, so there are $\binom{n}{4}$ points at which chords intersect within the circle.

2.2.21. **(a)** $\binom{52}{13}$

(b) Every rank must be present in the hand, and there are $\binom{4}{1} = 4$ ways to choose one card of any rank. Therefore, 4^{13} bridge hands have no pair.

(c) The rank of the pair can be chosen in 13 ways, and a pair of this rank in $\binom{4}{2}$ ways. The 11 remaining cards must be of all different ranks chosen from the 12 ranks different from the pair, and there are four choices of the suit from each rank. Altogether, there are $\binom{13}{1}\binom{4}{2}\binom{12}{11}4^{11}$ bridge hands with exactly one pair.

2.2.25. $Prob(9, 10, \text{ or } 11 \text{ heads}) = \dfrac{\binom{20}{9}+\binom{20}{10}+\binom{20}{11}}{2^{20}} \doteq 0.4966$, so it is just slightly better to bet against getting 9, 10, or 11 heads.

2.3.1. The left side f_{2n+1} counts the number of ways to tile a board of odd length, which requires at least one square tile. Now take cases that depend on the position of the rightmost square, and suppose that k dominoes are to its right. To the left of the rightmost square there is a board of length $(2n + 1) - (1 + 2k) = 2(n - k)$ that can be tiled in $f_{2(n-k)}$ ways. Summing over $k = 0,1,2, \ldots, n$ gives the right side of the identity.

2.3.7. ***Answer 1.*** There are $\binom{2n}{n}$ ways to place the n red tiles, with the remaining squares covered by blue tiles. This is the left side of the identity.

Answer 2. There are four types of columns: rr, bb, rb, and br, where the colors of the upper and lower tiles are listed in order. Suppose that there are k all-red columns of type rr. Since there are an equal number of red and blue tiles, there must also be k all-blue columns. The remaining $n - 2k$ columns can each be tiled with a red over a blue tile rb, or with a blue over a red tile br. The all red columns can be selected in $\binom{n}{k}$ ways, the all blue columns in $\binom{n-k}{k}$ ways, and the remaining $n - 2k$ columns can be tiled in 2^{n-2k} ways. Summing over k gives the right side of the identity.

2.3.11. The bowl series can be modeled with block walking through Pascal's triangle, so there are $\binom{100}{50}$ paths that represent the sequence of wins and losses. The Catalan number

$$C_{50} = \frac{1}{50+1}\binom{100}{50}$$

is the number of paths for which West never trailed East in games won. Assuming that every path is equally likely, the probability of this is 1 in 51.

2.3.13. Let $B = B(2n + 2, n + 1)$ be the point in row $2n + 2$ crossed by diagonal $n + 1$. The Catalan number C_{n+1} counts the paths from $A(0,0)$ to B that never cross the vertical line through A and B. However, that path can return to the line, say, for the first time at the point $F(2r + 2, r + 1)$ where $0 \leq r \leq n$. This means the path from $P(1,0)$ to $Q(2r + 1, r)$ never crosses the vertical line from P to Q, so there are C_r of these paths. There are also C_{n-r} paths from F to B. We see that there are $C_r\,C_{n-r}$ paths from A to B that first return to the line AB at the point F in row r. Summing over r gives the desired sum.

2.3.19. ***Answer 1.*** There are n choices of the block to support the pole, and 2^{n-1} subsets of the remaining $n - 1$ blocks on which to anchor guy wires. This is the right side of the identity.

Answer 2. There are $\binom{n}{k}$ ways to choose k blocks to used in the arrangement, and the one supporting the flag can be chosen in k ways. Summing over k gives the left side of the identity.

2.3.20. There are f_{n+1} tilings of a $1 \times (n + 1)$ board by squares and dominoes. Cases can be taken according to the number of dominoes in the tiling. If there are two or more dominoes, the penultimate domino can cover cells $j - 1$ and j, $2 \le j \le n - 1$. The first $j - 2$ cells can be tiled in f_{j-2} ways and the last (right most) domino can begin by covering any of the $n - j$ cells $j + 1$ through n. Thus the number of tilings with two or more dominoes is $\sum_{j=2}^{n-1} (n - j)f_{j-2}$. If the tiling has just one domino, it can begin at any of the n cells 1 through n, so there are n tilings with one domino. Finally, there is one tiling with all squares and no dominoes. Adding the cases accounts for all of the tilings and proves the identity.

2.4.1. **(a)** $13!/(2!2!2!)$

(b) View bat as a single *letter*, so there are $11!/(2!2!2!)$ words.

(c) $\dfrac{13!}{2!2!2!} - \dfrac{12!}{2!2!}$

2.4.3. **(a)** $\binom{9}{3,3,3}$

(b) $\binom{3}{1,1,1}^3 = (3!)^3 = 6^3$

2.4.5. **(a)** $27{,}720 = 6930 + 11{,}550 + 9240$

(b) The last step to (3,5,4) must be from (2,5,4), or (3,4,4), or (3,5,3).

(c) The last step to $(a,b,c,\ldots,z)$ must be from $(a-1, b, c, \ldots, z)$ or $(a, b - 1, c, \ldots, z) \ldots$, or $(a,b,c,\ldots,z - 1)$.

2.4.9. Let $x_1 = y_1 + 2$, then $y_1 + x_2 + x_3 + x_4 = 13, y_1 \ge 0, x_2 \ge 0, x_3 \ge 0, x_4 \ge 0$. Hence there are $\left(\!\binom{4}{13}\!\right) = \binom{16}{13} = 560$ solutions.

2.4.13. We have the equation $x_a + x_b + 3x_r = 10$ in nonnegative variables that specify the number of pies of each type. Clearly $x_r \in \{0, 1, 2, 3\}$, giving the four equations $x_a + x_b = 10, x_a + x_b = 7, x_a + x_b = 4, x_a + x_b = 1$, which have altogether $11 + 8 + 5 + 2 = 26$ solutions.

2.5.1. First, give each child one candy. The remaining 12 candies that can be distributed as 6 pairs in

$$\left(\!\binom{3}{6}\!\right) = \binom{6+3-1}{6} = \binom{8}{6} = \frac{8 \cdot 7}{2!} = 28$$

ways.

2.5.3. **(a)** $\left(\!\binom{10}{6}\!\right) = \binom{6+10-1}{6} = \binom{15}{6}$

(b) In how many ways can an order be placed for six ice cream cones at a store offering 10 flavors?

(c) $x_1 + x_2 + \cdots + x_{10} = 6, x_i \geq 0$

2.5.9. There are 2^m ways to distribute m distinct objects to two distinct recipients, including the two distributions in which one recipient is assigned all m objects. By the subtraction principle, there are $2^m - 2$ onto distributions.

2.5.17. When $x = k$, any positive integer, we have $(k)_n/n! = \binom{k}{n}$. Therefore, the polynomial

$$p(x) = x^m - \sum_{n=0}^{m} T(m,n)\frac{(x)_n}{n!}$$

when x is any positive integer. But the only polynomial with infinitely many zeros is the 0 polynomial, so $p(x) = 0$, which proves the identity.

2.6.1. **(a)** $7!$

(b) The men can be seated in $3!$ ways, and then the women seated between the men in $4!$ ways, giving $3! \cdot 4!$ seatings.

(c) Seat all but Mr. Smith in $6!$ ways, and then seat Mr. Smith to either the left or right of Mrs. Smith in 2 ways, giving $6! \cdot 2$ seatings.

(d) Seat everyone but Alice in $6!$ ways, and then seat Alice in any one of the five seats not adjacent to George. Therefore, there are $6! \cdot 5$ seatings.

2.6.5. **(a)** First seat person #1 in 2 ways, either at the left or right of any of the identical sides. The other 7 people can be seated around the table in $7!$ ways, so there are $2 \cdot 7!$ seatings.

(b) There are now 3 ways—the left, middle, or right seat—to seat person #1, and $11!$ ways to seat the remaining people. This gives $3 \cdot 11!$ seatings.

2.6.7. There are $n - 1$ choices for $\pi(1)$ and $(n-1)!$ ways to permute the remaining $n - 1$ elements. Thus, $(n-1)(n-1)!$ permutations do not fix element 1.

2.6.11. **(a)** Choose k elements in $\binom{n}{k}$ ways, and derange them in D_k ways. Therefore, there are $\binom{n}{k} D_k$ permutations of $[n]$ with $n-k$ fixed points and k deranged points.

(b) The $n!$ permutations of $[n]$ can be partitioned into disjoint subsets determined by the number, k, of points that are deranged by the partition, where $0 \leq k \leq n$. This gives the identity by part (a) and the addition principle.

Chapter 3

3.2.1. **(a)** $\sum_{n=0}^{8} \binom{8}{n} 9^n = (1+9)^8 = 10^8$

(b) $\sum_{n=0}^{8} (-1)^n \binom{8}{n} 11^n = (1-11)^8 = (-10)^8 = 10^8$

3.2.3. **(a)** $\binom{7}{3} 2^4 = 35 \cdot 16 = 560$

(b) 0, since no term involving w^3x^4 appears in the expansion.

(c) $\binom{5}{3} (2^3)(-1)^2 = 80$

3.2.5. [*Hint*: Use Pascal's identity several times.]

3.2.11. Choose $w = 1$ and $x = -1$ in the identity proved Problem 3.2.10(a).

3.2.17. Take the k-fold derivative $\partial^k/(\partial x_1 \partial x_2 \cdots \partial x_k)$ of the multinomial formula and then set $x_1 = x_2 = \cdots = x_k = 1$.

3.3.3. **(a)** $a_n = 3^n$

(b) $b_n = -\frac{1}{2}$

(c) $h(x) = \dfrac{1+x}{2-8x+6x^2} = \dfrac{1}{1-3x} - \dfrac{1}{2-2x} = f(x) + g(x)$ so the sequence is $c_n = a_n + b_n = 3^n - \frac{1}{2}$.

3.3.5. **(a)** The derivative of the geometric series is

$$\frac{1}{(1-x)^2} = \sum_{k=1}^{\infty} kx^{k-1}$$

so the desired formula is obtained by a multiplication by x.

(b) $\dfrac{x^3}{(1-5x)^2} = \dfrac{x^2}{5}\dfrac{(5x)}{(1-(5x))^2} = \dfrac{x^2}{5}\sum_{k=1}^{\infty} k(5x)^k = \sum_{k=1}^{\infty} k5^{k-1}x^{k+2} = x^3 + 10x^4$ $+ 75x^5 + \cdots$ so the sequence is $a_0 = a_1 = a_2 = 0, a_k = (k-2)5^{k-3}$, $k \geq 3$.

(c) $\sum_{k=0}^{\infty} 3k2^k x^k = 3\sum_{k=0}^{\infty} k(2x)^k = \dfrac{3(2x)}{(1-(2x))^2} = \dfrac{6x}{(1-2x)^2}$

3.3.11. **(a)** $(-1)_n = (-1)(-2)\cdots(-1-n+1) = (-1)^n(1)(2)\cdots(n) = (-1)^n n!$

(b) $(-\alpha)_n = (-\alpha)(-\alpha-1)\cdots(-\alpha-n+1)$
$= (-1)^n(\alpha)(\alpha+1)\cdots(\alpha+n-1) = (-1)^n(\alpha+n-1)_n$

3.3.13. $\dfrac{\left(\frac{1}{3}\right)\left(-\frac{2}{3}\right)\left(-\frac{5}{3}\right)\left(-\frac{8}{3}\right)\cdots\left(\frac{1}{3}-n+1\right)}{n!} = \dfrac{(-1)^{n-1}2\cdot 5\cdot 8\cdot\cdots\cdot(3n-4)}{3^n n!}$

3.3.17. Replace x with $-4x$ in equation (3.34).

3.4.3. **(a)** $(k-1)^3 = k^3 - 3k^2 + 3k - 1$ by the binomial theorem.

(b) $n^3 = \sum_{k=1}^{n}(k^3-(k-1)^3) = 3\sum_{k=1}^{n}k^2 - 3\sum_{k=1}^{n}k + \sum_{k=1}^{n}1 = 3\sum_{k=1}^{n}k^2$
$-3\dfrac{n(n+1)}{2} + n$

$$\therefore \sum_{k=1}^{n}k^2 = \frac{n^3}{3} + \frac{n^2+n}{2} - \frac{n}{3} = \frac{2n^3+3n^2+n}{6} = \frac{n(n+1)(2n+1)}{6}$$

3.4.7. $(1+x+x^2+\cdots)(x^3+x^4+x^5+\cdots)(x+x^2+x^3+x^4) =$ $x^4(1-x^4)/(1-x)^3$.

3.4.9. **(a)** $(x+x^4+x^9+\cdots)(x+x^8+x^{27}+\cdots)(x^2+x^3+x^5+x^7+x^{11}\cdots)$

(b) Three solutions

(c) Seven solutions

3.4.15. **(a)** $h(x) = f_P(x)f_N(x)f_D(x)f_Q(x)f_H(x)$

$$= \left(\sum_{r\geq 0}x^r\right)\left(\sum_{s\geq 0}x^{5s}\right)\left(\sum_{t\geq 0}x^{10t}\right)\left(\sum_{u\geq 0}x^{25u}\right)\left(\sum_{v\geq 0}x^{50v}\right)$$

$$= \frac{1}{1-x}\frac{1}{1-x^5}\frac{1}{1-x^{10}}\frac{1}{1-x^{25}}\frac{1}{1-x^{50}}$$

3.4.21. The 5 ways are 5B+1R, 3B+2R+1G, 3B+1R+2G, 1B+2R+3G, 1B+1R+4G.

3.5.1. $\left[\dfrac{x^n}{n!}\right]f(x) = \left[\dfrac{x^n}{n!}\right]\sum_{m\geq 0}h_m x^m = \left[\dfrac{x^n}{n!}\right]\sum_{m\geq 0}m!h_m\left(\dfrac{x^m}{m!}\right) = n!h_n = n![x^n]f(x)$

3.5.3. **(a)** $\sum_{n=0}^{\infty}(-1)^n\frac{x^{2n}}{(2n)!} = \cos x$

(b) $\sum_{n=0}^{\infty}(-1)^n\frac{x^{2n+1}}{(2n+1)!} = \sin x$

3.5.9. $\left(\frac{x^2}{2!}+\frac{x^3}{3!}+\frac{x^5}{5!}+\frac{x^8}{8!}\right)e^{3x}$

3.5.13. **(a)** $(2n)!! = (2n)(2n-2)\cdot\cdots\cdot(2) = 2^n n!$

(b) $\sum_{n\geq 0}(2n)!!\frac{x^n}{n!} = \sum_{n\geq 0}2^n n!\frac{x^n}{n!} = \sum_{n\geq 0}(2x)^n = \frac{1}{1-2x}$

3.5.19.
$$\left[\frac{x^{10}}{10!}\right](e^x-1)^5 = \left[\frac{x^{10}}{10!}\right](e^{5x}-5e^{4x}+10e^{3x}-10e^{2x}+5e^x-1)$$
$$= 5^{10} - 5\cdot 4^{10} + 10\cdot 3^{10} - 10\cdot 2^{10} + 5$$

3.5.27. **(a)** $[x^{10}](x+x^3+x^5+\cdots)^2 = [x^{10}]\frac{x^2}{(1-x^2)^2} = [x^{10}]\sum_{n=0}^{\infty}nx^{2n} = 5$

(b) $\left[\frac{x^{10}}{10!}\right](\sinh x)^2 = \left[\frac{x^{10}}{10!}\right]\frac{\cosh 2x - 1}{2} = \frac{2^{10}}{2} = 2^9 = 512$

Chapter 4

4.2.3. [*Hint*: Use the DIE method and consider subsets of $[n]$, adding or deleting the element 1.]

4.2.5. **(a)** By Example 4.7, the sum is $(-1)^n T(m,n)$, which is zero for $m < n$.
(b) By Example 4.7, the sum is $(-1)^n T(n,n) = (-1)^n n!$.

4.2.9. **(a)** By Example 4.7, the sum is

$$T(n+1,n) = \binom{n+1}{2}T(n,n) = \frac{(n+1)n}{2}n! = \frac{n(n+1)!}{2}$$

(b) [*Hint*: Distribute $n+1$ distinct objects to a subset of k distinct recipients chosen from a numbered group of n recipients.]

4.2.13. Let $t_n = (-1)^n s_n$; then $t_n = (-1)^n s_n = (-1)^n(-s_{n-3}) = (-1)^{n-3}s_{n-3} = t_{n-3}$.

4.3.1. **(a)** 1090 **(b)** 1,000,000 – 1090 = 998,910

4.3.3. 357

4.3.5. 871

4.3.9. $10! - (7! + 9! + 8!) + (6! + 5! + 7!) - 4! = 3{,}225{,}696$

4.3.15. $\binom{n}{0} - \binom{n}{1} + \binom{n}{2} - \binom{n}{3} + \binom{n}{4} + \cdots + (-1)^n \binom{n}{n} = 0$. This is the result of equation (4.5).

4.3.17. **(a)** $\varphi(24) = \varphi(2^3 3^1) = 24\left(1 - \frac{1}{2}\right)\left(1 - \frac{1}{3}\right) = 24\left(\frac{1}{2}\right)\left(\frac{2}{3}\right) = 8,$

$$\frac{5^8 - 1}{24} = \frac{390{,}624}{24} = 16{,}276$$

(b) $\varphi(17) = 17 - 1 = 16, \frac{2^{16} - 1}{17} = \frac{65{,}535}{17} = 3855$

4.4.1. **(a)** $(7 \cdot 9)(6 \cdot 8)(5 \cdot 7)(4 \cdot 6) = (7)_4 (9)_4$

(b) $(7)_4 (9)_4 / 4!$

4.4.3. $R(x, \boldsymbol{B}) = 1 + 6x + 9x^2 + 4x^3$

$4! - 6 \cdot 3! + 9 \cdot 2! - 4 \cdot 1! = 24 - 36 + 18 - 4 = 2$

4.4.9. **(a)** $1 + 7x + (2 \cdot 4 + 3)x^2 = 1 + 7x + 11x^2$

(b)
$$\begin{aligned} R(x, \boldsymbol{B}) &= R(x, \boldsymbol{B} - \boldsymbol{S}) + xR(x, \boldsymbol{B_S}) \\ &= (1 + 3x)^2 + x(1 + 2x) = 1 + 7x + 11x^2 \end{aligned}$$

4.4.11. $R(x, \boldsymbol{B}) = (1 + 3x + x^2)^2 + x(1 + 2x)^2 = 1 + 7x + 15x^2 + 10x^3 + x^4$

4.6.9. **(a)** 8, 2 **(b)** 10, 4 **(c)** 10, 3

4.6.10. **(a)** $r_0(\boldsymbol{B}) = 1,\ r_1(\boldsymbol{B}) = 4,\ r_2(\boldsymbol{B}) = 4,\ r_3(\boldsymbol{B}) = 1,\ r_4(\boldsymbol{B}) = 0$

$1 \cdot 4! - 4 \cdot 3! + 4 \cdot 2! - 1 \cdot 1! = 7$

(b) $r_0(\boldsymbol{B}) = 1,\ r_1(\boldsymbol{B}) = 6,\ r_2(\boldsymbol{B}) = 10,\ r_3(\boldsymbol{B}) = 4,\ r_4(\boldsymbol{B}) = 0$

$1 \cdot 4! - 6 \cdot 3! + 10 \cdot 2! - 4 \cdot 1! = 4$

(c) $r_0(\boldsymbol{B}) = 1,\ r_1(\boldsymbol{B}) = 5,\ r_2(\boldsymbol{B}) = 7,\ r_3(\boldsymbol{B}) = 3,\ r_4(\boldsymbol{B}) = 0$

$1 \cdot 4! - 5 \cdot 3! + 7 \cdot 2! - 3 \cdot 1! = 5$

4.6.11. There are $\binom{2n}{n} = (2n)(2n - 1)/2 = 2n^2 - n$ ways to place two rooks on the board that include the n disallowed cases in which both rooks are in the same row and the $n - 1$ disallowed cases in which both rooks are in the same column. Therefore there are $2n^2 - n - n - (n - 1) = 2n^2 - 3n + 1$ nonattacking placements of two rooks.

Chapter 5

5.2.1. Any tiling of a board of length n corresponds to a seating for which a square represents a return to the same seat and a domino represents a switch of adjacent seats.

5.2.7. **(b)**
$$\begin{aligned} F_0 + F_2 + F_4 + \cdots + F_{2m} &= 0 + (F_3 - F_1) + (F_5 - F_3) + (F_7 - F_5) \\ &\quad + \cdots + (F_{2m+1} - F_{2m-1}) \\ &= -F_1 + F_{2m+1} = F_{2m+1} - 1. \end{aligned}$$

5.2.9. **(a)**
$$\begin{aligned} L_0 + L_1 + L_2 + L_3 + \cdots + L_m &= (L_2 - L_1) + (L_3 - L_2) + (L_4 - L_3) \\ &\quad + (L_5 - L_4) + \cdots + (L_{m+2} - L_{m+1}) \\ &= -L_1 + L_{m+2} = -1 + L_{m+2} \end{aligned}$$

5.2.13. $L_{-k} = (-1)^k L_k$. Using mathematical induction, the formula is valid for $k = 1$ and $k = 2$. Then

$$\begin{aligned} L_{-(k+2)} &= L_{-k} - L_{-(k+1)} = (-1)^k L_k - (-1)^{k+1} L_{k+1} = (-1)^{k+2}(L_k + L_{k+1}) \\ &= (-1)^{k+2} L_{k+2} \end{aligned}$$

5.2.17. **(b)**
$$\begin{aligned} 5F_n + L_n &= 5\frac{\varphi^n - \hat{\varphi}^n}{\sqrt{5}} + (\varphi^n + \hat{\varphi}^n) = (1 + \sqrt{5})\varphi^n + (1 - \sqrt{5})\hat{\varphi}^n \\ &= 2\varphi^{n+1} + 2\hat{\varphi}^{n+1} = 2L_{n+1} \end{aligned}$$

5.2.21. **(a)** Use mathematical induction. For $n = 1$, the formula is valid since $F_0 = 0$ and $F_1 = F_2 = 1$. Now assume that the formula is valid for some $n \geq 1$.

Then $\begin{bmatrix} 1 & 1 \\ 1 & 0 \end{bmatrix}^{n+1} = \begin{bmatrix} 1 & 1 \\ 1 & 0 \end{bmatrix}\begin{bmatrix} 1 & 1 \\ 1 & 0 \end{bmatrix}^n = \begin{bmatrix} 1 & 1 \\ 1 & 0 \end{bmatrix}\begin{bmatrix} F_{n+1} & F_n \\ F_n & F_{n-1} \end{bmatrix} =$

$$\begin{bmatrix} F_{n+1} + F_n & F_n + F_{n-1} \\ F_{n+1} & F_n \end{bmatrix} = \begin{bmatrix} F_{n+2} & F_{n+1} \\ F_{n+1} & F_n \end{bmatrix}$$

(b) $F_{n+1}F_{n-1} - F_n^2 = \det \begin{bmatrix} F_{n+1} & F_n \\ F_n & F_{n-1} \end{bmatrix} = \det \begin{bmatrix} 1 & 1 \\ 1 & 0 \end{bmatrix}^n$

$$= \left(\det \begin{bmatrix} 1 & 1 \\ 1 & 0 \end{bmatrix}\right)^n = (-1)^n$$

5.3.3. **(a)** $\alpha = \dfrac{1 + i\sqrt{11}}{2}$ and $\beta = \dfrac{1 - i\sqrt{11}}{2}$

(b) $c_1\left(\dfrac{1 + i\sqrt{11}}{2}\right)^n + c_2\left(\dfrac{1 - i\sqrt{11}}{2}\right)^n$

5.3.5. **(a)** $\alpha = \beta = 1$ **(b)** $c_1 + c_2 n$

5.3.7. $(E - 1)^2(c_0 + c_1 n) = (E - 1)(c_0 + c_1(n + 1) - c_0 - c_1 n) = (E - 1)(c_1)$
$= c_1 - c_1 = 0$

5.3.11. $C(x) = x^2 - a_1 x - a_2 = (x - \alpha)(x - \beta) = x^2 - (\alpha + \beta)x + \alpha\beta$, so equating like coefficients of powers of x gives $\alpha + \beta = a_1$ and $\alpha\beta = -a_2$.

5.3.15. [*Hint*: Show $h_n = 2h_{n-1} + h_{n-2}$, and relate to the Pell numbers.]

5.3.21. **(a)** We suggest using the Binet formula for the Pell number $P_n = (\alpha^n - \beta^n)/2\sqrt{2}$, where $\alpha = 1 + \sqrt{2} \approx 2.4$ and $\beta = 1 - \sqrt{2} \approx -0.4$. Therefore

$$\left|\frac{\beta}{\alpha}\right| < 1 \therefore \lim_{n\to\infty} \left|\frac{\beta}{\alpha}\right|^n = 0$$

5.4.1. **(a)** $\alpha = \sqrt{3}, \beta = -\sqrt{3}$, and $\gamma = 2$ **(b)** $c_1(\sqrt{3})^n + c_2(-\sqrt{3})^n + c_3 2^n$

5.4.3. **(a)** It is possible. The fifth-order annihilating operator $C(E)$ may have a third-order divisor that annihilates the sequence.

(b) Yes. If $P(E)$ is any second-order operator, the seventh-order operator $P(E)C(E)$ continues to annihilate the sequence.

5.4.7. $c_0 2^n + c_1 \binom{n}{1} 2^{n-1} + d_0(-1)^{n-2} + d_1 \binom{n}{1}(-1)^n + d_2 \binom{n}{2}(-1)^{n-2}$

5.4.15.
$$\begin{aligned} T(E)u_n &= (P(E)S(E) - Q(E)R(E))u_n \\ &= S(E)P(E)u_n - Q(E)R(E)u_n \\ &= -S(E)Q(E)v_n + Q(E)S(E)v_n = 0. \end{aligned}$$
An analogous calculation shows that $T(E)v_n = 0$.

5.4.19. $h_{n+4} - 3h_{n+3} - 6h_{n+2} + 3h_{n+1} + h_n = 0$ since all four sequences have eigenvalues $\{\varphi^3, \hat{\varphi}^3, -\varphi, -\hat{\varphi}\}$.

5.5.1. **(a)** h_n is annihilated by the operator $(E - 3)(E + 2)(E - 1)$.

(b) $h_n = 3^n - 1$

(c) $h_{n+2} = 4h_{n+1} - 3h_n$

5.5.3. **(a)** $h_n = 3^n - 1$ **(b)** $h_n = 18 \cdot 2^n - 18 \cdot 3^n + 6n3^n$

5.5.7. **(a)** The $(n + 1)$st polygon adds $1 + (r - 2)\,n$ dots to number of the dots given by $c_n^{(r)}$.

(b) The associated homogeneous recurrence $g_n = g_{n-1}$ is annihilated by $E - 1$, and the nonhomogeneous term $q_n = 1 + (r - 2)n$ is a first-degree polynomial annihilated by $(E - 1)^2$.

(c) By part (b), $p_n^{(r)}$ is a quadratic polynomial $p_n^{(r)} = c_0 + c_1 \binom{n}{1} + c_2 \binom{n}{2}$. Now solve for the constants using the initial conditions.

5.5.11. **(a)** $h_n = h_{n-1} + 2n + 1,\ n \geq 2$. Solving, we find $h_n = 1 + 3n + n(n - 1) = (n + 1)^2$.

(b) In each row, there are $n + 1$ choices for the number of white squares that are to the left in the row. This gives $(n + 1)^2$ tilings.

5.6.1. The OGF is $f(x) = 1/(1 - x - 2x^2)$.

5.6.5. The OGF is $f(x) = (1 - x)/(1 - 3x - x^2 + x^3)$.

5.6.11. Since $F_{n+2} = F_{n+1} + F_n$, subtracting the second and third series from the first series shows that $(1 - x - x^2)f_F(x) = F_0 + (F_1 - F_0)x = 0 + (1 - 0)x = x$. Solving the equation for f_F gives $f_F(x) = x/(1 - x - x^2)$.

5.6.13. **(a)** $(1 - x^2 - x^3)f_p(x) = p_0 + (p_2 - p_1)x^2 = 3 + (2 - 3)x^2 = 3 - x^2$. Thus $f_p(x) = (3 - x^2)/(1 - x^2 - x^3)$.

(b) $(1 - x - x^2 - x^3)f_T(x) = T_0 + (T_1 - T_0)x + (T_2 - T_1 - T_0)x^2 = 0 + (0 - 0)x + (1 - 0 - 0)x^2 = x^2$. Thus $f_T(x) = x^2/(1 - x - x^2 - x^3)$.

5.6.21. **(a)** $f'(x) = \sum_{n\geq 1} na_n x^{n-1} = a_1 + \sum_{n\geq 2} na_n x^{n-1} = 1 + \sum_{n\geq 2} a_{n-1}x^{n-1}$
$+ x\sum_{n\geq 2} a_{n-2}x^{n-2} = f(x) + xf(x)$

(b) $f'(x)/f(x) = 1 + x$ so $\log f(x) = c + x + (x^2/2)$. Then $\log f(0) = \log 1 = 0 = c$ so $f(x) = e^{x+(x^2/2)}$.

Chapter 6

6.2.3. $p(n) = \binom{n}{0} + 3\binom{n}{1} + 3\binom{n}{2} = 1 + 3n + 3\frac{n(n-1)}{2} = \frac{3n^2+3n+2}{2}$

6.2.9. If not all of the coefficients c_j are 0, we may reasonably assume that k is the largest index for which $c_k \neq 0$. The polynomial then has the form $p(n) = c_k n^k + \hat{p}(n)$, where $\hat{p}(n)$ is a polynomial of degree at most $k-1$. Since $p(n)$ is the 0 polynomial, all of its coefficients are 0, so $c_k = 0$, a contradiction. Therefore all of the coefficients c_j are 0.

6.2.15. Using equations (6.10) and (6.14), we obtain

$$x^k = \sum_{i=0}^{k} \left\{ {k \atop i} \right\} (x)_i = \sum_{i=0}^{k} \left\{ {k \atop i} \right\} \sum_{j=0}^{i} (-1)^{i-j} \left[{i \atop j} \right] x^j$$
$$= \sum_{j=0}^{k} (-1)^j \left(\sum_{i=0}^{k} (-1)^i \left\{ {k \atop i} \right\} \left[{i \atop j} \right] \right) x^j$$

Comparing coefficients of x^j gives the result.

6.2.17. The formula is easily seen to be true for $j = 0$ and $j = 1$ for every $k \geq 0$. Now suppose that the formula holds for some $j \geq 1$ and all $k \geq 0$. Then, using the triangle identity for the Stirling numbers of the second kind, we obtain

$$\left\{ {j+1+k+1 \atop j+1} \right\} = (j+1)\left\{ {j+k+1 \atop j+1} \right\} + \left\{ {j+k+1 \atop j} \right\}$$
$$= (j+1)\left\{ {j+k+1 \atop j+1} \right\} + \sum_{i=0}^{j} i\left\{ {k+i \atop i} \right\} = \sum_{i=0}^{j+1} i\left\{ {k+i \atop i} \right\}$$

so the identity is true by mathematical induction.

6.2.25. [*Hint*: The sum $\sum_{j=1}^{k} \left[{k \atop j} \right]$ counts all ways to seat k people around identical circular tables, where j is the numbers of tables used for a seating. Either combine or separate persons 1 and 2.]

6.3.1. $2E = 2\left(\frac{1}{2} + \frac{1}{4} + \frac{1}{6} + \cdots\right) = 1 + \frac{1}{2} + \frac{1}{3} + \cdots = H$, so $D = H - E = E$. But $D > E$ since each term of D is strictly larger than the corresponding term of E. This is a contradiction, so the assumption that H is finite is false.

6.3.5. Take the derivative of the generating function to get

$$\sum_{n=1}^{\infty} nH_n x^{n-1} = \frac{1}{(1-x)^2} - \frac{\ln(1-x)}{(1-x)^2} = \sum_{n=1}^{\infty} nx^{n-1} + \sum_{n=0}^{\infty}\left(\sum_{k=0}^{n} H_k\right)x^n$$

Equating coefficients of x^{s-1} gives the formula.

6.3.9. $\sum_{n=1}^{k} \frac{2n+1}{n(n+1)} = \sum_{n=1}^{k}\left(\frac{1}{n} + \frac{1}{n+1}\right) = H_k + H_{k+1} - 1 = 2H_k + \frac{1}{k+1} - 1 = 2H_k - \frac{k}{k+1}$

6.3.15. $E_{\text{Jack}}(1,10) = 10 + H_{10} \approx 10 + \gamma + \ln 10.5 = 10 + 0.577\ldots + 2.351 = 12.928\ldots$

$E_{\text{Jill}}(10,1) = 1 + 10H_1 = 11$

6.4.1. $B_5(t) = 5\int\left(t^4 - 2t^3 + t^2 - \frac{1}{30}\right)\,dt = t^5 - \frac{5}{2}t^4 + \frac{5}{3}t^2 - \frac{1}{6}t + C$

where the constant of integration C is determined by the equation

$$0 = \int_0^1 B_5(t)\,dt = \int_0^1 \left(t^5 - \frac{5}{2}t^4 + \frac{5}{3}t^3 - \frac{1}{6}t + C\right)dt = \frac{1}{6} - \frac{1}{2} + \frac{5}{12} - \frac{1}{12} + C$$

$= 0 + C$

That is, $C = 0 = B_5$.

6.4.5.
$$B_n(t+h) = \left[\frac{x^n}{n!}\right]\frac{xe^{x(t+h)}}{e^x - 1} = \left[\frac{x^n}{n!}\right]\left(\frac{xe^{xt}}{e^x-1}e^{xh}\right)$$
$$= \left[\frac{x^n}{n!}\right]\left(\frac{xe^{xt}}{e^x-1}\right)(e^{xh}) = \left[\frac{x^n}{n!}\right]\left(\sum_{r=0}^{\infty} B_r(t)\frac{x^r}{r!}\right)\left(\sum_{s=0}^{\infty} h^s\frac{x^s}{s!}\right)$$
$$= \left[\frac{x^n}{n!}\right]\sum_{n=0}^{\infty}\left(\sum_{k=0}^{n}\binom{n}{k}B_k(t)h^{n-k}\right)\frac{x^n}{n!} = \sum_{k=0}^{n}\binom{n}{k}B_k(t)h^{n-k}$$

6.5.1. C3 =(C$1+1)*C2+($A3-C$1)*B2

6.5.3. Choose a nonempty subset $A \subseteq [k]$ and let B be its complement. If $\pi = a_1a_2\cdots a_{k-j}b_j\cdots b_2b_1$ where $a_i \in A$, $a_1 > a_2 > \cdots > a_{k-j}$ and $b_i \in B$, $b_j > \cdots > b_2 > b_1$, then π is a permutation of $[k]$ with exactly one ascent except in the k cases for which $A = \{k, k-1, \ldots, j+1\}$ for $j = 0, 1, \ldots, k-1$. Since there are $2^k - 1$ nonempty subsets A, there are $2^k - 1 - k$ permutations of $[k]$ with exactly one ascent.

6.5.7. [Hints: **(a)** Use a committee selection model and ask: "How many ways can a committee of size n, either with or without a chair, be chosen from a club of m men and one woman?"; **(b)** use the identity from part (a).]

6.6.3. **(a)** 1,999,280,893

(b) 4.10025×10^{12}

6.6.5. $(1+x^5+x^{10}+\cdots)(1+x^{10}+x^{20}+\cdots)(1+x^{25}+x^{50}+\cdots)=$
$\dfrac{1}{(1-x^5)(1-x^{10})(1-x^{25})}$

6.6.7. Any partition of n that is the sum of k odd integers $a_1, a_2, \ldots, a_j$, $1, \ldots, 1$, where $a_j \geq 3$ corresponds to the partition of $n-k$ with the form $a_1-1, a_2-1, \ldots, a_j-1$ that has no more than k even parts.

6.6.15. Each partition $a_1+a_2+a_3=n$ of n with three parts can be uniquely matched to the partition $(n-a_3)+(n-a_2)+(n-a_1)=2n$ of $2n$ into three parts, each smaller than n.

6.6.19. For example, in Maple, the commands

```
f:=(1-x)*(1-x^2)*(1-x^3);
g:=1/(1-x)^3/6+1/(1-x)^2/4+1/(1-x^2)/4+1/(1-x^3)/3;
simplify (f(x)*g(x));
```

return a 1.

6.7.1. A mountain range corresponds to a sequence of n positive ones (upstrokes) and n negative ones (downstrokes) with positive partial sums, so there are C_n mountain ranges.

6.7.3. Change will always be available if at least as many people with \$5 bills are in front of those needing change. There are 12! ways for the people to line up, of which $(6!)^2\ C_6$ ways have no problem making change. This gives the probability $(6!^2 C_6)/12! = 1/(6+1) = \frac{1}{7}$.

6.7.7. The given sequence corresponds to a path from A to A' with point A' one unit to the right of the 45° line through A. Any cyclic shift is a path of n steps along the extended path created by translates of the starting path by the vector A to A'. Only the cyclic shift beginning at the uppermost point B of the 45° line for which all further points lie below the 45° line gives a path with all positive partial sums.

Chapter 7

7.2.1. **(a)** $\{2,3\}$ **(b)** $\{1,2\}$ **(c)** $\{2,-2,3\}$

7.2.5. **(a)** $\mathcal{O}(v_n^2) \leq \binom{3+2-1}{2} = 6$ and $\mathcal{O}(t_{2n}+v_n^2) \leq 2+6=8$

(b) $\mathcal{E}(v_n^2) = \{4,9,-4,6,-6\}, \mathcal{O}(v_n^2) \leq 5$
$\mathcal{E}(t_{2n}+v_n^2) = \{4,9,-4,6,-6\}, \mathcal{O}(t_{2n}+v_n^2) \leq 5$

7.2.7. $\mathcal{E}(h_n) = \{2,1,1,1\}, \mathcal{O}(h_n) = 4$

7.2.13. Since $u_n^t = (c_0+c_1 n)^t \alpha^{nt} = p_k(n)(\alpha^t)^n$, where $p_k(n)$ is a polynomial in n of degree t, we see that u_n^t is annihilated by the operator $(E-\alpha^t)^{t+1}$.

7.3.1. **(a)** $(E+5)(E^2-3E+7)$
(b) $(E-1)^3(E^2-3)$
(c) $(E-1)(E-2)(E^3-E^2+2)$
(d) $(E-1)^2(E+1)(E-3)(E^2+E-4)$
(e) $(E^2-E-1)^2$

7.3.5. **(a)** $C(E)^2$
(b) $C(E)^2D(E)^2$

7.3.7. **(a)** $h_n = 2h_{n-1} + 3h_{n-2} + 4^n + 1$
(b) $h_n = -\frac{1}{4} + \frac{16}{5}\cdot 4^n - \frac{33}{40}(-1)^n - \frac{9}{8}\cdot 3^n$

7.4.1. The spectra of all of the terms $F_n^2, F_nF_{n+1}, F_{2n}$, and $(-1)^n$ that appear in the formulas are included in the spectrum $\{\varphi^2, \hat{\varphi}^2, -1\}$, so all of the terms are annihilated by $(E+1)(E-\varphi^2)(E-\hat{\varphi}^2) = (E+1)(E^2-3E+1)$. Since each equation (a), (b), (c) and (d) holds for $n = 0,1,2$, each identity is proved.

7.4.5. **(a)** $\{1, 2, 3\}$
(b) $s_n = \frac{1}{2} - 2\cdot 2^n + \frac{3}{2}\cdot 3^n = \frac{1}{2}(1 - 2^{n+2} + 3^{n+1})$
(c) $s_n = d_0 + d_1t_n + d_2t_{n+1}$, so the initial values $s_0 = 0, s_1 = 1, s_2 = 6$ gives $s_n = \frac{1}{2} + 3t_n - \frac{1}{2}t_{n+1}$

7.4.7. **(a)** If 1 and α were the eigenvalues for the operator $C(E) = E^2 - aE - b$, this would mean that $C(x) = x^2 - ax - b = (x-1)(x-\alpha) = x^2 - (1+\alpha)x + \alpha$, so $\alpha = -b$ and $1+\alpha = a$. This shows that $a + b = 1$, a contradiction.

7.4.9. **(a)** The eigenvalues of F_{n+1} are $\{\varphi, \hat{\varphi}\}$, so by Theorem 7.36, the sum $s_n = \sum \binom{n}{j} F_{1+j}$ is a recurrence sequence with the eigenvalues $\{1+\varphi, 1+\hat{\varphi}\} = \{\varphi^2, \hat{\varphi}^2\}$. Therefore, there are constants such that $s_n = c_1F_{2n} + c_2F_{2n+1}$. Since $s_0 = 1$ and $s_1 = 2$, we see that $s_n = F_{2n+1}$.

7.4.13. $F_2^2 + F_4^2 + \cdots + F_{2n}^2 = \frac{1}{5}(F_{4n+2} - 2n - 1)$

7.4.15. $\sum_{j=0}^{n} j^3 = \frac{1}{4}(n^4 + 2n^2 + n^2) = \left(\frac{n(n+1)}{2}\right)^2$

7.4.21. The recurrence relation has the form $s_{n+2} = as_{n+1} + bs_n$ so the initial conditions give us the equations $24 = 3a + 0,\ 198 = 24a + 3b$. Solving the linear system shows that $a = 8$ and $b = 2$.

Chapter 8

8.2.3. **(a)** k **(b)** 12 **(c)** 60, the least common multiple of 4, 5, and 6

8.2.11. $e = (1)(2)(3)(4)(5), \rho = (1\ 2\ 3\ 4)(5), \rho^2 = (13)(24)(5), \rho^3 = (1\ 4\ 3\ 2)(5)$

8.2.13. $G_{x_i} = \{e\}$ for $i = 1, 2, 3, 4$ and $G_{x_5} = G = C_4$

8.2.17. $G_{x_1} = G_{x_3} = \{e, \varphi_1\}, G_{x_2} = G_{x_4} = \{e, \varphi_3\}, G_{x_5} = D_4$

8.3.1. **(a)** $\langle f \rangle = \{ef, \varphi_1 f\} = \{ef, \varphi_2 f\} = \{ef, \varphi_3 f\}$

(b) $G_f = \{e, \rho, \rho^2\}$

(c)

	e	ρ	ρ^2
e	e	ρ	ρ^2
φ_1	φ_1	φ_3	φ_2

	e	ρ	ρ^2
e	e	ρ	ρ^2
φ_2	φ_2	φ_1	φ_3

	e	ρ	ρ^2
e	e	ρ	ρ^2
φ_3	φ_3	φ_2	φ_1

8.3.3. **(a)** $X_e = X, X_{\rho^i} = \emptyset, \quad i = 1, 2, 3, 4; \qquad X_{\varphi_i} = \{x_j\}, \quad j = 1, 2, 3, 4, 5$

(b) $G_{x_i} = \{e, \varphi_i\}, \quad i = 1, 2, 3, 4, 5$

(c) $\dfrac{1}{|D_5|} \sum_{\gamma \in D_5} |X_\gamma| = \dfrac{1}{10}(5 + 0 + 0 + 0 + 0 + 1 + 1 + 1 + 1 + 1) = 1$

(d) $\dfrac{1}{|D_5|} \sum_{x \in X} |G_x| = \dfrac{1}{10}(2 + 2 + 2 + 2 + 2) = 1$

8.3.7. [*Hint*: The group of actions is induced by the symmetries of the cyclic group $C_2 = \{e, \rho\}$.] **(a)**, **(b)** There are 36 nonequivalent colorings of the table.

8.4.3. The set $\{e, \varphi_1, \varphi_2, \varphi_3\}$ is not a group since it lacks closure. The correct group of the floating triangle is the dihedral group D_3. The cycle index is $Z = \frac{1}{6}\left(t_1^3 + 2t_3^1 + 3t_1^1 t_2^1\right)$. There are $\frac{1}{6}(3^3 + 2 \cdot 3 + 3 \cdot 3 \cdot 3) = \frac{60}{6} = 10$ nonequivalent 3-colorings.

8.4.5. **(a)** 14 **(b)** 4

8.4.7. **(a)** The centers of the faces of a regular tetrahedron are the vertices of regular tetrahedron.

(b) Letting $t_i = 4, i = 1, 2, 3$ in the cycle index shows that there are

$$Z(4, 4, 4) = \frac{1}{12}(4^4 + 3 \cdot 4^2 + 8 \cdot 4 \cdot 4) = \frac{256 + 48 + 128}{12} = 36$$

different molecules.

INDEX

Combinatorial Reasoning: An Introduction to the Art of Counting, First Edition. Duane DeTemple and William Webb.

Printed and bound by CPI Group (UK) Ltd, Croydon, CR0 4YY
16/09/2026
14963427-0001